FinTech and Robotics Advancements for Green Finance and Investment

Muhammad Arslan
Open Polytechnic of New Zealand, New Zealand

Alimshan Faizulayev
KIMEP University, Kazakhstan

IGI Global
Scientific Publishing
Publishing Tomorrow's Research Today

Vice President of Editorial Melissa Wagner
Director of Acquisitions Mikaela Felty
Director of Book Development Jocelynn Hessler
Production Manager Mike Brehm
Cover Design Phillip Shickler

Published in the United States of America by
IGI Global Scientific Publishing
701 East Chocolate Avenue
Hershey, PA, 17033, USA
Tel: 717-533-8845
Fax: 717-533-7115
Website: https://www.igi-global.com E-mail: cust@igi-global.com

Library of Congress Cataloging-in-Publication Data

Names: Arslan, Muhammad, 1989- editor. | Faizulayev, Alimshan, 1986- editor.
Title: Fintech and robotics advancements for green finance and investment / edited by: Muhammad Arslan, Alimshan Faízulayev.
Description: Hershey, PA : IGI Global Scientific Publishing, [2025] | Includes bibliographical references and index. | Summary: "The primary objective of this book is to explore scholarly research on the latest financial and robotics technological advancements (such as AI, fintech startups, big data, IoT, cryptocurrencies, robo-advisors, machine learning, and blockchain applications, among others) in the world of sustainable finance, green finance, green innovation, and green investment"-- Provided by publisher.
Identifiers: LCCN 2024058254 (print) | LCCN 2024058255 (ebook) | ISBN 9798337303550 (hardcover) | ISBN 9798337303567 (paperback) | ISBN 9798337303574 (ebook)
Subjects: LCSH: Fintech. | Artificial intelligence--Financial applications. | Clean energy investment.
Classification: LCC HG173 .F534 2025 (print) | LCC HG173 (ebook) | DDC 332--dc23/eng/20250320
LC record available at https://lccn.loc.gov/2024058254
LC ebook record available at https://lccn.loc.gov/2024058255
British Cataloguing in Publication Data
A Cataloguing in Publication record for this book is available from the British Library.

Table of Contents

Detailed Table of Contents

the effect of digitalization on the world economy was discussed in detail. The need for global ethical rules and standards was presented as well. The study is purely theoretical and descriptive.

Cryptocurrencies have emerged as a cornerstone of the digital transformation in the global economy during the last decades, introducing decentralized mechanisms that challenge the dominance of traditional financial systems. This innovation enhances transparency, security, and accessibility, while making financial systems more inclusive and efficient. Since the first inception of Bitcoin in 2009, the cryptocurrency landscape has expanded exponentially. By 2023, the combined market capitalization of over 22,000 cryptocurrencies exceeded $1 trillion, demonstrating their significant influence on financial markets. This study examines the dual nature of cryptocurrencies—assessing their transformative potential and inherent risks. Using case studies, historical data, and technological advancements, it provides a balanced perspective on how cryptocurrencies can reshape financial systems, bridge economic inclusion gaps, and drive innovation.

The fast acceleration of digital innovations has disrupted the banking system, putting conventional financial institutions under pressure. This study examines the fluctuating relationship between emerging Fintech startups and traditional banks in terms of opportunities for collaboration, regulatory challenges, and emerging trends. Initially competitive, this relationship has shifted to one of collaboration, as banks harness Fintech innovations to enhance digital efficiency. The research will also consider new banking models, including digital-first approaches and "Banking-as-a-Service," and how these changes influence customer demands for continuous and personalized services. Ultimately, the study will highlight the profound impact of Fintech and traditional bank integration on the modern financial ecosystem, particularly regarding cybersecurity and risk management.

This study examines the disruptive impact of FinTech technologies on financial behaviour, with a focus on transparency, personalisation and sustainability in digital financial services. It examines adoption trends and their impact on financial systems. Using a quantitative methodology, the analysis employs techniques such as PCA, MCA, ANOVA and logistic regressions, based on data collected through a survey of diverse consumers. The findings indicate that transparency and ease of use are key factors driving FinTech adoption, while sustainability criteria remain secondary for most users. This research enriches the literature by linking technology adoption theories with financial applications, and offers practical recommendations to enhance user trust and promote sustainable practices in digital financial ecosystems.

This study investigates the transformative potential of green fintech solutions in mitigating financed emissions, synthesizing insights through an integrative literature review approach. Green fintech which is powered by advanced technologies such as blockchain, artificial intelligence, big data analytics, and digital platforms, emerges as a critical driver of sustainability within financial systems. The study offers innovative solutions, including digital ESG analytics, green investment tools, crowdfunding platforms, and digital payment systems, to address the complex challenge of aligning financial activities with net-zero objectives. The study further identifies enabling factors which enhance transparency, efficiency, and accountability in green finance. However, these potentials are met with significant challenges, including limited data availability and standardization, high energy consumption of underlying technologies, regulatory and compliance uncertainties, greenwashing concerns, and unequal access to capital and technology.

The Chapter revolves around the attraction that the blockchain-backed decentralized financial products have created in the investment market. These new-gen decentralized finances would be Defi, Metaverse coins, Stablecoins, Cryptocurrency, smart contracts, and privacy coins. The influence of social media and high internet penetration levels have made the retail and partial investors very well aware of many alternate financial investment products. Though there is an increase seen in the Stock Market investments as well a drastic increase in the alternate market is seen. In recent times major stock markets have created indexes separately. It is the icing on the cake with the highest attention of the globe. The regulations have put the Defi Assets under unpredictable volatility but the attraction towards cryptocurrency and other digital asset gain still exists. The study aims to identify the relationship between the increase in investment in the stock market and the alternative.

Digital transformation in finance has increasingly become a key component of businesses' strategic plans as they strive to stay competitive. The effects of digital transformation encompass various aspects, including its influence on the risk-taking processes, on financial performances, its role in fostering green innovation, and on improving Environmental, Social, and Governance performance of businesses. For this purpose, this study seeks to examine the effect of finance digitalization and digital transformation on the financial structure and decision-making of businesses. To achieve this, a comprehensive review and evaluation of the existing literature and findings on the topic will be conducted, aiming to clarify how digitalization and transformation in finance influence business performance. Additionally, it provides practitioners, policymakers, and researchers with a wide-ranging overview of the literature on these concepts, highlighting areas where further development is needed.

Sustainability is the way forward. In order to achieve the sustainable growth, technology is a pivot. Leveraging Artificial Intelligence (AI), Blockchain, Big Data, Cloud Computing and related tech tools there is an applied tech lever, known as FinTech. FinTech is a tech revolution which has disrupted the business landscape. It has transformed the economy by enabling the institutions, firms, consumers and society at large. The chapter dwells on the role of FinTech in creating sustainable development. Further, the chapter covers aspects like FinTech architecture and ecosystem, impact of FinTech in realising Sustainable Development Goals (SDGs) and the significance of FinTech as an enabler of ESG theme and philosophy. Use cases of FinTech in promoting sustainability are documented. The chapter outlines the challenges of FinTech in terms of risks and ethical issues. Also, the chapter establishes the need for a sustainable FinTech ecosystem in the wake of its impact, potential, associated risks and ethical challenges.

To account for nonlinearities in the relationship between financial and technological development on income inequality, this chapter uses a nonlinear panel ARDL model that allows for heterogeneity within countries. The results show that in the long run, financial and technological development significantly impact income inequality in SSA countries. However, the impact of positive technology on income inequality is negative, while both positive and negative financial shocks have a positive impact on income inequality. The study indicates the importance of fiscal and financial policies of developing economies that should be implemented to achieve inclusive growth, improve welfare, and reduce income distribution in the SSA region.

A systematic literature review (SLR) of 120 peer-reviewed articles published between 2010 and 2024, examining the role of advanced technologies in sustainable

Preface

INTRODUCTION

The growing demand for sustainable development and environmentally responsible business practices has placed immense pressure on financial markets to adapt and evolve. Concurrently, rapid advancements in financial technology (FinTech) and robotics are transforming the landscape of finance, creating new opportunities and challenges for aligning investment strategies with environmental, social, and governance (ESG) principles. This book, *FinTech and Robotics Advancements for Green Finance and Investment*, offers a timely and valuable contribution to this emerging interdisciplinary field.

As editors of *FinTech and Robotics Advancements for Green Finance and Investment*, we are honored to present this comprehensive volume that bridges the frontiers of finance, technology, and sustainability. The accelerating integration of digital technologies into financial systems has opened up remarkable possibilities for transforming not only how financial services are delivered, but also how they contribute to global sustainability goals. From blockchain and artificial intelligence to robotic process automation and green digital platforms, this book explores the evolving interplay of FinTech and robotics within the broader context of green finance and investment.

Our objective in assembling this edited volume is to capture both the depth and breadth of current research and practice in this rapidly evolving domain. The chapters presented herein reflect diverse perspectives, geographies, and methodological approaches, unified by a shared interest in sustainable innovation. As technology continues to disrupt traditional paradigms, we believe it is critical for scholars, practitioners, and policymakers to understand both the opportunities and risks involved. The chapters aim to inform ongoing discussions around the role of digital technologies in promoting financial inclusion, transparency, carbon neutrality, and environmental, social, and governance practices.

This book would not have been possible without the contributions of our international panel of authors, each of whom has brought unique expertise and intellectual rigor to their chapters. We are equally grateful to our institutional affiliations—Open Polytechnic of New Zealand and KIMEP University—for their unwavering support of our academic endeavors. We extend our appreciation to the reviewers, researchers, and technical editors who helped shape the final form of this volume, and we hope this book serves as a valuable resource for your teaching, research, and practice in sustainable digital finance.

CHAPTER OVERVIEW

Chapter 1: Does Blockchain Adoption Related to Cryptocurrency Returns: Evidence from Bitcoin and Ethereum

Carlos Munoz Valdez and Halil Kiymaz explore the relationship between smart contract growth and cryptocurrency returns, offering empirical insights into blockchain adoption's impact on Ethereum versus Bitcoin.

Chapter 2: Thematic Analysis of Blockchain and Artificial Intelligence in Sustainable Finance

Muhammad Arslan and co-authors present a systematic review that maps the emerging themes of blockchain and AI integration into sustainable finance using bibliometric and qualitative tools.

Chapter 3: To the Moon and Beyond: How Nano-Entrepreneurs and Digital Currency are Revolutionizing the Future of Business in the Space!

This visionary chapter by Mohammad Rashed Hasan Polas and colleagues investigates how small-scale entrepreneurs leverage digital currencies to participate in the space economy.

Chapter 4: Blockchain, RPA, and AI as Main Elements of Digitalization: Benefits, Challenges, and Economic Impact

Maya Katenova and Mira Nurmakhanova assess the transformative elements of digitalization, detailing both their economic benefits and associated global challenges.

Chapter 5: The Transformative Role of Cryptocurrencies in Modern Finance

ILIMSAN FEYZULLAH offers a critical look at how cryptocurrencies challenge conventional financial structures while exploring the risks and innovations they introduce.

Chapter 6: Bank Challenges Facing the Emergence of FinTech: From Competition to Collaboration

Bilel Jarraya and Adel Alresheedi examine how banks are shifting from resisting to embracing FinTech, exploring the implications for digital-first strategies and customer-centric finance.

Chapter 7: FinTech and Financial Behavior: Determinants of Adoption and Sustainability

Ismaili Safae analyzes empirical data to uncover what drives FinTech adoption, emphasizing transparency and user experience over sustainability in most consumer choices.

Chapter 8: Assessing the Transformative Potential of Green Fintech Solutions

Abdulkadri Alabi synthesizes literature to propose how green fintech can reduce financed emissions while identifying the regulatory and operational hurdles impeding progress.

Chapter 9: Decentralized Financial Assets: An Attraction Beyond Stock Market Investments

Anand Patil and co-authors dissect the allure of decentralized finance for retail investors, noting its rise alongside traditional investment avenues.

Chapter 10: The Effects of Digital Transformation on Financial Decisions and Structures

Aysegul Kirkpinar presents a review-based investigation into how digital transformation reshapes financial decision-making and performance, especially in green innovation.

Chapter 11: FinTech for Sustainability: A Treatise

Batani Raghavendra Rao and collaborators provide a holistic exploration of FinTech's potential to drive sustainability, anchored in real-world use cases and policy insights.

Chapter 12: Investigating Asymmetric Impacts of Financial and Technological Development on Income Inequality in Sub-Saharan Africa

Hassan Lunku and Muhsin Danga analyze income inequality dynamics in Sub-Saharan Africa using a nonlinear econometric model, providing regional policy implications.

Chapter 13: AI in Sustainable Investing: A Dual-Edged Sword

Muhammad Nauman offers a systematic literature review that emphasizes the benefits and ethical dilemmas posed by AI in sustainable investment strategies.

Chapter 14: Assessing Financial Performance and Competitive Efficiency of Banks: The Impact of ESG and FinTech in Islamic Finance-Oriented Countries

Alimshan Faizulayev, Muhammad Arslan, and Aizada Zhakupova present empirical findings on the integration of ESG and FinTech in Islamic finance, exploring performance and competition metrics in QISMUT+3 nations.

CONCLUSION

As the world stands at the nexus of digital and green revolutions, the ideas and innovations explored in this volume provide a foundation for envisioning finance systems that are not only more efficient and inclusive but also more sustainable and equitable. Each chapter in this book contributes uniquely to our understanding of how advanced technologies—ranging from blockchain and AI to digital currencies and green fintech tools—are redefining the rules of engagement in finance. Together, these works form a compelling narrative that speaks to both the promise and responsibility of digital transformation.

We recognize that technological advancement alone cannot deliver sustainable outcomes. It must be accompanied by thoughtful policy, inclusive frameworks, and ethical innovation. In that spirit, we encourage future research to address the real-world implications of these technologies, including their socioeconomic impacts, regulatory risks, and potential for creating or resolving inequalities.

Finally, we hope that this volume will not only inform but inspire. Whether you are a scholar, policymaker, financial practitioner, or student, we invite you to reflect on the lessons, challenges, and opportunities presented here. The journey toward a more resilient, green, and tech-enabled financial ecosystem is underway—and it is our collective responsibility to shape its course with foresight, integrity, and collaboration.

Muhammad Arslan
Open Polytechnic of New Zealand, New Zealand

Alimshan Faizulayev
KIMEP University, Kazakhstan

Chapter 1
Does Blockchain Adoption Related to Cryptocurrency Returns:
Evidence From Bitcoin and Ethereum

Carlos Humberto Munoz Valdez
IPADE Business School, Mexico

Halil Kiymaz
https://orcid.org/0000-0003-4182-3129
Rollins College, USA & KIMEP University, Kazakhstan

ABSTRACT

This study investigates the connection between the creation of smart contracts in the Ethereum Virtual Machine and the returns of Ethereum and Bitcoin. The analysis reveals that while there is no connection between Bitcoin and smart contracts created on the Ethereum platform, a significant connection exists for Ethereum. Blockchain technology, which underpins cryptocurrencies, has garnered significant attention for its potential applications beyond financial transactions, including supply chain management, healthcare, and digital identity verification. We argue that blockchain adoption would increase the application of blockchain in general. We use the growth of smart contracts as a proxy for blockchain adoption. The findings suggest that the volume growth of smart contracts is correlated with the return on Ethereum but not on Bitcoin. This research contributes to understanding blockchain technology's impact on cryptocurrency performance and offers insights for academics and practitioners interested in the evolving landscape of digital assets.

DOI: 10.4018/979-8-3373-0355-0.ch001

1. INTRODUCTION

Blockchain technology has been one of the most interesting developments in finance in recent years. Major accounting firms, the Big Four, are developing applications using blockchain technology and promoting its acceptance (Atik and Kelten, 2021). Visualizing these technologies' full impact on the world is not feasible since their adoption is still in the early stages. That is why there is increasing interest from practitioners and academics alike in better understanding the possibilities and applications of these technologies. Central Banks, technology, and financial companies are at the vanguard of researching the advantages and challenges of applying blockchain technology and its potential (Kapadia, 2020). Increased awareness about blockchain technology is mainly due to cryptocurrencies, a world in constant flux, where hype and unsupported claims are common (Deane and Fines, 2023). Here lies the need for academic research to support this ever-changing phenomenon in its initial stages, where technological development will continue, and a more comprehensive regulatory framework is overdue.

Cryptocurrencies have gained attention for the technology behind them, and one technology that has generated a massive interest in the business community is blockchain. Cryptocurrencies rely on blockchain to keep track of all transactions and ownership to keep the system's integrity. This technology has promising applications beyond cryptocurrencies and is also the driver behind Web 3.0, a system in which "data are interoperable, decentralized, and controlled by individual users rather than centralized companies" (Harvey et al., 2022, p. 6).

While blockchain is an intriguing technology for businesses, companies are still exploring how to leverage its potential and overcome challenges such as ensuring data security and protecting the information stored on the blockchain. A survey published by the CFA Institute stated that "the principles underpinning DLT and blockchain technology are sound and could enhance the day-to-day functioning of economic exchanges through reduced friction and enhanced efficiency" (Deane and Fines, 2023, p. 21). However, the same respondents expressed concern regarding compatibility issues with legacy systems and potential attacks on system vulnerabilities.

Now, considering that cryptocurrencies rely on blockchain technology and that part of the awareness of blockchain is due to cryptocurrencies, is there a relationship between both that may lead cryptocurrencies to mature as an asset? Does the perception of cryptocurrencies limit blockchain adoption? This research studies the literature regarding blockchain and its application and whether its adoption has affected the performance of major cryptocurrencies like Bitcoin or Ethereum. Our study examines whether a practical application of blockchain technology like Ethereum smart contracts impacts the return of Bitcoin/Ethereum. Our study would contribute to and complement previous research in the field from an academic perspective. It would also provide perspectives on some of the functionalities blockchain technology supports and its impact on cryptocurrencies as an investment alternative.

2. LITERATURE REVIEW

We started our research by explaining blockchain technology and its mechanics. Then, we review its applications in smart contracts and cryptocurrencies. Blockchain technology has widespread adoption across various industries, influencing financial markets, including cryptocurrencies like Bitcoin and Ethereum. This literature review analyzes academic research and empirical studies to explore how blockchain adoption affects the returns of Bitcoin and Ethereum. Kapadia (2020) defines blockchain as an incorruptible digital ledger of economic transactions that can be programmed to record financial transactions and virtually everything of value. It is a secure distributed database that preserves ordered lists of records, known as blocks, which cannot be changed. Among the characteristics of blockchain are security, a shared database, and keeping records in chronological order registered in linked blocks, creating a chain of data. Blockchain also offers information transparency, live data synchronization, and immutable records that prevent fraud. The blockchain design does not allow retroactively altering records, so it is an immutable system and temper resistance technology (Chong et al., 2019). Blockchain is also known as the "next-generation internet," and it will have a lasting impact on the economy and how individuals interact with each other (Trivedi et al., 2021). The World Economic Forum (WEF) estimates that by the end of 2025, 10 percent of the world's GDP will be registered on blockchain platforms (Chong et al., 2019). Blockchain offers transparency, verifiability, and immutability to commercial transactions. It also offers security and resilience, individual data protection, and promotes cooperation among participants (Wigley and Cary, 2018).

As a developing technology, blockchain may revolutionize how digital operations are recorded, retrieved, and shared. It operates as a decentralized ledger of peer-to-peer transactions without third-party intermediaries (Liang et al., 2021). It is considered an anonymous, secure, transparent, efficient, and traceable system. However, it is important to note that anonymity means the sending and receiving parties are not identifiable, while transparency allows anyone to track all transactions stored in the blocks. Blockchain is also known as "trustless" technology. It replaces trust in a specific party with trust in consensus algorithms and the system's code (Kostić and Sedej, 2022). Blockchain is a distributed ledger without a third party that verifies and registers information within that ledger. "Blockchain resembles a fully distributed, decentralized system that captures and stores an immutable chronological log of every transaction among actors on its peer-to-peer network" (Chong et al., 2020, p. 1309). Blockchain is designed to encourage the active participation of actors to contribute computational power to the verification and recording of information. This process ensures the integrity of transactional data in an immutable and transparent manner.

Blockchain is a technology utilized by Distributed Ledger Technologies (DLT). The need to store records of transactions and activities has been present since the dawn of human history, but the methods employed by individuals and institutions have changed dramatically in recent years (Atik and Kelten, 2021). There are three categories of blockchains based on control and access permissions: public, private, and consortium. In a public blockchain, all members have access, share information, and collaboratively control the blockchain. In contrast, a private blockchain requires permission from a third party, introducing some degree of centralization. The consortium category is governed by the parties participating in the consensus algorithms (Atik and Kelten, 2021). Both private and consortium blockchains necessitate some level of trust (Kostić and Sedej, 2022). For instance, Bitcoin, the first cryptocurrency, employs blockchain technology to resolve the issue of double spending without relying on an authority or a third party. It also provides increased record capacity, enhanced security, and improved processing times for transactions (Creed and Northard, 2021).

Cryptocurrencies and blockchain, as emerging technologies, are at an early stage of achieving widespread adoption by users worldwide. In 2023, there were approximately 400,000 daily BTC transactions (Statista, 2023), which remains a small volume compared to other payment methods like Visa or Mastercard. A significant barrier to the widespread adoption of cryptocurrencies is their high volatility. There have been numerous debates surrounding cryptocurrencies: Are they currencies like the dollar or the euro? Are they investment assets? Or are they merely a series of codes that entertain technology enthusiasts? Cryptocurrencies lack the fundamentals of stocks; for instance, there are no tangible assets backing them, and they do not generate a continuous cash flow. These conditions, or lack thereof, complicate the process of determining a fair value for each cryptocurrency.

Contrary to more traditional assets that can be appraised by discounted cash-flows, multiples, or assets, cryptocurrencies do not possess the characteristics required for these valuation methods. Instead, researchers have proposed alternative methods like network valuation, production cost, or scarcity-based approach (Deane & Fines, 2023). Nevertheless, a cryptocurrency like Bitcoin is sometimes compared to gold as a store of value due to its scarcity, limited additional supply, and even as a refuge during times of crisis. Both assets also share that their prices are mainly set by supply and demand (Ammous, 2018). In a 2022 survey in the United States, 16% of 9,989 respondents in four waves between January and December of 2022 mentioned cryptocurrencies as one of the most used financial products, comparable to stocks and insurance (Figure 1). Multiple answers were possible. During the survey, the question was phrased as follows: “Which of these financial products and investments do you currently use/own?”

There is no doubt that cryptocurrencies are a recurring topic among retail investors, but the challenge for widespread adoption lies in achieving higher participation from institutional investors, who require legal certainty and proven technology because of their fiduciary obligations (Deane & Fines, 2023).

Cryptocurrencies have attracted significant attention due to the technology behind them, particularly blockchain, which has sparked substantial interest in the business community. Cryptocurrencies depend on blockchain to track all transactions and ownership, ensuring the integrity of the system. This technology holds promising applications beyond cryptocurrencies and serves as the driving force behind Web 3.0 (Harvey et al., 2022).

Figure 1. Most used financial products in the US 2022

Most Used Financial Products in the US 2022

This figure provides the most commonly used financial products in the US. This would highlight how commonly cryptocurrencies have become.

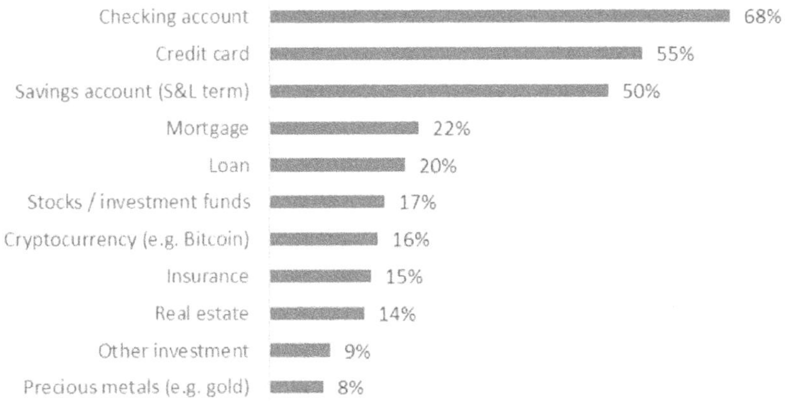

Note. Statista (2022).

This figure provides the most commonly used financial products in the US. This would highlight how common cryptocurrencies have become.

Now, considering that cryptocurrencies rely on blockchain technology and that part of the awareness of blockchain stems from cryptocurrencies, is there a relationship between the two that may lead to the maturation of cryptocurrencies as an asset? Does the perception of cryptocurrencies hinder blockchain adoption? This research examines the literature on blockchain and its applications, as well as whether its adoption has influenced the performance of major cryptocurrencies like Bitcoin and Ethereum. From an academic perspective, due to the relatively short lifespan of cryptocurrencies and blockchain applications, there is a unique opportunity to contribute to and enhance previous research in the field. For practitioners, it offers an intriguing view of some functionalities supported by blockchain technology and its impact on cryptocurrencies as an investment alternative.

How Blockchain Works

Blockchains have a genesis block that is not linked to any previous block. Once a block is filled, it connects to a new one using a hash value that identifies the connection and sequence among different blocks (Atik and Kelten, 2021). Users who wish to connect to a specific blockchain utilize nodes or connected computing devices. Every new block must be approved or rejected by other nodes. Consensus algorithms govern users; these agreements outline the rights of blockchain users. Examples of consensus algorithms include Proof of Work, Proof of Stake, Delegated Proof of Stake, Proof of Elapsed Time, Practical Byzantine Fault Tolerance, Delegated Byzantine Fault Tolerance, Proof of Weight, Proof of Burn, Proof of Capacity, Proof of Importance, Proof of Activity, and Directed Acyclic Graphs (Atik and Kelten, 2021, p. 498).

Each algorithm has diverse levels of security, decentralization, difficulty, and level of participation, and each algorithm requires different energy levels.

Proof of Work (POW) is a consensus mechanism used for Bitcoin, the earliest verification method. It depends on computer power to confirm a transaction and add it to the chain's most recent block of data (Folkinshteyn and Lennon, 2016). Computer power is used to solve mathematical problems involving a hash of the operations waiting to be included in the blockchain and an additional number called a *nonce*, connecting with the previous block's hash. It is an iterative process in which computers try to find the right number. This process is designed to consume time, hence the definition of proof of work. This time-consuming process is a barrier to modifying and rewriting historical records, not only because of the time it will require to do it but also because of the complexity of correctly connecting with the hash of the previous work. A corrupt participant in the system will be required to control at least 51% of the computer power in it to surpass the processing capacity of the remaining nodes and modify historical information, this is known as a "51% attack" (Coyne and McMickle, 2017). The set of transactions that will have the highest probability of solving the mathematical puzzle first is the one that attracts more computer power to work on it.

Proof of Stake (POS) is another common consensus mechanism used by Ethereum. Instead of computer power, POS requires confirming stake positions in the blockchain (number of Ethers for Ethereum). The size of the position is directly proportional to the voting power to validate transactions (Folkinshteyn and Lennon, 2016). This characteristic makes POS less resource-demanding than POW, which has been criticized for the computer power it requires and its environmental side effects. The higher the stake of a participant in the system, the higher the incentive to preserve the integrity of the data, which would be the carrot. If a participant registers an invalid transaction, a fine might be charged against his position, which would be the stick (Coyne and McMickle, 2017).

Mining is part of the process of creating new blocks; it is resource-intensive, and its complexity makes fraudulent transactions difficult. Once a new block is accepted, the information is shared among all nodes to include the latest block in the current chain. Every new accepted block will contain a hash, a unique digital fingerprint, of the block preceding it (Creed and Northard, 2021).

Blockchain General Applications

Blockchain technology is utilized across diverse industries beyond cryptocurrencies. Its P2P technology has enhanced internet security (Creed and Northard, 2021). Blockchain technology applies to cryptocurrencies and various financial activities. For instance, it can be employed in "supply chain management, voting in elections, healthcare records management, identity management systems, access control systems, decentralized notary services, and tourism organizations" (Atik and Kelten, 2021, p. 499). Integrating blockchain technology into supply chain processes can provide accurate and comprehensive information for sustainable reporting. Other services take advantage of the permanent nature of blockchain to notarize documents, issue and transfer licenses or digital art, enable faster post-trade settlement of securities, and even facilitate secure communications and website certification verification (Meiklejohn and Cary, 2018).

Blockchain can also verify the authenticity of the pharmaceutical supply chain for drugs and facilitate the exchange and monitoring of medical records among physicians and hospitals. In finance, blockchain can support transactions that require security, trust, and consensus (Liang et al., 2021). Different applications include the storage of medical insurance information and healthcare data, such as medical records, diagnosis certificates, and patient-owned medical apps (Liang et al., 2021). Other studies have reported various blockchain applications, including "e-government, food supply information security, energy systems, sharing economy, health care, education, online education, intellectual property management, banking, personal identity, asset ownership, supply chains, and the prevention of unethical business transactions" (Yang and Hwang, 2020, p. 2).

Smart Contracts

Blockchain enables independent parties to create, manage, synchronize, and update common share records. It also facilitates the decentralized management of data and enforcement through algorithms known as smart contracts. Kostić and Sedej (2022) point out that a smart contract is a method for recording obligations that trigger other commitments and operate automatically. The benefits of blockchain's distributed data management, consensus mechanisms, and automated execution via smart contracts primarily apply to transactions that can only be conducted through blockchain, as well as to highly standardized, verifiable, and codifiable transactions. Smart contracts consist of a set of covenants programmed into the blockchain code, executing these covenants when predefined conditions are fulfilled. This execution is automatic and deterministic, consistently resulting in the correct outcome. The immutability feature of blockchain protects contract terms from manipulation or alteration. A smart contract can also store a digital asset prior to exchange; for example, it can withhold payment to a vendor until the customer has confirmed full receipt under agreed conditions. In this scenario, the shipment is not released until the funds have been transferred from the customer to the blockchain first. The future of digital contracts and digital asset management may significantly rely on blockchain's evolution and smart contracts' adoption (Coyne and McMickle, 2017). Smart contract technology is secure and transparent, as cryptography safeguards information, and coded processes operate automatically (Rijanto, 2021).

Smart contracts can automatically execute financial transactions and enforce obligations among all parties without the need for a middleman or associated costs. They enhance speed and simplify processes, thereby reducing licensing, ticketing costs, and overhead expenses. Simultaneously, data integrity during transmission is preserved, and transactions are executed only when programmed conditions are met. Furthermore, all parties involved in the transaction can access recorded information, significantly decreasing the likelihood of errors (Kapadia, 2020). Another intriguing application of smart contracts is in supply chain finance, which mitigates manual operational risks. When the parties share contractual information on a decentralized ledger accessible to the bank, customer, and supplier, payments can be processed according to specific times and conditions programmed into the smart contracts.

Another example of smart contract use would be estate planning. An individual can create a smart contract to manage his/her will imprinted in the code of a blockchain. The contract automatically transfers all the testator's money and assets to the beneficiary when the person dies. Additionally, the testator can include limits or conditions like triggering the contract when the beneficiary reaches a certain age (Trivedi et al., 2021).

Blockchain Advantages

Blockchain technology offers numerous significant advantages that are transforming various industries and processes. The first major benefit is its decentralized nature and enhanced security features. By eliminating the need for central authorities, blockchain creates a distributed system where data is stored across multiple nodes, making it extremely difficult for hackers to manipulate or compromise the network. Advanced cryptography further ensures the security of all transactions and stored data. Another advantage pertains to transparency and traceability. Every transaction is recorded in a public ledger visible to all network participants, establishing an unbroken chain of information from origin to destination. This complete audit trail significantly reduces the potential for fraud and simplifies verifying the authenticity of products and transactions throughout supply chains. Additional benefits include a high level of data integrity. Once information is recorded on the blockchain, it cannot be altered or deleted, creating a permanent and tamper-proof record. This feature is particularly valuable for maintaining legal documents, contracts, and other sensitive information that requires absolute certainty of authenticity.

Smart contracts represent a revolutionary advancement enabled by blockchain technology. These self-executing contracts have their terms directly written into code, automatically enforcing agreements when predetermined conditions are met. This automation reduces the need for intermediaries, significantly decreases transaction costs, and accelerates execution times. Additionally, cost reduction is a significant benefit organizations can achieve through blockchain implementation. Businesses can operate more efficiently and allocate resources to other critical areas by eliminating intermediary fees and reducing administrative overhead. The automation of various processes further contributes to these cost savings.

Ultimately, blockchain systems enhance speed and efficiency, delivering significant operational advantages. In contrast to traditional systems that function only during business hours, blockchain networks operate continuously, allowing for near-instant transaction processing. This capability dramatically reduces settlement times from days to mere minutes and eliminates the need for lengthy reconciliation processes. Blockchain technology enables innovative business models that were previously unattainable; it supports the creation of decentralized applications, allows for new forms of asset tokenization, and opens up opportunities for novel fundraising methods such as Initial Coin Offerings (ICOs) and Security Token Offerings (STOs).

Blockchain Disadvantages

As an emerging technology, blockchain has some disadvantages: storage and execution costs, its intensive use of electricity, and the side effects of global warming, security issues, and cybercrime. Blockchain technology is costly, and its hardware requires substantial energy. Energy consumption represents a significant drawback of blockchain technology, particularly for proof-of-work systems. The mining process needed to validate transactions and maintain network security demands enormous amounts of electricity, leading to considerable environmental concerns. This high energy usage contributes to ecological issues and results in substantial operational costs that may be unsustainable in the long term (Atik & Kelten, 2021).

From an operational perspective, the speed of blockchain is increasing, but traffic is rising as well. The scalability limitations of blockchain technology pose a significant challenge for widespread adoption. Current blockchain networks struggle to process large volumes of transactions efficiently, with major networks like Bitcoin managing only a few transactions per second compared to traditional payment systems that can handle thousands. This bottleneck creates delays and increases transaction costs during periods of high network activity. Access to blockchain platforms is becoming more congested as adoption accelerates. It is essential to address the need to develop greater capacity (Kapadia, 2020).

One critical and sensitive issue is the carbon footprint that blockchain technology contributes to the environment. While countries struggle to agree on solutions for the global warming crisis, blockchain has significant progress to make in participating in a greener economy. Addressing this problem is essential for the widespread adoption of this technology (Kapadia, 2020).

Another challenge for blockchain is the regulatory framework. Regulatory uncertainty continues to hinder blockchain adoption across various industries. The lack of clear regulatory frameworks in many jurisdictions creates hesitation among potential adopters, particularly in the finance and healthcare sectors, where compliance is crucial. Organizations must navigate complex and sometimes contradictory regulations while implementing blockchain solutions. In most countries, it is not straightforward to determine how to address issues such as conflict, fraud, terrorism financing, and cybercrime (Atik & Kelten, 2021). Blockchain technology holds significant promise for the banking industry, but its vulnerabilities could limit its application, especially in more sophisticated markets like Western Europe and the United States. Additionally, blockchain requires adequate monitoring of international standards (Kapadia, 2020). All efforts in this direction will contribute to accelerating the adoption of blockchain technology.

The current configuration of blockchain platforms makes them inefficient compared to other payment systems like Visa or Mastercard. The time required to validate a transaction poses a significant disadvantage for widespread adoption. Ultimately, the size of the information and the technology's storage capacity will present a critical limit to its expansion. For instance, a single network node needs more than 170GB of capacity (Liang et al., 2021). Other challenges that blockchain technology encounters include hard forks, compatibility among different chains, and security issues such as a 51% attack. These vulnerabilities necessitate constant monitoring and security measures to mitigate potential risks.

Despite the potential benefits of implementing blockchain technology, it still encounters numerous obstacles to widespread adoption. In accounting, for instance, public blockchains clash with the need for information confidentiality; the inability to modify historical data and the limitations surrounding transaction verification are traits that, according to some authors, render blockchain undesirable for business accounting (Coyne & McMickle, 2017). Additional challenges include cost, efficiency, regulation, security, and system scalability, among others. For example, utilizing Ethereum is twice as expensive as the cloud services offered by Amazon SWF (Poonam et al., 2020).

In sectors like supply chains, blockchain is still in its early stages, with implementation costs and technological complexity presenting significant barriers to its expansion. Challenges such as integration with legacy IT systems, scalability, regulatory issues, and substantial investment in the necessary computing power hinder blockchain from gaining momentum (Rijanto, 2021). Very few start-ups are able to implement blockchain on a larger scale. Despite the interest in leveraging blockchain to address market inefficiencies and reduce costs and delays, the technology's nascent stage suggests a long road ahead before blockchain becomes standard in the business world. Blockchain continues in its phase of experimentation and improvement, aiming to eventually unlock its full potential (Yang, 2019).

Blockchain Adoption

Blockchain technology differs from the existing infrastructure of the financial system in that even compliance with regulatory obligations presents a limitation to its implementation. Blockchain adoption can be likened to the development of transmission control protocol/internet protocol (TCP/IP), which revolutionized computer networking but took many years after its initial introduction in 1972 to achieve widespread acceptance (Lee et al., 2019, p. 3). Positive sentiment towards blockchain adoption can lead to increased returns in Bitcoin and Ethereum. Analyzing social media and news sentiment supports the assertion that investor perceptions of blockchain adoption influence cryptocurrency price movements.

Blockchain technology has seen widespread adoption across various industries, impacting financial markets, including cryptocurrencies such as Bitcoin and Ethereum. Research consistently indicates that enterprise blockchain adoption follows a distinct pattern influenced by several key factors. Guo and Liang (2016) report that financial institutions were early adopters, primarily motivated by potential reductions in transaction processing and settlement costs. A comprehensive study by Holotiuk et al. (2019) identified four critical factors influencing enterprise adoption: organizational readiness and technical infrastructure, regulatory environment and compliance requirements, network effects, industry standardization, and perceived benefits relative to implementation costs.

This literature review analyzes academic research and empirical studies to explore how blockchain adoption affects the returns of Bitcoin and Ethereum. Positive sentiment toward blockchain adoption leads to increased returns for both Bitcoin and Ethereum. Sentiment analysis from social media and news supports the assertion that investor perception of blockchain adoption influences cryptocurrency price movements. Despite the advantages of blockchain, there are challenges and obstacles to its adoption in specific sectors. For instance, Xu et al. (2019), Chang et al. (2020), and Mishra and Kaushik (2021) argue that advancements in the banking sector are relatively difficult. Moreover, AlShamsi et al. (2022) note that the adoption of blockchain in organizations is still in its early stages, with limited studies in this area. They report that, first, most prior studies have focused on the technological factors affecting blockchain adoption. Second, much of the existing literature on blockchain adoption emphasizes supply chains while overlooking business adoption, particularly in the banking sector. Additionally, the authors report that most studies regarding organizational decisions on blockchain adoption have relied on structural equation modeling (SEM) to examine the relationships among various adoption factors. As blockchain is an emerging technology, research on its adoption in banking remains quite limited. Among the few studies, for example, Kawasmi et al. (2020) examine how blockchain is adopted in the global banking industry and identify three categories of adoption factors: supporting, hindering, and circumstantial. Conversely, Saheb and Mamaghani (2021) qualitatively affirm that technological, organizational, and environmental factors significantly impact adopting blockchain in the banking sector compared to other industries. Finally, Ozturan et al. (2019) focus on the level of technology readiness within the banking industry.

Blockchain introduces a decentralized ledger that stores the complete transaction history across numerous controlled-access computers within a bank, replicating the breadcrumb trail of banking activities (Butt et al. 2019; Uddin et al. 2021; Bellucci et al. 2022). It is impossible to delete or alter this activity history. Blockchain executes various core workflows in finance and banking, including record-keeping, cybersecurity, currencies, debts, and equity management (Bellucci et al., 2022). In a recent study, Lu et al. (2024) argue that blockchain, as an emerging technology, has attracted attention in both academic and economic fields; however, its adoption remains limited within the banking sector in Taiwan. The authors investigate the critical factors affecting blockchain adoption from an organizational perspective within the banking industry. Their methodology employs hybrid methods to address gaps in the literature, utilizing the technology-organization-environment framework as a basis and blending relevant factors into a comprehensive framework to identify pertinent evaluation criteria. The findings indicate that organizational and technological aspects are primary considerations for enhancing and promoting the effectiveness of blockchain adoption. This study proposes valuable strategies to encourage blockchain adoption in Taiwan's banking sector.

Blockchain is a groundbreaking technology; however, it has not achieved widespread adoption. It also requires a significant financial investment, presenting a risk similar to that associated with any investment in innovative technologies. The pool of experts in the field remains limited, with only a handful of companies integrating blockchain into their daily operations (Liang et al., 2021). A common challenge in adopting innovative technologies like blockchain is security; the potential for cyberattacks and other unknown threats hampers further implementation (Trivedi et al., 2021).

Although blockchain remains in an experimental phase, it also shows signs of being a “relatively mature technology.” One of its most distinctive qualities is enabling independent parties who do not necessarily trust each other to conduct business transparently and maintain reliable records with added oversight (Kostić and Sedej, 2022). However, this interest does not exempt companies from uncertainties related to financial, technical, and operational aspects.

Despite the trustless nature of blockchains, a certain level of trust in the consensus rules and the system's underlying code is necessary. Additionally, permitted blockchains require trust in a “gatekeeper” who manages the platform and all other participants within the system. That is why “some level of familiarity and ex-ante trust needs to be established between partners” (Kostić and Sedej, 2022, p. 127). Furthermore, collaboration among parties is critical for successfully adopting blockchain technology, as it is a multi-party system that creates interdependence and follows a sequence of user interactions. Lastly, it is essential to note that once blockchain technology is adopted, it will dramatically influence how companies organize the flow of information and activities to fully leverage the technology's characteristics.

Blockchain has evolved through three distinct stages. The first generation was Bitcoin, a cryptocurrency. The second generation moved beyond monetary transactions to include smart contracts and blockchain-supported services. We are likely at the beginning of the third generation, which is embracing blockchain as a universal platform for research, healthcare, entertainment, government, bookkeeping, and more. The adoption of blockchain may have significant implications for sensitive topics such as democracy. Blockchain could be utilized for voting, storing results, and increasing the visibility of information to the public (Trivedi et al., 2021).

Bitcoin

Based on Nakamoto's 2008 paper, Bitcoin became a reality, and its demand surged in 2013, establishing it as the largest cryptocurrency in the world (Baur and Dimpfl, 2021). However, a significant portion of the population remains unaware of what BTC is; for instance, in El Salvador, when BTC was adopted as legal tender in September 2021, 90% of the population had no idea what Bitcoin was (Czeczeli and Vilonya, 2022). The Bitcoin market operates entirely electronically and peer-to-peer, without a central authority. Transactions are validated by a network of nodes that verify the integrity of the most recent operations against the record of total transactions forming a blockchain. These operations are documented in the ledger, and the latest information is broadcast to other nodes in the network.

Bitcoin encompasses various technological components, including asymmetric public-key cryptography, proof of work (PoW) algorithms, and peer-to-peer (P2P) network technology. It represents a distinctive, digital, decentralized form of currency that does not rely on a trusted third party (Huber and Sornette, 2022).

The total supply of Bitcoins is capped at 21 million, and new Bitcoins must be mined. Miners must present a valid transaction that is confirmed through mathematical computations, using a specific number called a nonce. A block is accepted once the nonce is validated with a hash function. This underscores another difference from fiat currency, where a central bank has the power to print and create additional money (Czeczeli and Vilonya, 2022).

Due to the lack of an asset backing Bitcoin's value, its prices are influenced by supply and demand. Additionally, the marginal cost of production or extraction can affect Bitcoin's price. If the demand for transactions increases faster than the available supply, it is reasonable to expect that the price of Bitcoin will rise. Some studies suggest that past values influence Bitcoin prices in the short term and that blockchain technology does not determine Bitcoin prices (Chen, 2021).

Bitcoin prices are sensitive to user announcements and events aimed at regulating its use and access. Examples include efforts by China or Iran to limit Bitcoin usage, the adoption of Bitcoin as legal tender in El Salvador, Elon Musk's endorsement of Bitcoin as a payment method, and his subsequent withdrawal over concerns about Bitcoin's environmental impact, all of which have affected Bitcoin's price by over 10% in some instances (Czeczeli and Vilonya, 2022). In contrast, hacker attacks do not notably influence Bitcoin prices because the attacks do not damage or compromise its infrastructure; instead, they are largely attempts at theft.

According to data from early 2018, the ten largest mining pools in the world (including F2Pool, AntPool, and BitFury) control approximately 80% of the entire Bitcoin network's capacity. Furthermore, the system is maintained by an authorized group of developers who wield significantly more influence within the Bitcoin ecosystem than many miners. This group of developers arises from the periodic need to update and resolve bugs in Bitcoin's underlying software. Additionally, the developers can alter the commission policy and greatly impact conflict resolution through hard forks. Regarding Bitcoin's anonymity, despite its claim to be an anonymous system, achieving complete anonymity in practice is impossible (Pestunov, 2021).

There is ongoing debate about the bubble that Bitcoin has created in the financial world. According to Huber and Sornette (2022), Bitcoin has initiated one of the largest bubbles in human history; it represents a breakthrough that combines existing technologies with novel applications, known as "combinatorial evolution," and it will have significant socio-economic impacts as a disruptive technological innovation. The rise in Bitcoin prices has spurred a speculative wave, leading to a series of bubbles that have supported the existence of Bitcoin. Bitcoin is a successful case following a number of failed attempts like DigiCash, Hashcash, or Bitgold.

Bitcoin represents a significant social aspect, functioning as a technology that coordinates the actions of miners, investors, and developers. This last group is responsible for the protocol that governs Bitcoin. When participants adhere to this protocol, the network's consensus sustains Bitcoin's relevance. Nodes that diverge from the network's agreement can circumvent the protocol, leading to a "hard fork"- a replicated version of the original protocol that adopts new rules for operating and recording transactions. Since the launch of the first crypto-exchange in 2010, Bitcoin has gone through multiple boom-bust cycles, each marked by increased volume and new price peaks (Figure 2). These cycles align with the Gartner Hype Cycle definition, where every bull market exhibits an exponential price surge followed by a dramatic crash (Huber and Sornette, 2022). Such bubbles have contributed to the spread of Bitcoin technology and heightened its visibility.

Figure 2. Bitcoin's historical price in USD (logarithmic scale).

Bitcoin's Historical Price in USD (Logarithmic Scale).

Note. Source: investing.com (2023)

Bubbles are fueled by positive feedback. This represents a "self-reinforcing feedback loop of herding behavior" (Huber and Sornette, 2022, p. 128). As a result of the proliferation of these bubbles, Bitcoin has broadened its adopter base. Developers, entrepreneurs, and speculators have engaged with these cycles with great excitement, committing financial resources that reinforced these loops and garnered mainstream media attention. These bubbles, driven by speculation, have propelled the Bitcoin ecosystem into existence (Huber and Sornette, 2022). This marks the first time a technological breakthrough has leveraged Hype Cycles as a necessary process for adoption and diffusion, reinforcing the idea of bubbles as catalysts for technological revolutions. It is reasonable to anticipate a new wave of enthusiasm in Bitcoin's history that will draw more users with a diminished aversion to risk due to the hype surrounding cryptocurrencies.

Ethereum

In 2014, Vitalik Buterin published the white paper "Ethereum: A Next-Generation Smart Contract and Decentralized Application Platform," marking a milestone in the practical use of cryptocurrencies. Referring to Satoshi's grand experiment, this paper highlights the concept of a work-based blockchain that facilitates consensus on the chronological order of transactions. It also proposed one of Ethereum's most important characteristics: the ability to create smart contracts using programming-contracts that automatically reallocate digital assets based on pre-specified parameters (Buterin, 2014).

Ethereum has a utility token, Ether, which is used as a payment method to conduct operations within the network. Users pay with Ethers to process transactions on the Ethereum Virtual Machine, the environment for Ethereum smart contracts. These smart contracts are reliable, immutable, and recorded on a distributed-ledger blockchain platform (Routledge and Zetlin-Jones, 2022). The Ethereum blockchain consists of an ever-growing set of blocks. Ethereum transactions require payment in Ethers, referred to as gas; it serves as a fee for operating on the network. Users set the limit they are willing to spend on gas for a specific transaction. Gas limits and gas prices are integral components of smart contract execution. Gas prices represent a conversion rate for Ethers; the higher the conversion rate, the more appealing it becomes for miners to process the transaction. Miners are responsible for executing the code within a smart contract (Bistarelli et al., 2020). Ethereum smart contracts are coded in a programming language called Solidity and are executed by the Ethereum Virtual Machine. Smart contracts carry out a series of steps outlined in clauses and covenants; they consist of three components:

The code of a program embodies contractual logic, comprising the messages that the program can receive, representing the events that activate the contract, and the methods that trigger the reactions anticipated by the contractual logic. (Pinna et al., 2019, p. 78197)

Once the terms of a smart contract are executed, they cannot be altered, and the parties involved in the transaction must honor their commitments. Ethereum is a general platform for smart contracts, but there are examples of more specialized platforms such as R3 Corda, Hyperledger Fabric, Quantum, and Achain. Smart contracts can be categorized based on their functionality: notary contracts that register agreements between parties, wallet contracts that serve as deposits, library contracts that support functionally different contracts, financial contracts for monetary value management, and game contracts used for digital collections and lotteries.

3. MODEL AND DATA

The following are a series of constructs that will underpin the hypotheses and research model of the present study:

Blockchain "is an incorruptible digital ledger of economic transactions that can be programmed to record not just financial transactions but virtually everything of value" (Kapadia, 2020, p. 130).

Cryptocurrency is defined as "a chain of digital signatures that creates an electronic coin" (Nakamoto, 2008, p. 2). This study will focus exclusively on information regarding Bitcoin and Ethereum, as they represent nearly two-thirds of the total cryptocurrency market.

Smart contracts serve as a means to record obligations, triggering further obligations that operate automatically. The advantages of blockchain's distributed data management, consensus mechanisms, and automated execution via smart contracts are primarily applicable to transactions solely managed on the blockchain, as well as to highly standardized, verifiable, and codifiable transactions (Kostić and Sedej, 2022, p. 125). We utilize smart contracts as a practical example of the application of cryptocurrencies and the actual use of blockchain technology.

Despite the large number of active cryptocurrencies in the world, Bitcoin and Ethereum account for nearly two-thirds of the total capitalization value in the cryptocurrency markets. Furthermore, the availability of historical data to analyze their performance makes them ideal candidates for supporting this research. These two cryptocurrencies have allowed blockchain technology to evolve beyond Satoshi Nakamoto's original concept from 2008. If the size and liquidity of Bitcoin continue to increase, price volatility will likely stabilize within a lower range (Folkinshteyn and Lennon, 2016). Based on this, the hypotheses of this research are:

H_1/H_2 The growth in volume of smart contracts coded in the Ethereum network is positively correlated to the return on Bitcoin (and Ethereum).

Figure 3. Research model

Research Model

This figure outlines our research model, where we hypothesize that the growth in Ethereum's smart contract volume will be related to Ethereum and Bitcoin returns.

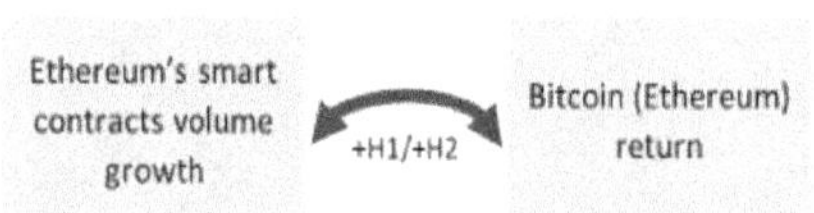

This figure illustrates our research model. We hypothesize that the growth in Ethereum's smart contract volume will correlate with the returns of Ethereum and Bitcoin.

Our study focuses on connecting smart contracts coded in Ethereum with the performance of both Bitcoin and Ethereum. Cryptocurrencies have generated an awareness effect surrounding the technology behind them. Although not all blockchain activity directly impacts cryptocurrencies, certain functions, such as smart contracts coded in Ethereum, may influence their performance. To assess the practical use of cryptocurrency blockchain technology, we use the number of smart contracts coded in the Ethereum network as a proxy. According to Davis's TAM theory, the perceived usefulness of technology is the most essential factor for its adoption and actual use. While cryptocurrencies have faced scrutiny due to their perceived lack of practical use, this perception has shifted with the introduction of functionalities like smart contracts on the Ethereum network. Ethereum is regarded as a general platform for programming smart contracts, with other more specialized platforms like Fabric or Corda also available. However, Ethereum uniquely utilizes its token (Ethers) for operation, reflecting a direct impact on a major cryptocurrency. Hypotheses H1 and H2 propose that smart contracts are related to the returns of Bitcoin and Ethereum. This study measures the correlation between the growth rate of smart contracts created on the Ethereum Virtual Machine (EVM) and the returns on Bitcoin and Ethereum. Smart contract information can be accessed at dune.com (Dune, 2023), which extracts data from EVM. While Bitcoin lacks a function comparable to Ethereum, it remains the most significant cryptocurrency in terms of market capitalization and general awareness. Price information for both Ethereum and Bitcoin is available at Coinmarketcap.com (Coinmarketcap, 2023). Price returns are computed using logarithmic returns, and Eviews software is employed for these calculations.

4. EMPIRICAL RESULTS

Smart contract data obtained from dune.com shows the daily creation of contracts since Ethereum's launch on August 7, 2015. Our sample includes 2,824 observations. From August 7, 2015, to April 30, 2023, 59,610,160 contracts were created. We organized our data to include both Bitcoin and Ethereum, enabling a consistent analysis period.

We analyze the data from a daily growth rate perspective to explore a potential relationship between smart contracts and the returns of BTC and ETH. We consider the daily changes (returns) in BTC and ETH prices, with all growth rates computed using logarithmic formulas. The results of this analysis provide a clearer view of the story (Table 1). First, the average growth of smart contracts is 0.00519 (0.519%), significantly higher than the returns of BTC and ETH, which are 0.00165 and 0.00231, respectively. Next, the minimum value of smart contract growth is a positive 0.0001, while the logarithmic returns for BTC and ETH include negative values of -0.46473 and -1.30290, respectively. In other words, smart contracts do not decrease, whereas the returns of BTC and ETH can. This illustrates the difference between the transactional purpose of cryptocurrencies, where market values fluctuate based on supply and demand, resulting in positive or negative swings to balance positions. In the case of smart contracts, this serves as a functional characteristic of Ethereum. As more individuals and companies utilize smart contracts to conduct their business, this number will continue to grow, provided that the generic platform EVM is used. This study is limited to EVM; however, other more specialized platforms like Fabric and Corda also support the functional use of blockchain technology. The standard deviation of smart contract growth is 0.02798, while the BTC price return is 0.03852 and the ETH price return is 0.06300, confirming the more erratic trend in BTC and ETH prices previously discussed. Another interesting result is the negative skewness of BTC and ETH returns, whereas growth remains positive for smart contracts (Figure 3). Finally, all three parameters exhibit a Kurtosis level well beyond 3, indicating fat tails.

This table provides descriptive statistics for smart contracts growth rate, Bitcoin, and Ethereum returns. The sample dates are from August 8, 2015, to April 30, 2023.

Table 1. Descriptive statistics: Smart contracts growth, BTC and ETH returns

Metric	SmartC Cum Growth	BTC Price Return	ETH Price Return
Mean	0.00519	0.00165	0.00231
Median	0.00130	0.00181	0.00044
Maximum	1.31661	0.22512	0.41044
Minimum	0.00010	-0.46473	-1.30290
Std. Dev.	0.02798	0.03852	0.06300
Skewness	37.619	-0.750	-3.135
Kurtosis	1,720.196	14.124	73.249
Observations	2,823	2,823	2,823

Figure 4. Histogram smart contracts growth, bitcoin and ethereum returns

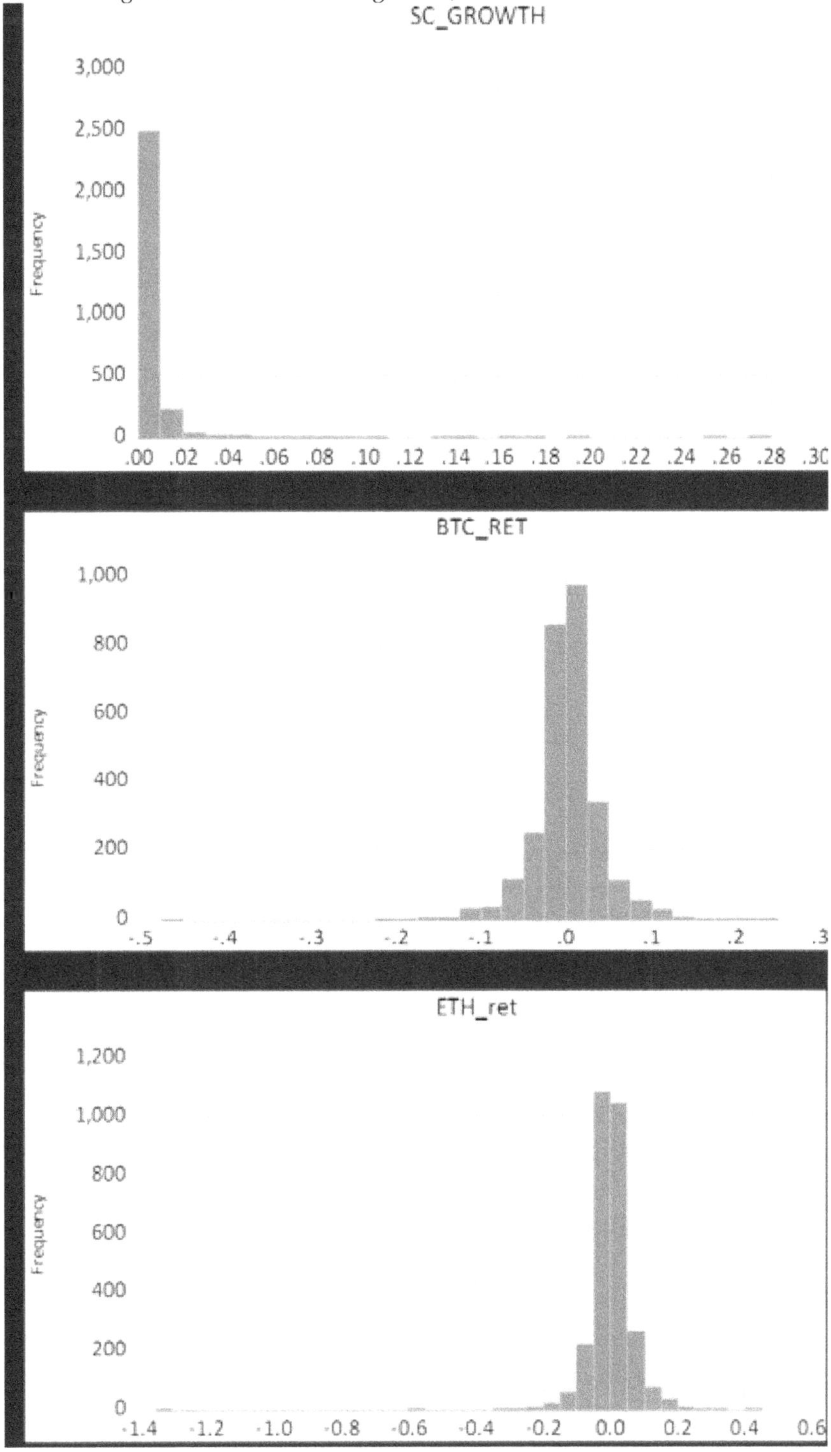

This figure provides a histogram for Smart Contracts growth rate, Bitcoin and Ethereum returns during the sample period of August 8, 2015, and April 30, 2023.

Finally, the correlation analysis of smart contract growth with BTC and ETH returns reveals mixed results (Table 2). The correlation between smart contracts and ETH is -0.3303, indicating that as the growth rate of smart contracts increases, the return rate of ETH decreases. In terms of Bitcoin, its correlation is nearly 0, specifically -0.0201, but this indicator has a p-value above 0.05; therefore, it is not significant.

This table reports the correlation analysis of smart contract, Bitcoin, and Ethereum returns.

Table 2. Correlation analysis smart contracts growth, BTC and ETH returns

Correlation (Probability)	**SC Cum. Growth**	**BTC Return**	**ETH Return**
SC Cum Growth	1.00000		

BTC Price Return	-0.02014	1.00000	
	(0.2848)	-----	
ETH Price Return	-0.33032	0.55737	1.00000
	(0.0000)	(0.0000)	-----
Covariance Analysis: Ordinary			

Our analysis indicates that we do not support our first hypothesis regarding a correlation between the volume growth of smart contracts on the Ethereum network and Bitcoin returns. However, we observe that the volume growth of smart contracts on the Ethereum network is negatively correlated with Ethereum's returns, which contradicts our expectations. Several explanations could account for why the correlation between smart contracts and Bitcoin is not significant, while it is significantly negative for Ethereum. Both Bitcoin and Ethereum have a speculative component, a trait absent in smart contracts. Unlike smart contracts, which may have expiration dates, Bitcoin and Ethereum do not; once mined, these cryptocurrencies remain permanently within the network. Regarding expiration dates, the information available for smart contracts includes the number of contracts created in the Ethereum Virtual Machine (EVM), but we lack data on how many of those contracts are currently active. In the case of Ethereum's negative correlation, its return rate may exhibit a trend similar to Bitcoin's as Ethereum evolves as a crypto asset. Lastly, and possibly the most critical reason, is that blockchain technology has garnered attention as a business solution independent of the fortunes of cryptocurrencies. This suggests that the applications of blockchain technology will continue to expand as its adoption increases.

Another potential explanation for why BTC returns are not related to the growth of smart contracts may be its transactional nature. BTC was designed as a payment method, aiming to serve as an alternative to fiat currencies rather than offering specific functionalities like smart contracts with their tokens, which were introduced in ETH seven years later. While BTC relies on blockchain technology for bookkeeping purposes, it does not utilize its own blockchain to develop additional capabilities. Moreover, the limited supply of BTC, both mined and unmined, poses a significant constraint on the adoption of blockchain technology. Once the maximum limit of twenty-one million BTC is reached, its blockchain will focus on tracking and recording all transactions among users.

Most recent examples, such as the creation of BTC ETFs approved for institutional investors, also highlight the transactional nature of BTC. This would be an interesting research topic to explore whether the creation of this ETF, and more importantly, an increased participation of institutional investors, have affected the patterns found in this study.

During this study, we reference other distributed ledger technologies such as Corda and Fabric. One limitation of this study is that it does not account for the potential growth of blockchain adoption through these platforms, which may influence how solution providers and users choose to enhance blockchain adoption. ETH serves as a versatile platform that can be an intriguing starting point for companies looking to apply blockchain technology to their processes. However, as a general solution, it has limitations regarding what a company can develop on such a platform compared to its actual needs.

5. SUMMARY AND CONCLUSIONS

The emergence of cryptocurrencies in 2008, highlighted by Nakamoto's paper, created a completely new arena for investments and speculation. However, Nakamoto's innovations also brought various solutions to the business world, including the use of blockchain technology to securely and accurately record information. Due to this shared origin, the first part of this study aims to explore a possible connection between blockchain technology and cryptocurrencies. There is a significant opportunity to develop a theory of finance and technology within this field. Nevertheless, this potential connection is analyzed through the creation of smart contracts on the

Ethereum platform, practical applications of blockchain technology, and cryptocurrency as an incentive, specifically Ethers.

In analyzing a potential connection between smart contracts and cryptocurrencies, the information considered is the growth in the number of contracts created on the Ethereum Virtual Machine and the daily returns of Bitcoin and Ethereum. This analysis begins on August 7, 2014, when the first contract was coded in the EVM. Our analysis indicates that there is no connection between the growth of smart contracts and the return of Bitcoin. We observe a negative correlation between the growth of smart contracts and the return of Ethereum. Finally, we find a strong correlation between the returns of Bitcoin and Ethereum.

This lack of connection between smart contracts and BTC, along with a negative correlation with ETH, may be explained by several factors. The information about smart contracts from EVM pertains to the number of contracts created without indicating how many are currently active. Additionally, smart contracts serve a functional purpose that is independent of market fluctuations; they will be created and executed regardless of whether the crypto market is in a bullish or bearish trend. Moreover, Bitcoin and Ethereum carry a speculative nature that is unrelated to business operations. It is also important to consider that the Ethereum network is a general platform for coding smart contracts, adhering to the rules of the broader network. It is an open system, and all participants must follow Ethereum's protocol. For certain businesses, this technology may not meet their specific needs, such as those requiring a need-to-know basis, notary services, or managing permissioned users. In these cases, alternative blockchain solutions like Fabric, which specializes in general business, and Corda, which focuses on banking systems subject to mandatory regulations, may be more suitable. Additionally, regarding ETH, the negative correlation can be attributed to the fact that as ETH matures as a crypto asset, it is likely to show lower returns, similar to what has occurred with BTC.

Blockchain technology applications have the opportunity to thrive regardless of faith in cryptocurrencies, and cryptocurrencies will eventually face a heavier regulatory framework, but they are a method of payment, unit of measure, or store of value, not a tool for business solutions. This lack of connection with BTC and limited connection with ETH may be excellent news for both parties, as well as blockchain technology applied to smart contracts and cryptocurrency performance.

A particular field of study is the practical use of blockchain technology. As mentioned in this research, there is limited knowledge about what blockchain can offer as a business solution and, more importantly, how to implement it. Combining finance and technology represents a demanding task, but as other technologies like the Internet and artificial intelligence continue to dominate many headlines, this combination of finance and technology resembles a long-term marriage.

For practitioners, several implications can be drawn. The lack of connection between smart contracts and Bitcoin, along with a slight negative correlation with Ethereum returns, is promising news for companies investing in blockchain technology for their operations. If Ethereum's volatility increases its separation from practical uses of the Ethereum Virtual Machine, smart contracts will function without the distorting effects of Ethereum's challenges. Concerning the potential regulatory burden that cryptocurrencies may encounter in the near future, their impact on the Ethereum Virtual Machine might be limited, but further study will be necessary to validate that assertion.

Exploring and learning about the adoption of blockchain technology is a valuable field of study. As technology advances and expertise in this area increases, more resources will emerge to enhance understanding. For instance, this study focuses exclusively on the Ethereum Virtual Machine, which serves as a generic platform. Investigating the development of specialized platforms offers significant opportunities for academic inquiry. As this field continues to evolve, additional data sources will become available. A major limitation of this study was the insufficient information regarding smart contracts and their behavior since their inception.

REFERENCES

AlShamsi, M., Al-Emran, M., & Shaalan, K. (2022). A systematic review on blockchain adoption. *Applied Sciences (Basel, Switzerland), 12*(9), 4245. DOI: 10.3390/app12094245

Atik, A., & Kelten, G. S. (2021). Blockchain technology and its potential effects on accounting: A systematic literature review. *Istanbul University Journal of the School of Business Administration, 50*(2), 495–515. DOI: 10.26650/ibr.2021.50.806870

Baur, D. G., & Dimpfl, T. (2021). The volatility of Bitcoin and its role as a medium of exchange and a store of value. *Empirical Economics, 61*(5), 2663–2683. DOI: 10.1007/s00181-020-01990-5 PMID: 33424101

Bellucci, M., Bianchi, D. C., & Manetti, G. (2022). Blockchain in accounting practice and research: A systematic literature review. *Meditari Accountancy Research, 30*(7), 121–146. DOI: 10.1108/MEDAR-10-2021-1477

Bistarelli, S., Mazzante, G., Micheletti, M., Mostarda, L., Sestili, D., & Tiezzi, F. (2020). Ethereum smart contracts: Analysis and statistics of their source code and opcodes. *Internet of Things : Engineering Cyber Physical Human Systems, 11*, 100198. DOI: 10.1016/j.iot.2020.100198

Buterin, V. (2014). Ethereum: A next-generation smart contract and decentralized application platform. [White Paper]. Ethereum.org.

Butt, T. A., Iqbal, R., Salah, K., Aloqaily, M., & Jararweh, Y. (2019). Privacy management in social internet of vehicles: Review, challenges and blockchain-based solutions. *IEEE Access : Practical Innovations, Open Solutions, 7*, 79694–79713. DOI: 10.1109/ACCESS.2019.2922236

Chang, V., Baudier, P., Zhang, H., Xu, Q., Zhang, J., & Arami, M. (2020). How blockchain can impact financial services—The overview, challenges and recommendations from expert interviewees. *Technological Forecasting and Social Change, 158*, 120166. DOI: 10.1016/j.techfore.2020.120166 PMID: 32834134

Chen, Y. (2021). Empirical analysis of Bitcoin price. *Journal of Economics and Finance, 45*(4), 692–715. DOI: 10.1007/s12197-021-09549-5

Chong, A. Y. L., Lim, E. T. K., Hua, X., Zheng, S., & Tan, C. (2019). Business on chain: A comparative case study of five blockchain-inspired business models. *Journal of the Association for Information Systems, 20*(9), 1308–1337. DOI: 10.17705/1jais.00568

Coinmarketcap. (2023, May). *Bitcoin historical data.* https://coinmarketcap.com/currencies/bitcoin/historical-data/

Coyne, J. G., & McMickle, P. L. (2017). Can blockchains serve an accounting purpose? *Journal of Emerging Technologies in Accounting*, *14*(2), 101–111. DOI: 10.2308/jeta-51910

Creed, R. R., & Northard, A. (2021). The role of blockchain technology in cryptocurrencies. *Journal of New Business Ideas and Trends*, *19*(3), 8–13.

Czeczeli, V., & Vilonya, M. (2021). Exchange rate developments of cryptocurrencies based on event study analysis. *Public Finance Quarterly (0031-496X)* 67, no. 2: 231–47. https://doi.org/.DOI: 10.35551/PFQ_2022_2_5

Davis, F. D. (1986). *A technology acceptance model for empirically testing new end-user information systems: Theory and results.* [Doctoral study, Sloan School of Management, Massachusetts Institute of Technology].

Davis, F. D. (1989). Perceived usefulness, perceived ease of use, and user acceptance of information technology. *Management Information Systems Quarterly*, *13*(3), 319–340. DOI: 10.2307/249008

Davis, F. D., Bagozzi, R. P., & Warshaw, P. R. (1989). User acceptance of computer technology: A comparison of two theoretical models. *Management Science*, *35*(8), 982–1003. DOI: 10.1287/mnsc.35.8.982

Deane, S., & Fines, O. (2023). *Cryptoassets: Beyond the hype*. CFA Institute. DOI: 10.56227/23.1.1

Dune. (2023, February). *Daily SM on EVM.*https://dune.com/queries/4104463

Fenn, J., Raskino, M., & Burton, B. (2017). Understanding Gartner's Hype Cycles. *Gartner. Inc.*, *G00251964*, 35.

Folkinshteyn, D., & Lennon, M. (2016). Braving Bitcoin: A Technology Acceptance Model (TAM) analysis. *Journal of Information Technology Case and Application Research*, *18*(4), 220–249. DOI: 10.1080/15228053.2016.1275242

Guo, Y., & Liang, C. (2016). Blockchain application and outlook in the banking industry. *Financial Innovation*, *2*(1), 1–12. DOI: 10.1186/s40854-016-0034-9

Harvey, C. R., Zeid, T. A., Draaisma, T., Luk, M., Neville, H., Rzym, A., & Van Hemert, O. (2022). *An investor's guide to crypto.* [SSRN Scholarly Paper. Rochester, NY]. https://doi.org/DOI: 10.2139/ssrn.4124576

Holotuik, F., & Moormann, J. (2019). Dimensions, Success Factors, and Obstacles of the Adoption of Blockchain Technology. *Australasian Conference on Information Systems*

Huber, T. A., & Sornette, D. (2022). Boom, bust, and Bitcoin: Bitcoin-bubbles as innovation accelerators. *Journal of Economic Issues (Taylor & Francis Ltd)* 56, no. 1: 113–36. https://doi.org/.DOI: 10.1080/00213624.2022.2020023

Investing.com. (2023, May). *Bitcoin historical data.*https://www.investing.com/crypto/bitcoin/historical-data

Kapadia, J. M. (2020). Blockchain technology: Application in the financial industry. *Scholedge International Journal of Management & Development*, *7*(8), 130–135. DOI: 10.19085/sijmd070801

Kawasmi, Z., Gyasi, E. A., & Dadd, D. (2020). Blockchain adoption model for the global banking industry. *Journal of International Technology and Information Management*, *28*(4), 112–154. DOI: 10.58729/1941-6679.1436

Kiymaz, H., & Munoz Valdez, C. H. (2023). Blockchain in Fintech: A catalyst for disrupting the financial world. In Baker, H. K., Filbeck, G., & Black, K. (Eds.), *The Emerald Handbook on Fintech: Reshaping Finance.* Emerald Publishing.

Kostić, N., & Sedej, T. (2022). Blockchain technology, inter-organizational relationships, and management accounting: A synthesis and a research agenda. *Accounting Horizons*, *36*(2), 123–141. DOI: 10.2308/HORIZONS-19-147

Lee, C. C., Kriscenski, J. C., & Lim, H. S. (2019). An empirical study of behavioral intention to use blockchain technology. *Journal of International Business Disciplines*, *14*(1), 1–21.

Liang, T. P., Kohli, R., Huang, H.-C., & Li, Z.-L. (2021). What drives the adoption of the blockchain technology? A fit-viability perspective. *Journal of Management Information Systems*, *38*(2), 314–337. DOI: 10.1080/07421222.2021.1912915

Litan, A., Leow, A., & Kandaswamy, R. (2022). Hype cycle for blockchain and Web3. *Gartner. Inc.*, *G00770861*, 83.

Liu, L., & Tsyvinski, A. (2021). Risks and returns of cryptocurrency. *Review of Financial Studies*, *34*(6), 2689–2727. DOI: 10.1093/rfs/hhaa113

Meiklejohn, S., & Cary, N. (2018). Introduction. In *The future is decentralized* (pp. 4–8). United Nations Development Programme.

Mishra L, Kaushik V (2021) Application of blockchain in dealing with sustainability issues and challenges of the financial sector. Journal of Sustainable Finance and Investment. . 2021. 19408 05.DOI: 10. 1080/ 20430 795

Nakamoto, S. (2008). Bitcoin: A peer-to-peer electronic cash system. https://bitcoin .org/en/bitcoin-paper

Ozturan, M., Atasu, I., & Soydan, H. (2019). Assessment of blockchain technology readiness level of banking industry: Case of Turkey. *International Journal of Business and Management*, *4*(12), 1–13.

Pestunov, A. I. (2021). Cryptocurrencies and blockchain: Potential applications in government and business. *Problems of Economic Transition*, *62*(4–6), 286–297. DOI: 10.1080/10611991.2020.1968751

Pinna, A., Ibba, S., Baralla, G., Tonelli, G., & Marchesi, M. (2019). A Massive analysis of Ethereum smart contracts empirical study and code metrics. *IEEE Access : Practical Innovations, Open Solutions*, *7*, 78194–78213. DOI: 10.1109/ ACCESS.2019.2921936

Rijanto, A. (2021, November 8). Blockchain technology adoption in supply chain finance. *Journal of Theoretical and Applied Electronic Commerce Research*, *16*(7), 3078–3098. DOI: 10.3390/jtaer16070168

Routledge, B., & Zetlin-Jones, A. (2022). Currency stability using blockchain technology. *Journal of Economic Dynamics & Control*, *142*, 104155. DOI: 10.1016/j. jedc.2021.104155

Saheb T, & Mamaghani, FH. (2021). Exploring the barriers and organizational values of blockchain adoption in the banking industry. Journal of High Technology and Management Research, 32(2):100417ey.

Statista. (2023, October). *Most Used Financial Products in the US 2022*https://www -statista-com.ezproxy.rollins.edu/forecasts/997144/most-used-financial-products -in-the-us

Statista. (2023, October). *Number of daily transactions on the blockchain of Bitcoin from January 2009 to August 8, 2023,*https://www-statista-com.ezproxy.rollins.edu/ statistics/730806/daily-number-of-bitcoin-transactions/

Trivedi, S., Mehta, K., & Sharma, R. (2021). Systematic literature review on application of blockchain technology in e-finance and financial services. *Journal of Technology Management & Innovation*, *16*(3), 88–100. DOI: 10.4067/S0718-27242021000300089

Uddin MA, Stranieri A, Gondal I, & Balasubramanian V. (2021). A survey on the adoption of blockchain in IoT: challenges and solutions. Blockchain Research Applications. . bcra. 2021. 100006.DOI: 10. 1016/j

Wigley, R., & Cary, N. (2018). *The future is decentralized.* United Nations Development Programme.

Xu, M., Chen, X., & Kou, G. (2019). A systematic review of blockchain. *Financial Innovation*, *5*(1), 1–14. DOI: 10.1186/s40854-019-0147-z

Yang, Y.-J., & Hwang, J.-C. (2020). Recent development trend of blockchain technologies: A patent analysis. *International Journal of Electronic Commerce Studies*, *11*(1), 1–12. DOI: 10.7903/ijecs.1931

Chapter 2
Thematic Analysis of Blockchain and Artificial Intelligence in Sustainable Finance

Muhammad Arslan
https://orcid.org/0000-0001-5046-7627
Open Polytechnic, New Zealand & Te Pūkenga New Zealand Institute of Skills and Technology, New Zealand

Alimshan Faizulayev
https://orcid.org/0000-0002-0212-0538
KIMEP University, Kazakhstan

Muhammad Nauman
https://orcid.org/0009-0005-0301-5982
University of Punjab, Pakistan

Zazli Lily Wisker
Eastern Institute of Technology, New Zealand

ABSTRACT

In recent years, sustainable finance has emerged as a central concept at the convergence of finance and the Sustainable Development Goals (SDGs). This chapter synthesizes and systematically reviews the existing literature on blockchain technology and artificial intelligence in sustainable finance. The authors gathered the relevant studies from Google Scholar, Scopus, and Web of Science databases. The authors applied co-occurrence mapping and thematic analysis with the help of VOS viewer and NVivo software. In this chapter, based on a systematic literature

DOI: 10.4018/979-8-3373-0355-0.ch002

review of 509 studies, the findings revealed five thematic areas (i.e., ESG measurement and disclosure; tracking and trading of carbon footprint; renewable energy and circular economy; transparency, security, governance, and compliance; and social and ethical aspects) attracting research interest. The authors of this chapter also identified several challenges faced by sustainable finance in the contemporary business environment. The authors outlines, with a few remarks, general trends for the use of AI and BT in sustainable finance.

1. INTRODUCTION

In the current technological era, regenerative finance (ReFi) and environmental, social, and governance (ESG) have gained popularity, which shows the importance and significance of sustainability in the business landscape (Schletz et al., 2023). Due to this popularity, firms are under immense pressure to utilize technological resources in a responsible and efficient manner. In recent years, investments in sustainability have also increased, which shows the behavior and attitude of investors toward sustainability. Additionally, investors are prioritizing and considering non-financial factors in their decision-making. The sustainable finance market has grown rapidly, and there are no signs of slowing down (Kumar, Sharma, Rao, Lim, & Mangla, 2022). Even during the COVID-19 pandemic, global borrowing involving green bonds, loans, and equity increased 100 times since 2012. In 2020, the United Nations estimated the need for $5 to $7 trillion annually to achieve the Sustainable Development Goals (SDGs), while ESG-related funds will exceed $53 trillion by 2025 (Craig, 2021). Regenerative finance has the potential to drive change and create value for society and investors by channeling capital investments towards sustainability initiatives. In addition, it also has the potential to reshape the financial industry, play a critical role in advancing global SDGs, and make the financial industry more inclusive and sustainable. The UN has 17 SDGs that are crucial to fostering sustainable development globally. Sustainable finance has great potential; however, it still has a long way to go. The financial sector can play an imperative role by transforming our economies into sustainable ones (Alonso & Marqués, 2019; Chayjan, Bagheri, Kianian, & Someh, 2020), which in turn will bring prosperity to the society. Nonetheless, greenwashing practices have increased skepticism among investors about social and environmental information. This has undermined investors' confidence in sustainable finance and led to reduced participation and unsatisfied demand (Lammens, 2020). In addition, it has also discouraged asset managers from designing and developing sustainable products (such as ESG; socially responsible investing (SRI); sustainable; or impact), which ultimately reduced investment in sustainability.

In the realm of finance, blockchain technology (BT) and artificial intelligence (AI) have emerged as powerful tools for driving sustainable practices and advancing environmental and social goals. The origins of AI can be traced to the mid-20th century, with its formal conceptualization emerging from the Dartmouth workshop in 1956, where it was theorized that machines could be designed to simulate several characteristics of human intelligence and learning (Moodaley & Telukdarie, 2023). Since then, AI has evolved significantly, predominantly over the past decade, signifying ample improvements in computational capabilities that align with the increasing emphasis on sustainability reporting. However, the deployment of AI systems presents challenges, predominantly in terms of high computational resource consumption, which raises concerns regarding sustainability (OECD, 2022). Consequently, it is imperative to measure and mitigate the adverse environmental impacts associated with AI development and application to ensure its alignment with sustainability objectives (OECD, 2022). BT and AI can help businesses achieve their sustainability goals by optimizing resource usage, identifying and implementing sustainable practices, and developing sustainable products (Arslan, 2024; Arslan, Faizulayev, Abeuova, & Kayani, 2024). Wang and Sun (2024) conducted a study by taking a sample from 51 countries for the period 2000-2019. They found a significant positive relationship between AI and green innovation. They also documented that AI could enhance environmental innovation. Scholars acknowledged the capabilities of AI and BT in addressing and navigating societal issues such as sustainability. In the presence of complex business environments (Boskov & Drakulevski, 2017; Falcone & Sica, 2019; Nishant, Kennedy, & Corbett, 2020), a cautious approach is needed to mitigate the risks associated with implementation of AI and BT (Arslan, 2024; Arslan et al., 2024). Stanislas and Constanze (2020) conducted a survey and found that most retail investors are interested in sustainable investments. They further argued that investors were ready to accept financial trade-offs for a positive social and environmental impact. However, the greenwashing and lack of reliability are the culprits behind low sustainable investments. A report published by the 2 Degrees Investing Initiative (2020) reveals that a significant number of financial actors who made sustainability claims were not able to provide any evidence and were even not complying with existing regulatory guidance. Such greenwashing practices reduce the investors' confidence and trust in sustainable investments. Hence, the question arises about the extent to which BT and AI can play a role in reducing greenwashing practices and achieving the SDGs through sustainable finance. In contemporary technological era, regenerative finance (ReFi) and environmental, social, and governance (ESG) factors have become imperative in shaping financial decision-making. The financial sector faces challenges in ensuring transparency, accountability, and efficiency as firms increasingly prioritize sustainability. Artificial intelligence (AI) and blockchain technology (BT) have emerged as transformative tools capable of addressing these challenges by enhancing data integrity, automation, and decision-making processes (Arslan et al., 2024). Several studies have explored the role of digital finance and emerging technologies in sustainability (Kumar et al., 2022; Nishant et al., 2020). A recent review by Schletz et al. (2023) provides a broad discussion on blockchain and regenerative finance in one hand. On the other hand, another systematic review by Hu, Zhu, Zhao, and Xu (2023) examines AI's role in corporate sustainability. However, these reviews do not specifically analyze the intersection of AI and blockchain in sustainable finance through a structured thematic approach. This chapter fills this gap by conducting a systematic thematic analysis of AI and blockchain applications in sustainable finance. This chapter aims to answer the following questions: How have AI and blockchain been utilized in sustainable finance? What key themes emerge from existing literature on these technologies? What challenges and opportunities do AI and blockchain present for ESG reporting, carbon trading, and financial governance? Our contribution is threefold i.e., (i) The authors systematically categorize AI and blockchain applications into five key themes in sustainable finance; (ii) The authors provide recommendations for financial institutions, regulators, and policymakers to harness AI and blockchain for financial sector development; and (iii) The authors identify research gaps and propose future empirical studies to validate AI and blockchain's effectiveness in sustainability. The structure of the chapter is as follows: Section 1 discusses the introduction. Section 2 presents the literature review. Section 3 presents the methodology, while Section 4 presents the results and discussion. At the end, Section 5 presents the conclusion, recommendation, significance, and future research directions of the chapter.

2. LITERATURE REVIEW

Sustainable finance weaves environmental, social, and governance (ESG) considerations into financial decision-making, leading to more long-term investments in sustainable economic activities and projects (see Figure 1). Unlike traditional finance, which focuses on profit-driven financing, sustainable finance focuses on ESG risk management, sustainable impact investing, impact first investing, and sustainable philanthropy. Sustainable finance increases brand reputation and value, provides greater access to green bonds and loans, avoids legal and financial risks, and helps in retaining employees and customers.

Figure 1. Sustainable finance

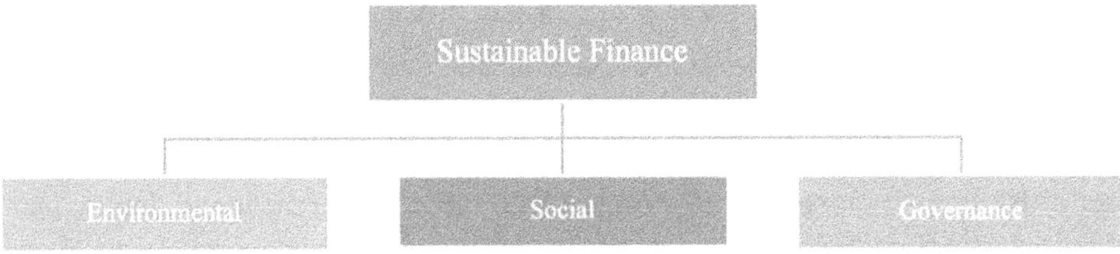

Source: developed by the researcher(s)

Scholars acknowledged the importance of sustainable finance in delivering climate and sustainability objectives (Hu et al., 2023; Mavlutova et al., 2023). Hu et al. (2023) argued that sustainable Finance refers to investment strategies that incorporate ESG criteria to support climate goals, social responsibility, and ethical governance. Schletz et al. (2023) argued that regenerative finance (ReFi) is a concept within sustainable finance that focuses on circular economy principles and equitable capital distribution to create positive environmental and social change. It is also important to meet European international climate commitments and achieve the EU's green deal. The core concepts of sustainability finance involve ESG integration, impact investing, green financing, and social responsibility. Sustainability finance helps in channeling private investment into the transition to a climate-neutral, climate-resilient, resource-efficient, and fair economy, which helps in addressing environmental challenges such as climate change, resource depletion, and pollution. Sustainability finance also ensures that economic growth is inclusive and benefits all segments of society. In existing literature, the term green finance is widely used. Green finance (such as green bonds, green-tagged loans, green investment funds, and climate risk insurance) is a branch of financial activity that encourages bilateral engagement between the environment, finance, and investment. Green finance focuses more on climate issues such as energy efficiency and pollution reduction and excludes social and economic aspects. Artificial intelligence (AI) is technology that enables computers and digital devices to learn, read, write, create, analyze, and make data-driven decisions, widely used in predictive analytics, financial modeling, and ESG scoring (Wang et al., 2024). A blockchain technology (BT) is a decentralized distributed database or ledger shared among a computer network's nodes that ensures data transparency, security, and traceability, particularly beneficial for carbon credit trading, compliance monitoring, and fraud prevention (Yermack, 2017). A blockchain (BT) is a distributed database or ledger shared among a computer network's nodes. Scholars have argued that blockchain and artificial intelligence can analyze large amounts of data, which can help identify areas for improvement and foster sustainability practices (Nishant et al., 2020; Otto & Thornton, 2023). In a similar vein, other scholars also argued that new technologies (such as blockchain and artificial intelligence) are the most disruptive technologies in the business landscape (Arslan, 2024; Ljepava, 2022; Păvăloaia & Necula, 2023) that can help in assessing and tracking the sustainability of businesses and their suppliers. However, the impact of BT and AI on sustainability depends on the development and deployment of these technologies. With the potential to offer a circular economy for food and consumer electronics, BT and AI would be saving more than $200 billion annually by 2030. Scholars argued for the need for stakeholders' engagement, clear policies and regulations, and integration of existing systems to get the most benefits from BT and AI (Al-Sartawi, 2022; Andrianaivo & Kpodar, 2011; Craig, 2021). Scholars also agreed that blockchain and artificial intelligence are disruptive technologies (Arslan, 2024; Christensen, Raynor, Dyer, & Gregersen, 2011), and these technologies can also help firms in manage and report their financial and non-financial performance and ensure transparency and data integrity. For example, blockchain technology can minimize the risks of errors and fraud and ensure the security and transparency of financial transactions.

Correspondingly, artificial intelligence can automate financial reporting and analysis, which leads to efficiency. The degradation of the physical environment and the climate crisis are complex problems. However, BT and AI offer unprecedented opportunities for streamlining data collection and analysis, allowing companies and regulators to make better-informed decisions, automate tedious processes, and adapt swiftly to dynamic market conditions to achieve the SDGs. Nevertheless, companies across industries are busy leveraging sustainable technologies not only to enhance their reputation and boost their attractiveness to investors but also to drive innovation and create long-term value.

Sustainable tech innovations, such as renewable energy solutions and waste reduction initiatives, enable companies to minimize their environmental footprint, foster inclusive growth, and uphold ethical standards. The existing literature shows a rise in awareness about BT and AI in sustainable finance to achieve the SDGs (Al-Sartawi, Al-Okaily, Hannoon, & Khalid, 2022; Al-Sartawi, 2022). Scholars (such as Türegün (2019); Al-Sartawi (2022); Arslan (2024); and Arslan and Faizulayev (2024)) highlighted the need to explore the impact of BT and AI on financial reporting, ESG, green finance, and sustainability. Environmental, social, and governance (ESG), also called sustainability, refers to a set of criteria used to evaluate firms' commitment to sustainable products and services. It is often used interchangeably with the term sustainable finance, which refers to investment decision- making in sustainable economic activities and projects and consideration of ESG factors.

In 2004, United National introduced the concept of ESG in a report, "Who Care Winds: Connecting the Financial Market to a Changing World." Since then, it has received evolving attention from academia, researchers, and regulators. In fact, the number of ESG-related publications have increased to sixfold from 2004-2008 to 2018-2022 (Li, Wang, Sueyoshi, & Wang, 2021). The question arises: how can AI and BT promote ESG? ESG is heavily dependent on data, and AI and BT can help in collecting and analyzing ESG data, which in turn can aid in fostering sustainability. AI can help financial institutions visualize future landscapes through the simulation of ESG scenarios, which will help in the development of strategies and managing risks (Arslan et al., 2024; Li, 2003). AI can also reduce repetitive tasks and compliance checks through automation. Similarly, BT can help in promoting the transparency, integrity, and security of ESG data, and this supports more agile and resilient operations, which in turn leads to responsible and sustainable business practices. The environmental degradation and climate crises pose significant challenges; however, AI and BT offer innovative solutions for advancing data collection and analysis. Blockchain distributed ledger technology reduces the involvement of multiple intermediaries (Wang & You, 2022), reduces costs, streamlines the financial workflow (Arslan et al., 2024), minimizes delays, and makes it more susceptible to fraud (Lagoarde-Segot, 2019; Makarov & Schoar, 2022). Blockchain and artificial intelligence provide flexibility to adapt to market changes and provide more equitable and inclusive financial systems that are crucial in achieving the SDGs and minimizing disparities. Machine learning and artificial intelligence systems are changing the landscape of the financial sector with the introduction of new rules of competition and modifying the existing connections among parties (Kabza, 2020; Ma, 2023). The potential of these technologies is therefore a great challenge for the economy and the financial system; it calls for new regulations, and that is why it is tremendously imperative to consider the consequences of new technologies for the financial sector, especially in sustainable finance. Existing literature suggests a rise in awareness of the encompassing role of blockchain and artificial intelligence in sustainable development investments. Nevertheless, the studies are still very limited in exploring the role of BT and AI in sustainable finance. This chapter synthesizes and systematically reviews the existing literature on blockchain technology and artificial intelligence in sustainable finance.

3. METHODOLOGY

This chapter aimed to synthesize and systematically review the existing literature on blockchain technology (BT) and artificial intelligence (AI) in sustainable finance. Particularly, the authors aimed to explore the benefits and challenges of BT and AI in sustainable finance through systematic review and thematic analysis. A systematic literature review is secondary research that involves a clearly articulated search process and selection of literature and focuses on gathering and synthesizing evidence. Existing studies (such as Arslan et al. (2024); Nightingale (2009)) recommend the use of systematic literature review. Selection bias may occur in systematic review when exclusion and inclusion criteria are not clear and are restricted (Nightingale, 2009). Therefore, it is critical to clearly state the aims and objectives of the review and develop the inclusion and exclusion criteria. In addition, systematic review requires the use of search strategies (i.e., sensitive or specific) for the inclusion of all relevant studies. This chapter used a sensitive search strategy because a specific search strategy may find more relevant studies but can miss certain important studies. After conducting the systematic literature review, the authors used thematic analysis. It emphasizes identifying, analyzing, and interpreting patterns of meaning within qualitative data (Braun & Clarke, 2006). These techniques are appropriate here since the authors try to explore themes from existing literature. The authors used Nvivo11, Endnote X9, Excel spreadsheets, and VOS viewer for organization, information extraction, thematic analysis, and for the constructing and visualization of bibliometric networks within data.

It is difficult to draw an archetypal sample from existing studies. Therefore, the authors used Google Scholar, Web of Science (WoS), and Scopus databases for searching and collecting relevant studies from existing literature. Each of these bibliometric databases has its own benefits and limitations. For example, Web of Science and Scopus provide curated bibliometrics and metadata but are limited to their indexes and may face issues of bias for using their selective indexing criteria (Martín-Martín, Orduna-Malea, Thelwall, & López-Cózar, 2018). Google Scholar has the largest citation-based coverage, having over 300 million records (Delgado-López-Cózar & Martín-Martín, 2019), and provides a highly inclusive and multidisciplinary bibliometric crawler-based search system (Martín-Martín et al., 2018). However, metadata such as abstracts or references is not included in Google Scholar, and researchers need to perform manual web scraping. In addition, the quality of selected papers can vary significantly due to a lack of curation. The authors removed all restrictions on Google Scholar and performed keyword searches such as artificial intelligence, blockchain, and sustainable finance to find relevant studies. The authors then applied the rank-order approach and stopped looking for further studies when three sequential studies were found extraneous. The authors further restricted the period from 2000 to 2023. Although blockchain technology gained reputation after Bitcoin's introduction in 2008, AI applications in finance predate this period. AI-driven financial forecasting models and algorithmic trading systems were already established in the early 2000s (Li, 2003). Hence, the authors include studies from 2000 onward to capture the full evolution of these technologies in sustainable finance. A little over 600 articles were selected by looking into the abstract and/or full text[1]. It is also possible that a rank-based approach excluded some relevant studies (Arslan, 2020a), therefore the authors repeated the same search criteria with close synonyms such as machine learning and big data in ESG, green finance, regenerative finance, etc. and found another 353 articles. The authors performed the same search criteria on the Web of Science and Scopus databases and selected 209 articles. Several articles emerged in all three databases and subsequently removed from the sample. In addition, the authors also removed reports and other irrelevant studies, which brought the final sample to 509 studies. The use of the rank-order approach provided a saturated sample, and additional articles will not add value to thematic analysis. After collecting the data, the authors performed co-occurrence mapping of keywords in existing literature with the help of VOS viewer (see Figure 2). Figure 2 presents the co-occurrence mapping of keywords in existing literature, analyzed in VOS viewer[2] and focuses on blockchain and artificial intelligence in sustainability finance. Co-occurrence refers to the frequency with which two or more words appear together in a corpus of text, such as a collection of web pages or documents. In simple terms, when two words frequently appear together, they are likely to have some kind of association or be semantically related (Su & Lee, 2010). Co-occurrence mapping is an effective tool for understanding knowledge structures and research trends. Similarly, the co-occurrence network mapping in Figure 2 reveals the relationship of various keywords (such as blockchain and artificial intelligence) in existing sustainability finance literature and helps in identifying networks of important themes (terms), their influence, and understanding the nexus among them.

Figure 2. Co-occurrence mapping of keywords in literature

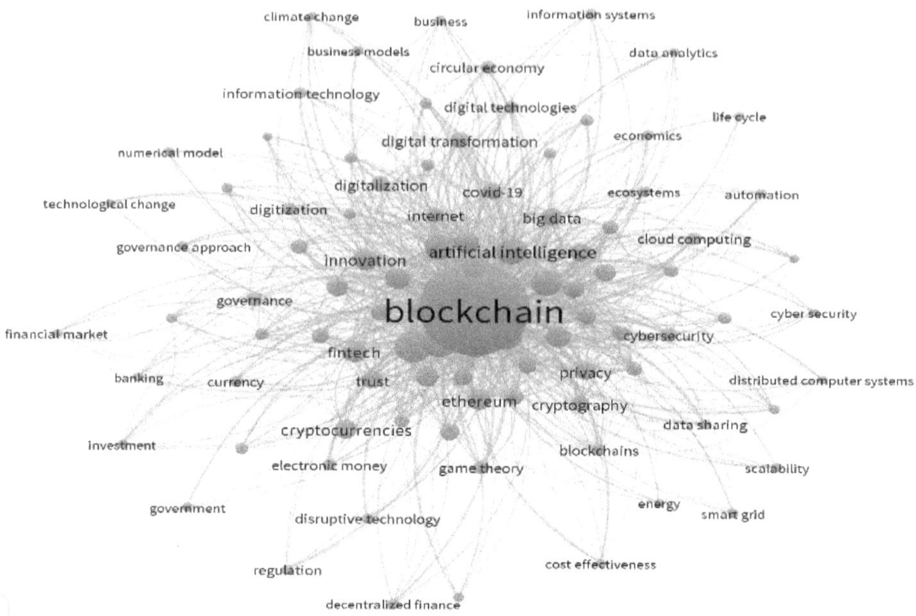

Source: Developed by the researcher(s) in VOS viewer

Figure 2 reveals the occurrence and influence of blockchain and artificial intelligence, among several other words such as innovation, trust, privacy, cost effectiveness, scalability, and ecosystems, among others. This analysis is important to get semantic understanding, and it also helps understand the topic association of related words and phrases. In line with the study of Arslan and Alqatan (2020), the authors applied open, axial, and selective coding for thematic analysis. Open coding created several codes and helped curtail the data into shorter codes. Axial coding helped in finding the common codes and identifying relationships among them to create subcategories. At the end, selective coding revealed the aggregate themes by looking into all the categories and subcategories. The authors refined the themes using open, axial, and selective coding (Braun & Clarke, 2006) to ensure a comprehensive categorization. Using VOSviewer and NVivo, the authors identified five major themes based on keyword co-occurrence and content analysis (see Figure 3):

- ESG Measurement and Disclosure
- Tracking and Trading of Carbon Footprint
- Renewable Energy and Circular Economy
- Transparency, Security, Governance, and Compliance
- Social and ethical aspects

In addition to the five themes, the authors also identified the challenges faced by sustainable finance in the contemporary business environment.

Figure 3. Thematic analysis

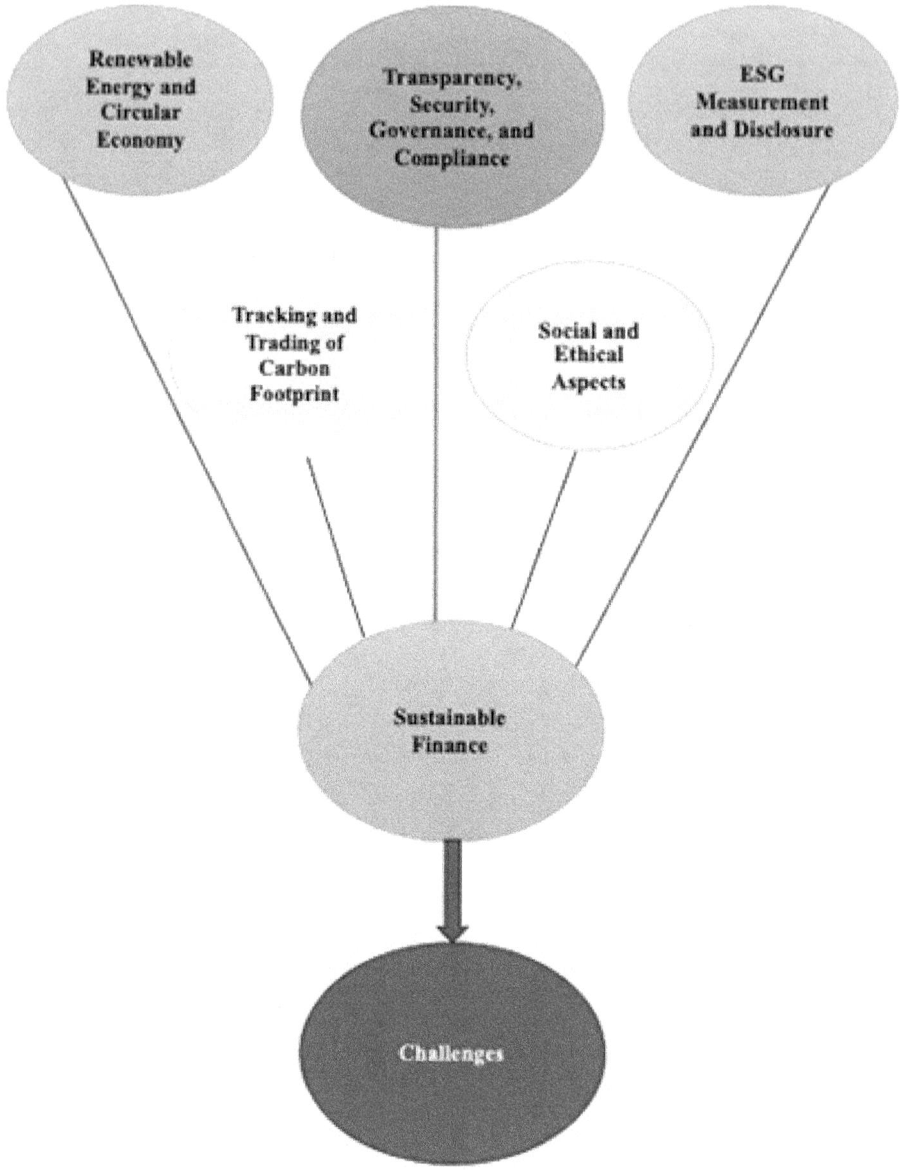

Source: developed by the researcher(s)

4. FINDINGS

This section presents the findings of this chapter. The findings of this chapter provide a comprehensive analysis of the intersection between artificial intelligence (AI), blockchain technology (BT), and sustainable finance. The systematic review identified five thematic areas that encapsulate the core contributions of these emerging technologies: (i) ESG measurement and disclosure, (ii) carbon footprint tracking and trading, (iii) renewable energy and circular economy, (iv) transparency, security, governance, and compliance, and (v) social and ethical considerations. Each of these themes represents critical aspects of sustainable finance and reflects how BT and AI can drive meaningful change in financial markets. Additionally, this chapter identifies key challenges that hinder the effective deployment of these technologies.

4.1. The Role of Blockchain Technology and Artificial Intelligence in Sustainable Finance

This section presents the benefits of using blockchain technology and artificial intelligence in sustainable finance.

4.1.1. ESG Measurement and Disclosure

The growing call for transparency in ESG reporting has driven the adoption of advanced technologies such as AI and blockchain to improve measurement and disclosure processes. These technologies offer significant improvements in accuracy, efficiency, and credibility, focusing on the fundamental limitations of traditional ESG reporting methods, which often rely on manual data collection and subjective assessments (Ahmad, Yaqub, & Lee, 2024). ESG disclosure, or ESG index, helps investors learn about ESG performance of the companies and also the associated trends. More and more firms are disclosing ESG data to attract investors who are interested in sustainable investments. ESG information is critical to determining the firm's environmental, social, and governance policies. BT and AI are very helpful in recording, tracking, developing, and analyzing the ESG performance or indexes, such as the green sentiment index, of the companies. It was not possible to conduct sentimental analysis; however, it is possible due to emerging technologies. Briere and Ramelli (2020) developed a green sentiment index to capture the investors' appetite for environmental responsibility. Several scholars have used machine learning and other AI tools to perform textual analysis on ESG data and ESG issues that firms are trying to address. For example, Goloshchapova and Poon (2019) used text mining to explore corporate social responsibility (CSR) disclosure by taking a sample of 15 industrialized countries that were included in MSCI Europe. Similarly, Raman and Bang (2020) used a pertained language model for the analysis of corporate earnings calls and to detect historic trends in ESG. AI also helps investors analyze real-time changes in ESG data and make decisions accordingly. Clarkson and Ponn (2020) used a computer-based textual analysis of the CSR reports of US companies. They further used classified algorithms for forecasting and analyzing future ESG disclosures. Chang and Lee (2024) used the machine learning selection method for forecasting earnings per share and the impact of SDGs on corporate performance. Reig-Mullor and Garcia-Bernabeu (2022) suggested combining AHP and TOPSIS techniques and furthering the development of a neutrosophic model to assess ESG performance. Bouyé and Menville (2020) examined the sovereign ESG ratings with the help of regression and principal component regression. Scholars also developed language models for better analysis of financial data. For example, Huang and Wang (2023) developed a large language model called FinBERT for better analysis of financial texts. This clearly shows the benefits of using AI for analysis and tracking ESG and other financial data. In the governance aspect of ESG, firms try to monitor and assess the potential ESG risks that help ensure compliance with relevant policies and laws. AI can automate ESG data, which can help regulators implement ESG policies and also help businesses track and increase their ESG practices (Chen, Wang, & Jin, 2022; Fan & Wu, 2022; Sokolov, Mostovoy, Ding, & Seco, 2021). In addition, blockchain ensures ESG data integrity, reducing greenwashing risks and enhancing investor confidence (Sokolov et al., 2021). The fintech ecosystem can optimize operations, prepare financial budgets, ensure the accuracy of transactions, and resolve auditing issues that reduce irregularities and unethical behavior by company executives (Yadav, Kar, & Kashiramka, 2022). It also helps in assessing the social impact of corporate decisions, for example, plant closures and layoffs, and improving overall governance practices (De Lucia, Pazienza, & Bartlett, 2020; Turunen, 2021). Blockchain technology, characterized by its decentralized and immutable ledger system, provides a secure and transparent mechanism for recording and verifying ESG data (Moodaley & Telukdarie, 2023). This technological framework can effectively mitigate risks associated with greenwashing while enhancing stakeholder trust. Concurrently, AI algorithms possess the capability to process extensive datasets from diverse sources, enabling pattern identification, risk prediction, and the generation of actionable insights aimed at improving ESG performance (Moodaley & Telukdarie, 2023). Furthermore, the integration of AI into sustainability initiatives has the potential to facilitate culturally relevant organizational processes and individual practices that contribute to reduced consumption of natural resources and energy (Nishant et al., 2020). Nevertheless, challenges persist in accurately assessing the impact of AI-driven interventions and addressing associated cybersecurity risks (Nishant et al., 2020). The convergence of AI and blockchain technologies holds transformative potential for ESG measurement and disclosure. AI-driven analytics can extract insights from complex datasets, while blockchain ensures the integrity and transparency of reported data. This synergy enhances the reliability, efficiency, and credibility of ESG reporting, thereby enabling organizations to make more informed decisions and contribute meaningfully to sustainability efforts. By leveraging AI's analytical prowess alongside blockchain's transparency and security features, organizations can provide stakeholders with verifiable and trustworthy information regarding their environmental and social impact. This integration fosters greater accountability and trust, ultimately advancing global sustainability initiatives.

4.1.2. Tracking and Trading of Carbon Footprint

The reduction in carbon footprint is one of several UN SGDs. The carbon footprint is one of the reasons for climate change globally, and several countries have recognized the need to track and reduce the carbon footprints. Consequently, the global transition towards a sustainable future requires a comprehensive approach to carbon accounting and management. Traditional carbon tracking systems often suffer from issues of transparency, reliability, and efficiency, which incumber effective global carbon reduction efforts and policy implementation. Nevertheless, emerging technologies, such as AI and Blockchain, offer transformative solutions to these challenges, enabling accurate carbon accounting and secure carbon trading mechanisms. These technologies can help in reducing energy costs and reducing carbon footprints, which will help in achieving UN SDGs. Zhong and Zhong (2024) took a sample from 66 countries and found that the impact of AI on carbon emissions diverges from country to country. Similarly, Arya and Bachheti (2024) argued that AI helps in monitoring, predicting, and reducing the emission of greenhouse gases (GHGs), which helps in a cleaner environment.

In recent years, carbon tracking has become much more transparent with blockchain integration. The tracking of carbon footprints provides accountability and offers opportunities to quantify and control carbon emissions for countries, which in turn helps the regulators develop individual policies for reducing carbon footprints. It will also help them comply with established caps and make enforcement easier. BT makes traceability possible, while artificial intelligence can analyze large amounts of data and draw patterns. BT can also collect a complete and accurate record of data and make it publicly available. Consequently, it can help with accountability and meeting emission targets. Carbon credit is one of the initiatives that can attract firms to reduce their carbon footprint and meet emission targets. Scholars argued that AI can help in predicting market trends through automation and intelligence investments (Cao, 2020; Ferreira, Gandomi, & Cardoso, 2021), which will help in decision-making (Baulkaran & Jain, 2023; Hakala, 2019; Katterbauer & Moschetta,

2022). Scholars (such as de la Barcena Grau (2021)) suggested the use of machine learning for social impact. Faccini and Matin (2023) invested the climate risk in stock prices and found that only risks stemming from the U.S. climate-policy debate are priced. Scholars found that ESG indexes were not affected during COVID, while market indexes decreased with the number of cases and deaths (Mousa, Saleem, & Sági, 2021). Though the carbon markets have immense potential for advancing climate action and mobilizing finance, their true potential is limited due to issues such as fragmentation, a lack of transparency, and siloed approaches.

Artificial Intelligence has the potential to revolutionize carbon accounting by automating data extraction, analysis, and validation across diverse sources. Natural Language Processing (NLP) can unlock a wealth of alternative datasets, including financial transactions, shipping records, and industrial reports, thereby providing a more comprehensive and real-time understanding of carbon emissions and their sources (Oladeji & Mousavi, 2023). By leveraging machine learning algorithms, AI can identify discrepancies, detect fraudulent claims, and improve emission forecasting, significantly enhancing the accuracy and transparency of carbon accounting systems (Cowls, Tsamados, Taddeo, & Floridi, 2023). This data-driven approach fosters informed decision-making, allowing governments and businesses to implement more effective carbon management strategies. Blockchain technology, on the other hand, is pivotal in ensuring the integrity, security, and efficiency of carbon trading systems. The decentralized, immutable, and transparent nature of blockchain can mitigate issues related to double counting, fraud, and lack of standardization in carbon credit markets. By embedding smart contracts and compliance mechanisms, blockchain facilitates automated verification and enforcement of carbon credit transactions, reducing the reliance on intermediaries and minimizing administrative costs (Dunkel, 2018). Furthermore, blockchain enables the seamless exchange of digital carbon credits across national and subnational boundaries, fostering global partnerships and ensuring a fair, verifiable, and liquid carbon market. The convergence of AI and Blockchain technology holds the potential to address the complex challenges associated with carbon tracking and trading. AI-driven insights enhance the precision and accountability of carbon measurement, while blockchain ensures the security and transparency of carbon markets. Together, these technologies can create a synergistic ecosystem that not only optimizes carbon accounting but also accelerates global efforts toward carbon neutrality. Governments, businesses, and policymakers must prioritize investment in these technologies to establish robust carbon management frameworks that drive meaningful climate action.

4.1.3. Renewable Energy and Circular Economy

The convergence of AI and blockchain technology provides a transformative opportunity to advance renewable energy systems and enhance a circular economy (Adelakun, 2024). There is a consensus among global and regional leaders about the need for decentralization of renewable energy technology to achieve the SDGs. This can be addressed through the development and implementation of trading mechanisms. Energy and sustainable finance are critical for socio-economic development (Arslan, Abeuova, & Alqatan, 2021; Burney, 1995; Orazayeva & Arslan, 2024), and any unfavorable setbacks to the energy sector can have devastating consequences for socio-economic development (Chopra & Khanna, 2015). Though few countries have started prioritizing renewable energy to meet the low carbon and clean energy targets, the majority of countries are still facing issues such as unfair energy prices, lack of access to credit, unavailability of skilled manpower, lack of required technology, and infrastructure (Babu et al., 2020; Yaqoot, Diwan, & Kandpal, 2016). Thus, a quest for affordable, secure, equitable, reliable, resilient, and adequate renewable energy supplies is critical for achieving the UN SDGs (Chopra & Khanna, 2015). Scholars have argued that blockchain and artificial intelligence can play a critical role in the development and implementation of renewable energy (Arslan et al., 2024; Delina, 2023; Ma, 2023; Mavlutova et al., 2023). Distributed energy resources and peer-to-peer energy trading can support the exchange of surplus energy among parties connected to grids. Blockchain technology can store real-time data for multiple transactions and optimize the grid's capacity while minimizing waste. AI technology can also help in the design and production of energy-efficient vehicles, thereby providing better charging and battery management. Some companies (such as BMW and Siemens) are already using AI technology for sustainable mobility and the production of energy-efficient vehicles (Gailhofer et al., 2021). This provides an opportunity for the researchers to explore technological innovations (Bolton, Raven, & Mintrom, 2021). The AI and blockchain have the potential to enhance the traceability of materials in supply chains, optimize energy distribution networks, and accelerate the transition towards sustainable resource management. AI can analyze vast datasets and identify patterns that plays a critical role in predictive maintenance of renewable energy systems, lowering operational costs, reducing downtime, and prolonging the lifespan of critical infrastructure. The integration of AI into big data, machine learning and predictive modeling can lead to greater efficiency, reliability, and sustainability (Pachot & Patissier, 2022).

Blockchain technology and artificial intelligence can also help in achieving a circular economy. Blockchain enables more efficient and cleaner economic transactions for the participants and also helps in achieving harmony among the environment, society, and economy. Blockchain technology provides a decentralized, transparent, and secure platform for tracking energy production, consumption, and trading which supplements AI-driven innovations. This is typically critical in energy trading, where blockchain-based smart contracts can facilitate real-time transactions, reducing inefficiencies, and improving market transparency. In renewable energy, one of the most promising applications of blockchain is peer-to-peer (P2P) energy trading. Decentralized blockchain platforms fosters energy independence and resilience by empowering communities and individuals to generate, store, and trade renewable energy. Additionally, blockchain fosters greater accountability in energy markets, enhances the traceability of energy sources, and ensuring that consumers can verify the origin and sustainability credentials of the electricity they use (Hu et al., 2023). The reuse of materials, circular material design, integration of AI, data collection, and visualization can help in the development of more sustainable products with smart production (Ghoreishi & Happonen, 2020). Similarly, Wilson and Paschen (2022) argued that different forms of AI provide significant benefits in the reverse logistics process. Both BT and AI can provide untapped opportunities for renewable energy and accelerate the shift toward a circular economy that can reshape the economy into one that is regenerative, resilient, and fit for the long term. The integration of AI and blockchain technology will be instrumental in addressing global energy challenges, reducing environmental impact, and promoting a decentralized and more equitable energy landscape.

4.1.4. Transparency, Security, Governance, and Compliance

Traditional finance is very vulnerable to fraud, scams, and cyberattacks (Bulyga, Vasyukova, Grigoryeva, & Dimov, 2022; Chris, 2023). Scholars argued that security is of utmost concern in the financial sector due to the sensitive nature of data and the large amount of money at stake. One of the core benefits of blockchain technology for sustainable finance is transparency and security. The lack of transparency and security can lead to mistrust among stakeholders and thus affect the efficient functioning of financial markets (Arslan et al., 2024; Shkolnyk, Kozmenko, Nowacki, & Mershchii, 2020). The convergence of AI and blockchain technologies holds the potential to revolutionize various sectors, including governance, security, compliance, and supply chain management. Organizations can unlock new levels of efficiency, security, and trust through integration of AI's data processing and decision-making capabilities with Blockchain's secure and transparent ledger system (Chen, Bao, Li, & Lin, 2023; Liu, He, Dong, & Cao, 2023). It is reported that a huge investment is needed to achieve the SDGs and sustainability, therefore, it is crucial to retain the trust of stakeholders in sustainable financial investments. Blockchain ensures transparency with publicly accessible ledgers that contain data about every transaction and are accessible to all participants in the network (Piazza, 2017; Yermack, 2017), thus providing authorities' oversight and mitigating unlawful and manipulative business practices. Another advantage of blockchain is its immutability. Once a transaction is recorded in the form of a block, it becomes impervious to alteration or deletion. Firms can improve product quality, increase efficiency, and ensure transparency and security by using emerging technologies such as AI and BT (Arslan et al., 2024; Jiang, Li, Si, & Su, 2022; Palomares et al., 2021; Trivedi, Mehta, & Sharma, 2021; Wilson et al., 2022). BT can control fraud and manipulation through the execution of smart contracts (Yermack, 2017). Similarly, artificial intelligence can automate processes and reduces costs while increasing efficiency. AI can also be used to execute smart contracts and analyze any abnormalities in real time (Dai & Vasarhelyi, 2017; Marshall & Lambert, 2018). Additionally, the integration of AI and Blockchain can enhance cybersecurity measures by creating decentralized identity management systems and improving data privacy protocols. Scholars have argued that AI can provide transparency to predictive analysis (Meskó, Hetényi, & Győrffy, 2018) and reduce human bias in the process. However, some scholars argue that AI systems may not be totally free from bias and can depend on the advancement of technology and data itself (Arslan, 2024). Scholars also argued that blockchain and artificial intelligence can improve compliance and regulatory checks. BT and AI can automate transactions in a secure way, and these emerging technologies can help regulators monitor and track compliance. Additionally, machine learning algorithms can prevent illicit activities such as terrorist financing and money laundering and can process large amounts of data quickly. AI and BT systems can also reduce the risk of regulatory fines and enhance the integrity of financial systems. This is very important for sustainable finance, where investors can track the performance of their investments in a transparent way.

4.1.5. Social and Ethical Aspects

Scholars argued that AI and BT can help financial institutions develop more transparent, inclusive, and interpretable models; however, key considerations should be given to reduce biases against any social group (Hoepner & Rogelj, 2021; Lacoste, Luccioni, Schmidt, & Dandres, 2019). It is crucial to develop transparent AI and BT systems, especially in the financial sector, where decisions have significant implications (Shkolnyk et al., 2020; Shkolnyk, Melnyk, Havrysh, & Ivanchenko, 2019). Consequently, ethical considerations are vital to the successful adoption of AI and BT systems (Rane, 2023a, 2023b; Shkolnyk et al., 2020). Thus, a critical examination of their social and ethical implications becomes increasingly essential. On the other hand, scholars have argued that a cautious approach needs to be exercised when relying on BT and AI technologies (Arslan, 2024; Arslan et al., 2024). Biased and unfair AI and BT systems can lead to unfair outcomes, which in turn can affect the creditability of financial systems. It is true that AI and BT have certain advantages when it comes to protecting data through encryption and real-time data monitoring (Arslan et al., 2024; Kalubanga & Mbekeka, 2024; Zhang, Sun, Yang, & Wang, 2020). However, discreet information can be exposed when using more and more data to train AI and BT algorithms. Scholars agreed that AI and BT can reduce biases and unfairness with the effective development of algorithms (Moeuf et al., 2020). Notwithstanding such advancements, the ethical challenges associated with AI and blockchain technology must not be overlooked. On the contrary, the prevalence of errors in facial recognition can marginalize some segments of society even further (Cichosz, Wallenburg, & Knemeyer, 2020). In line with the perspectives of organizational theory, technologies have changed the social landscape with the introduction of new concepts such as digital discretion, automated discretion, and artificial discretion (Busch & Henriksen, 2018; Young, Bullock, & Lecy, 2019; Zouridis, Van Eck, & Bovens, 2020). The integration of AI and blockchain can unlock innovative applications, such as smart contracts, predictive analytics, and decentralized decision-making. Thus, it has the potential to enhance accountability, minimize fraud risks, and empower individuals with greater control over their data (Ruel & Njoku, 2021). AI and BT systems can be misused by relevant authorities and organizations; therefore, a need arises to establish principles for their use. The transformative technologies can change the decision-making process by altering human judgments, which in turn can provide equality and inclusivity and reduce exploitation (Arslan, 2020b; Kim, Andersen, & Lee, 2022). Regulatory bodies around the globe have increased efforts to develop ethical guidelines to mitigate the associated financial risks with AI and BT systems.

4.2. Challenges Faced by Sustainable Finance

In recent years, sustainable finance has become attractive to many investors, and investments are growing in sustainable financial products such as green equities, green funds, green bonds, and sustainability-backed loans (Al-Sartawi, 2022; Krueger, Sautner, & Starks, 2020; Memdani, 2020). However, awareness is still limited due to the markets' immaturity and flaws. Additionally, the return on investment is lower for sustainable projects, and some investors lose interest in them. There are several challenges in the use of AI and BT for sustainable finance, and these emerging technologies poses several challenges. Scholars argued that governments, firms, investors, and academics are not currently focusing on AI and BT applications for environmental applications. In addition, the liquidity of sustainable finance projects is limited compared to other industries; therefore, highly demanded projects may be overvalued. Scholars also found high entry barriers for green bonds, and these bonds are not sold in fractions like other bonds. This limits the investments from the retail investors and poses significant entry challenges. Another challenge in sustainability finance is the lack of technical knowledge and skills for effective use of BT and AI, thus changing the labor demands (Al-Sartawi, Razzaque, & Kamal, 2021; Shihadeh, 2020). Several countries are acting individually to oversee AI and BT and are thus struggling to keep up with the pace and rapid advancements. The imperfect data also poses a significant challenge, and it leads to a lack of trust among AI and BT applications. Scholars also argued that most of claims made by sustainable projects are not supported with sufficient evidence, which in turn reduces trust in the projects. Another challenge is the lack of the necessary infrastructure for BT and AI applications. Developed countries have the required infrastructure and are leading the deployment of AI and BT. However, developing countries are still struggling due to a lack of the required infrastructure. The effectiveness of AI and BT applications in sustainable finance depends on data quality, transparency, and system security. However, data quality varies among firms, and this poses data quality issues and further impacts the quality and applicability to a wide array of use cases. In addition, data security and privacy also pose significant challenges. Scholars highlighted the need for regulations from multiple perspectives (Haenlein & Kaplan, 2019).

5. CONCLUSION AND RECOMMENDATIONS

This chapter aimed to synthesize and systematically review the existing literature on blockchain technology (BT) and artificial intelligence (AI) in sustainable finance. Particularly, the authors aim to explore the benefits and challenges of BT and AI in sustainable finance through systematic review and thematic analysis. This chapter identified five archetypical thematic areas (i.e., ESG measurement and disclosure; tracking and trading of carbon footprint; renewable energy and circular economy; transparency, security, governance, and compliance; and social and ethical aspects) attracting research interest. This chapter offers valuable insights into each theme and contributes to a more nuanced understanding of the area.

In conclusion, the integration of blockchain technology and artificial intelligence offers uncapped opportunities for sustainable finance and the betterment of society, which in turn can lead to achieving the SGDs. Blockchain and artificial intelligence are the most disruptive technologies in modern times, and these have enormous potential to address environmental externalities and help combat climate change. The findings also reveal some challenges, such as lack of awareness, quality of data, data security, and ethical issues, that sustainable businesses face in the contemporary business environment. Nonetheless, it is imperative to emphasize the benefits of BT and AI and involve stakeholders in preparation of policies and decision-making for avoiding adverse outcomes. Collaborative policymaking offers a prudent approach to confirming prevailing collaboration in dynamic landscape. The combination of BT and AI enhances ESG disclosure, helps in its measurement, and thus takes ambiguity, abuse, and fraud out of the process. These technologies also help in the traceability and trading of carbon footprints, supplement the decision-making process, improve production processes, and reduce energy consumption, which contribute to more sustainable practices. In addition, BT and AI provide seamless cross-border transactions and immutable data storage, which in turn fosters transparency and trust among stakeholders in the financial sector. These technologies also offer opportunities for the development of inclusive products by efficient resource management. This chapter finds that retail investors are willing to trade profit for sustainable investments which can drive positive social and environmental impacts. However, there is a lack of compliance, and sustainable investments are not well supported with sufficient evidence. The findings of this chapter recommend more regulatory oversight on sustainable investments and actions from governments to promote sustainable investing.

5.1 Implications of the Chapter

This chapter has significant implications for both academic research and practical applications in sustainable finance. By identifying key thematic areas where AI and blockchain technology can enhance financial sustainability, this chapter provides valuable insights for policymakers, financial institutions, and investors. From a policy perspective, the findings emphasize the need for regulatory frameworks that facilitate the integration of emerging technologies in sustainable finance while ensuring ethical considerations and data privacy protections. From a practical standpoint, the chapter highlights how AI and blockchain can improve transparency, efficiency, and accountability in financial markets, thereby attracting more investment into sustainable initiatives. From an academic perspective, this chapter contributes to the growing body of literature on financial technology (FinTech) and sustainable finance by offering a systematic review and identifying research gaps that future studies can address. It underscores the need for interdisciplinary research to explore the intersection of finance, technology, and sustainability.

5.2 Limitations and Future Research Directions

This chapter uses a systematic literature review and thematic analysis and suggests exploring further the latent opportunities of evolving technologies such as blockchain and artificial intelligence for comprehensive analysis of ESG and sustainability through quantitative data. Future studies can provide a comprehensive analysis by adopting circular economy principles and incorporating sustainable development theories. The continuing advancement of AI and Blockchain will present new opportunities and challenges for businesses, policymakers, and regulators. Though the integration of these technologies can drive innovation and efficiency, it is imperative to address the concerns related to scalability, interoperability, and ethical considerations. Future research and development efforts should focus on refining AI algorithms, enhancing Blockchain scalability, and establishing comprehensive regulatory frameworks to ensure responsible and sustainable adoption.

REFERENCES

Adelakun, N. O. (2024). Future trends in artificial intelligence for energy Management. *Information Matters, 4*(1).

Ahmad, H., Yaqub, M., & Lee, S. H. (2024). Environmental-, social-, and governance-related factors for business investment and sustainability: A scientometric review of global trends. *Environment, Development and Sustainability, 26*(2), 2965–2987. DOI: 10.1007/s10668-023-02921-x PMID: 36714213

Al-Sartawi, A., Al-Okaily, M., Hannoon, A., & Khalid, A. A. (2022). Financial technology: literature review paper. *Artificial Intelligence for Sustainable Finance and Sustainable Technology: Proceedings of ICGER 2021 1*, 194-200.

Al-Sartawi, A. M. M., Razzaque, A., & Kamal, M. M. (2021). *Artificial Intelligence Systems and the Internet of Things in the Digital Era:Proceedings of EAMMIS 2021* (Vol. 239): Springer Nature. DOI: 10.1007/978-3-030-77246-8

Al-Sartawi, M. (2022). *Artificial Intelligence for Sustainable Finance and Sustainable Technology*. Springer International Publishing. DOI: 10.1007/978-3-030-93464-4

Alonso, A., & Marqués, J. M. (2019). Innovación financiera para una economía sostenible (Financial Innovation for a Sustainable Economy). *Banco de España Occasional Paper*(1916).

Andrianaivo, M., & Kpodar, K. (2011). ICT, financial inclusion, and growth: Evidence from African countries.

Arslan, M. (2020a). Corporate social sustainability in supply chain management: A literature review. *Journal of Global Responsibility, 11*(3), 233–255. DOI: 10.1108/JGR-11-2019-0108

Arslan, M. (2020b). Mechanisms of labour exploitation: The case of Pakistan. *International Journal of Law and Management, 62*(1), 1–21. DOI: 10.1108/IJLMA-07-2018-0145

Arslan, M. (2024). Digitalisation in Accounting: Is It a Risk or an Opportunity? In *Impact of Digitalization on Reporting, Tax Avoidance, Accounting, and Green Finance* (pp. 112-133): IGI Global.

Arslan, M., Abeuova, D., & Alqatan, A. (2021). Corporate Social Responsibility and Institutional Investors: Evidence from Emerging Markets. *Pakistan Journal of Commerce and Social Sciences, 15*(1), 31–57.

Arslan, M., & Alqatan, A. (2020). Role of institutions in shaping corporate governance system: Evidence from emerging economy. *Heliyon*, *6*(3), e03520. DOI: 10.1016/j.heliyon.2020.e03520 PMID: 32181393

Arslan, M., Faizulayev, A., Abeuova, D., & Kayani, U. N. (2024). Exploring the Risks, Challenges, and Opportunities of Using Digital Financial Technologies for Green Finance. *Impact of Digitalization on Reporting, Tax Avoidance, Accounting, and Green Finance*, 226-242.

Arya, A., Bachheti, A., Bachheti, R. K., Singh, M., & Chandel, A. K. (2024). Role of Artificial Intelligence in Minimizing Carbon Footprint: A Systematic Review of Recent Insights. *Biorefinery and Industry 4.0: Empowering Sustainability*, 365-386.

Babu, T. S., Vasudevan, K. R., Ramachandaramurthy, V. K., Sani, S. B., Chemud, S., & Lajim, R. M. (2020). A comprehensive review of hybrid energy storage systems: Converter topologies, control strategies and future prospects. *IEEE Access : Practical Innovations, Open Solutions*, *8*, 148702–148721. DOI: 10.1109/ACCESS.2020.3015919

Baulkaran, V., & Jain, P. (2023). Who uses robo-advising and how? *Financial Review*, *58*(1), 65–89. DOI: 10.1111/fire.12324

Bolton, M., Raven, R., & Mintrom, M. (2021). Can AI transform public decision-making for sustainable development? An exploration of critical earth system governance questions. *Earth System Governance*, *9*, 100116. DOI: 10.1016/j.esg.2021.100116

Boskov, T., & Drakulevski, L. (2017). Addressing the role of risk management and digital finance technology on financial inclusion. *Quality - Access to Success*, *18*(161), 113–115.

Bouyé, E., & Menville, D. (2020). The convergence of sovereign environmental, social and governance ratings. *Social and Governance Ratings (December 21, 2020)*.

Braun, V., & Clarke, V. (2006). Using thematic analysis in psychology. *Qualitative Research in Psychology*, *3*(2), 77–101. DOI: 10.1191/1478088706qp063oa

Briere, M., & Ramelli, S. (2020). *Personal values, responsible investing and stock allocation*

Bulyga, E. S., Vasyukova, L. K., Grigoryeva, V. I., & Dimov, R. P. (2022). Digital transformation of the financial market: Risks and opportunities. *Наука Красноярья*, *11*(2), 49–67. DOI: 10.12731/2070-7568-2022-11-2-49-67

Burney, N. A. (1995). Socioeconomic development and electricity consumption A cross-country analysis using the random coefficient method. *Energy Economics*, *17*(3), 185–195. DOI: 10.1016/0140-9883(95)00012-J

Busch, P. A., & Henriksen, H. Z. (2018). Digital discretion: A systematic literature review of ICT and street-level discretion. *Information Polity*, *23*(1), 3–28. DOI: 10.3233/IP-170050

Cao, L. (2020). AI in finance: A review. *Available atSSRN* 3647625.

Chang, A., Lee, T. S., & Lee, H. M. (2024). Applying sustainable development goals in financial forecasting using machine learning techniques. *Corporate Social Responsibility and Environmental Management*, *31*(3), 2277–2289. DOI: 10.1002/csr.2694

Chayjan, M. R., Bagheri, T., Kianian, A., & Someh, N. G. (2020). Using data mining for prediction of retail banking customer's churn behaviour. *International Journal of Electronic Banking*, *2*(4), 303–320. DOI: 10.1504/IJEBANK.2020.114770

Chen, J. E., Bao, F., Li, C., & Lin, Y. (2023). The application and Ethics of Artificial Intelligence in Blockchain: A bibliometric-content analysis. [JGIM]. *Journal of Global Information Management*, *31*(7), 1–32. DOI: 10.4018/JGIM.323656

Chen, L., Wang, Y., & Jin, S. (2022). How green credit guidelines policy affect the green innovation in china? [EEMJ]. *Environmental Engineering and Management Journal*, *21*(3).

Chopra, S. S., & Khanna, V. (2015). Interconnectedness and interdependencies of critical infrastructures in the US economy: Implications for resilience. *Physica A*, *436*, 865–877. DOI: 10.1016/j.physa.2015.05.091

Chris, K. (2023). Scams cost Kiwis almost $200m - Ministry of Business, Innovation and Employment (MBIE). nzherald.co.nz. Retrieved from https://www.nzherald.co.nz/nz/scams-cost-kiwis-almost-200m-ministry-of-business-innovation-and-employment-report/7BA6OADPBBGNLDXU765DEXUFQM/

Christensen, C. M., Raynor, M. E., Dyer, J., & Gregersen, H. (2011). *Disruptive Innovation: The Christensen Collection (The Innovator's Dilemma, The Innovator's Solution, The Innovator's DNA, and Harvard Business Review article" How Will You Measure Your Life?")(4 Items)*. Harvard Business Press.

Cichosz, M., Wallenburg, C. M., & Knemeyer, A. M. (2020). Digital transformation at logistics service providers: Barriers, success factors and leading practices. *International Journal of Logistics Management*, *31*(2), 209–238. DOI: 10.1108/IJLM-08-2019-0229

Clarkson, P. M., Ponn, J., Richardson, G. D., Rudzicz, F., Tsang, A., & Wang, J. (2020). A textual analysis of US corporate social responsibility reports. *Abacus, 56*(1), 3–34. DOI: 10.1111/abac.12182

Craig, C. D. (2021). Sustainable finance starts with data Symposium conducted at the meeting of the World Economic Forum Retrieved from https://www.weforum.org/agenda/2020/01/sustainable-finance-starts-with-data/

Dai, J., & Vasarhelyi, M. A. (2017). Toward blockchain-based accounting and assurance. *Journal of Information Systems, 31*(3), 5–21. DOI: 10.2308/isys-51804

de la Barcena Grau, A. M. (2021). Social Impact Returns. Filling the Finance Gap with Data Value. *Risk Management,* •••, 125–143.

De Lucia, C., Pazienza, P., & Bartlett, M. (2020). Does good ESG lead to better financial performances by firms? Machine learning and logistic regression models of public enterprises in Europe. *Sustainability (Basel), 12*(13), 5317. DOI: 10.3390/su12135317

Degrees Investing Initiative. (2020). *EU Retail Funds' Environmental Impact Claims Do Not Comply with Regulatory Guidance*. Paris, France: 2 Degrees Investing Initiative. Retrieved from https://2degrees-investing.org/resource/marketing-claims/

Delgado-López-Cózar, E., & Martín-Martín, A. (2019). El Factor de Impacto de las revistas científicas sigue siendo ese número que devora la ciencia española:¿ hasta cuándo? *Anuario ThinkEPI, 13.*

Delina, L. L. (2023). Fintech RE in a global finance centre: Expert perceptions of the benefits of and challenges to digital financing of distributed and decentralised renewables in Hong Kong. *Energy Research & Social Science, 97*, 102997. DOI: 10.1016/j.erss.2023.102997

Faccini, R., Matin, R., & Skiadopoulos, G. (2023). Dissecting climate risks: Are they reflected in stock prices? *Journal of Banking & Finance, 155*, 106948. DOI: 10.1016/j.jbankfin.2023.106948

Falcone, P. M., & Sica, E. (2019). Assessing the opportunities and challenges of green finance in Italy: An analysis of the biomass production sector. *Sustainability (Basel), 11*(2), 517. DOI: 10.3390/su11020517

Fan, G., & Wu, X. (2022). Going Green: The Governance Role of Environmental Regulations on Firm Innovation and Value. *Singapore Management University School of Accountancy Research Paper*(2023-163).

Ferreira, F. G., Gandomi, A. H., & Cardoso, R. T. (2021). Artificial intelligence applied to stock market trading: A review. *IEEE Access : Practical Innovations, Open Solutions*, *9*, 30898–30917. DOI: 10.1109/ACCESS.2021.3058133

Gailhofer, P., Herold, A., Schemmel, J. P., Scherf, C.-S., de Stebelski, C. U., Köhler, A. R., & Braungardt, S. (2021). *The role of artificial intelligence in the European green deal*. European Parliament Luxembourg.

Ghoreishi, M., & Happonen, A. (2020). Key enablers for deploying artificial intelligence for circular economy embracing sustainable product design: Three case studies*AIP Publishing*. Symposium conducted at the meeting of the AIP conference proceedings

Goloshchapova, I., Poon, S.-H., Pritchard, M., & Reed, P. (2019). Corporate social responsibility reports: Topic analysis and big data approach. *European Journal of Finance*, *25*(17), 1637–1654. DOI: 10.1080/1351847X.2019.1572637

Haenlein, M., & Kaplan, A. (2019). A brief history of artificial intelligence: On the past, present, and future of artificial intelligence. *California Management Review*, *61*(4), 5–14. DOI: 10.1177/0008125619864925

Hakala, K. (2019). *Robo-advisors as a form of artificial intelligence in private customers' investment advisory services*. Aalto University. Retrieved from https://aaltodoc.aalto.fi/items/e279201f-e5f0-455d-9222-2aa94aa240e4

Hoepner, A. G., & Rogelj, J. (2021). Emissions estimations should embed a precautionary principle. *Nature Climate Change*, *11*(8), 638–640. DOI: 10.1038/s41558-021-01109-3

Hu, S., Zhu, Q., Zhao, X., & Xu, Z. (2023). Digital Finance and Corporate Sustainability Performance: Promoting or Restricting? Evidence from China's Listed Companies. *Sustainability (Basel)*, *15*(13), 9855. DOI: 10.3390/su15139855

Huang, A. H., Wang, H., & Yang, Y. (2023). FinBERT: A large language model for extracting information from financial text. *Contemporary Accounting Research*, *40*(2), 806–841. DOI: 10.1111/1911-3846.12832

Jiang, Q., Li, J., Si, H., & Su, Y. (2022). The impact of the digital economy on agricultural green development: Evidence from China. *Agriculture*, *12*(8), 1107. DOI: 10.3390/agriculture12081107

Kabza, M. (2020). Artificial intelligence in financial services–benefits and costs. In *Innovation in Financial Services* (pp. 183–198). Routledge. DOI: 10.4324/9781003051664-14

Kalubanga, M., & Mbekeka, W. (2024). Compliance with government and firm's own policy, reverse logistics practices and firm environmental performance. *International Journal of Productivity and Performance Management, 73*(5), 1427–1478. DOI: 10.1108/IJPPM-09-2022-0463

Katterbauer, K., & Moschetta, P. (2022). A robo-advisory framework for Islamic and Environmental, Social and Governance (ESG) compliance–A benchmark study on the S&P 500 stock index. *The Journal of Muamalat and Islamic Finance Research*, 34-46.

Kim, S., Andersen, K. N., & Lee, J. (2022). Platform government in the era of smart technology. *Public Administration Review, 82*(2), 362–368. DOI: 10.1111/puar.13422

Krueger, P., Sautner, Z., & Starks, L. T. (2020). The importance of climate risks for institutional investors. *Review of Financial Studies, 33*(3), 1067–1111. DOI: 10.1093/rfs/hhz137

Kumar, S., Sharma, D., Rao, S., Lim, W. M., & Mangla, S. K. (2022). Past, present, and future of sustainable finance: Insights from big data analytics through machine learning of scholarly research. *Annals of Operations Research*, •••, 1–44. PMID: 35002001

Lacoste, A., Luccioni, A., Schmidt, V., & Dandres, T. (2019). Quantifying the carbon emissions of machine learning. *arXiv preprint arXiv:1910.09700.*

Lagoarde-Segot, T. (2019). Sustainable finance. A critical realist perspective. *Research in International Business and Finance, 47*, 1–9. DOI: 10.1016/j.ribaf.2018.04.010

Lammens, H. (2020). *EU Retail Funds' Environmental Impact Claims Do Not Comply with Regulatory Guidance Analysis of a sample of 230 funds against the criteria of the EU Multi-Stakeholder Dialogue on Environmental Claims*. Retrieved from https://policycommons.net/artifacts/1802403/eu-retail-funds-environmental-impact-claims-do-not-comply-with-regulatory-guidance-analysis-of-a-sample-of-230-funds-against-the-criteria-of-the-eu-multi-stakeholder-dialogue-on-environmental-claims/2534047/

Li, S. (2003). Future trends and challenges of financial risk management in the digital economy. *Managerial Finance, 29*(5/6), 111–125. DOI: 10.1108/03074350310768797

Li, T.-T., Wang, K., Sueyoshi, T., & Wang, D. D. (2021). ESG: Research progress and future prospects. *Sustainability (Basel), 13*(21), 11663. DOI: 10.3390/su132111663

Liu, W., He, Y., Dong, J., & Cao, Y. (2023). Disruptive technologies for advancing supply chain resilience. *Frontiers of Engineering Management, 10*(2), 360–366. DOI: 10.1007/s42524-023-0257-1

Ljepava, N. (2022). AI-enabled marketing solutions in Marketing Decision making: AI application in different stages of marketing process. *TEM Journal, 11*(3), 1308–1315. DOI: 10.18421/TEM113-40

Ma, J. (2023). How digital finance promotes renewable energy consumption in China? *Environmental Science and Pollution Research International, 30*(46), 102490–102503. DOI: 10.1007/s11356-023-29504-7 PMID: 37667128

Makarov, I., & Schoar, A. (2022). Cryptocurrencies and Decentralized Finance (DeFi). *Brookings Papers on Economic Activity, 2022*(1), 141–215. DOI: 10.1353/eca.2022.0014

Marshall, T. E., & Lambert, S. L. (2018). Cloud-based intelligent accounting applications: Accounting task automation using IBM watson cognitive computing. *Journal of Emerging Technologies in Accounting, 15*(1), 199–215. DOI: 10.2308/jeta-52095

Martín-Martín, A., Orduna-Malea, E., Thelwall, M., & López-Cózar, E. D. (2018). Google Scholar, Web of Science, and Scopus: A systematic comparison of citations in 252 subject categories. *Journal of Informetrics, 12*(4), 1160–1177. DOI: 10.1016/j.joi.2018.09.002

Mavlutova, I., Spilbergs, A., Verdenhofs, A., Kuzmina, J., Arefjevs, I., & Natrins, A. (2023). The Role of Green Finance in Fostering the Sustainability of the Economy and Renewable Energy Supply: Recent Issues and Challenges. *Energies, 16*(23), 7712. DOI: 10.3390/en16237712

Memdani, L. (2020). Demonetisation: A move towards cashless economy in India. *International Journal of Electronic Banking, 2*(3), 205–211. DOI: 10.1504/IJEBANK.2020.111428

Meskó, B., Hetényi, G., & Győrffy, Z. (2018). Will artificial intelligence solve the human resource crisis in healthcare? *BMC Health Services Research, 18*(1), 1–4. DOI: 10.1186/s12913-018-3359-4 PMID: 30001717

Moeuf, A., Lamouri, S., Pellerin, R., Tamayo-Giraldo, S., Tobon-Valencia, E., & Eburdy, R. (2020). Identification of critical success factors, risks and opportunities of Industry 4.0 in SMEs. *International Journal of Production Research, 58*(5), 1384–1400. DOI: 10.1080/00207543.2019.1636323

Moodaley, W., & Telukdarie, A. (2023). Greenwashing, sustainability reporting, and artificial intelligence: A systematic literature review. *Sustainability (Basel), 15*(2), 1481. DOI: 10.3390/su15021481

Mousa, M., Saleem, A., & Sági, J. (2021). Are ESG shares a safe haven during COVID-19? Evidence from the Arab region. *Sustainability (Basel), 14*(1), 208. DOI: 10.3390/su14010208

Nightingale, A. (2009). A guide to systematic literature reviews. *Surgery (Oxford), 27*(9), 381–384. DOI: 10.1016/j.mpsur.2009.07.005

Nishant, R., Kennedy, M., & Corbett, J. (2020). Artificial intelligence for sustainability: Challenges, opportunities, and a research agenda. *International Journal of Information Management, 53*, 102104. DOI: 10.1016/j.ijinfomgt.2020.102104

OECD. (2022). *Measuring the environmental impacts of artificial intelligence compute and applications: The AI footprint*. OECD Publishing.

Orazayeva, A., & Arslan, M. (2024). Social responsibility portfolio optimisation in the context of emerging markets. *International Journal of Managerial and Financial Accounting, 16*(3), 266–281. DOI: 10.1504/IJMFA.2024.139513

Otto, M., & Thornton, J. (2023). The Role Of Digital Finance And Technological Innovation In The Growth Of The Green Economy. *Qo 'qon universiteti xabarnomasi, 7*, 19-22.

Palomares, I., Martínez-Cámara, E., Montes, R., García-Moral, P., Chiachio, M., Chiachio, J., & Fernández, B. (2021). A panoramic view and swot analysis of artificial intelligence for achieving the sustainable development goals by 2030: Progress and prospects. *Applied Intelligence, 51*(9), 6497–6527. DOI: 10.1007/s10489-021-02264-y PMID: 34764606

Păvăloaia, V.-D., & Necula, S.-C. (2023). Artificial intelligence as a disruptive technology—A systematic literature review. *Electronics (Basel), 12*(5), 1102. DOI: 10.3390/electronics12051102

Piazza, F. S. (2017). Bitcoin and the blockchain as possible corporate governance tools: Strengths and weaknesses. *Bocconi Legal Papers, 9*, 125.

Raman, N., Bang, G., & Nourbakhsh, A. (2020). Mapping ESG trends by distant supervision of neural language models. *Machine Learning and Knowledge Extraction, 2*(4), 453–468. DOI: 10.3390/make2040025

Rane, N. (2023a). Contribution of ChatGPT and other generative artificial intelligence (AI) in renewable and sustainable energy. *Available atSSRN* 4597674. DOI: 10.2139/ssrn.4597674

Rane, N. (2023b). Role of ChatGPT and similar generative artificial intelligence (AI) in construction industry. *Available atSSRN* 4598258. DOI: 10.2139/ssrn.4598258

Reig-Mullor, J., Garcia-Bernabeu, A., Pla-Santamaria, D., & Vercher-Ferrandiz, M. (2022). Evaluating ESG corporate performance using a new neutrosophic AHP-TOPSIS based approach. *Technological and Economic Development of Economy, 28*(5), 1242–1266-1242–1266.

Ruel, H., & Njoku, E. (2021). AI redefining the hospitality industry. *Journal of Tourism Futures*, *7*(1), 53–66. DOI: 10.1108/JTF-03-2020-0032

Schletz, M., Constant, A., Hsu, A., Schillebeeckx, S., Beck, R., & Wainstein, M. (2023). Blockchain and regenerative finance: Charting a path toward regeneration. *Frontiers in Blockchain*, *6*, 1165133. DOI: 10.3389/fbloc.2023.1165133

Shihadeh, F. (2020). Online payment services and individuals' behaviour: new evidence from the MENAP. *International journal of electronic banking, 2*(4), 275-282.

Shkolnyk, I., Kozmenko, O., Nowacki, R., & Mershchii, B. (2020). Dependence of the state of public finances on their transparency and the level of corruption in a country. *Economics & Sociology (Ternopil)*, *13*(4), 281–296. DOI: 10.14254/2071-789X.2020/13-4/18

Shkolnyk, I., Melnyk, T., Havrysh, Y., & Ivanchenko, A. (2019). Local finance transparency in Ukraine. *Public and Municipal Finance*, *8*(1), 73–82. DOI: 10.21511/pmf.08(1).2019.06

Sokolov, A., Mostovoy, J., Ding, J., & Seco, L. (2021). Building machine learning systems for automated ESG scoring. *The Journal of Impact and ESG Investing*, *1*(3), 39–50. DOI: 10.3905/jesg.2021.1.010

Stanislas, D., Constanze, B., & Thierry, S. (2020). *Survey of French and German retail investors' sustainability objectives*. Paris, France: 2degrees investing. Retrieved from https://2degrees-investing.org/wp-content/uploads/2020/03/A-Large-Majority-of-Retail-Clients-Want-to-Invest-Sustainably-Final.pdf

Su, H.-N., & Lee, P.-C. (2010). Mapping knowledge structure by keyword co-occurrence: A first look at journal papers in Technology Foresight. *scientometrics, 85*(1), 65-79.

Trivedi, S., Mehta, K., & Sharma, R. (2021). Systematic literature review on application of blockchain technology in E-finance and financial services. *Journal of Technology Management & Innovation*, *16*(3), 89–102. DOI: 10.4067/S0718-27242021000300089

Türegün, N. (2019). Impact of technology in financial reporting: The case of Amazon Go. *Journal of Corporate Accounting & Finance*, *30*(3), 90–95. DOI: 10.1002/jcaf.22394

Turunen, J. (2021). *Exploring the corporate financial and social performance relationship with net impact method.* Tampere University. Retrieved from https://trepo.tuni.fi/bitstream/handle/10024/134449/978-952-03-2137-6.pdf?sequence=2&isAllowed=y

Wang, Q., Sun, T., & Li, R. (2024). Does artificial intelligence promote green innovation? An assessment based on direct, indirect, spillover, and heterogeneity effects. *Energy & Environment*, 0958305X231220520.

Wang, Y., & You, J. (2022). [Retracted] The Operation Mode of Agricultural Supply Chain Finance Using Blockchain. *Computational Intelligence and Neuroscience*, *2022*(1), 3338030. PMID: 36188687

Wilson, M., Paschen, J., & Pitt, L. (2022). The circular economy meets artificial intelligence (AI): Understanding the opportunities of AI for reverse logistics. *Management of Environmental Quality*, *33*(1), 9–25. DOI: 10.1108/MEQ-10-2020-0222

Yadav, H., Kar, A. K., & Kashiramka, S. (2022, September). Artificial intelligence adoption for FinTech industries-an exploratory study about the disruptions, antecedents and consequences. In *Conference on e-Business, e-Services and e-Society* (pp. 3-12). Cham: Springer International Publishing.

Yaqoot, M., Diwan, P., & Kandpal, T. C. (2016). Review of barriers to the dissemination of decentralized renewable energy systems. *Renewable & Sustainable Energy Reviews*, *58*, 477–490. DOI: 10.1016/j.rser.2015.12.224

Yermack, D. (2017). Corporate governance and blockchains. *Review of Finance*, *21*(1), 7–31.

Young, M. M., Bullock, J. B., & Lecy, J. D. (2019). Artificial discretion as a tool of governance: A framework for understanding the impact of artificial intelligence on public administration. *Perspectives on Public Management and Governance*, *2*(4), 301–313. DOI: 10.1093/ppmgov/gvz014

Zhang, Y., Sun, J., Yang, Z., & Wang, Y. (2020). Critical success factors of green innovation: Technology, organization and environment readiness. *Journal of Cleaner Production*, *264*, 121701. DOI: 10.1016/j.jclepro.2020.121701

Zhong, J., Zhong, Y., Han, M., Yang, T., & Zhang, Q. (2024). The impact of AI on carbon emissions: Evidence from 66 countries. *Applied Economics*, *56*(25), 2975–2989. DOI: 10.1080/00036846.2023.2203461

Zouridis, S., Van Eck, M., & Bovens, M. (2020). Automated discretion. *Discretion and the quest for controlled freedom*, 313-329.

KEY TERMS AND DEFINITIONS

Artificial Intelligence: Is the process of simulating human intelligence through machines and computer systems.

Blockchain: Refers to an advanced database mechanism that allows secure and transparent information sharing within a network.

Thematic Analysis: Is a method of analysing qualitative data to identify themes through coding.

Financial Technology: Refers to the use of new technologies such as artificial intelligence, blockchain, cloud computing, and big data (referred to as ABCD) to provide financial services along with traditional financial methods.

Green Finance: Refers to financial products and services that promote and adopt sustainable practices

ENDNOTES

1 613 articles/reports/monographs were selected.

2 VOSviewer is a software tool for constructing and visualizing bibliometric networks. More details https://www.vosviewer.com

Chapter 3 To the Moon and Beyond: How Nano-Entrepreneurs and Digital Currency Are Revolutionizing the Future of Business in Space!

Mohammad Rashed Hasan Polas
https://orcid.org/0000-0002-6080-1075
Asia Pacific University of Technology and Innovation, Malaysia

Md. Shahbub Alam
https://orcid.org/0009-0002-0880-9173
Sonargaon University, Bangladesh

Mohammad Falahat
Asia Pacific University of Technology and Innovation, Malaysia

Md. Tajbir Husain
Sonargaon University, Bangladesh

Md. Omar Faruque Molla
Sonargaon University, Bangladesh

ABSTRACT

As the space economy gains momentum, fueled by advancements in technology and a growing interest in resource exploration beyond Earth, new opportunities for diverse economic participation are emerging. This chapter explores the intersection of the space economy, digital currency, and nano-entrepreneurship small-scale, often individual-driven businesses with minimal capital investment. By examining how digital currency can democratize access to the space economy, this chapter

DOI: 10.4018/979-8-3373-0355-0.ch003

proposes that even nano-entrepreneurs can play a significant role in this frontier. It investigates the potential for micro-investments in space ventures, the benefits of decentralized finance for nano-entrepreneurs, and the sustainability implications of these interactions. By leveraging case studies and theoretical frameworks, this chapter provides insights into how digital currency could revolutionize the space economy and offers a roadmap for nano-entrepreneurs looking to enter this high-potential but traditionally high-barrier field.

1.0 INTRODUCTION

In the era of technological advancements are reshaping economic landscapes, the convergence of the space economy and digital currency presents vast opportunities for fostering sustainable nano-entrepreneurship (Agarwal et al., 2021; Ferreira et al., 2023; Dzharbulov et al., 2024). As humanity expands beyond Earth, there is an increasing demand for decentralized financial systems and agile entrepreneurial models (Kang, 2024). Nano-entrepreneurs like small-scale innovators leveraging digital tools and occupying emerging technological spaces are poised to shape the trajectory of this evolving ecosystem by promoting inclusive growth, resource efficiency, and economic resilience (Nikraftar et al., 2022). This chapter explores how the fusion of space-based commerce and blockchain-powered financial systems can unlock new frontiers for sustainable micro-enterprise development within the global economy (Ferreira et al., 2023; Dzharbulov et al., 2024).

The space economy encompasses all activities and resources that generate value and benefits from space exploration, commercial ventures, and operational usage of space (Agarwal et al., 2021; Ferreira et al., 2023). Over the past two decades, the space industry has experienced significant growth, moving beyond its traditional domination by federal agencies such as NASA, the European Space Agency (ESA), and Roscosmos. Private-sector companies like SpaceX, Blue Origin, and Virgin Galactic have introduced groundbreaking space technologies that not only reduce the costs of accessing space but also create new opportunities for commercial participation (Crawford, 2016). According to research by the Space Foundation, the global space market reached a value exceeding $469 billion in 2021, with projections indicating it will surpass $1 trillion by 2040 (Dzharbulov et al., 2024). Modern satellite technologies remain a key driver of growth in the space sector, alongside advancements in asteroid exploration, planetary settlement initiatives, and space tourism. These technological developments have opened the door for small-scale space enterprises and decentralized participation in the space economy. Historically, government agencies and large corporations were the primary players in space-related activities due to the high financial requirements, advanced infrastructure, and specialized expertise needed (Arnone, 2024).

The creation of nano-entrepreneurship which features small-scale low-capital business ventures represents an entirely new concept (Korobova et al., 2024). Through digital currency combined with blockchain technology financial barriers to entry have lowered which allows nano-entrepreneurs to invest in space-related activities through micro-investments as well as DeFi mechanisms and tokenized assets (Jora et al., 2023). The accelerating use of digital currencies worldwide creates significant effects on businesses operating in space-related markets. The combination of cryptos and blockchain presents a secure financial system that eliminates borders and decentralizes financial operations thereby improving space commerce access. Delegated financial platforms will likely become the funding source for satellite launches while enabling space research and interplanetary monetary exchanges according to upcoming market developments. Digital currency will become a tool for expanding space economy accessibility because it will enable nano-entrepreneurs and small business owners together with researchers to enter projects needing significant financial investments (Kang, 2024).

Digital currency exists as electronic money which runs separate from typical banking structures yet relies on blockchain systems for operation (Lal & Jatav, 2024). The cryptocurrencies Bitcoin and Ethereum alongside stablecoins have shown capabilities for performing secure transparent transactions without border restrictions. These characteristics apply directly to space economic activities because they enable borderless financial operations which involve supply chain funding and digital ownership in space initiatives (Nikraftar et al., 2022). The main hurdle in traditional space investment involves needing large financial support and fulfilling all regulatory criteria. The execution of digital currency through Decentralized Finance (DeFi) helps eliminate traditional restrictions by establishing peer-to-peer acquisitions and fractional estate ownership along with crowd-funding options. Through blockchain technology platforms investors may execute decentralized funding operations and compose smart contracts with tokenized space resources which enable less affluent investors to join limited-access space economic segments (Martin & Sunley, 2017).

Individuals together with small groups establish new businesses by investing minimal funds through digital platforms while utilizing technological innovations toward their development (Saito et al., 2019). The widespread use of this concept began with e-commerce but its importance for the space economy is currently on the rise. Nano-entrepreneurs would conduct activities that include satellite servicing and space-based 3D printing as well as data analytics for space exploration and virtual space experience delivery. De-centralized funding mechanisms which use digital currency enable nano-entrepreneurs to explore new business prospects without needing extensive initial capital outlays (Sun, 2024).

The chapter addresses digital currency together with nano-entrepreneurship and the space economy by examining financial innovations which reduce entry hurdles for tiny market participants. Digital currency along with blockchain technology function as potential financial solutions which enable nano-entrepreneurs to participate in sustainable space-related business ventures. The chapter examines questions that appear in the subsequent section.

1. How can digital currency democratize access to space economy ventures for nano-entrepreneurs?
2. What are the specific ways in which nano-entrepreneurs could contribute to the space economy, particularly regarding sustainable practices and micro-investment opportunities?
3. What are the potential environmental and economic impacts of increased nano-entrepreneurial activity in the space economy?
4. What regulatory frameworks and policies would support the responsible integration of digital currency and nano-entrepreneurship in the space economy?

Therefore, digital currency serves as a transformative financial tool within space economics because it requires us to develop equal opportunities and innovative solutions. This research presents nano-entrepreneurs with an action plan to use blockchain technology and decentralized financial frameworks for active space venture participation (Tong & Jiayou, 2021). This study seeks to steer national policy development while promoting frameworks which ensure viable financial approaches for space trade activities. The transition to space commercial activities and planetary exploration requires accessible finances to define industry progress (Valente et al., 2025). Digital currency creates an opportunity between large-scale space investments and small business startups to discover fresh economic boundaries that build a sustainable space economy for every participant. Rapid space economy expansion generates various chances from investment to innovation and modern entrepreneurship (Tong & Jiayou, 2021). Financial barriers act as a major obstacle for business creators who want to enter this market. Crypto payments through decentralized financial systems offer entrepreneurial solutions that help nano-entrepreneurs enter space-related activities regardless of traditional financial barriers. The analysis of digital currency's role in making space investments accessible to all users contributes essential understanding to the combination of technology alongside finance and entrepreneurship which drives a modern space economy (Valente et al., 2025).

2.0 THE SPACE ECONOMY: OPPORTUNITIES AND CHALLENGES

The space economy, a rapidly expanding sector, refers to the economic activities that revolve around the use and commercialization of outer space (Aglietti, 2020). The field has long been dominated by government space agencies like NASA, the European Space Agency (ESA), and Roscosmos, but in recent years, private companies such as SpaceX, Blue Origin, and Virgin Galactic have entered the scene, marking a shift in the way space-related activities are conducted (Allam et al., 2022). This dynamic transformation is creating new opportunities for participation across various sectors, ranging from satellite technology to space tourism and space mining. The space economy encompasses a broad spectrum of industries that include satellite communications, Earth observation, space travel, and the extraction of resources from celestial bodies like asteroids and the Moon (Hart & Halden, 2020).

One of the most significant components of the space economy is satellite technology. Satellites play a crucial role in communication systems, navigation (GPS), weather forecasting, and surveillance (Hermoso et al., 2022). They enable a wide range of applications, including broadband internet access in remote areas, monitoring of natural disasters, and supporting global positioning systems. The development and deployment of these technologies have paved the way for the commercialization of space (Marzaleh et al., 2022). Another growing area is space tourism, with private companies making significant strides in enabling non-professional astronauts to visit outer space for leisure. Companies like SpaceX, Virgin Galactic, and Blue Origin have developed space travel programs that offer suborbital flights, orbital missions, and potentially even lunar expeditions in the near future. These activities not only serve as the foundation for commercial spaceflight but also open up new avenues for revenue generation (Caprotti et al., 2017).

A particularly fascinating aspect of the space economy is space mining, the potential extraction of valuable resources like rare metals and water from asteroids and the Moon (Hopkinson et al., 2018). This component of the space economy has the capacity to redefine resource distribution on Earth, particularly in light of dwindling natural resources. Minerals such as platinum, gold, and even water could be sourced from outer space, reducing the pressure on Earth's limited resources (Porru et al., 2020). Although the technology to extract these resources is still in its infancy, the economic impact of space mining could be transformative. Beyond these well-known sectors, the space economy also includes Earth observation, space manufacturing, space exploration, and infrastructure development, all of which contribute to the growth and expansion of space as a commercial enterprise (Lin et al., 2022).

Despite its promise, the space economy faces numerous barriers to entry that make it difficult for new players, especially small-scale enterprises or nano-entrepreneurs, to participate (Nair et al., 2021). One of the most significant barriers is the capital intensity of space-related ventures. Building spacecraft, satellites, and launching infrastructure requires substantial investment. Launching a single satellite can cost hundreds of millions of dollars, and the research and development (R&D) required creating new technologies for space exploration or exploitation often requires years of investment and testing (Chen et al., 2017). Space-related ventures often involve highly complex systems, and even the smallest mistake can result in the loss of an expensive payload or jeopardize the success of a mission. For small companies or individual entrepreneurs, the capital requirements are often far beyond their reach.

In addition to financial constraints, the technological complexity of space ventures presents a considerable challenge. Space exploration, satellite deployment, and resource extraction from celestial bodies require advanced expertise in aerospace engineering, robotics, artificial intelligence, and other highly specialized fields (Tan et al., 2022). The engineering and technical hurdles involved in designing and operating spacecraft, rovers, satellites, and space habitats are immense. These technologies are often at the cutting edge of human capabilities, and developing new solutions often requires the collaboration of large teams of scientists and engineers. Consequently, the technical barriers to entry are high, particularly for those who lack access to the necessary resources and expertise (Jayasooria, 2016).

Moreover, regulatory and legal constraints add further complexity to the space economy. The outer space environment is governed by international treaties, such as the Outer Space Treaty of 1967, which establishes principles for space exploration and use (Secinaro et al., 2022). These treaties are aimed at ensuring that space is used for peaceful purposes and that space-related activities benefit all of humankind. National space agencies, such as the U.S. Federal Aviation Administration (FAA) or the European Space Agency (ESA), impose strict regulations on the commercial use of space (Crawford, 2016). These regulations often require companies to obtain specific licenses for launching payloads, conducting space missions, and using orbital slots. Legal challenges may also arise regarding property rights and claims over resources harvested from celestial bodies like the Moon or asteroids. These complexities make it difficult for new entrants, especially small companies or individual entrepreneurs, to navigate the regulatory landscape and gain the necessary approvals for their ventures.

Another challenge lies in the monopolization of the space economy by a few large companies and state-run agencies. Major space agencies like NASA and ESA control much of the space infrastructure, and private companies such as SpaceX and Boeing are often the dominant players in the commercial spaceflight and satellite sectors (Weinzierl, 2018). These large companies have access to vast resources, well-established networks, and the technological capabilities needed to conduct ambitious space missions. The high entry barriers created by these established players make it challenging for new firms to compete in terms of technology, infrastructure, and market share. This monopolization often means that small-scale entrepreneurs may struggle to find a niche in the space economy (Valente et al., 2025).

Nevertheless, the rise of digital currencies and decentralized finance (DeFi) presents a new set of opportunities for small-scale participation in the space economy. One of the primary advantages of digital currencies is their ability to democratize access to investment opportunities (Pomeroy et al., 2019). In traditional models, the high cost of entry into the space economy often restricts participation to wealthy investors or large corporations (Fiore & Elvis, 2025). However, digital currencies and DeFi platforms enable micro-investments, allowing nano-entrepreneurs and small investors to participate in space ventures. This is particularly important in the context of space mining and satellite technology, where crowdfunding or token-based investments could enable individuals or small firms to invest in space projects without needing large capital outlays (Fiore & Elvis, 2025).

Furthermore, space-related data including satellite imagery, Earth observation data, and communications data represents a valuable asset that nano-entrepreneurs can leverage. Small businesses and startups specializing in data analytics, AI-driven insights, and predictive models can take advantage of this wealth of information to develop new applications and services (Krugman, 2018). For instance, satellite data can be used for climate change modeling, disaster management, agricultural planning, and urban development. The increasing availability of open-source satellite data presents a unique opportunity for small firms to provide value-added services, such as geospatial analysis or environmental monitoring, without having to own or operate satellites themselves (Vittori et al., 2025).

Another opportunity for small-scale participation exists in satellite component manufacturing. As the demand for smaller, cheaper, and more efficient satellites grows, there is an increasing market for CubeSats small, low-cost satellites that can be launched into space at a fraction of the cost of traditional satellites (Foreman Campins, 2023). Nano-entrepreneurs could potentially capitalize on this growing sector by providing components, design services, or even fully operational CubeSat missions. The ability to manufacture and launch small-scale satellites opens up numerous entrepreneurial possibilities, including providing satellite communication services to underserved areas, conducting scientific research, or offering Earth observation capabilities to commercial clients (Lee et al., 2012).

Moreover, space-related consulting and educational services present another avenue for nano-entrepreneurs to engage with the space economy (Caprara, 2020). Given the rapid growth of the space sector, there is a growing demand for specialized knowledge in areas such as space law, commercial space policies, and satellite technology. Small consulting firms or independent entrepreneurs could offer services related to regulatory compliance, market analysis, and business development in the space sector. Additionally, the space economy offers potential for education and training programs focused on space-related technologies and industries. By developing educational resources, workshops, or even online courses, nano-entrepreneurs could provide valuable insights and training for the next generation of space professionals (Kutter & Sowers, 2016). While the space economy presents significant opportunities for innovation and growth, it is also fraught with challenges, particularly for small-scale players. The high capital requirements, technological complexity, and regulatory constraints make it difficult for nano-entrepreneurs to gain a foothold in the industry (Eastwood et al., 2017). However, the increasing availability of digital currencies, decentralized finance, and new technologies is creating pathways for smaller enterprises to enter the space economy. By focusing on niche areas like satellite component manufacturing, data analytics, consulting, and micro-investments, nano-entrepreneurs can contribute to the growth of the space economy while overcoming traditional barriers to entry (Polanyi, 2018).

3.0 DIGITAL CURRENCY AS AN ENABLER FOR NANO-ENTREPRENEURSHIP

In recent years, digital currency has emerged as a transformative force, not only reshaping traditional financial systems but also offering new possibilities for entrepreneurs, particularly those operating on a small scale (Auplat, 2012). Nano-entrepreneurship, often characterized by minimal capital investment, individual-driven businesses, and limited resources, has historically faced numerous barriers to entry, including access to capital, financial inclusion, and the challenges of navigating complex banking systems (Keister & Sanches, 2023). Digital currencies, especially cryptocurrencies, and the technologies supporting them, such as blockchain, provide innovative solutions that can democratize access to economic opportunities and empower nano-entrepreneurs. By enabling faster, cheaper, and more secure transactions, digital currencies create a level playing field, offering small-scale entrepreneurs the chance to tap into industries like space commerce, global markets, and technology-driven startups (Haque & Shoaib, 2023).

3.1 Overview of Digital Currency and Blockchain Technology

Digital currencies, primarily in the form of cryptocurrencies like Bitcoin, Ethereum, and others, are currencies that exist only in digital form (Verma et al., 2023). Unlike traditional fiat currencies (such as USD or EUR), which are issued and regulated by central banks, digital currencies are typically decentralized, operating on a technology called blockchain. Blockchain is a distributed ledger technology that records all transactions made with a particular cryptocurrency in a transparent, immutable, and secure manner. It removes the need for intermediaries like banks, enabling peer-to-peer transactions without relying on centralized authorities (Akhtar et al., 2019).

The decentralized nature of digital currencies provides several key advantages over traditional financial systems (Verma et al., 2023). First, they eliminate the need for third-party intermediaries, such as banks or payment processors, which typically charge fees and may limit access based on geography, financial status, or creditworthiness. Second, digital currencies offer faster transaction speeds, particularly for international transfers, which traditionally take several days when processed through conventional banking systems. Additionally, blockchain ensures transparency and security, providing nano-entrepreneurs with a reliable, cost-effective way to manage transactions and investments (Hashemi et al., 2020).

3.2 Lowering Barriers to Entry for Nano-Entrepreneurs

One of the primary challenges that nano-entrepreneurs face is access to capital. Small-scale businesses often struggle to secure financing from traditional banks or venture capital firms due to a lack of credit history, collateral or sufficient business networks (Ronaghi, 2023). Digital currencies provide a promising alternative by enabling micro-investments and crowdfunding in decentralized spaces. Through Initial Coin Offerings (ICOs), Security Token Offerings (STOs), and Decentralized Finance (DeFi) platforms, nano-entrepreneurs can raise capital directly from a global pool of investors without the need for traditional financial intermediaries (Truby, 2018).

For example, DeFi platforms offer a wide range of financial services, including lending, borrowing, trading, and saving, all conducted without the need for banks. These platforms use smart contracts—self-executing contracts with the terms of the agreement directly written into code to ensure that transactions are transparent, secure, and executed automatically once specific conditions are met. By leveraging DeFi, nano-entrepreneurs can access liquidity, fund their ventures, and create financial products that suit their specific needs, all while avoiding the constraints and high costs typically associated with conventional financing routes (Ronaghi, 2023).

Moreover, digital currencies allow entrepreneurs to tap into a global market, breaking down geographical barriers that traditionally limited small businesses (Ronaghi, 2023). A nano-entrepreneur based in a developing country, for instance, could easily access a global investor base through digital currency platforms. These entrepreneurs can also offer their products or services to international customers by accepting cryptocurrencies as payment, avoiding the fees and delays associated with traditional international banking systems. This global reach creates new avenues for small businesses to expand and thrive, even in industries like space commerce, where traditional financing models might be out of reach (Truby, 2018).

3.3 Enhancing Financial Inclusion and Reducing Transaction Costs

For nano-entrepreneurs, especially those in regions with limited access to banking services, financial inclusion has been a persistent problem. According to the World Bank, over 1.7 billion people worldwide remain unbanked, with many lacking access to basic financial services such as savings accounts, loans, and insurance (Naheem, 2019). Digital currencies, however, can provide financial services to anyone with internet access, bypassing the need for traditional banking infrastructure (Efanov & Roschin, 2018). In regions with weak banking systems or unstable currencies, digital currencies can provide a stable and reliable alternative. Cryptocurrencies like Bitcoin or stablecoins (cryptocurrencies pegged to the value of traditional assets like the US dollar) offer a hedge against inflation and currency devaluation, which is particularly important in emerging markets where local currencies may be subject to extreme volatility. Nano-entrepreneurs operating in such markets can leverage digital currencies to protect their savings, conduct transactions, and access global markets without relying on local banking systems that may be unstable or inaccessible (Naheem, 2019).

In addition to improving financial access, digital currencies can significantly reduce transaction costs. Traditional banking systems and payment processors often charge high fees for domestic and international transactions, particularly for small-scale transactions (Lu et al., 2021). For nano-entrepreneurs, these fees can quickly erode their profits. By using digital currencies, small businesses can reduce or eliminate transaction fees, especially in cross-border transactions, where traditional fees are typically the highest. This is particularly advantageous for entrepreneurs involved in niche markets like space commerce or technology startups, where even small reductions in costs can make a significant impact on overall profitability (Ertz & Boily, 2019).

3.4 Empowering Nano-Entrepreneurs with Smart Contracts and Decentralized Platforms

Another significant advantage of digital currencies is the ability to utilize smart contracts, which are automated agreements executed directly on the blockchain (Zhang et al., 2022). These contracts remove the need for intermediaries, such as lawyers, notaries, or payment processors, streamlining the process of entering into agreements and ensuring that transactions are executed securely and automatically when predefined conditions are met. For nano-entrepreneurs, this offers a powerful tool to automate business processes, reduce administrative costs, and ensure transparency in all dealings (Tasatanattakool & Techapanupreeda, 2018). For example, a nano-entrepreneur in the space economy could enter into a smart contract with a satellite provider for data services, where payment is automatically processed upon delivery of data. Similarly, a nano-entrepreneur involved in space tourism could create a smart contract for booking services, ensuring that payments are made once specific conditions (e.g., flight schedule confirmation) are met. This automation not only reduces the risk of fraud but also streamlines the customer experience, providing businesses with a reliable and efficient way to operate (Sabry, 2019).

Furthermore, decentralized platforms that utilize blockchain and digital currencies offer nano-entrepreneurs the ability to create and manage businesses without relying on centralized authorities (Gamage et al., 2020). This is particularly relevant in industries like space commerce, where traditional funding sources, regulatory frameworks, and market access points are often controlled by a few large players. By using decentralized platforms, small-scale entrepreneurs can collaborate with others, share resources, and access global markets without the need for intermediaries or gatekeepers. This fosters a more inclusive, innovation-driven environment, where ideas can flourish regardless of the size or financial standing of the participants (Komalavalli et al., 2020).

3.5 Implications for Nano-Entrepreneurs in the Space Economy

In the context of the space economy, digital currencies offer several unique opportunities for nano-entrepreneurs. As the space sector becomes increasingly commercialized, small-scale businesses can participate in space ventures, such as satellite component manufacturing, space tourism, and space mining, by leveraging the advantages of digital currencies (Giungato et al., 2017). For instance, space mining, a sector that involves extracting valuable resources from asteroids or the Moon, requires massive capital investment. However, through tokenized investments, nano-entrepreneurs can pool funds from smaller investors globally, enabling them to invest in space mining projects. Similarly, decentralized platforms could allow for the creation of space-based decentralized applications (dApps), providing services such as satellite data analysis or space tourism bookings, where nano-entrepreneurs can enter the market without needing to develop expensive infrastructure (Dong et al., 2023). Moreover, tokenization allows for fractional ownership of space assets, such as satellites or spacecraft, enabling small investors and nano-entrepreneurs to share in the profits generated by these ventures. By issuing digital tokens that represent shares in space-related assets, entrepreneurs can create new funding models that reduce the risk of entry and make participation in the space economy more accessible to small players (Niranjanamurthy et al., 2019).

3.6 Challenges and Risks of Digital Currency for Nano-Entrepreneurship

While digital currencies offer significant opportunities for nano-entrepreneurs, they are not without their challenges. The volatility of cryptocurrencies, legal uncertainties, and potential regulatory hurdles present risks for small-scale businesses (Chen et al., 2018). The value of cryptocurrencies like Bitcoin can fluctuate dramatically, which might result in sudden losses for nano-entrepreneurs who hold large amounts of digital currency. Furthermore, the lack of a clear regulatory framework in many countries raises concerns about the legality of using digital currencies for business transactions, especially in emerging markets (Ramadoss, 2022). Additionally, there is a risk of fraud and cybersecurity threats. While blockchain technology is generally secure, digital currency exchanges and wallets can be vulnerable to hacks, putting entrepreneurs' funds at risk. Nano-entrepreneurs must be aware of these risks and take steps to protect their assets, such as using secure wallets, conducting transactions on reputable platforms, and staying informed about the latest cybersecurity best practices (Chen et al., 2018; Ramadoss, 2022).

Therefore, digital currencies and blockchain technology present significant opportunities for nano-entrepreneurs, particularly in emerging sectors like the space economy. By lowering barriers to entry, improving financial inclusion, reducing transaction costs, and enabling decentralized business models, digital currencies can empower small-scale businesses to innovate and grow (Laroiya et al., 2020). However, nano-entrepreneurs must be aware of the challenges and risks associated with digital currencies and take appropriate precautions to protect their ventures. As the digital currency ecosystem continues to evolve, it is likely that nano-entrepreneurs will play an increasingly important role in the global economy, unlocking new frontiers in space commerce and other high-potential industries (Valdeolmillos et al., 2020).

4.0 THE ROLE OF NANO-ENTREPRENEURS IN THE SPACE ECONOMY

Nano-entrepreneurship represents a specialized form of entrepreneurship that focuses on ventures operating at the nano and micro-scale, often in high-tech and innovation-driven industries (Glaser et al., 2016). This type of entrepreneurship emphasizes technological innovation, niche market engagement, and the ability to scale with precision, making it particularly relevant in sectors such as nanotechnology, micro-manufacturing, and biotechnology (Jahangir & Bulut, 2024). As such, nano-entrepreneurs are highly specialized, embracing the complexity and risks involved in developing cutting-edge technologies for targeted applications. Their impact is notable across multiple industries, with their small-scale operations often driving significant advancements in technology, economic growth, and job creation (Sørensen et al., 2024).

4.1 Characteristics of Nano-Entrepreneurship

Nano-entrepreneurship refers to ventures that operate on a smaller scale, focusing on highly specialized markets that require advanced technological expertise (Jahangir & Bulut, 2024). The defining characteristics of nano-entrepreneurship include a strong emphasis on innovation, a high degree of agility, and a focus on niche markets. Nano-entrepreneurs are often at the forefront of research and development (R&D) activities, creating products or services based on complex scientific and engineering principles at microscopic scales. The size and nature of their operations allow them to pivot and adapt quickly to changes in technology and market demand, which is a significant advantage in high-tech sectors (Sørensen et al., 2024).

Innovation and technological advancement are at the heart of nano-entrepreneurship, and these entrepreneurs typically engage in R&D to create new products or improve existing ones (Sørensen et al., 2024). The complexity of their work requires not only a deep understanding of science and engineering but also a comprehensive knowledge of market dynamics. Furthermore, nano-entrepreneurs often rely on collaborative networks with universities, research institutions, and other start-ups to access resources, funding, and knowledge. These networks foster innovation, accelerate time-to-market, and provide essential support for market entry (Glaser et al., 2016).

4.2 Examples of Nano-Entrepreneurial Contributions

Nano-entrepreneurs are making significant contributions across various high-tech and specialized sectors, demonstrating the broad potential of nano-scale innovation. Below are several examples that highlight their impact:

Nano-Formulations for Drug Delivery: In the pharmaceutical industry, nano-entrepreneurs have pioneered the development of advanced nano-formulations to improve drug delivery systems. These formulations enhance targeting capabilities and overcome physiological barriers, improving treatment efficacy for conditions like respiratory diseases. By utilizing the unique properties of nanomaterials, these ventures have opened new markets, demonstrating the transformative potential of nano-entrepreneurship in the health tech sector (Jeevanandam et al., 2016; Musari et al., 2023).

Sustainable Food Packaging Solutions: Nano-entrepreneurs are revolutionizing the food packaging industry by developing nano-packaging materials that improve the shelf life and quality of food. For example, using nano-clay in polymer composites enhances the barrier properties of packaging, making it more resistant to moisture and gases. This innovation meets both sustainability goals and consumer demands for longer-lasting food products, highlighting the role of nano-entrepreneurs in addressing environmental and market challenges (Guillard et al., 2018).

Enhanced Barrier Properties in Materials: Nano-entrepreneurs have also contributed to advancements in materials science. For instance, incorporating nano-zinc oxide in biopolymer films has significantly improved mechanical durability and environmental resistance, making it ideal for use in packaging and construction. These advancements showcase the potential of nano-entrepreneurs to develop high-performance materials that meet industry standards (Sanchez-Garcia et al., 2010).

Innovations in Renewable Energy: Nano-entrepreneurs are playing a key role in the renewable energy sector by developing advanced materials for energy generation and storage. For example, nano-materials are being used to enhance the efficiency of solar panels and reduce production costs. This innovation aligns with global efforts to shift toward more sustainable energy solutions (Elia et al., 2021).

Advanced Robotics for Medical Applications: In healthcare, nano-entrepreneurs are pioneering the use of microrobots for surgical applications. These technologies enable highly precise and minimally invasive procedures, improving patient outcomes. The intersection of nanotechnology and robotics in the medical field highlights the interdisciplinary nature of nano-entrepreneurship (Xie, 2016).

These examples demonstrate the broad and impactful contributions of nano-entrepreneurs across diverse sectors, from pharmaceuticals and food packaging to renewable energy and healthcare. By addressing specific industry challenges, nano-entrepreneurs are driving technological advancements and reshaping traditional markets.

4.3 Opportunities for Nano-Entrepreneurs in the Space Economy

The space economy is rapidly expanding, creating numerous opportunities for nano-entrepreneurs to innovate and contribute to the industry's growth. As the space sector evolves, the demand for miniaturized, high-precision components is increasing, opening new avenues for nano-entrepreneurs to apply their expertise. The following are key opportunities for nano-entrepreneurs in the space economy:

Micro-Manufacturing Solutions: Nano-entrepreneurs can capitalize on the growing need for micro-manufacturing, which involves producing small-scale components with high precision for space applications. Micro-manufacturing is essential for creating miniaturized devices such as sensors, communication systems, and scientific instruments that are integral to space missions. Advanced materials, such as polymer nanocomposites, offer enhanced mechanical and thermal properties that are crucial for aerospace applications. These innovations help reduce the weight of spacecraft, improving performance and sustainability (Qin, 2018).

Resource Sharing and Collaboration: One of the major opportunities for nano-entrepreneurs in the space economy is collaboration. Smaller firms can collaborate with larger entities, such as space agencies and private aerospace companies, to share resources, technologies, and research facilities. Collaborative initiatives allow nano-entrepreneurs to engage in high-stakes projects while reducing financial risks and leveraging shared expertise. These partnerships can accelerate innovation, improve efficiency, and provide opportunities for smaller firms to contribute to large-scale space projects (Bailey-Haine et al., 2014).

Sustainable Supply Chain Innovation: Sustainability is a top priority in the space economy, and nano-entrepreneurs can play a vital role in optimizing supply chains to reduce carbon footprints and enhance resource efficiency. By developing sustainable materials, such as biodegradable composites, and leveraging digital technologies and data analytics, nano-entrepreneurs can create more efficient logistics systems for space missions. This focus on sustainability not only aligns with global environmental goals but also presents significant market opportunities for entrepreneurs focused on green technologies (Gao et al., 2017).

Addressing Technical Challenges through Innovation: The challenge of long-term space exploration such as in-situ resource utilization (ISRU) present unique opportunities for nano-entrepreneurs. ISRU refers to the process of using local resources on celestial bodies, such as the Moon or Mars, to produce life-supporting materials like water or oxygen. Nano-entrepreneurs can innovate in resource extraction, processing, and utilization technologies, enabling more sustainable and self-sufficient space missions (O'Cass & Wetzels, 2018).

Market Entry into Growing Aerospace Ventures: As private companies increasingly enter the space exploration market, nano-entrepreneurs have the opportunity to contribute to emerging aerospace ventures. The rise of satellite manufacturing, small launch vehicles and commercial space travel creates demand for specialized components and services that nano-entrepreneurs can provide. Small firms have the ability to adapt quickly to market shifts, enabling them to offer innovative solutions that larger companies may overlook (McGuire & Islam, 2015).

Therefore, nano-entrepreneurs are strategically positioned to capitalize on the growing space economy. Through micro-manufacturing, collaboration, sustainable supply chain innovation, and addressing technical challenges, nano-entrepreneurs can significantly contribute to the development of the space sector. As the space economy continues to expand, nano-entrepreneurs will be key players in driving technological advancements and shaping the future of space exploration.

5.0 SUSTAINABILITY IMPLICATIONS IN THE DIGITAL CURRENCY-SPACE ECONOMY NEXUS

The integration of digital currencies into space economy ventures presents several critical sustainability implications, particularly as they relate to environmental stewardship and socioeconomic inclusivity. The inclusion of digital currencies in small-scale space projects can help foster sustainable development by creating financial ecosystems that promote local businesses and space-related activities. These developments support the broader goal of circular economies, which aim to maximize resource efficiency and reduce waste (Pfister, 2024).

5.1 Key Implications of Digital Currencies in Space Economy Sustainability

The role of digital currencies in fostering sustainability in the space economy is multifaceted. On one hand, local digital currencies, when adopted, can empower communities to participate in space ventures by providing the financial means to engage in projects that contribute to sustainability. For instance, digital currencies can enable small businesses involved in space-related activities to access financing, thus supporting local economic development and innovation in space technologies (Balogun, 2024).

However, there are concerns regarding the environmental costs of digital currencies, particularly related to the energy consumption of cryptocurrency mining. Studies have shown that the high energy demands of cryptocurrencies like Bitcoin could hinder sustainability goals, with considerable carbon emissions tied to these activities. This necessitates a reconsideration of regulatory frameworks to address these challenges. Policymakers and industry stakeholders must prioritize the development of energy-efficient technologies and adopt sustainable practices in the cryptocurrency space to mitigate the negative environmental impacts (Chorzempa, 2021).

On the other hand, Central Bank Digital Currencies (CBDCs) hold promise for transforming financial systems toward sustainability. CBDCs can make digital currencies more accessible to businesses and promote the circular economy by facilitating sustainable production and consumption. The transparency and security inherent in CBDCs may also lead to greater trust in digital currencies, promoting broader adoption and more efficient use of resources in space-related projects (Balogun, 2024).

5.2 Circular Economy Models and Digital Currencies

Circular economies are closely aligned with the principles of digital currencies. A circular economy focuses on resource efficiency, waste reduction, and maximizing the use of existing resources (Ronaghi & Fallahi, 2024). Digital currencies can play a crucial role in facilitating these principles within space ventures by reducing transaction costs, increasing transparency, and providing traceable financial systems for space projects. Cryptocurrencies, in particular, can help fund and incentivize sustainable practices such as recycling in spaceports or promoting resource-efficient farming technologies on Mars (Ozili, 2022). The long-term integration of digital currencies into the space economy, with a focus on sustainability, could further promote resource efficiency and reduce waste. By supporting decentralized financial systems, digital currencies could contribute to reducing the ecological footprint of space missions while enhancing the economic resilience of space projects (Ronaghi & Fallahi, 2024).

5.3 Case Studies: Digital Currency for Resource Efficiency and Circular Economy in Space

Real-world examples highlight the potential for digital currencies to contribute to the circular economy and sustainability within the space sector. NASA's use of blockchain technology to manage supply chains and track materials in space missions is one such example. Blockchain allows for more transparent tracking of resources, leading to better allocation and utilization (Siderius & Zink, 2023). This can enhance the efficiency of space missions by ensuring that resources are managed optimally. Additionally, the potential for digital currencies to support sustainable agriculture technologies in Mars mission planning through in-situ resource utilization (ISRU) proposals exemplifies their role in fostering innovation in space sustainability. Similarly, digital currencies could facilitate the recycling of materials at spaceports by creating rewards systems that incentivize sustainable practices. These applications showcase how digital currencies can align with the principles of a circular economy while advancing space missions (Kouhizadeh et al., 2020).

5.4 Environmental and Social Impacts of Expanding Participation in Space Economy

Broadening participation in space economy initiatives through digital currencies brings environmental and social implications that must be carefully considered (Ge et al., 2021). Environmental concerns primarily stem from the energy-intensive nature of digital currency mining, which can contribute to high carbon emissions (Toivonen, 2022). However, alternatives such as digital community currencies (DCCs) present opportunities to mitigate these impacts by focusing on localized transactions that reduce transportation costs and energy consumption. On the social front, the expansion of digital currencies in space-related projects offers the potential for greater inclusivity and equitable access to economic opportunities (Bernhardt & Pollak, 2016). Digital currencies can break down financial barriers for underserved populations, allowing them to participate in global space ventures (Al-Emran, 2023). This could significantly improve socioeconomic equity, especially in regions with limited access to traditional banking services. By facilitating financial inclusion, digital currencies could empower underrepresented communities and help foster digital literacy, thus contributing to a more resilient and inclusive space economy (Ge et al., 2021).

The integration of digital currencies into the space economy holds immense potential for fostering sustainability across both environmental and social dimensions. While challenges remain, particularly in managing the environmental impact of cryptocurrency mining, the ongoing exploration of digital currencies and their regulatory frameworks will play a pivotal role in shaping the future of space economy sustainability. Embracing digital currencies can contribute to resource efficiency, transparency, and inclusivity, thus enabling a more sustainable and equitable future for space ventures and the communities they impact (Al-Emran, 2023).

6.0 REGULATORY AND POLICY CONSIDERATIONS

The integration of digital currency into the space economy represents a revolutionary shift in global finance, offering decentralized and efficient payment mechanisms for space-based enterprises (Al-Emran, 2023). As commercial space activities expand, financial transactions beyond Earth will require secure and legally compliant frameworks. However, the absence of established financial regulations for extraterrestrial commerce poses significant challenges. Issues such as jurisdictional ambiguity, cybersecurity risks, taxation, and transaction validation must be addressed to create a sustainable financial infrastructure. This section explores the key regulatory challenges that must be overcome to facilitate the seamless integration of digital currency into the space economy (Chen et al., 2017).

6.1 Key Regulatory Challenges Associated with Integrating Digital Currency in the Space Economy

The key regulatory challenges to integrating digital currency into the space economy can be broken down into several categories, each presenting its own unique hurdles.

a. Jurisdictional Ambiguities and Legal Frameworks

The legal governance of space activities is primarily dictated by international treaties such as the Outer Space Treaty (1967), which lacks explicit provisions for financial regulations. This lack of a clear regulatory framework creates significant challenges in determining which country's laws govern digital transactions in space. The absence of a global authority overseeing space-based financial operations leads to uncertainties, complicating the enforcement of contracts, resolution of disputes, and ensuring legal accountability in extraterrestrial commerce (Ochigbo et al., 2024).

b. Cross-Border and Multilateral Regulatory Coordination

Space-based digital transactions inherently involve multiple jurisdictions, which requires international cooperation for effective regulation. Different countries have varying policies on digital currencies, leading to inconsistencies in their legal treatment. The lack of a unified regulatory framework makes it difficult to ensure compliance across different legal systems, increasing the risk of conflicts between nations. Establishing a globally recognized financial authority for space commerce is crucial to harmonizing regulations and facilitating seamless economic transactions (Beck, 2016).

c. Cybersecurity and Data Privacy Risks

The security of digital transactions in space presents significant concerns. Space-based financial systems are highly vulnerable to cyberattacks, hacking, and data breaches. Unlike Earth-based financial institutions that operate within regulated cybersecurity frameworks, space commerce lacks specific regulations to ensure data protection and user privacy. The enforcement of cybersecurity measures in a space environment is also technologically challenging due to communication limitations and the risks associated with remote data management. To ensure secure financial transactions in space, robust encryption, decentralized ledger technologies, and international cybersecurity standards are essential (Wylde et al., 2022).

d. Stability of Financial Infrastructure and Transaction Validation

Unlike traditional banking systems, space-based financial transactions must operate within a decentralized framework, which poses challenges for financial stability. Cryptocurrency volatility further complicates space commerce by introducing financial risks for businesses and investors. Additionally, the physical constraints of space, such as high-latency communication and radiation exposure, make transaction validation and settlement difficult. Developing resilient blockchain networks and decentralized financial protocols tailored for extraterrestrial environments is essential for ensuring the stability and reliability of digital currency transactions in space (Schwarcz, 2018).

e. Anti-Money Laundering (AML) and Counter-Terrorism Financing (CTF) Compliance

The anonymity of blockchain-based digital currencies creates challenges in preventing illicit financial activities, such as money laundering and terrorism financing. Space-based financial transactions may be exploited by malicious actors to bypass traditional AML and CTF regulations. Additionally, verifying the identities of financial participants in an extraterrestrial environment is complex due to the lack of clear regulatory oversight. Implementing stringent compliance mechanisms and international cooperation will be essential in preventing financial crimes in space commerce (Bebris & Mukkamala, 2018).

f. Taxation and Regulatory Compliance

Determining taxation policies for space-based digital transactions is a complex issue, as national tax laws are designed for terrestrial economies. The absence of clear tax jurisdiction raises concerns about revenue collection, regulatory evasion, and market distortions. Space entrepreneurs may exploit legal loopholes to avoid taxation, creating unfair competitive advantages. Governments and international regulatory bodies must develop tax policies that balance incentives for space commerce with equitable revenue collection mechanisms to ensure a fair and transparent financial ecosystem (Machfuzhoh & Pratiwi, 2021).

g. Interoperability between Space and Earth-Based Financial Systems

The integration of digital currencies into the space economy requires seamless interoperability with traditional banking systems on Earth. However, differences in regulatory policies, transaction protocols, and technological infrastructures present major challenges. Real-time financial transactions between Earth and space are affected by communication delays, network instability, and compatibility issues. Standardizing financial protocols and creating space-compatible payment solutions will be essential for ensuring smooth financial interactions between terrestrial and extraterrestrial economies (Machfuzhoh & Pratiwi, 2021).

h. Intellectual Property Rights and Smart Contracts

Blockchain technology enables the execution of smart contracts, automating financial agreements without intermediaries. However, the legal recognition and enforcement of smart contracts in space remain uncertain. Disputes over digital assets and intellectual property rights in extraterrestrial commerce lack clear arbitration mechanisms. Without internationally accepted legal frameworks, enforcing financial agreements and resolving contractual disputes in space will be challenging. Establishing cross-border legal recognition of smart contracts and blockchain-based agreements will be crucial for fostering trust and stability in space-based financial transactions (Bebris & Mukkamala, 2018).

i. Environmental and Ethical Considerations

The environmental impact of digital currency operations in space must be considered, particularly the energy-intensive nature of cryptocurrency mining. Space-based mining and blockchain processing could contribute to increased resource consumption and potential environmental hazards. Additionally, ethical concerns such as financial accessibility and inclusivity in the space economy must be addressed to prevent economic disparities. Regulatory frameworks should ensure that space-based digital finance aligns with sustainability goals and ethical business practices to promote a responsible and equitable financial system (Naheem, 2019).

j. Future Policy Development and Governance Frameworks

To address the regulatory challenges of integrating digital currency into the space economy, international policymakers must develop comprehensive governance structures. A globally recognized financial regulatory body dedicated to space commerce is essential to provide oversight, ensure compliance, and mitigate risks. Collaboration between governments, private enterprises, and international organizations will be crucial in shaping policies that balance innovation with regulatory stability. Proactive policy development will be necessary to adapt to emerging financial technologies and create a secure, stable, and legally compliant financial ecosystem for the space economy (O'Cass & Wetzels, 2018).

6.2 Policy Suggestions to Facilitate Nano-Entrepreneurship in the Space Economy

The emergence of the space economy presents new opportunities for nano-entrepreneurs—small-scale businesses leveraging advanced technology and digital finance to participate in extraterrestrial commerce (Porru et al., 2020). Digital currencies, blockchain-based smart contracts, and decentralized financial systems provide innovative financial mechanisms that can enable nano-entrepreneurs to operate in space-related industries. However, for these small businesses to thrive, policymakers must address challenges related to safe and equitable access, risk management, and sustainability. This section presents key policy recommendations to support the integration of nano-entrepreneurship into the space economy while ensuring a secure, inclusive, and environmentally responsible framework (Ozili, 2022).

a. Establishing Inclusive Access to Space-Based Markets

Ensuring equitable access to the space economy is crucial for fostering nano-entrepreneurship. Policymakers should create financial support programs, including micro-grants, investment incentives, and space-focused incubators, to help small businesses enter the market. Partnerships between governments, private sector players, and academic institutions can facilitate knowledge transfer, training, and technological access for aspiring space entrepreneurs. Additionally, international organizations should develop regulatory frameworks that prevent monopolization by large corporations, ensuring fair competition and market participation for small businesses (Schwarcz, 2018).

b. Implementing Decentralized Financial Infrastructure for Space Commerce

To enable nano-entrepreneurs to transact seamlessly in the space economy, policymakers must support the development of decentralized financial systems. Blockchain technology can provide transparent, secure, and efficient cross-border payments, reducing reliance on traditional banking infrastructure. Regulatory bodies should promote interoperability between space-based digital currencies and terrestrial financial systems while ensuring compliance with anti-money laundering (AML) and counter-terrorism financing (CTF) regulations. Facilitating low-cost and frictionless financial transactions will allow small businesses to engage in space commerce without excessive financial burdens (Tan et al., 2022).

c. Strengthening Cybersecurity and Data Protection Regulations

Space-based digital transactions are highly susceptible to cybersecurity threats, including hacking, fraud, and data breaches. Governments and international regulators must establish cybersecurity standards tailored to space commerce, focusing on encryption protocols, multi-signature verification, and fraud prevention mechanisms. Nano-entrepreneurs should have access to secure financial networks and cyber risk insurance programs to mitigate potential financial losses. Collaborative efforts between space agencies, financial regulators, and technology firms can drive innovation in cybersecurity solutions for extraterrestrial financial transactions (Tasatanattakool & Techapanupreeda, 2018).

d. Developing Risk Mitigation Strategies for Space-Based Startups

Nano-entrepreneurs operating in the space economy face significant financial and operational risks, including technological failures, regulatory uncertainties, and market volatility. Policymakers should establish dedicated space insurance mechanisms to protect small businesses from unexpected financial losses. Public-private partnerships can provide risk-sharing models, enabling small entrepreneurs to access capital while minimizing exposure to unpredictable space-related risks. Additionally, the development of standardized legal frameworks for contract enforcement in space commerce will enhance investor confidence and business stability (Valdeolmillos et al., 2020).

e. Enforcing Fair Trade and Ethical Business Practices

To promote a balanced and competitive space economy, governments must enforce fair trade policies that prevent anti-competitive behavior and monopolistic practices by large space corporations. Transparent procurement policies, fair access to space infrastructure, and open-source technology initiatives can empower nano-entrepreneurs to participate in space ventures. Establishing global regulatory oversight bodies will ensure that space commerce remains an inclusive and ethical business environment where small enterprises can thrive alongside major industry players (Schwarcz, 2018).

f. Promoting Sustainable Space Operations

Sustainability is a critical aspect of the space economy, and policies must encourage environmentally responsible business practices. Governments should implement strict regulations on space debris management, energy-efficient technologies, and sustainable resource utilization. Incentives for green innovation, such as tax breaks for eco-friendly space startups and funding for sustainable propulsion systems, will encourage nano-entrepreneurs to adopt responsible practices. Furthermore, blockchain-based smart contracts can be utilized to track environmental compliance, ensuring that businesses operate within sustainability guidelines (Vittori et al., 2025).

g. Establishing Legal Recognition of Digital Assets and Smart Contracts in Space

As nano-entrepreneurs increasingly rely on digital currencies and blockchain-based contracts, policymakers must establish clear legal frameworks for their recognition and enforcement. International agreements should define the legal status of digital assets in space transactions, ensuring that blockchain-based financial agreements are legally binding and dispute-resolution mechanisms are in place. The development of a global space commerce regulatory authority will help streamline legal recognition across jurisdictions, providing legal certainty (Zhang et al., 2022).

7.0 FUTURE DIRECTIONS FOR RESEARCH AND PRACTICE

As the space economy continues to evolve, new opportunities and challenges emerge for both researchers and practitioners. While substantial progress has been made, significant knowledge gaps remain in understanding the long-term financial implications and risk factors associated with space-based digital currencies and nano-entrepreneurship. This section explores potential areas for future research, offers practical suggestions for nano-entrepreneurs, and discusses the broader implications for the future of the space economy.

7.1 Identification of Knowledge Gaps and Future Research Areas

There is a pressing need for further research to better understand the financial mechanisms that will drive the space economy. Specifically, future studies could focus on developing advanced financial models that account for the complexities of space-based markets, such as cross-border digital currency transactions, financial volatility, and currency stability in the space environment. Research into the economic feasibility of space-based businesses, especially small-scale nano-entrepreneurial ventures, is also vital (Kutter & Sowers, 2016). Risk assessment models tailored to space commerce must be refined to address unique factors such as the unpredictability of space missions, technology failures, and legal ambiguities. Furthermore, long-term economic studies exploring the sustainability and scalability of space commerce, including how space-based businesses can thrive while maintaining fiscal health, will be crucial. These research directions will inform both policymakers and industry leaders about the financial dynamics shaping the space economy (Kouhizadeh et al., 2020).

7.2. Practical Suggestions for Nano-Entrepreneurs Interested in Participating in the Space Economy

For nano-entrepreneurs aiming to participate in the space economy, several practical considerations must be taken into account. First, acquiring a strong foundation in emerging technologies, such as blockchain, artificial intelligence, and space communication systems, will be crucial for understanding and leveraging the digital currency ecosystem. Entrepreneurs should focus on enhancing their technical, financial, and operational knowledge to navigate the complexities of space commerce effectively (Haque & Shoaib, 2023). Developing partnerships with established space agencies, technology providers, and financial institutions will provide access to necessary resources and networks. Nano-entrepreneurs should also prioritize building resilient business models that are adaptable to the uncertainties and risks associated with space ventures. This includes focusing on sustainable practices, preparing for potential financial volatility, and ensuring that their business plans align with international regulatory frameworks. By equipping themselves with the right skills, resources, and strategies, nano-entrepreneurs can capitalize on the opportunities presented by the burgeoning space economy (Hermoso et al., 2018).

7.3. Implications for the Future of the Space Economy, Digital Currency, and Nano-Entrepreneurship

Looking ahead, the integration of digital currencies into the space economy will have transformative implications. The widespread adoption of decentralized financial systems could enable more fluid and secure transactions across international and extraterrestrial borders. As space commerce becomes more accessible, nano-entrepreneurship will likely become a key driver of innovation, allowing small businesses to play a significant role in space-related industries (Hart & Halden, 2020). However, this shift also presents challenges related to regulatory oversight, cybersecurity, and financial stability that must be addressed through international cooperation and governance frameworks. The future of the space economy will depend on developing a robust and secure financial ecosystem that fosters inclusivity and encourages sustainable growth. By addressing these challenges and focusing on the development of supportive infrastructure, the space economy has the potential to become a thriving, dynamic, and global marketplace, where digital currencies and nano-entrepreneurship can work in tandem to shape the future of humanity's ventures beyond Earth (Guillard et al., 2018).

8.0 CONCLUSION

The intersection of digital currency, nano-entrepreneurship, and the space economy presents a transformative paradigm with the potential to redefine how small-scale actors participate in frontier markets. This chapter has examined how decentralized finance mechanisms and blockchain-based currencies could democratize access to the high-barrier space sector, thereby enabling nano-entrepreneurs to engage meaningfully in space ventures. The chapter reveals a strong alignment between digital financial innovation and the need for inclusivity, sustainability, and resource optimization in space-based economic activities.

From a theoretical standpoint, this chapter contributes to emerging scholarship in several key ways. First, it expands the discourse on space economy by integrating concepts from entrepreneurship theory, decentralized finance (DeFi), and sustainability studies. Second, it offers a novel conceptual framework connecting nano-entrepreneurship with space commerce, previously dominated by state and corporate actors. This new lens provides a foundation for future academic inquiry into micro-level participation in space ventures, including the dynamics of capital formation, financial risk, and technology adoption in extraterrestrial markets. Third, the chapter introduces sustainability as a central theoretical dimension of the digital-space economy nexus, inviting further exploration of circular economy models and low-impact business practices beyond Earth.

On the practical side, several implications arise for policymakers, business practitioners, educators, and space agencies. Governments and international organizations must consider creating inclusive regulatory frameworks that allow nano-entrepreneurs to participate in space commerce safely and equitably. This includes legal recognition of digital currencies, international treaties governing financial activity in space, and public-private partnerships to support space-tech innovation at the grassroots level. For entrepreneurs, this chapter underscores the need to develop technical proficiency in blockchain, digital wallets, and decentralized investment platforms as foundational tools for accessing space markets. Universities and incubators should also focus on building capacity among aspiring space entrepreneurs through interdisciplinary training in finance, space technology, and sustainability.

Moreover, fostering a space economy that is both inclusive and sustainable requires strategic investment in technological infrastructure, such as micro-satellite networks, distributed ledgers for space transactions, and AI-enabled financial platforms tailored to small-scale users. Such infrastructure will not only reduce the entry barrier for nano-entrepreneurs but also ensure efficient, low-impact resource utilization critical for sustaining life and commerce in space. The integration of nano-entrepreneurship into the space economy, powered by digital currencies, also opens new avenues for addressing global development goals. It offers potential for equitable wealth distribution, employment generation, and sustainable growth, even in underrepresented regions of Earth. As the space economy becomes more accessible, the democratization of this frontier can help ensure that the benefits of extraterrestrial exploration are shared more widely across nations and socioeconomic groups.

Therefore, the future of the space economy lies not only in rockets and celestial exploration but also in inclusive digital infrastructures and entrepreneurial ecosystems that support diverse participation. Digital currencies and nano-entrepreneurs, though small in scale, could play a disproportionately large role in shaping a resilient and sustainable space economy. To fully harness this potential, continued interdisciplinary research, strategic policy development, and proactive stakeholder engagement will be essential. As humanity ventures "to the moon and beyond," it is imperative to ensure that the economic opportunities of space are open to all fueling not just innovation and profit, but also shared progress and planetary stewardship.

REFERENCES

Agarwal, J. D., Agarwal, M., Agarwal, A., & Agarwal, Y. (2021). Economics of cryptocurrencies: Artificial intelligence, blockchain, and digital currency. In *Information for efficient decision making: big data, blockchain and relevance* (pp. 331-430).

Aglietti, G. S. (2020). Current challenges and opportunities for space technologies. *Frontiers in Space Technologies, 1*, 1. DOI: 10.3389/frspt.2020.00001

Akhtar, F., Li, J. P., Heyat, M. B. B., Quadri, S. L., Ahmed, S. S., Xiao, Y. U. N., & Haq, A. U. (2019, December). Potential of blockchain technology in digital currency: a review. In *2019 16th International Computer Conference on Wavelet Active Media Technology and Information Processing* (pp. 85-91). IEEE. DOI: 10.1109/ICCWAMTIP47768.2019.9067546

Al-Emran, M. (2023). Beyond technology acceptance: Development and evaluation of technology-environmental, economic, and social sustainability theory. *Technology in Society, 75*, 102383. DOI: 10.1016/j.techsoc.2023.102383

Allam, Z., Sharifi, A., Bibri, S. E., Jones, D. S., & Krogstie, J. (2022). The metaverse as a virtual form of smart cities: Opportunities and challenges for environmental, economic, and social sustainability in urban futures. *Smart Cities, 5*(3), 771–801. DOI: 10.3390/smartcities5030040

Arnone, G. (2024). Security and Privacy in the Digital Currency Space. In *Navigating the World of Cryptocurrencies: Technology, Economics, Regulations, and Future Trends* (pp. 63–77). Springer Nature Switzerland. DOI: 10.1007/978-3-031-69176-8_7

Auplat, C. A. (2012). Radical innovation and policy-making: nanotechnology public R&D funding in the USA and the EU. *International Journal of Innovation and Regional Development 5, 4*(3-4), 281-292.

Bailey-Hainer, B., Beaubien, A., Posner, B., & Simpson, E. (2014). Rethinking library resource sharing: New models for collaboration. *Interlending & Document Supply, 42*(1), 7–12. DOI: 10.1108/ILDS-12-2013-0038

Balogun, A. A. (2024). Legal Tender in the Age of Central Bank Digital Currencies. In *Global Developments in Central Bank Digital Currency* (pp. 160–169). IGI Global. DOI: 10.4018/979-8-3693-5588-6.ch011

Bebris, A. V., & Mukkamala, R. R. (2018). Anti-Money Laundering and Counter-Terrorism Financing Methodology for Cryptocurrency Exchanges in the European Union.

Beck, T. (2016). Regulatory cooperation on cross-border banking–progress and challenges after the crisis. *National Institute Economic Review*, *235*(1), R40–R49. DOI: 10.1177/002795011623500114

Bernhardt, T., & Pollak, R. (2016). Economic and social upgrading dynamics in global manufacturing value chains: A comparative analysis. *Environment and Planning A. Environment & Planning A*, *48*(7), 1220–1243. DOI: 10.1177/0308518X15614683

Caprara, G. (2020). From the Lunar Station to the Space Economy. In *A History of the Italian Space Adventure: Pioneers and Achievements from the XIVth Century to the Present* (pp. 353–370). Springer International Publishing. DOI: 10.1007/978-3-319-73987-8_13

Caprotti, F., Cowley, R., Datta, A., Broto, V. C., Gao, E., Georgeson, L., Herrick, C., Odendaal, N., & Joss, S. (2017). The New Urban Agenda: Key opportunities and challenges for policy and practice. *Urban Research & Practice*, *10*(3), 367–378. DOI: 10.1080/17535069.2016.1275618

Chen, G., Xu, B., Lu, M., & Chen, N. S. (2018). Exploring blockchain technology and its potential applications for education. *Smart Learning Environments*, *5*(1), 1–10. DOI: 10.1186/s40561-017-0050-x

Chen, Z., Dong, J., & Ren, R. (2017). Urban underground logistics system in China: Opportunities or challenges? *Underground Space*, *2*(3), 195–208. DOI: 10.1016/j.undsp.2017.08.002

Chorzempa, M. (2021). China, the United States, and central bank digital currencies: How important is it to be first? *China Economic Journal*, *14*(1), 102–115. DOI: 10.1080/17538963.2020.1870278

Crawford, I. A. (2016). The long-term scientific benefits of a space economy. *Space Policy*, *37*, 58–61. DOI: 10.1016/j.spacepol.2016.07.003

Dong, S., Abbas, K., Li, M., & Kamruzzaman, J. (2023). Blockchain technology and application: An overview. *PeerJ. Computer Science*, *9*, e1705. DOI: 10.7717/peerj-cs.1705 PMID: 38077532

Dzharbulov, T., Nikolaeva, I., & Gryaznov, S. (2024, May). Creation of Unified Space for Digital Transfers and Payments in National Currencies. In *International Scientific Conference "Digital Transformation of the Economy: Challenges, Trends, New Opportunities"* (pp. 155-164). Cham: Springer Nature Switzerland. DOI: 10.1007/978-3-031-65659-0_20

Eastwood, J. P., Biffis, E., Hapgood, M. A., Green, L., Bisi, M. M., Bentley, R. D., Wicks, R., McKinnell, L.-A., Gibbs, M., & Burnett, C. (2017). The economic impact of space weather: Where do we stand? *Risk Analysis*, *37*(2), 206–218. DOI: 10.1111/risa.12765 PMID: 28230267

Efanov, D., & Roschin, P. (2018). The all-pervasiveness of the blockchain technology. *Procedia Computer Science*, *123*, 116–121. DOI: 10.1016/j.procs.2018.01.019

Elia, A., Kamidelivand, M., Rogan, F., & Gallachóir, B. Ó. (2021). Impacts of innovation on renewable energy technology cost reductions. *Renewable & Sustainable Energy Reviews*, *138*, 110488. DOI: 10.1016/j.rser.2020.110488

Ertz, M., & Boily, É. (2019). The rise of the digital economy: Thoughts on blockchain technology and cryptocurrencies for the collaborative economy. *International Journal of Innovation Studies*, *3*(4), 84–93. DOI: 10.1016/j.ijis.2019.12.002

Ferreira, J. P., El Khantouti, I., Bhattacharya, A., Mommerency, M., & Rondoni, F. (2023). The space-faring Africa: Commercial space industry and its readiness for innovation-driven investment. *New Space*, *11*(1), 32–43. DOI: 10.1089/space.2021.0026

Fiore, F., & Elvis, M. (2025). Space science & the space economy. *arXiv preprint arXiv:2502.20922*.

Foreman Campins, L. (2023). *NewSpace and the european space economy* (Master's thesis, Universitat Politècnica de Catalunya).

Gamage, H. T. M., Weerasinghe, H. D., & Dias, N. G. J. (2020). A survey on blockchain technology concepts, applications, and issues. *SN Computer Science*, *1*(2), 114. DOI: 10.1007/s42979-020-00123-0

Gao, D., Xu, Z., Ruan, Y. Z., & Lu, H. (2017). From a systematic literature review to integrated definition for sustainable supply chain innovation (SSCI). *Journal of Cleaner Production*, *142*, 1518–1538. DOI: 10.1016/j.jclepro.2016.11.153

Ge, T., Hao, X., & Li, J. (2021). Effects of public participation on environmental governance in China: A spatial Durbin econometric analysis. *Journal of Cleaner Production*, *321*, 129042. DOI: 10.1016/j.jclepro.2021.129042

Giungato, P., Rana, R., Tarabella, A., & Tricase, C. (2017). Current trends in sustainability of bitcoins and related blockchain technology. *Sustainability (Basel)*, *9*(12), 2214. DOI: 10.3390/su9122214

Glaser, A., Ben Slimane, S., Auplat, C., & Coeurderoy, R. (2016). Enabling nanotechnology entrepreneurship in a French context: Towards a holistic theoretical framework. *Journal of Small Business and Enterprise Development*, *23*(4), 1009–1031. DOI: 10.1108/JSBED-10-2015-0139

Guillard, V., Gaucel, S., Fornaciari, C., Angellier-Coussy, H., Buche, P., & Gontard, N. (2018). The next generation of sustainable food packaging to preserve our environment in a circular economy context. *Frontiers in Nutrition*, *5*, 121. DOI: 10.3389/fnut.2018.00121 PMID: 30564581

Haque, M. A., & Shoaib, M. (2023). The digital currency in India: Challenges and prospects. *BenchCouncil Transactions on Benchmarks,Standards and Evaluations,3*(1), 100107.

Hart, O. E., & Halden, R. U. (2020). Computational analysis of SARS-CoV-2/ COVID-19 surveillance by wastewater-based epidemiology locally and globally: Feasibility, economy, opportunities and challenges. *The Science of the Total Environment*, *730*, 138875. DOI: 10.1016/j.scitotenv.2020.138875 PMID: 32371231

Hashemi Joo, M., Nishikawa, Y., & Dandapani, K. (2020). Cryptocurrency, a successful application of blockchain technology. *Managerial Finance*, *46*(6), 715–733. DOI: 10.1108/MF-09-2018-0451

Hermoso, V., Carvalho, S. B., Giakoumi, S., Goldsborough, D., Katsanevakis, S., Leontiou, S., Markantonatou, V., Rumes, B., Vogiatzakis, I. N., & Yates, K. L. (2022). The EU Biodiversity Strategy for 2030: Opportunities and challenges on the path towards biodiversity recovery. *Environmental Science & Policy*, *127*, 263–271. DOI: 10.1016/j.envsci.2021.10.028

Hopkinson, P., Zils, M., Hawkins, P., & Roper, S. (2018). Managing a complex global circular economy business model: Opportunities and challenges. *California Management Review*, *60*(3), 71–94. DOI: 10.1177/0008125618764692

Jahangir, R., & Bulut, M. (2024). Nano Entrepreneurship and Saving-Based Finance Concept in Waqf Literature: A Systematic Review and Future Research Direction. *Islamic Financial Institutions from the Early Modern Period to the 20th Century*, 37-77.

Jayasooria, D. (2016). Sustainable development goals and social work: Opportunities and challenges for social work practice in Malaysia. *Journal of Human Rights and Social Work*, *1*(1), 19–29. DOI: 10.1007/s41134-016-0007-y

Jeevanandam, J., San Chan, Y., & Danquah, M. K. (2016). Nano-formulations of drugs: Recent developments, impact and challenges. *Biochimie, 128*, 99–112. DOI: 10.1016/j.biochi.2016.07.008 PMID: 27436182

Jora, O. D., Roşca, V. I., Iacob, M., Murea, M. M., & Nedef, M. . (2023). Small and medium enterprises shooting for the stars: what matters, besides size, in outer space economy?. *Management & Marketing, 18*(1).

Kang, K. Y. (2024). Digital currency and privacy. *Theoretical Economics, 19*(1), 131–167. DOI: 10.3982/TE5081

Keister, T., & Sanches, D. (2023). Should central banks issue digital currency? *The Review of Economic Studies, 90*(1), 404–431. DOI: 10.1093/restud/rdac017

Komalavalli, C., Saxena, D., & Laroiya, C. (2020). Overview of blockchain technology concepts. In *Handbook of research on blockchain technology* (pp. 349–371). Academic Press. DOI: 10.1016/B978-0-12-819816-2.00014-9

Korobova, O. V., Egorchenkova, N. B., & Dzyubenko, Y. S. (2024). Linguistic Representation of the Concept "Digital Currency" in the System of Remote Investments of the Russian Federation. In *Remote Investment Transactions in the Digital Age: Perception, Techniques, Law Regulation* (pp. 55–61). Springer Nature Switzerland. DOI: 10.1007/978-3-031-51536-1_6

Kouhizadeh, M., Zhu, Q., & Sarkis, J. (2020). Blockchain and the circular economy: Potential tensions and critical reflections from practice. *Production Planning and Control, 31*(11-12), 950–966. DOI: 10.1080/09537287.2019.1695925

Krugman, P. (2018). How the economy organizes itself in space: A survey of the new economic geography. In *The economy as an evolving complex system II* (pp. 239–262). CRC Press. DOI: 10.1201/9780429496639-10

Kutter, B. F., & Sowers, G. F. (2016). Cislunar-1000: transportation supporting a self-sustaining Space Economy. In *AIAA SPACE 2016* (p. 5491). DOI: 10.2514/6.2016-5491

Lal, M., & Jatav, A. (2024). Digital Currencies Impact on Development of International Trade and Economy Growth. In *Digital Currencies in The New Global World Order* (pp. 191–215). Springer Nature Singapore. DOI: 10.1007/978-981-97-4810-5_13

Laroiya, C., Saxena, D., & Komalavalli, C. (2020). Applications of blockchain technology. In *Handbook of research on blockchain technology* (pp. 213–243). Academic press. DOI: 10.1016/B978-0-12-819816-2.00009-5

Lee, E. K. S., Zhao, S. X., & Xie, Y. (2012). Command and control cities in global space-economy before and after 2008 geo-economic transition. *Chinese Geographical Science*, *22*(3), 334–342. DOI: 10.1007/s11769-012-0533-1

Lin, H., Wan, S., Gan, W., Chen, J., & Chao, H. C. (2022, December). Metaverse in education: Vision, opportunities, and challenges. In *2022 IEEE International Conference on Big Data (Big Data)* (pp. 2857-2866). IEEE. DOI: 10.1109/BigData55660.2022.10021004

Lu, W., Wu, L., Zhao, R., Li, X., & Xue, F. (2021). Blockchain technology for governmental supervision of construction work: Learning from digital currency electronic payment systems. *Journal of Construction Engineering and Management*, *147*(10), 04021122. DOI: 10.1061/(ASCE)CO.1943-7862.0002148

Machfuzhoh, A., & Pratiwi, R. (2021). The effect of understanding taxation regulations, tax rates, tax sanctions, tax socialization, fiscus services and online services on the level of compliance with MSME taxpayers. *Jurnal Riset Akuntansi Terpadu*, *14*(1).

Martin, R., & Sunley, P. (2017). The post-Keynesian state and the space economy. In *Economy* (pp. 459–473). Routledge. DOI: 10.4324/9781351159203-21

Marzaleh, M. A., Peyravi, M., & Shaygani, F. (2022). A revolution in health: Opportunities and challenges of the Metaverse. *EXCLI Journal*, *21*, 791. PMID: 35949490

McGuire, S., & Islam, N. (2015). Indigenous technological capabilities, emerging market firms and the aerospace industry. *Technology Analysis and Strategic Management*, *27*(7), 739–758. DOI: 10.1080/09537325.2014.959482

Musari, K., Fajariyah, N. A., & Khotimah, K. (2023). Understanding the Behavior of Qardhul Hassan Borrowers in Nano-Entrepreneurship: A Field Experiment of Islamic Nanofinance. In *Strategies and Applications of Islamic Entrepreneurship* (pp. 21-33). IGI Global. DOI: 10.4018/978-1-6684-7519-5.ch002

Naheem, M. A. (2019). Exploring the links between AML, digital currencies and blockchain technology. *Journal of Money Laundering Control*, *22*(3), 515–526. DOI: 10.1108/JMLC-11-2015-0050

Nair, M. M., Tyagi, A. K., & Sreenath, N. (2021, January). The future with industry 4.0 at the core of society 5.0: Open issues, future opportunities and challenges. In *2021 international conference on computer communication and informatics (ICCCI)* (pp. 1-7). IEEE.

Nikraftar, T., Hosseini, E., & Mohammadi, E. (2022). The factors influencing technological entrepreneurship in nanotechnology businesses. *Revista de Gestão, 29*(1), 76–99. DOI: 10.1108/REGE-02-2021-0029

Niranjanamurthy, M., Nithya, B. N., & Jagannatha, S. J. C. C. (2019). Analysis of Blockchain technology: Pros, cons and SWOT. *Cluster Computing, 22*(S6), 14743–14757. DOI: 10.1007/s10586-018-2387-5

O'Cass, A., & Wetzels, M. (2018). Contemporary issues and critical challenges on innovation in services. *Journal of Product Innovation Management, 35*(5), 674–681. DOI: 10.1111/jpim.12464

Ochigbo, A. D., Tuboalabo, A., Labake, T. T., Buinwi, U., Layode, O., & Buinwi, J. A. (2024). Legal frameworks for digital transactions: Analyzing the impact of Blockchain technology. *Finance & Accounting Research Journal, 6*(7), 1205–1223. DOI: 10.51594/farj.v6i7.1313

Ozili, P. K. (2022). Circular economy and central bank digital currency. *Circular Economy and Sustainability, 2*(4), 1501–1516. DOI: 10.1007/s43615-022-00170-0 PMID: 35434721

Pfister, C. (2024). Monetary sovereignty in the digital currency era. In *Digital Assets and the Law* (pp. 35–66). Routledge.

Polanyi, K. (2018). The economy as instituted process. In *The sociology of economic life* (pp. 3–21). Routledge. DOI: 10.4324/9780429494338-2

Pomeroy, C., Calzada-Diaz, A., & Bielicki, D. (2019). Fund me to the moon: Crowdfunding and the new space economy. *Space Policy, 47*, 44–50. DOI: 10.1016/j.spacepol.2018.05.005

Porru, S., Misso, F. E., Pani, F. E., & Repetto, C. (2020). Smart mobility and public transport: Opportunities and challenges in rural and urban areas. [English edition]. *Journal of Traffic and Transportation Engineering, 7*(1), 88–97. DOI: 10.1016/j.jtte.2019.10.002

Qin, Y. (2018). Integrated, multidisciplinary approaches for micro-manufacturing research, and new opportunities and challenges to micro-manufacturing. *Proceedings of the Institution of Mechanical Engineers. Part N, Journal of Nanomaterials, Nanoengineering and Nanosystems, 232*(1), 5–21. DOI: 10.1177/2397791417750350

Ramadoss, R. (2022). Blockchain technology: An overview. *IEEE Potentials, 41*(6), 6–12. DOI: 10.1109/MPOT.2022.3208395

Ronaghi, M. H. (2023). A contextualized study of blockchain technology adoption as a digital currency platform under sanctions. *Management Decision, 61*(5), 1352–1373. DOI: 10.1108/MD-03-2022-0392

Ronaghi, M. H., & Fallahi, A. (2024). How does the use of cryptocurrency affect circular economy practices in Iranian manufacturing companies? *Cleaner Engineering and Technology, 23*, 100835. DOI: 10.1016/j.clet.2024.100835

Sabry, S. S., Kaittan, N. M., & Majeed, I. (2019). The road to the blockchain technology: Concept and types. [PEN]. *Periodicals of Engineering and Natural Sciences, 7*(4), 1821–1832. DOI: 10.21533/pen.v7i4.935

Saito, K., Hatta, S., & Hanada, T. (2019). Digital currency design for sustainable active debris removal in space. *IEEE Transactions on Computational Social Systems, 6*(1), 127–134. DOI: 10.1109/TCSS.2018.2890655

Sanchez-Garcia, M. D., Lopez-Rubio, A., & Lagaron, J. M. (2010). Natural micro and nanobiocomposites with enhanced barrier properties and novel functionalities for food biopackaging applications. *Trends in Food Science & Technology, 21*(11), 528–536. DOI: 10.1016/j.tifs.2010.07.008

Schwarcz, S. L. (2018). Secured transactions and financial stability: Regulatory challenges. *Law and Contemporary Problems, 81*, 45.

Secinaro, S., Brescia, V., Lanzalonga, F., & Santoro, G. (2022). Smart city reporting: A bibliometric and structured literature review analysis to identify technological opportunities and challenges for sustainable development. *Journal of Business Research, 149*, 296–313. DOI: 10.1016/j.jbusres.2022.05.032

Siderius, T., & Zink, T. (2023). Markets and the future of the circular economy. *Circular Economy and Sustainability, 3*(3), 1569–1595. DOI: 10.1007/s43615-022-00196-4 PMID: 36035512

Sørensen, F., Friis Jensen, J., & Laura Dam, A. (2024). Developing innovation practices of nature tourism entrepreneurs. *Scandinavian Journal of Hospitality and Tourism, 24*(3), 332–360. DOI: 10.1080/15022250.2024.2440556

Sun, Y. (2024). The Impact and Opportunities of Digital Currency Development. *Advances in Economics and Management Research, 12*(1), 781–781. DOI: 10.56028/aemr.12.1.781.2024

Tan, T. F., Li, Y., Lim, J. S., Gunasekeran, D. V., Teo, Z. L., Ng, W. Y., & Ting, D. S. (2022). Metaverse and virtual health care in ophthalmology: Opportunities and challenges. *Asia-Pacific Journal of Ophthalmology, 11*(3), 237–246. DOI: 10.1097/APO.0000000000000537 PMID: 35772084

Tasatanattakool, P., & Techapanupreeda, C. (2018, January). Blockchain: Challenges and applications. In *2018 international conference on information networking (ICOIN)* (pp. 473-475). IEEE.

Toivonen, A. (2022). Sustainability dimensions in space tourism: The case of Finland. *Journal of Sustainable Tourism, 30*(9), 2223–2239. DOI: 10.1080/09669582.2020.1783276

Tong, W., & Jiayou, C. (2021). A study of the economic impact of central bank digital currency under global competition. *China Economic Journal, 14*(1), 78–101. DOI: 10.1080/17538963.2020.1870282

Truby, J. (2018). Decarbonizing Bitcoin: Law and policy choices for reducing the energy consumption of Blockchain technologies and digital currencies. *Energy Research & Social Science, 44*, 399–410. DOI: 10.1016/j.erss.2018.06.009

Valdeolmillos, D., Mezquita, Y., González-Briones, A., Prieto, J., & Corchado, J. M. (2020). Blockchain technology: a review of the current challenges of cryptocurrency. In *Blockchain and applications:International congress* (pp. 153-160). Springer International Publishing. DOI: 10.1007/978-3-030-23813-1_19

Valente, M., Caviggioli, F., & Agostini, L. (2025). Space Economy and Sustainability: A Systematic Review. *Sustainable Development (Bradford)*, sd.3383. DOI: 10.1002/sd.3383

Verma, J., Sharma, J., & Gupta, M. (2023). Digital currency and blockchain technology in the financial world. In *Perspectives on Blockchain Technology and Responsible Investing* (pp. 216–225). IGI Global. DOI: 10.4018/978-1-6684-8361-9.ch010

Vittori, D., Panniello, U., Messeni Petruzzelli, A., & Cupertino, F. (2025). Reaching for the stars: Analyzing market entry in the space economy. *Industry and Innovation*, •••, 1–37. DOI: 10.1080/13662716.2025.2468923

Weinzierl, M. (2018). Space, the final economic frontier. *The Journal of Economic Perspectives, 32*(2), 173–192. DOI: 10.1257/jep.32.2.173

Wylde, V., Rawindaran, N., Lawrence, J., Balasubramanian, R., Prakash, E., Jayal, A., Khan, I., Hewage, C., & Platts, J. (2022). Cybersecurity, data privacy and blockchain: A review. *SN Computer Science, 3*(2), 127. DOI: 10.1007/s42979-022-01020-4 PMID: 35036930

Xie, S. (2016). Advanced robotics for medical rehabilitation. *Springer tracts in advanced robotics, 108*(1), 357.

Zhang, P., Cheng, Q., Zhang, M., & Luo, X. (2022). A group covert communication method of digital currency based on blockchain technology. *IEEE Transactions on Network Science and Engineering*, *9*(6), 4266–4276. DOI: 10.1109/TNSE.2022.3196816

KEY TERMS AND DEFINITIONS

Nano-Entrepreneurship: A form of entrepreneurship that involves starting and managing very small-scale businesses, often with limited capital, resources, and workforce typically run by a single person or a small team using digital tools or innovative business models.

Space Economy: The collection of economic activities related to the research, development, production, and use of equipment and technologies that contribute to space exploration, satellite services, and commercial ventures in outer space.

Digital Currency: A type of money that exists only in electronic form. It is not tangible like paper money or coins and can be used to buy goods or services, especially online. Examples include cryptocurrencies like Bitcoin and government-issued digital currencies (CBDCs).

Space Business Innovation: The process of creating new ideas, technologies, or business models specifically aimed at enhancing or creating businesses related to the space sector—such as satellite communication, space tourism, or lunar resource extraction.

Decentralized Finance (DeFi) in Space: A futuristic concept that involves using blockchain and decentralized financial systems in space-related operations, allowing financial transactions and services (like lending or investing) to occur without traditional banks, possibly supporting autonomous interplanetary economies.

Chapter 4
Blockchain, RPA, and AI as Main Elements of Digitalization:
Benefits and Challenges and the Effect on the World Economy

Maya Katenova
https://orcid.org/0009-0008-7311-8049
KIMEP University, Kazakhstan

Mira Nurmakhanova
https://orcid.org/0000-0002-1399-7773
KIMEP University, Kazakhstan

ABSTRACT

The study explores three main elements of digitalization such as Artificial Intelligence, Blockchain, and Robotic Process Automation. In addition, the study properly defines the issue of digitalization and focuses on its benefits and disadvantages. All benefits and disadvantages of digitalization were carefully presented in the study. Moreover, the effect of digitalization on the world economy was discussed in detail. The need for global ethical rules and standards was presented as well. The study is purely theoretical and descriptive.

DOI: 10.4018/979-8-3373-0355-0.ch004

1. BACKGROUND

The era of digitalization has started already today. According to the International Institute of Applied System Analysis (2018), every day 207 billion emails are sent, 8.8 billion YouTube videos are watched, 4.2 billion Google searches are conducted, and 152 million Skype calls are made. The digitization process has been incorporated not only into our lives but into our businesses as well. Business digitalization has evolved from simple adoption of digital tools (e.g., email, websites, e-commerce platforms) to the integration of more complex technologies like Artificial Intelligence and automation. Digitalization in business refers to the integration of digital technologies into all aspects of a business's operations, fundamentally changing how businesses operate and deliver value to customers. The process has accelerated rapidly with advancements in technologies such as Artificial Intelligence (AI), Blockchain, Robotic Process Automation (RPA), and machine learning. These technologies are not only enhancing business efficiency but also reshaping the very way companies interact with consumers, manage resources, and develop products and services.

What is digitalization? Digitalization is a term, which describes the phenomenon of adopting digital technologies in business and society worldwide. It means the growing use of information and communication technology in every area of our lives. It has indiscriminately impacted all levels of our society and influenced the way we do our business today. Digitalization has created many challenges as well as excellent opportunities in the workplace and in our lives. Moreover, the term also covers the affiliated changes in the connectivity of individuals, organizations, and objects (Urbach & Röglinger, 2019). Digital businesses are expanding at a very rapid pace, with new mobile and intangible digital goods and services becoming available to consumers daily. Businesses are leveraging digital platforms to expand their global reach, enhance customer experiences, and streamline internal processes. With the rise of the digital economy, companies are adopting cloud computing, big data analytics, and Internet of Things devices to gain deeper insights and improve decision-making processes. This evolution allows businesses to operate more flexibly and adapt to market demands faster, contributing to overall growth and competitiveness. Furthermore, as governments worldwide push for digitalization, businesses are increasingly integrating digital technologies to align with national and international digitalization programs, ensuring compliance and access to emerging markets.

Digitalization and technological advances have brought new opportunities, but they also present new challenges today. Digitalization of society promotes continuity and business vitality but requires a change in the culture and business ethics of companies (Garczarek & Steuer, 2019; Rezer, 2021). Thus, digitalization has advantages and disadvantages, which will be discussed in this book chapter carefully. The attention is focused on the greatest opportunities and challenges of a change through digitalization. In addition, digitalization requires new rules and standards to be developed. Such a transformation is a long-term process, which requires a lot of time and effort.

Which key benefits does digitalization offer? Digitalization offers several key benefits at both the macroeconomic and microeconomic levels. It enhances efficiency and productivity by automating repetitive tasks with technologies like Artificial Intelligence (AI) and Robotic Process Automation (RPA), reducing labor costs and human errors. This also allows businesses to optimize operations and forecast trends using data analytics (Strategic Finance, Institute of Management Accountants). Additionally, digitalization enables cost reduction and profit growth by automating tasks and reallocating resources to higher-value activities, while Artificial Intelligence-driven decision-making leads to more targeted products and services. The digital transformation also opens doors for global market expansion through e-commerce and digital marketing, boosting sales with minimal overhead. Moreover, it fosters innovation, providing businesses with tools like Artificial Intelligence, Blockchain, and data analytics to develop personalized products and maintain a competitive edge (Young et al., 2019). On a macroeconomic scale, digitalization contributes to GDP growth by improving labor efficiency and expanding services, prompting governments to support digital transformation programs.

Despite its many benefits, digitalization presents several challenges for businesses. Ethical concerns surrounding privacy, confidentiality, data security, and transparency arise as companies adopt technologies like Artificial Intelligence (AI) and blockchain. This shift also demands new ethical standards and cultural changes within organizations, as regulatory frameworks often lag behind technological advancements, creating potential risks Kirchschlaeger (2021). Furthermore, the integration of new technologies can be complex and costly, as companies may struggle to align digital tools with existing systems and train employees effectively. The fast pace of technological change requires constant updates and investments. Digitalization also increases cybersecurity risks, requiring robust security measures to protect sensitive data from attacks and breaches. Additionally, automation can lead to job displacement, particularly for low-skilled workers, while requiring businesses to invest in reskilling their workforce. The digital divide further exacerbates challenges, with smaller businesses, especially in developing regions, facing difficulties in accessing digital tools. Finally, businesses must navigate evolving regulatory and compliance requirements, such as data protection laws, which can create uncertainty and increase operational costs globally.

As a result, while digitalization offers tremendous opportunities for businesses to enhance efficiency, reduce costs, expand globally, and drive innovation, it also presents significant challenges, particularly around business ethics, security, workforce transformation, and regulatory compliance. Digital transformation is not just a technological shift but also a cultural and ethical one, requiring careful consideration and strategic planning for sustainable growth. Companies must adapt to these changes by creating new standards of ethics and addressing the accompanying social and technological issues. A well-designed management system can reduce the risk of inefficient usage of labor and financial resources. The question we face in this new age of digitization is "How can organizations adapt to digitalization and create new standards of ethics in the era of digitalization and innovation?". The paper adds to the literature by discussing all possible advantages and disadvantages of digitalization in a global arena and provides some useful advice on ethical standards and rules in the era of digitalization. In addition, the paper discusses the effect of digitalization on the world economy. The next chapter is dedicated to literature review.

2. LITERATURE REVIEW

It should be noted that digitalization has increased the tempo of daily life. "Speed-up society," with a faster pace through the day, fewer pauses, more multi-tasking, and more time pressures. It all started long ago, in 1969 when the internet was deployed. Today, one of the biggest impacts of digitalization on organizations is that information has become more accessible and transparent, which allows organizations to share more information with all employees—even those at lower levels of the hierarchy (Kuusisto, 2015). This easy accessibility resulted in an increased productivity, efficiency, and in greater demands at work and at home (Chesled, 2010). Innovative processes are strongly popularized today. They cover a variety of activities including Robotic Process Automation (RPA), predictive analytics, cloud-based accounting software, and digital auditing tools. Robotic Process Automation (RPA) uses software robots to handle repetitive tasks such as data extraction and invoice processing, freeing up human resources for judgment-based strategic analysis. Predictive analytics leverages historical data, statistics, and algorithms in order to forecast future trends and patterns, which can aid in processes such as budgeting and financial planning, for example. Cloud-based accounting software enables real-time collaboration and data access, facilitating remote work and reducing the need for physical infrastructure. Digital auditing tools enhance audit efficiency by automating testing procedures and improving accuracy in financial statements. These activities have advanced underlying processes such as journal entry, month-end close, accounts payable, purchase orders, invoicing, and vendor payments for management accountants (Vineet, 2023).

When it comes to the theory of digitalization, it should be noted that this theory is one of the most popular theories today. It is related to its relevance to the current stage of economic development. It is necessary to take into account all risks, which may arise in the future, the identification of which will contribute to its positive change. Guryanova et al. (2020) analyzed various sources on social and ethical issues of digitization of healthcare, data privacy, violation of existing social structures, inequality, and injustice. Taylor & Pagliari (2018) conducted a study on the usage of social media data for the research, and authors developed their own five-stage process. The digital revolution has been clearly led by the industry and the global digital platforms. Data and Artificial Intelligence (AI) are now being actively employed to make better decisions, optimize processes, customize services, predict patterns, and understand society better. Sustainable Development Goals have proved to be a suitable framework to catalyze and promote uses of Data and Artificial Intelligence to address global challenges (Vinuesa et al., 2020). A major challenge to modern society is the online science, digital technology, Artificial Intelligence, and Robotics Process Automation because they lead to the loss of autonomy, cybercrime, threats to child safety and national security. At the same time, digitalization is very effective in climate change and disaster mitigation and improvement of the connection between smart cities. The introduction of innovative IT promotes international cooperation in cultural, economic, social, and ethical spheres (Hantrais & Lenihan, 2021).

2.1. Advantages of Digitalization

Driven by the new era of technological revolution and global industrial change, digitalization was presented as an important opportunity for businesses worldwide to build key competencies and gain global competitive advantages. Existing studies on the impact of digitalization on businesses have been conducted mainly from the perspectives of Sustainable Development Goals, cost savings, and economic growth. For instance, Gao and Li et al. (2022) showed the positive effect of digital technology development on green total factor energy efficiency using panel data from 213 cities in China. Digital technology development helped to decrease costs and achieve efficiency. Borowski (2021) demonstrated that digitalization slightly improved energy efficiency and significantly reduced industrial energy consumption, which had an extremely positive effect on sustainable development. Myovella and Karacuka (2020) discussed the contribution of digitalization to economic growth globally and it was discovered that digitalization contributes positively to economic growth in both developed and emerging and less developed countries as well. However, Internet technologies contribute least to underdeveloped countries, while mobile communication technologies contribute most to economic growth in less developed countries. Boikova and Zeverte-Rivza (2020) mentioned that digitalization has become a key factor of competitiveness and is an important guarantee for economic growth in Europe. In addition, Zhang showed that digitalization changes the old international trade structure by influencing trade participants and trade services, thus, facilitating the development of services trade. Guryanova et al. (2020) outlined the primary elements of a digital economy and examined how they change the structure of a traditional economy and transform social relations. The authors strongly believe that socio-ethical problems should be taken into consideration when describing the future of digital transformation. There are many new opportunities with the rise of new jobs and new markets, rising environmental sustainability, improving ergonomics and the safety of production processes, etc. (Governo Italiano, n.d.). Digitalization is creating keen changes in the world of work. Castells (2010) mentioned that the key feature of wireless communication is not mobility but perpetual connectivity. Serkina et al. (2022) has mentioned the importance of digitalization in the business world today. Economic development facilitates the digitalization of businesses, economies, social systems, and even small individual businesses. Some researchers argue that digital transformation of a traditional society provides it with a competitive advantage (Dias & Vieira, 2013; Cochoy et al., 2020; Bonnaud, 2021). People would benefit from new technologies and experience new ethical improvements globally. The fact that these changes will mainly be driven by firms makes business ethics a central discipline in the discussion of ethical challenges of digital change. In order to achieve effective digitalization, it is required to abolish traditional systems and operating models, paper-based data centers, etc. (Sizganova et al., 2022). Digital documentation, technological innovations in the industrial and environmental sectors, and dematerialization make it possible to conserve natural resources and minimize the negative effect of human activity on the environment globally (Zhili et al., 2019; Chen et al., 2020).

2.2. Disadvantages of Digitalization

When it comes to challenges of digitalization, it should be noted that the proliferation of digital technologies within organizational contexts may facilitate new forms of exploiting users' 'free' labor, which creates further problems (Beverungen et al., 2015). In addition, their 'excess' resources and capacities should be taken into consideration (Bauer et al., 2015). It has also been suggested that digitalization promotes an illusory sense of resistance, which counteracts genuine forms of resistance (Ossewaarde M.et al., 2017), and that it has contributed to a large range of new workplace conflicts (Upchurch M.& Grassman R., 2016). Probably, higher unemployment rate and problems of further employment are associated mainly with digitalization. Instead of applying digital technologies for positive social and organizational transformation, firms employ digital technologies for the purpose of profit maximization only today (Trittin H., Schoeneborn D., 2017). Because of self interest, businesses ignore all issues and challenges associated with digitalization. Large internet and platform companies, who hold powerful monopolistic positions in the global arena, and who generate extraordinary rents are often seen as the main drivers of digitalization. These above-mentioned large corporations constitute 'information empires' and employ digitalization for their benefit only (Wu T., 2011). They control large parts of the internet, and thus, have gained unprecedented economic and political power (Flyverbom M., et al. 2019). Hence, there is a power asymmetry between these big tech companies and most other firms in the market (who commonly depend on services of those large corporations). It creates a new form of so-called "economic elite". In addition, studies indicate that these large companies actively employ low paid precarious labor force without adequate worker's rights, where workers have been surveilled and actively punished for attempts to restore such rights (De Stefano V., 2019). For example, Uber has attempted to avoid labor-related costs by defining its apps' users as 'independent contractors' rather than as regular employees; Amazon fired a number of their workers, who raised concerns about their working conditions at the large warehouse.

3. THREE MAIN ELEMENTS OF DIGITALIZATION

As was mentioned earlier, digitalization is a major part of our life nowadays. Though, the transformation of a traditional society into a digital one is a complex process. Technological innovations generate a growing need for proper digital processing of all data, correct and ethically driven storage, and transmission of information. Many different jobs are replaced by Artificial Intelligence today. RPA (Robotic Process Automation) is useful for repetitive and iterative tasks (Institute of Management Accountants, 2022). While the Blockchain is useful for storage purposes, Artificial Intelligence (AI) might replace current repetitive jobs people do today. All the above-mentioned technologies are essential parts of digitalization. This chapter explores the history, the current state and the impact of these three digitalization pillars: Artificial Intelligence (AI), Robotic Process Automation (RPA), and Blockchain on businesses.

3.1. Artificial Intelligence

The history of Artificial Intelligence (AI) dates back to the early 1950s, with one of the first successful AI programs written in 1951 by Christopher Strachey. Strachey's checkers program was a simple AI that could play the game against a human user. Within just a year, it could play a complete game of checkers at an average level. This early success laid the foundation for the exponential growth of AI over the following decades. In the years that followed, AI research expanded into more complex fields, including machine learning, neural networks, and natural language processing, leading to the development of technologies that began to permeate various aspects of human life.

Today, Artificial Intelligence has evolved far beyond its initial capabilities. Modern AI is integrated into a wide array of technologies that impact our daily lives. Homes have transformed into “smart homes” equipped with AI-driven devices like voice assistants and smart thermostats. Smartphones are no longer just communication tools but have become powerful AI-driven devices, with features like predictive text, personal assistants, and face recognition. AI is also embedded in vehicles, turning cars into “smart cars” capable of autonomous driving and advanced safety features. The use of AI in industries is vast, from healthcare and finance to retail and education, with machine learning, neural networks, and AI algorithms driving innovations in productivity and decision-making. AI technologies are also expected to raise global GDP by 14% by 2030, particularly boosting economies like the United States and China.

According to Rodrigo (2023), Artificial Intelligence (AI) is a branch of research in which computer systems are created to execute skills that are often associated with human beings. Some examples of these functions include recognizing faces or voices, playing chess, or driving a vehicle through traffic. Artificial Intelligence in Education (AIED) is a term that describes the application of AI in educational settings with the goal of enhancing the teaching, learning, or overall educational experience. It is possible that AIED will have a significant effect on human cognition as well as the development of the brain. Ever since Socrates argued that writing led to forgetting, people have held the belief that technology has had a substantial impact on human growth and intellect. Despite this fact, though, the repercussions are very likely to be complicated. For instance, one study has revealed that increased usage of Global Positioning System (GPS) navigation devices and applications leads to a deterioration in spatial memory, while another study has demonstrated that using smartphone alerts frees cognitive resources for other tasks. Both of these studies were conducted by the same researchers Rodrigo (2023). Holes (2022) mentioned the importance of Artificial Intelligence and AIED for adults as well as for children.

Artificial intelligence (AI) is profoundly transforming business by automating routine tasks, increasing efficiency, and allowing employees to focus on more strategic work. It enhances decision-making by analyzing large volumes of data and uncovering insights that might otherwise be overlooked. AI also improves customer interactions, optimizes supply chain management, and drives cost reductions across various areas by streamlining operations. Furthermore, it accelerates innovation, enables new business models, and disrupts traditional ones. AI plays a vital role in strengthening cybersecurity by detecting and responding to threats, while businesses that adopt these technologies early often gain a competitive advantage.

Looking forward, the development of Artificial Intelligence is expected to accelerate, with transformative effects on both the economy and society. AI and automation are predicted to significantly influence global trade by reducing the importance of labor costs in production processes. This could reshape international trade dynamics, especially as AI lowers barriers related to labor supply and production costs. In education, Artificial Intelligence has the potential to revolutionize learning by providing personalized learning experiences, improving educational outcomes, and democratizing access to knowledge. AI-enabled technologies will continue to evolve, driving advancements in fields like healthcare, where Artificial Intelligence could aid in diagnostics and personalized medicine, and in business, where it will further optimize operations. As Artificial Intelligence becomes more embedded in everyday life, it will further transform industries, enhance productivity, and play a central role in the future economy. The role of Artificial Intelligence is transformative and it creates the future economy, which is rapidly growing today.

3.2. Robotic Process Automation

Robotic Process Automation (RPA) is an emerging technology that combines elements of computer science, electronics, mechanical engineering, and information technology to automate repetitive tasks. The concept of automation itself has existed for many years, but the integration of robotics with software for business process automation is a more recent development. Robotic Process Automation leverages software "robots" to perform tasks traditionally carried out by human workers, such as data entry, processing transactions, or managing records. This technology gained significant attention in the last decade, driven by the increasing need to streamline operations, reduce costs, and improve accuracy across industries. However, empirical research and published articles on RPA have been relatively limited, indicating that this is still an evolving field with considerable potential for growth.

Robotic Process Automation (RPA) is revolutionizing business operations by automating repetitive, rule-based tasks, leading to significant improvements in efficiency and accuracy. By using software robots to handle routine processes such as data entry, invoice processing, and customer service inquiries, Robotic Process Automation frees up human employees to focus on more complex, value-added activities. This reduces operational costs, minimizes human error, and speeds up task execution. Robotic Process Automation also enhances scalability, as businesses can quickly deploy robots to manage fluctuating workloads without the need for additional personnel. Additionally, Robotic Process Automation fosters innovation by streamlining workflows and improving service delivery, making businesses more agile and competitive in an increasingly fast-paced environment. Currently, Robotic Process Automation is gaining traction in technology-driven sectors that handle large volumes of data, such as information technology, finance, and manufacturing. It allows organizations to automate routine, rule-based tasks, which do not require human intervention, significantly improving operational efficiency. For example, in industries like oil and gas, medicine, and education in Europe, Robotic Process Automation (RPA) has been successfully integrated into daily operations. RPA's use is particularly widespread in data processing and document management, where it can quickly handle large datasets and improve workflow efficiency. However, other sectors such as education, wholesale trade, and the chemical and manufacturing industries have been slower to adopt Robotic Process Automation, utilizing it to a lesser extent.

The future of Robotic Process Automation looks promising, as it is expected to continue expanding into new industries and applications. As the technology matures, Robotic Process Automation will likely evolve to integrate more advanced capabilities, such as Artificial Intelligence and Machine Learning, enabling more complex and intelligent automation. This could lead to Robotic Process Automation systems that not only execute predefined tasks but also make decisions based on data analysis, further reducing the need for human intervention. In the coming years, industries that are currently under-digitized, such as education and retail, may see more widespread adoption of RPA to optimize operations. With continuous advancements in Artificial Intelligence and data processing technologies, RPA could become an even more powerful tool for improving productivity, enhancing accuracy, and lowering operational costs across various sectors globally.

3.3. Blockchain

The concept of Blockchain technology emerged with the creation of Bitcoin in 2008 by an anonymous person (or group of people) under the pseudonym Satoshi Nakamoto. Blockchain was originally designed as the underlying technology for cryptocurrencies, providing a decentralized, secure, and immutable ledger for transactions without the need for intermediaries like banks or financial institutions. This innovation allowed peer-to-peer transactions to occur directly between parties, with each transaction being verified and recorded by a distributed network of computers (nodes). Over time, Blockchain's potential was recognized beyond cryptocurrencies, and it began to be explored for applications in various other fields, including finance, supply chain management, healthcare, and even government services.

Blockchain technology is significantly impacting businesses by providing a secure, transparent, and decentralized way to record transactions and manage data. It enhances trust and accountability by ensuring that once data is recorded, it cannot be altered without consensus, reducing the risk of fraud and errors. Blockchain streamlines operations by enabling faster, more efficient transactions, especially in industries like finance, supply chain management, and healthcare. It also reduces costs by eliminating intermediaries, enabling peer-to-peer transactions. Furthermore, Blockchain fosters innovation by supporting the creation of decentralized applications (dApps) and smart contracts, which automate processes and enhance business efficiency, ultimately transforming industries and creating new opportunities for businesses to operate more securely and transparently. Currently, blockchain technology is widely used in the financial sector for various applications such as digital assets, remittances, and online payments, all of which can take place without the need for banks or other financial institutions. Blockchain is also being applied in areas like smart contracts, where agreements can be automatically executed when pre-defined conditions are met, eliminating the need for intermediaries. Other sectors, such as the Internet of Things (IoT), reputation systems, and security services, are increasingly leveraging blockchain for its decentralized and immutable nature. The technology is valued for its ability to provide high reliability, honesty, and security, making it particularly attractive to businesses that need to guarantee trust in their transactions. Furthermore, blockchain's distributed architecture solves the problem of a single point of failure, ensuring the resilience of the system. However, the widespread adoption of blockchain is still in the early stages, and there are notable technical challenges that need to be addressed.

The future of blockchain holds great promise, especially in building the next generation of Internet systems. As blockchain technology matures, it is expected to become more scalable and secure, enabling it to handle a greater volume of transactions more efficiently. The development of blockchain applications will likely expand to encompass even more industries, including healthcare, voting systems, and legal documentation. However, key challenges like scalability and security still need to be overcome for blockchain to realize its full potential. Blockchain's ability to provide a decentralized, immutable ledger offers significant advantages for businesses, governments, and consumers alike, particularly in creating transparent, trustworthy, and automated systems. As solutions to current technical limitations emerge, blockchain may play an integral role in shaping the future digital economy and in enabling a more secure and decentralized Internet.

3.4. Advantages and Disadvantages of Digitalization

As it was shown earlier, all three pillars of digitalization, Artificial Intelligence (AI), Robotic Process Automation (RPA), and Blockchain, have significantly changed the landscape of business operations. Despite the numerous benefits, they present several challenges and issues that businesses must navigate. The topic is a global one, and it requires the attention of both academics and practitioners worldwide.

The table below presents the main advantages and disadvantages of digitalization for businesses globally.

Table 1. Advantages of digitalization

ADVANTAGES OF DIGITALIZATION
Cost Reduction and Profit Growth: Digitalization allows businesses to reduce operational costs and improve profit margins simultaneously. By automating tasks, businesses can reallocate resources to higher-value activities, which can lead to cost savings. Additionally, Artificial Intelligence and data-driven decision-making can lead to better-targeted products and services, resulting in higher revenue generation.
Enhanced Efficiency and Productivity: One of the key benefits of digitalization is the increased efficiency and productivity across business operations. For instance, the automation of repetitive tasks through technologies like Robotic Process Automation and Artificial Intelligence helps businesses cut down on labor costs and human errors while speeding up processes. Businesses can collect and store vast amounts of data more efficiently, leveraging advanced algorithms and machine learning models to optimize operations and forecast market trends.
Global Reach and Market Expansion: The digitalization of business practices enables companies to operate on a global scale. With online platforms, businesses can reach new customer segments and enter international markets with minimal overhead costs. For example, the use of e-commerce and digital marketing strategies allows businesses to target customers worldwide, contributing to increased sales.
Innovation and Competitive Advantage: The digital transformation of business processes fosters innovation, allowing companies to develop new products and services. Digital tools like Artificial Intelligence, Blockchain, and data analytics provide businesses with the opportunity to create personalized offerings, improving customer satisfaction and loyalty. Moreover, companies embracing digitalization are able to stay competitive in the fast-evolving global market.
Economic Growth and GDP Contribution: On a macroeconomic level, digitalization can significantly contribute to GDP growth. It enables economies to improve labor efficiency and expand services, directly impacting national economic performance. Governments recognize the potential of digitalization to foster economic growth and have initiated various programs to encourage digital transformation in their industries.
Improved Knowledge Acquisition: Digitalization has revolutionized knowledge acquisition by making learning accessible from virtually anywhere in the world. Online platforms, digital courses, and educational resources allow individuals to acquire new skills and knowledge at their own pace, regardless of their geographical location. This flexibility enables learners to tap into global expertise, access diverse educational content, and engage with a variety of learning styles. As a result, digitalization fosters continuous learning, empowering individuals and organizations to stay competitive and adaptive in an increasingly digital world.
Flexibility of Digitalization: Digitalization offers unparalleled flexibility, allowing businesses and individuals to adapt quickly to changing needs and circumstances. Whether it's working remotely, accessing cloud-based services, or using mobile apps, digital tools enable seamless adjustments in real-time.
Human Resources and Talent Management: AI tools can scan resumes, assess candidates' qualifications, and even conduct preliminary interviews, saving time for HR departments. Additionally, AI can track employee performance and identify areas for improvement or training, aiding in talent management and career development.

Table 2. Disadvantages of digitalization

DISADVANTAGES OF DIGITALIZATION
Ethical and Social Concerns: As businesses adopt digital technologies like Artificial Intelligence and blockchain, ethical concerns arise regarding privacy, data security, and transparency. The digitalization of society also requires a shift in business ethics and culture, as companies need to develop new ethical standards to address these concerns. Improvement of legislation related to the ethical standards of the information usage, is time consuming. Thus, the rapid development of technology sometimes outpaces the development of regulatory frameworks, creating potential risks and gaps in governance.
Technological Integration and Adaptation: Digitalization incorporates different approaches to information processes and transmission. Moreover, digitalization requires the development of heavy programs to process the information. Thus, adopting new technologies can be a complex and costly process. Companies may face difficulties in integrating new digital tools with legacy systems or in training employees to use these technologies effectively. Additionally, a systematic approach to learning and using the new technologies may be time consuming. The fast pace of technological change can leave companies struggling to keep up, leading to increased investment in constant upgrades and innovations.
Cybersecurity Risks: There is the increased risk of cyberattacks and data breaches. Businesses must invest in robust cybersecurity measures to protect sensitive data and avoid financial and reputational damage. As the frequency and sophistication of cyberattacks grow, companies must stay vigilant and adopt advanced security protocols.
Job Displacement and Workforce Transformation: The automation of repetitive tasks through Robotic Process Automation and Artificial Intelligence technologies can lead to job displacement, particularly for low-skilled workers. Although digitalization creates new job opportunities, it also requires businesses to reskill and upskill their workforce to remain competitive. Lack of qualified specialists in the field of digital technologies creates shifts in labor demands. This poses challenges for employees who need to adapt to new roles in the digital economy.
Digital Divide and Accessibility Issues: Not all businesses have the same level of access to digital technologies, leading to a digital divide. Smaller businesses, particularly those in developing countries or rural areas, may struggle to keep up with digital transformation due to limited resources and infrastructure. This inequality in access could exacerbate existing economic disparities and hinder inclusive growth. On the other hand, digitalization may exclude some customers from access to the market since the older generation has certain difficulties using online products.
Regulatory and Compliance Challenges: As digitalization evolves, businesses face challenges in complying with ever-changing regulations regarding data privacy, intellectual property, and digital governance. For instance, the introduction of new data protection laws has forced companies to reevaluate their data collection and usage practices. Constantly evolving legal frameworks can create uncertainty and increase compliance costs in the future.

4. THE EFFECT OF DIGITALIZATION ON WORLD ECONOMY

The main difficulty of digitalization is its ambiguous effect on society and the world economy. On one hand, digitalization is having a transformative impact on the global economy, offering numerous benefits across various sectors. On the other hand, digitalization also introduces significant challenges and negative consequences for the global economy. In this chapter we discuss the benefits and challenges posed by digitalization on the world economy.

One of the most significant positive effects is the reduction in costs and increased efficiency in business operations. As digital technologies, including Artificial Intelligence (AI), automation, and digital platforms, become more prevalent, they enable companies to streamline processes, reduce operational expenses, and improve productivity. For example, businesses can leverage Artificial Intelligence to automate repetitive tasks, freeing up human workers for higher-value activities. This leads to increased profits for firms, particularly for smaller businesses, which now have the tools to compete on a global scale with relatively low initial investments. Additionally, digital technologies facilitate access to global markets, allowing firms to engage in e-commerce, participate in the platform economy, and utilize digital payment systems, which are expanding the reach of businesses worldwide.

Furthermore, digitalization has the potential to boost economic growth, improve living standards, and enhance consumer welfare. The use of digital products and services, such as those offered by e-commerce and fintech platforms, reduces transaction costs and increases access to goods and services, making them more affordable for consumers globally. For instance, the reduction in the prices of information and communication technology (ICT) allows for more efficient data collection, innovation, and distribution, benefiting users with free or low-cost services. Additionally, Artificial Intelligence (AI) is facilitating advancements in both the public and private sectors, driving productivity, innovation, and service efficiency. With improved data analytics, businesses and governments can make better decisions, leading to the creation of new business models and the better allocation of resources.

Digitalization also holds the potential to address global economic challenges. Innovations in digital payments, e-commerce, and the platform economy are making it easier to reach underserved populations and create new economic opportunities, especially in emerging markets. This can foster inclusive growth, helping to bridge the digital divide and stimulate development in less economically advanced regions.

Despite the many benefits, digitalization also introduces significant challenges and negative consequences, particularly in terms of employment and income inequality. The automation of repetitive tasks, driven by Artificial Intelligence and Robotics, can lead to job displacement, especially in industries that rely mostly on manual labor or low-skilled workers. Jobs in sectors such as manufacturing, retail, and even traditional banking are particularly vulnerable to automation, resulting in a reduction in available positions and increasing unemployment rates in certain regions and industries. Although digitalization may create new opportunities in tech-driven sectors, the displacement of workers in traditional sectors is a growing concern. According to the International Monetary Fund, the popularity of Artificial Intelligence may possibly result in a huge gap between rich and poor countries across several different channels. If we discuss it in more details, an IMF (International Monetary Fund) study discovered that, given the fact that there are higher salaries in developed countries, the motivation to replace the labor with new technologies is much stronger there. Hence, we shall see higher demand for investment in Robotics rather than traditional aspects of the economy in those developed countries. According to the IMF (International Monetary Fund), it may result in the diversion of investments from emerging nations to advanced economies, "thus resulting in a transitional decline in GDP in the developing or emerging country". In addition, an IMF study discovered that the area where Artificial Intelligence is expected to have the most effect in the developing world is the labor market. The manpower of developing countries has been employed by developed countries for outsourcing purposes, in such industries as manufacturing and logistics, for example. Despite the benefits, Artificial Intelligence creates many disadvantages. AI productivity gains are mainly captured by developed countries and major tech firms, creating an asymmetry in the global arena. These risks widen the gap in incomes because leading countries get most of the benefits. And, emerging countries are left far behind today. Moreover, Artificial Intelligence may easily erode the competitive benefit of emerging economies reliant on cheap labor. As AI-fueled automation develops, it may reduce economic incentives for further trade. Further, it may halt progress towards narrowing the income gap. It should be noted that the majority of emerging countries need to create enough quality jobs for growing young populations. Artificial Intelligence is expected to automate millions of existing jobs, upending traditional growth models and development strategies. This tendency will possibly increase the rate of unemployment and inequality in the labor market. These significant threats require careful navigation further. Moreover, digitalization can exacerbate income and wealth inequality globally. As digital products and services become more widely available, they tend to benefit wealthier individuals who can afford the latest technologies and innovations. These individuals also spend a larger proportion of their income on digital goods, thus widening the gap between the rich and poor. On the other hand, individuals with lower incomes may struggle to keep up with the rising demand for digital skills and access to technology, which can hinder their ability to participate in the growing digital economy. The destruction of traditional business models, especially in industries like banking, is another significant downside of digitalization. In the financial sector, digital platforms and fintech innovations are displacing traditional banks, creating challenges for those institutions that have relied on established business models based on physical infrastructure and face-to-face interactions. While digitalization in banking reduces costs and improves efficiency, it also puts pressure on the stability of traditional financial institutions and their ability to compete with new digital-first entrants. The shift toward digital platforms is also reducing the informational frictions that once benefited universal banks, which could disintegrate their business models over time. According to recent research by SAS and Coleman Parkes, around 90% of bank managers have allocated budgets for Artificial Intelligence projects in 2025, which signals their commitment to the vast opportunities this tech unlocks — from fraud detection to -personalized customer experiences. The global digital banking market is projected to hit $79.4 billion by 2030. (https://innowise.com/blog/digital-banking-trends/)

The global business process automation market size was estimated at $13.7 billion in 2023, and it is expected to reach $41.8 billion by 2033. Such a global firm as Amazon employs machine learning (ML) as a component of its Artificial Intelligence business process automation to manage inventory administration, customer suggestions, and supply chain management concurrently. Through these implementations, Amazon slightly improves operational effectiveness and enhances customer satisfaction. KPMG incorporated IBM Watson to automate financial examinations today. (https://hbr.org/2018/01/artificial-intelligence-for-the-real-world)

Finally, the regulatory challenges associated with digitalization are substantial. Governments and policymakers face difficulties in keeping pace with the rapid development of digital technologies. Issues related to privacy, data protection, and cybersecurity are becoming increasingly important, and without proper regulations, there is a risk of creating loopholes that could lead to the exploitation of consumers and businesses. Additionally, digitalization may complicate the enforcement of monetary policies, financial stability, and regulatory frameworks, as digital platforms can bypass traditional regulatory controls, leading to greater financial volatility and risks.

In conclusion, the impact of digitalization on the world economy is both positive and negative. On one hand, it drives economic growth, increases productivity, and opens up new opportunities for businesses. Larger businesses have better opportunities rather than smaller ones. On the other hand, digitalization raises concerns about job displacement, increased inequality, and the destruction of traditional business models. As the digital economy continues to expand, it is crucial for governments and businesses to develop effective policies and strategies that mitigate the negative effects while maximizing the benefits of digital transformation. By addressing these challenges and investing in digital skills and inclusive growth, the global economy can harness the full potential of digitalization to improve living standards and create a more connected, prosperous future.

5. RATIONALE FOR ETHICAL STANDARDS AND ADOPTION OF GLOBAL PUBLIC POLICIES WORLDWIDE

Digital technologies provide excellent opportunities but also pose serious drawbacks, and their effects depend on the adoption of rules and policies at the national and global level. Rules and policies are strongly needed. Such an author as Acemoglu (2021) discussed several factors that pushed to boost demand for labour. The author discussed the interaction between labour market institutions protecting workers and the emergence of new technologies boosting human productivity, after World War II and the government's role as a major buyer of new technologies. The author strongly argues that public policy, funding and leadership are critical in shaping the future course of Artificial Intelligence developments. At the same time, he mentioned the need for international coordination, as countries will not be able to set the future direction of Artificial Intelligence, when "hundreds of thousands of researchers in other countries can still pursue surveillance and military applications of Artificial Intelligence technology and eagerly continue the trend towards labour-saving automation". This issue requires international coordination rather than local one. Appropriate policies and standards must be incorporated with respect to digitalization in the world arena. Active training and learning may help to build the skills that are needed to capture all benefits of digitalization. At the same time, increased education and a strong public safety net often are not enough to deal with some of the inclusion challenges in the labour market driven by digitalization. The issue of digitalization requires further investigation and research.

When it comes to business ethics in general, it should be noted that business ethics plays a crucial role in the sphere of business. Business ethics is a branch of applied ethics and deals with the question of how moral regulations and behavior should work in the sphere of business. It demonstrates the rules of a business, and it guides business people in their activities. Business ethics is subdivided into ethical aspects of firms, individuals, orders and consumption. Haider et al. (2021) mentioned that the central focus of digitalization was related to ethical issues. To address those concerns, the author proposed a five-step approach (**access, knowledge acquisition, adoption, innovation, and transformation**), which protects the confidentiality of information at the national level. At the same time, Quinn (2021) analyzed the ethical principles, standards and values, which affect a democratic digital society. According to his opinion, ethical principles are related not only to data transparency and trustworthiness but also to data decryption, data protection and confidentiality. It was discovered by the author that society's ethical reasoning should involve quantitative aspects as well as social and political issues. It should be noted that proper regulation will help to strengthen trust in Artificial Intelligence further. The COVID-19 crisis has accelerated digital transformation across the globe. The lockdown situation pushed much of the world online, and the use of digital technologies became indispensable to guarantee the continuity of public, private, and social life. This acceleration has amplified processes that have been under way for decades, resulting in the pervasive impact of digital technologies on many domains: from the individual level (engaging in online learning, working, and shopping) to entire nations (shifting towards digital economies, governance, society), from companies (chasing new business models, new services, new ways to deliver) to entire industries (moving to process automation, exploring benefits of artificial intelligence), from local authorities to national governments (offering more transparent and efficient governance, digital public services).While digital transformation is a global trend today and the world becomes increasingly digital (some estimate that around 70 percent of the new economic value globally will be created on digitally-enabled platforms in nearest future and the global digital transformation market is expected to grow more than twice by 2025-2026), and not everyone is affected in the same way. Digital transformation is not happening at the same pace in the world and the level of intensity in all countries. It was noticed that some countries are advancing more rapidly, while others are still in the early stages of adoption. The scope of digital transformation efforts varies from country to country, depending on factors such as economic, political and social, as well as the level of digital connectivity, skills, regulatory maturity or others. However, in 2022, around 3 billion people remain offline. Universal and meaningful connectivity remains a distant prospect for least developed countries (LDCs), where only 36 percent of the population used the Internet in 2022, compared with 66 percent globally. (https://digitalregulation .org/national-digital-transformation-strategy-mapping-the-digital-journey/)

Finland was the first European country with the highest score on the Digital Economy and Society Index (DESI), ranking first in the human capital component thanks to the advanced digital skills of its citizens in 2022. At the same time, Denmark ranked first for its connectivity. Estonia was the first in digital public services. Concurrently, Greece, Bulgaria, and Romania were the member states with the lowest scores in the digitalization of their economy and society. (https://www.statista.com/ statistics/1245595/eu-digitalization-level/)

Different countries are taking different approaches to address data ethics today. Being one of the largest Western European countries, Germany created its own Data Ethics Commission in 2018 to provide recommendations for the Federal Government's Strategy on Artificial Intelligence. It confirmed that its principles were based on the Constitution, European values, and its "cultural and intellectual history." Concurrently, in Canada, the advancement of Artificial Intelligence technologies and their use in public services has created a discussion about the issue of data ethics. Canada's recommendations are focused primarily on public service officials and processes. It provided certain guiding principles to ensure ethical usage of Artificial Intelligence and developed a comprehensive Algorithmic Impact Assessment online tool to help government officials explore Artificial Intelligence in a way that is "governed by clear values, ethics, and laws." In addition, The Korean Ministry of Science and Information and Communication Technology, in collaboration with the National Information Society Agency, released Special Ethics Guidelines for the Intelligent Information Society in 2018. Those guidelines are primarily built on the Robots Ethics Charter. It strongly supports developing Artificial Intelligence and robots, which do not have "antisocial" characteristics." Korean ethical policies are mainly focused on the adoption of robots into society, while emphasizing the necessity to balance protecting human dignity and the common good. When we compare three different countries' approaches, differences among these initiatives seem to be related to their unique traditions, institutional arrangements, and many other cultural aspects. Germany places emphasis on autonomous vehicles and it developed a rather comprehensive view on ethics. At the same time, Canada focuses its attention primarily on public officials. Contrary, South Korea established its rules and standards through the prism of robots. However, none of them clearly defines what data ethics is. None of them is meant to have any legal (juridical) aspect. Rather, they stipulate the principles of the information society. Discussion and debate on digitalization and ethics will continue for many years in future and we need to consider global universal standards and rules to guide the issue of digitalization (Artificial Intelligence, Robotic Process Automation, Blockchain). But the sooner we reach a certain consensus on key definitions, principles, and approaches, the easier our debates can turn into real actions. Data ethics are equally important for all stakeholders such as government, businesses, and individuals. In addition, this issue should be discussed globally openly today. The process of global discussion will serve itself as a knowledge-sharing mechanism (Gromova et. al, 2021).

6. CONCLUSION AND FUTURE RESEARCH DIRECTIONS

This study made an attempt to define a digitalization in a new era as well as discuss its three main elements (Artificial Intelligence, Robotic Process Automation, Blockchain). This study is purely descriptive and it is mostly focused on the issue of digitalization globally in the business world today. This manuscript is primarily focused on advantages and disadvantages of digitalization in the global business arena today. Hence, this study contributes to the digitalization literature, by providing a clear understanding of it as well as its main benefits and disadvantages.

Digitalization is having a transformative impact on the global economy, offering numerous benefits across various sectors globally. The analysis made by us concluded that digitalization positively impacts businesses by reducing costs, saving time, and increasing efficiency. It also offers greater flexibility than traditional processes. It has the advantage of great flexibility compared to the traditional organization of processes. Moreover, it contributes to constant product improvement. One of the most significant positive effects is the reduction in costs and increased efficiency in business operations. Digitalization has the potential to boost economic growth, improve living standards, and enhance consumer welfare worldwide. The use of digital products and services, such as those offered by e-commerce and fintech platforms, reduces transaction costs and increases access to goods and services, making them more affordable for consumers globally.

However, digitalization also introduces significant challenges and negative consequences for the business world today. There is a serious problem associated with the lack of qualified specialists in the field of digital technologies, which created shifts in labor demands. Not all businesses have the same level of access to digital technologies, leading to a digital divide. And, the absence of the internet in remote regions can create certain problems in the future, but these issues can be further addressed with governmental support at both regional and local levels. Digitalization is mostly popular in developed countries, but it should be promoted in developing and emerging economies as well. Such an asymmetry may further create additional problems. Therefore, in order to prevent such problems, digitalization should strive to benefit all stakeholders equally. As it was mentioned earlier, rules and ethical standards related to digitalization must be developed in a global arena, not only in selected countries. Overall, there are several issues, which must be addressed in the global arena today. Rules and standards, policies and strategies related to digitalization must be presented and followed on a global arena today.

We carefully described the effect of digitalization on the world economy today. The impact of digitalization on the world economy is both positive and negative at the same time. It significantly drives economic growth, increases productivity, and opens up new opportunities for all types of businesses, particularly smaller firms, to compete globally. Digitalization helps to promote all types of goods and services indiscriminately worldwide. On the other hand, digitalization raises concerns about job displacement, increased inequality, and the destruction of traditional business models. As the digital economy continues to expand, it is crucial for governments and businesses to develop effective policies and strategies, which will mitigate the negative effects while maximizing the benefits of global digital transformation worldwide. To strengthen this particular argument, as it was mentioned, we carefully discussed all possible benefits and challenges of digitalization on businesses as well as on world economy and provided our own recommendations in terms of ethical rules and standards needed in a global arena today. Recommendations are very useful and ethical rules and standards must be developed and presented in a global arena today. The effect of digitalization on the global economy was carefully discussed and recommendations were carefully provided as well.

To conclude this manuscript, we present both practical and theoretical implications of the study. The practical implication of this particular study is the importance of digitalization on governmental level globally. Digitalization mostly benefits businesses globally and it will help to transform our society further. It affects the global economy and all businesses will benefit from it if we overcome and solve its main disadvantages mentioned in this study. The issue of digitalization must be studied further in a global arena today. The theoretical implication of this study is further development of this research topic. The topic can be interesting both for academicians and practitioners worldwide. The further study can take place in developing and developed countries concurrently, digitalization progress can be studied in detail further. It will be interesting to see a quantitative study on the usefulness of digitalization in a global arena. The further study may present how digitalization affected firms' profit, efficiency or market value in different countries. Comparative study may add value to existing literature. In addition, it would be interesting to conduct an in-depth interview with business people from different countries/continents about the usefulness of digitalization and its' practical implications. Future research should aim at a broader understanding of the digitalization phenomenon, therefore, our research can be enriched by the analysis of related topics, such as digital disruption, digital convergence or digital divide. Digital divide is the greatest disadvantage of digitalization. As it was mentioned earlier, not all businesses/companies have the same level of access to digital technologies, leading to a digital divide. We mostly talk about those in developing countries or rural areas, who may struggle to keep up with digital transformation due to limited resources and poor infrastructure. This inequality in access could exacerbate existing economic disparities and hinder inclusive growth. Smaller firms do not have an access to digital technologies and large companies may benefit from this access and grow further. As a results, smaller firms will not be able to develop and grow. Future research should be based on numbers, interviews and questionnaires rather than theory, and the research should be broader and deeper. In addition, further research may create new theories or models of digitalization. Further study may provide new practical implications of digitalization globally and help to deal with the issue of digital divide. This study serves as a promising start for any further research in this area.

REFERENCES

Artificial Intelligence for Real World. (2025), Harvard Business Review, https://hbr.org/2018/01/artificial-intelligence-for-the-real-world

Bauer, R. M., & Gegenhuber, T. (2015). Crowdsourcing: Global Search and the Twisted Roles of Consumers and Producers. *Organization*, *22*(5), 661–681. DOI: 10.1177/1350508415585030 PMID: 27524930

Beverungen, A., Böhm, S., & Land, C. (2015). Free Labour, social media, Management: Challenging Marxist Organization Studies. *Organization Studies*, *36*(4), 473–489. DOI: 10.1177/0170840614561568

Boikova, T., Zeverte-Rivza, S., Rivza, P., & Rivza, B. (2021). The Determinants and Effects of Competitiveness: The Role of Digitalization in the European Economies. *Sustainability (Basel)*, *13*(21), 11689. DOI: 10.3390/su132111689

Bonnaud, O. (2021). Innovative strategy to meet the challenges of the future digital society. *Adv Technol Innov*, *6*, 106–116. DOI: 10.46604/aiti.2021.6724

Borowski, P. F. (2021). Digitization, Digital Twins, Blockchain, and Industry 4.0 as Elements of Management Process in Enterprises in the Energy Sector. *Energies*, *14*(7), 1885. DOI: 10.3390/en14071885

Chen, X., Despeisse, M., & Johansson, B. (2020). Environmental sustainability of digitalization in manufacturing: Area view. *Sustainability (Basel)*, *12*(24), 10298. DOI: 10.3390/su122410298

Cochoy, F., Licoppe, C., McIntyre, M. P., & Sörum, N. (2020). Digitalizing consumer society: Equipment and devices of digital consumption. *Journal of Cultural Economics*, *13*(1), 1–11. DOI: 10.1080/17530350.2019.1702576

De Stefano, V. (2019). Negotiating the Algorithm: Automation, Artificial Intelligence, and Labour Protection. *Comparative Labor Law & Policy Journal*, *41*(1), 15–46.

Dias, G. A., & Vieira, A. A. N. (2013). Big data: Emerging legal and ethical issues. *Cienc Inf*, *42*, 174–184.

Digital Regulatory Platform. (2025) https://digitalregulation.org/national-digital-transformation-strategy-mapping-the-digital-journey/

Flyverbom, M., Deibert, R., & Matten, D. (2019). The Governance of Digital Technology, Big Data, and the Internet: New Roles and Responsibilities for Business. *Business & Society*, *58*(1), 3–19. DOI: 10.1177/0007650317727540

Gao, D., Li, G., & Yu, J. (2022). Does digitization improve green total factor energy efficiency? Evidence from 213 Chinese cities. *Energy*, *247*, C. DOI: 10.1016/j.energy.2022.123395

Garzcarek, U., & Steuer, D. (2019) Approaching ethical guidelines for data scientists, *Springer, Cham,* 1, 151-169.

Governo Italiano. (2024) The Changing World of Work. Digitalization, automation and the future of work. Italy's contribution to the ILO centenary Retrieved from http://www.lavorochecambia.lavoro.gov.it/documenti/Documents/THE-CHANGING-WORLD-OF-WORK.pdf

Gromova, K., & Eferin, Y. (2021). *Ethics in the digital world: where we are now and what's next*. World Bank Blogs.

Guryanova, A. V., Smotrova, I. V., Makhovikov, A. E., & Koychubaev, A. S. (2020) Socio-ethical problems of the digital economy: challenges and risks, *Conference: Proceedings of the II International Scientific Conference GCPMED 2019 - Global Challenges and Prospects of the Modern Economic Development.*

Haider, S. A., Zeeshan, M., Irshad, M., Noman, S. M., Arshad, J., Ahmed Shah, S. M., Pervaiz, A., & Naseer, F. (2021). The inclusive analysis of ICT ethical issues on healthy society: A global digital divide approach. *Procedia Computer Science*, *183*, 801–806. DOI: 10.1016/j.procs.2021.03.001

Hantrais, L., & Lenihan, A. T. (2021). Social dimensions of evidence-based policy in a digital society. *Contemporary Social Science*, *16*(2), 141–155. DOI: 10.1080/21582041.2021.1887508

Innowise (2025), https://innowise.com/blog/digital-banking-trends/

International Institute for Applied Systems Analysis (IIASA) International Institute for Applied Systems Analysis (IIASA) (2018) *Digitalization will transform the global economy.*

International Monetary Fund. (2024) https://www.imf.org/external/pubs/ft/ar/2024/

Madakam, S., Vidyapeeth, K., & Jaiswal, K. (2019). The future digital workforce: Robotic Process Automation (RPA). *Journal of Information Systems and Technology Management*, •••, 16.

Marabelli, M., Vaast, E., & Li, J. L. (2021). Preventing the digital scars of COVID-19. *European Journal of Information Systems*, *30*(2), 176–192. DOI: 10.1080/0960085X.2020.1863752

Marinho, S., & Sánchez-García, P. (Eds.). (2022). *Trial and Error in Journalism and Communication Education Between the Classroom and Industry*. Cambridge Scholars Publishing.

McQueen, K. (2015). Ethical issues of knowledge organization in designing a metadata schema for the Leo Kottke archives. *Knowledge Organization, 42*(5), 332–338. DOI: 10.5771/0943-7444-2015-5-332

Myovella, G., Karacuka, M., & Haucap, J. (2020). Digitalization and economic growth: A comparative analysis of Sub-Saharan Africa and OECD economies. *Telecommunications Policy, 44*(2), 101856. DOI: 10.1016/j.telpol.2019.101856

Ossewaarde, M., & Reijers, W. (2017). The Illusion of the Digital Commons: False Consciousness" in Online Alternative Economies. *Organization, 24*(5), 609–628. DOI: 10.1177/1350508417713217

Pangrazio, L., & Sefton-Green, J. (2021). Digital rights, digital citizenship and digital literacy: What's the difference? *J New Approaches Educ Res, 10*(1), 15–27. DOI: 10.7821/naer.2021.1.616

Quinn, R. A. (2021). Artificial intelligence and the role of ethics. *Statistical Journal of the IAOS, 37*(1), 35–77. DOI: 10.3233/SJI-210791

Reinsch, W., & Caporal, J. (2020). *Key Trends in the Global Economy through 2030, Center for Strategic and International Studies*. CSIS.

Rezer, T. M. (2021). Social values of students in conditions of digitalization of education and COVID-19. *Integra Educativa, 30*(2), 176–192. DOI: 10.15507/1991-9468.103.025.202102.226-243

Serkina, Y., Novikova, Z., & Sukhorukih, Z. (2022). Digitalization processes vs. traditional ones: Ethical and environmental aspects. *Ethics in Science and Environmental Politics, 22*, 57–67. DOI: 10.3354/esep00202

Sizganova, E. Y., Kaydashova, A. K., & Ashin, A. A. (2022) Didactic effectiveness of distance learning in the context of digitalization of higher education, *Springer, Singapore,* 65:365-373.

Statista (2025) https://www.statista.com/statistics/1245595/eu-digitalization-level/

Sunil, A., & Nair, K. S. (2021). Going green and CSR: Evidence from Indian companies' best practices. *Academy of Strategic Management Journal, 2*(Special Issue), 1–12.

Taylor, J., & Pagliari, C. (2018). Mining social media data: How are research sponsors and researchers addressing the ethical challenges? *Research Ethics*, *14*(2), 1–39. DOI: 10.1177/1747016117738559

Trittin, H., & Schoeneborn, D. (2017). Diversity as Polyphony: Reconceptualizing Diversity Management from a Communication-Centered Perspective. *Journal of Business Ethics*, *144*(2), 305–322. DOI: 10.1007/s10551-015-2825-8

Upchurch, M., & Grassman, R. (2016). Striking with Social Media: The Contested (Online) Terrain of Workplace Conflict. *Organization*, *23*(5), 639–656. DOI: 10.1177/1350508415598248

Uskov, V. N., Bulat, P. V., & Arkhipova, L. P. (2014). Classification of gas-dynamic discontinuities and their interference problem. *Research Journal of Applied Sciences, Engineering and Technology*, *8*(22), 2248–2254. DOI: 10.19026/rjaset.8.1225

Vineet, J. (2023). *Innovative Processes in Finance and Accounting. Strategic Finance*. Institute of Management Accountants.

Vochozka, M. (2020) Digital transformation of the economy: challenges, trends and new opportunities, *Advances in Intelligent Systems and Computing, Springer, Cham,* 908, 96-100.

Walker, K. L. (2016). Surrendering information through the looking glass: Transparency, trust, and protection. *Journal of Public Policy & Marketing*, *35*(1), 144–158. DOI: 10.1509/jppm.15.020

Wu, T. (2011). *The Master Switch: The Rise and Fall of Information Empires*. Vintage.

Young, A., Selander, L., & Vaast, E. (2019). Digital organizing for social impact: Current insights and future research avenues on collective action, social movements, and digital technologies. *Information and Organization*, *29*(3), 100–157. DOI: 10.1016/j.infoandorg.2019.100257

Zheng, Z., Xie, S., Dai, H., & Wang, H. (2017) An Overview of Blockchain Technology: Architecture, Consensus, and Future Trends, *Conference:6th IEEE International Congress on Big Data.* DOI: 10.1109/BigDataCongress.2017.85

Zhili, D., Boqiang, L., & Chunxu, G. (2019). Development path of electric vehicles in China under environmental and energy security constraints. *Resources, Conservation and Recycling*, *143*, 17–26. DOI: 10.1016/j.resconrec.2018.12.007

KEY TERMS AND DEFINITIONS

Digitalization: An: adaptation of a system, process, etc. to be operated with the use of computers and the internet. "Digitalization allowed companies to sell goods without a physical presence"

Robotic Process Automation: (RPA): A form of business process automation technology that uses software robots to automate tasks performed by humans.

Blockchain: A system in which a record of transactions, especially those made in a cryptocurrency, is maintained across computers that are linked in a peer-to-peer network.

Artificial Intelligence: The theory and development of computer systems able to perform tasks normally requiring human intelligence, such as visual perception, speech recognition, decision-making, and translation between languages.

Innovation: the action or process of innovating. "Innovation is crucial to the continuing success of any organization"

Transformation: a marked change in form, nature, or appearance. "British society underwent a radical transformation"

World Economy: refers to the sum total of all economic activity taking place globally. It encompasses the production, distribution, and consumption of goods ...

Chapter 5
The Transformative Role of Cryptocurrencies in Modern Finance:
Opportunities, Risks, and Future Directions

Ilimsan Feyzullah
https://orcid.org/0009-0004-4734-8861
Akdeniz University, Antalya, Turkey

ABSTRACT

Cryptocurrencies have emerged as a cornerstone of the digital transformation in the global economy during the last decades, introducing decentralized mechanisms that challenge the dominance of traditional financial systems. This innovation enhances transparency, security, and accessibility, while making financial systems more inclusive and efficient. Since the first inception of Bitcoin in 2009, the cryptocurrency landscape has expanded exponentially. By 2023, the combined market capitalization of over 22,000 cryptocurrencies exceeded $1 trillion, demonstrating their significant influence on financial markets. This study examines the dual nature of cryptocurrencies—assessing their transformative potential and inherent risks. Using case studies, historical data, and technological advancements, it provides a balanced perspective on how cryptocurrencies can reshape financial systems, bridge economic inclusion gaps, and drive innovation.

DOI: 10.4018/979-8-3373-0355-0.ch005

INTRODUCTION

On the IV century on the island of "Yap", how stones were used to record and control the debts of the island's inhabitants. With the expansion of the tribe on this island, it became almost impossible to keep track of who owed who and how much they owed, which caused tension within the community, and as a solution, the island's older, authoritative elders suggested appointing someone responsible for keeping records of the exchange of stones, and one person was chosen for the job.

Unfortunately, such a system failed and did not solve their problems, as the registrar began to charge a separate fee for each registration of the exchange transaction and, under pressure from older authoritarian individuals, began to forcibly register false information. Eventually, the members of this tribe abandoned the unilateral registration scheme and adopted a new practice whereby anyone who reached an agreement would publicize the agreement to other islanders. In the end, the logic of the new working system - when the majority of the islanders agreed, the agreement was considered valid and fulfilled. (Fitzpatrick, 2001; Demidova, 2021)

Cryptocurrencies have emerged as a cornerstone of the digital transformation in the global economy during the last decades, introducing decentralized mechanisms that challenge the dominance of traditional financial systems. Operating on Blockchain technology, cryptocurrencies ensure the integrity of transactions through a distributed ledger system that eliminates reliance on centralized authorities. Since the first inception of Bitcoin in 2009, the cryptocurrency landscape has expanded exponentially.

Security risks are another persistent issue; high-profile breaches, such as the Mt. Gox exchange hack in 2014 or the more recent FTX collapse in 2022, had resulted in billions of dollars in losses, undermining trust in cryptocurrency ecosystems. General regulatory uncertainty further complicates the landscape.

Despite these hurdles, the global cryptocurrency market surpassed $1 trillion in 2023, with daily trading volumes averaging around $125 billion. This remarkable growth highlights the need to critically assess the role of cryptocurrencies as financial instruments within an evolving economic landscape and to explore their potential to transform global finance.

By analyzing case studies, historical data, and technological advancements, this research provides a balanced perspective on how cryptocurrencies can reshape financial systems, bridge gaps in economic inclusion, and drive innovation while addressing the obstacles that could hinder their progress.

The unconventional nature of cryptocurrencies has also attracted the attention of law enforcement agencies due to their association with alleged illegal activities. Despite this relatively small percentage, the public perception of cryptocurrencies as tools for money laundering and cyber crime remains a significant barrier to acceptance.

This work aims to provide a comprehensive analysis of cryptocurrencies, balancing their innovative potential with the risks and uncertainties they pose. by examining case studies, historical trends, and recent developments, this research explores how cryptocurrencies are reshaping economic systems, fostering financial innovation, and presenting unique challenges for regulators and market participants. Ultimately, this study seeks to contribute to the ongoing dialogue on the future of cryptocurrencies and their role in an increasingly digitized global economy.

Cryptocurrencies and Blockchain Fundamentals

Cryptocurrency is a set of concepts and technologies that form the basis of today's Electronic (Digital) Money ecosystem. The crypto ecosystem has evolved significantly since Bitcoin was introduced in 2009, encompassing a wide range of digital assets and decentralized applications. (Grasic & Vidnjevic, 2024).

There are several important definitions and differences to consider when addressing this issue:

a) Digital money is completely virtual and backed by nothing
b) Although the system is completely transparent, all transactions are strictly anonymous. Any outside user can see the entire chain of transactions of Cryptocurrencies at any time, but the people who carry out these transactions will remain completely anonymous due to the structure of the system keys
c) Realized Payments can never be canceled. This is one of the fundamental and most important features of Cryptocurrencies. Satoshi Nakamoto stated this definition in his own work as follows: "Since traditional payments can be canceled, the seller has to be always on the alert, which is why more information than necessary is always requested by the buyer" (Nakamoto, 2008)
d) Essentially the most important difference that started it all is that cryptocurrency does not have a single control center. This eliminates intermediaries and commissions, as well as transferring accounting records from centralized financial institutions to the autonomous computer network, thereby reducing the costs of doing business with cryptocurrency and preventing corruption and theft that can occur in intermediary institutions. In other words, by eliminating the need for intermediaries, the system provides an infrastructure where foreigners can do business with each other. (Nakamoto, 2008)

It is worth noting that "Bitcoin" as a Cryptocurrency was the first of its kind, while "Forbes" magazine, just a few years after Bitcoin's emergence, published articles that permanently embedded the now familiar term "Cryptocurrencies" in society.

Since Bitcoin and Cryptocurrencies are often identified and defined in the first articles written on this subject, and since Bitcoin was also one of the first Cryptocurrencies to succeed, the history of the emergence of the Cryptoasset phenomenon will start exactly from the creation of this digital instrument.

The long journey of Cryptocurrencies began on October 31, 2008. It is assumed that it started when the then unknown software programmer Satoshi Nakamoto published the article **"Peer-to-Peer Electronic Cash System"** on internet resources. (Nakamoto, 2008)

Despite the promising perspectives of Satoshi Nakamoto's proposal, the Bitcoin system did not gain much traction in the initial phase, being met with extreme skepticism by society. (Nakamoto, 2008)

Figure 1. Block chain creation scheme

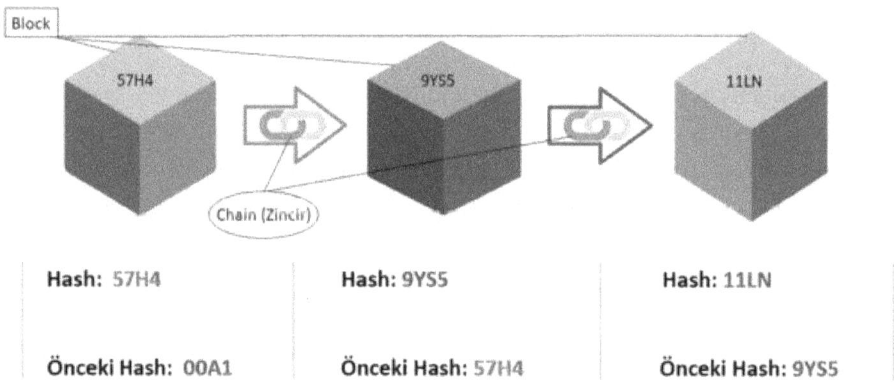

(By Author, 2025)

However, the system described by Satoshi Nakamoto had two unique features:

- Blockchain technology was unknown until then (Figure 1.),
- The existence of monetary incentives for computer owners in the online network to maintain the system. (Nakamoto, 2008)

Through trial and error, Satoshi Nakamoto and his followers were able to develop the system into what we know today. This development was not hindered by a number of known moments of crisis during its formation. In this context, 2014 is considered the worst year in the development and history of Cryptocurrencies to date.

Figure 2. Bitcoin price chart throughout history from 2012 to 2025,

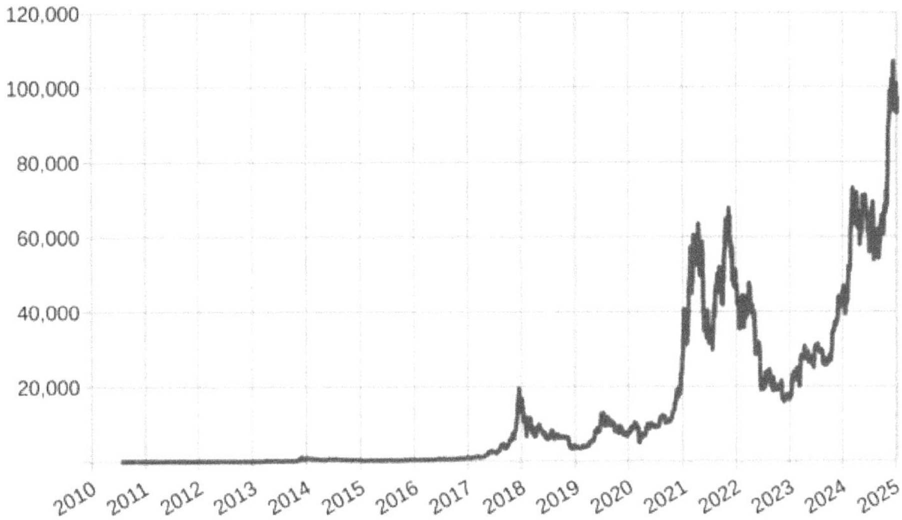

(Myfin Team, 2025)

Since then, over than thousands new Cryptocurrencies with the same or similar infrastructure and fundamentals have emerged on the market, but none have reached the success and value level of Bitcoin. The trend of Bitcoin prices between 2009 and 2021 is shown in Figure 2. As seen in the chart in 2021, Bitcoin renewed its historical peak at a record price.

As mentioned earlier, Block-Chain technology forms the basis of the ledger of Cryptocurrencies. As can be seen in Figure 1., a block chain is a long chain of blocks or transactions occurring simultaneously. As long as the system works, these block chains will be constantly added and grow. By creating a time-fixed transfer sequence for each transaction, the Block-Chain System allows users to monitor the account status of users making transactions at any given time, and also ensures the continuity of the entire system by providing the ability to review and view the data attached to Bitcoin or part of it, when it was created, consumed or received. The essence of blockchain technology is that the data subject to this process cannot be stored in a single environment/computer/source. This information is stored in a specific community network(s) and/or by installing special electronic wallets on the devices of individual computer owners. That is, in the form of special programs provided to users in the form of encrypted programs. These wallets/programs are used to manage users' Accounts. The creators of the “Block-Chain Revolution”, A. Tapscott and D. Tapscott, wrote in their work: “This system is an infinite digital transaction log that can be programmed and distributed to record not only financial transactions, but virtually anything of value”. (Tapscott and Tapscott, 2016)

The main reason and purpose for the creation of cryptocurrencies is to create and maintain an infrastructure where foreign parties can conduct business with each other, eliminating the need for an intermediary. Block-Chain technology has achieved this by transferring the most important function of keeping financial records from centralized and authoritarian institutions to a network of autonomous computers that form a system of trust in which tasks and transactions are distributed so that they are not controlled by any single institution.

At their core, cryptocurrencies are based on the idea of a universal ledger that is publicly available and is kept track of each other at the same time by high-performance computers running independently of each other.

By eliminating intermediaries with their commissions, cryptocurrency allows you to reduce the costs of doing business and also aims to ensure continuity by preventing the corruption that exists in intermediary structures.

The potential of this technology as a means of providing transparency and control extends far beyond the financial sphere, as it can eliminate any information intermediaries. In essence, this technology is a form of social organization and control mechanism that transfers control of cash flows and information from powerful and elite institutions/individuals to ordinary people.

It is envisaged that this system will provide great opportunities for ordinary people, giving them the right to independently manage their assets and unique abilities. Thus, it can be concluded that the Block-Chain technology is the central infrastructure of the Cryptocurrencies system, on which all its work is based.

Despite the fact that this technology has so far only been used in finance, its design is universal in nature and applicable to almost all areas of human activity. (Hayes and Khartit, 2022)

LITERATURE REVIEW

Theoretical Framework

The study draws on theories of decentralization, financial innovation, and risk management to analyze the role of cryptocurrencies. Decentralization theory emphasizes the elimination of intermediaries, while financial innovation theory highlights the disruptive potential of Blockchain technology.

Cryptocurrencies like Bitcoin and Ethereum have significantly transformed the financial landscape, challenging the dominance of traditional banking systems and introducing a revolutionary model of decentralized, transparent, and efficient transactions.

Blockchain, the foundational technology behind cryptocurrencies, functions as a distributed ledger where transactions are recorded immutably and transparently. This technology eliminates intermediaries, reducing costs and inefficiencies associated with traditional banking systems. For instance, cross-border payments, which traditionally involve high fees and lengthy settlement times, are now completed almost instantaneously with minimal costs using Blockchain solutions.

As known, cryptocurrencies operate on a fundamental technological innovation called Blockchain, a distributed ledger that allows for the secure, transparent, and immutable recording of transactions. This technology underpins all major cryptocurrencies, including Bitcoin, Ethereum, and newer Blockchain platforms such as Solana and Polkadot. Blockchain's primary feature is decentralization, meaning no single entity or group has control over the entire network. This ensures that transactions are secure, verifiable, and cannot be manipulated by central authorities or malicious actors.

Blockchain is based on a chain of cryptographic blocks, each containing a batch of transactions. Each block in the chain is linked to the previous one through a cryptographic hash, creating a sequence that is tamper-resistant. In the Bitcoin Blockchain, for example, once a transaction is confirmed by the network and added to a block, it becomes nearly impossible to alter that transaction due to the computational difficulty of modifying the entire chain. This process is what provides the security and transparency that makes Blockchain ideal for cryptocurrencies and other applications. (Nakamoto, 2008; Hayes and Khartit, 2022)

Review of Relevant Previous Studies

Previous research has explored the technological foundations of cryptocurrencies, their impact on financial systems, and associated risks. For instance, studies by Nakamoto and Buterin laid the groundwork for Bitcoin and Ethereum, respectively. More recent studies focus on DeFi, tokenization, and regulatory challenges. Unlike conventional fiat currencies managed by central banks and financial institutions, cryptocurrencies are digital assets that operate independently, leveraging Blockchain technology to ensure transaction integrity and security. This innovative model has disrupted traditional financial ecosystems and created opportunities for greater financial inclusivity, autonomy, and innovation. (Nakamoto, 2008; Hayes and Khartit, 2022; Buterin, 2013)

Key Concepts and Definitions

Bitcoin: The largest and most influential cryptocurrency by market capitalization. It was the first cryptocurrency, introduced in 2009, and continues to dominate the space both in terms of recognition and investment. (Berwick, 2025)
Ethereum: The Smart contract and DeFi Leader, introduced a groundbreaking innovation: the ability to execute smart contracts. (Riley, 2024)
Ethereum2.0: Ethereum's transition to Ethereum 2.0, an upgrade that includes the switch from a proof-of-work (PoW) to a proof-of-stake (PoS) consensus mechanism, is set to significantly improve its scalability, energy efficiency, and overall network security. (Riley, 2024)
Smart contracts: Ethereum smart contracts are decentralized applications operating autonomously on the blockchain. They manage assets often worth millions of dollars, executing tasks automatically when conditions are met. Imagine a vending machine but for digital agreements; you insert a condition, and out pops the action when that condition's satisfied. (Riley, 2024)
Blockchain: Blockchain is a digital ledger that's changing how we handle transactions. This is a ledger that everyone in a network has a copy of. In a blockchain, transactions are recorded in blocks. Every block has a unique identifier, this forms a secure chain. Unlike traditional systems where a central authority oversees everything, blockchain operates on a network of multiple nodes or computers. Each node has a copy of the ledger, so no single entity has control. This decentralization ensures that even if one node fails or is compromised, the integrity of the blockchain remains. (Riley, 2024)
Blockchain Scalability and Layer 1 and Layer 2 Solutions: Scalability has been one of the primary challenges for Blockchain networks. The more popular a Blockchain becomes, the more transactions it has to process, leading to potential congestion, slower transaction speeds, and higher fees. (Riley, 2024)
Decentralized Finance (DeFi): These studies suggest that Decentralized Finance (DeFi) is a blockchain-based financial system that operates without centralized intermediaries, offering services like payments, lending, and trading

through smart contracts, but it faces challenges such as regulatory compliance, security risks, and potential market manipulation. (Makarov and Schoar, 2022)
Tokenization: Representing real-world assets as digital tokens on Blockchain platforms. Digital tokens authenticated on blockchain ledgers offer the potential to make buying and selling assets much easier. (Whiffin, 2024; Riley, 2024)

CONSENSUS MECHANISMS: PROOF OF WORK, PROOF OF STAKE, AND OTHER ALTERNATIVES OF BLOCKCHAIN TECHNOLOGY

The security and integrity of Blockchain networks are maintained through consensus mechanisms, which ensure that all participants agree on the validity of transactions. Two of the most widely used consensus algorithms are Proof of Work (**PoW**), Proof of Stake (**PoS**), Delegated Proof of Stake (**DPoS**), Proof of authority (**Poa**). These mechanisms differ significantly in their approach to validating transactions and maintaining the network.

- **Proof of Work (PoW):** Bitcoin uses PoW as its consensus mechanism. In this system, miners compete to solve complex cryptographic puzzles in order to validate transactions and add them to the Blockchain. This process requires a significant amount of computational power, as miners must expend energy to solve these puzzles. The first miner to successfully solve the puzzle is rewarded with newly minted Bitcoins and transaction fees. However, PoW has drawn criticism for its high energy consumption. According to the Cambridge Centre for Alternative Finance, Bitcoin mining consumes more electricity annually than entire countries, such as Argentina or the Netherlands. This energy-intensive process has raised concerns about its environmental impact.
- **Proof of Stake (PoS):** Ethereum transitioned from PoW to PoS in 2022 with the launch of Ethereum 2.0, significantly reducing the network's energy consumption. In PoS, validators are chosen to validate transactions based on the amount of cryptocurrency they are willing to "stake" as collateral. Unlike PoW, which requires extensive computational resources, PoS is much more energy-efficient. In PoS systems, validators are incentivized to act honestly because if they validate fraudulent transactions, they risk losing their staked cryptocurrency. Ethereum's transition to PoS was expected to reduce its energy consumption by more than 99%, addressing one of the most significant criticisms of Blockchain technology.
- **Delegated Proof of Stake (DPoS):** This variant of PoS allows stakeholders to vote for a small group of validators, known as delegates, to secure the network

and validate transactions. This system is faster and more efficient than PoW, as fewer participants are responsible for maintaining the Blockchain. EOS and TRON are notable Blockchain platforms that use DPoS. This mechanism is designed to improve scalability and transaction speed, making it suitable for high-throughput applications like gaming or social media platforms.

- **Proof of authority (Poa):** Poa is often used in private or enterprise Blockchains where a trusted set of validators is responsible for validating transactions. Unlike PoS or PoW, Poa does not rely on staking or mining; instead, validators are selected based on their reputation and authority. Poa is faster and more scalable, making it ideal for use cases such as supply chain management and private financial networks. For instance, Vechain uses Poa for its supply chain tracking solutions.

(Wendl et al., 2022; Bitsgap Team, 2024; The Cambridge Centre for Alternative Finance, 2025; Criddle, 2021; Awati, 2024)

BLOCKCHAIN SCALABILITY AND LAYER 1 AND LAYER 2 SOLUTIONS

Scalability has been one of the primary challenges for Blockchain networks. The more popular a Blockchain becomes, the more transactions it has to process, leading to potential congestion, slower transaction speeds, and higher fees. For example, Bitcoin can only handle about 7 transactions per second (TPS), while Ethereum handles around 30 TPS. This is far from sufficient for global-scale applications. To address scalability issues, Blockchain developers have turned to Layer 2 solutions and other techniques to improve throughput. (Oku, 2025)

Second-Layer Scalability Solutions

Layer-2 Solutions: These are technologies built on top of the main Blockchain (**Layer-1**) to improve transaction speeds and reduce costs. A prominent example is the Lightning Network for Bitcoin. It allows users to open payment channels off-chain, where they can make multiple transactions without recording each one on the main Blockchain. After completing the transactions, the final balance is settled on the Bitcoin Blockchain. This reduces congestion and transaction fees, enabling faster and cheaper payments. (Cong at al., 2023)

Ethereum also uses Layer 2 solutions, such as Optimistic Rollups and ZK-Rollups, to address scalability. These solutions bundle multiple transactions into a single one and submit it to the main Ethereum chain, improving throughput and lowering energy fees. Rollups are considered one of the most promising solutions for scaling Ethereum without compromising on security. Second-layer scalability solutions, operate on top of the existing blockchain, processing transactions off-chain to lessen the burden on the main blockchain and boost scalability. (Cong at al., 2023)

Figure 3. Blockchain types

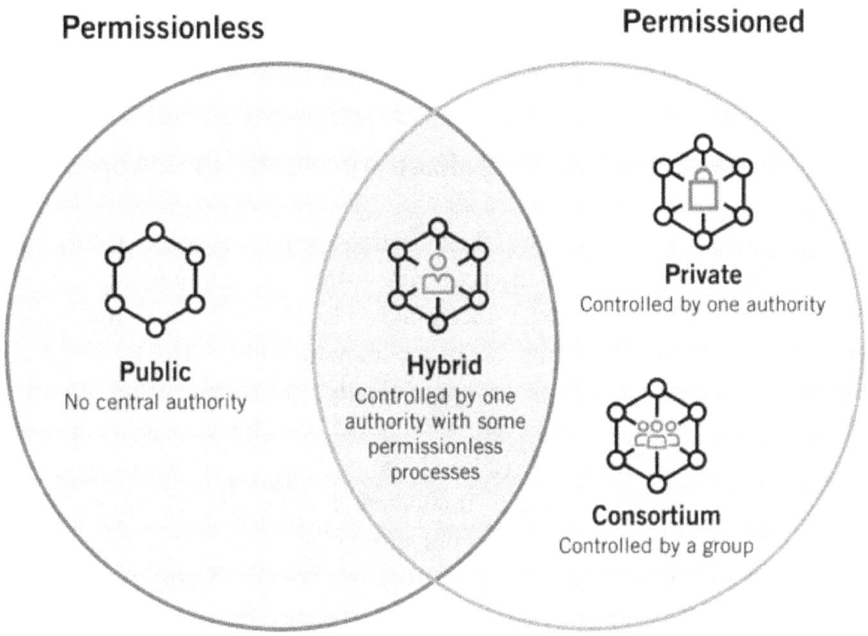

(Wegrzyn and Wang, 2021)

Block-Chain Types

1. **Public Blockchains:** Public blockchains allow all nodes of the blockchain to have equal rights to access the blockchain, create new blocks of data, and validate blocks of data. To date, public blockchains are primarily used for exchanging and mining cryptocurrency such as Bitcoin, Ethereum, and Litecoin.
2. **Private (or Managed) Blockchains:** Private blockchains, which may also be referred to as managed blockchains, are permissioned blockchains controlled by a single organization. Some examples of private blockchains are the business-to-business virtual currency exchange network Ripple and Hyperledger.

3. **Consortium Blockchains:** Consortium blockchains are permissioned blockchains governed by a group of organizations, rather than one entity, as in the case of the private blockchain. Consortium blockchains, therefore, enjoy more decentralization than private blockchains, resulting in higher levels of security.
4. **Hybrid blockchains:** Hybrid blockchains are blockchains that are controlled by a single organization, but with a level of oversight performed by the public blockchain, which is required to perform certain transaction validations. An example of a hybrid blockchain is IBM Food Trust, which was developed to improve efficiency throughout the whole food supply chain. (*Wegrzyn and Wang, 2021)*

Smart Contracts and Decentralized Applications (DApps)

As we already pointed out, one of the most transformative aspects of Blockchain technology is the ability to create smart contracts—self-executing contracts where the terms of the agreement are written directly into code. These contracts automatically execute when predefined conditions are met, eliminating the need for intermediaries. For instance, a smart contract could automatically transfer ownership of a house when the buyer's payment is received, without the need for a notary or real estate agent. Ullah, F., & Al-Turjman, F. (2021).

Ethereum, the first Blockchain to implement smart contracts, has become the dominant platform for decentralized applications (DApps), which are applications that run on a Blockchain. These dApps cover a wide range of industries, including finance, gaming, supply chain management, and more. Some popular examples of dApps include Uniswap, a decentralized exchange (DEX) that allows users to trade cryptocurrencies without a central authority, and MakerDaO, a decentralized lending platform that allows users to borrow funds against their cryptocurrency holdings. (Lim, 2022; Urusov at al., 2024; Chaleenutthawutat al., 2024).

The rise of Decentralized Finance (DeFi), powered by smart contracts, has been one of the most significant trends in the cryptocurrency space. DeFi platforms offer traditional financial services, such as lending, borrowing, and trading, without the need for banks or other intermediaries. DeFi applications are built on Blockchains like Ethereum, Binance Smart chain (BSC), and avalanche. In 2021, the total value locked (TVL) in DeFi reached over $100 billion, highlighting the growing importance of decentralized financial systems. However, DeFi platforms also come with risks, including smart contract vulnerabilities and the potential for hacks, as demonstrated by the high-profile attack on the Poly Network in 2021, where hackers exploited a vulnerability in the system to steal over $600 million. (Cointelegraph Team, 2023; Chainalysis Team, 2021)

Tokenization and NFT's

Another significant development in the Blockchain space as we see before is tokenization, the process of representing real-world assets as digital tokens on a Blockchain. Tokenization enables assets like real estate, art, stocks, and commodities to be divided into smaller, tradable units, making them accessible to a broader range of investors. For example, real estate platforms like RealT allow investors to buy fractional ownership of properties, providing exposure to the real estate market without the need for large capital outlays.

Figure 4. Asset tokenization: Unlocking new possibilities for the enterprise.

(Lewis, 2024)

In addition to tokenizing traditional assets, Blockchain has enabled the creation of non-fungible tokens (NFTs), unique digital assets that represent ownership or proof of authenticity of digital or physical items. NFTs have gained significant popularity in the art, music, and gaming industries, allowing artists to monetize their work directly on Blockchain platforms. In 2021, the NFT market reached over $41 billion in sales, with high-profile NFT sales, such as Beeple's digital artwork "Everydays: The First 5000 Days," which sold for $69 million at Christie's auction house, drawing mainstream attention to the space. (Brown, 2021)

NFTs also have potential applications beyond digital art. For instance, Nba Top Shot is a platform that allows fans to buy, sell, and trade officially licensed basketball highlights as NFTs. Similarly, NFTs are being explored for use in identity verification, event ticketing, and even real estate transactions. (Brown, 2021)

Figure 5. How does a tokenized transaction work,

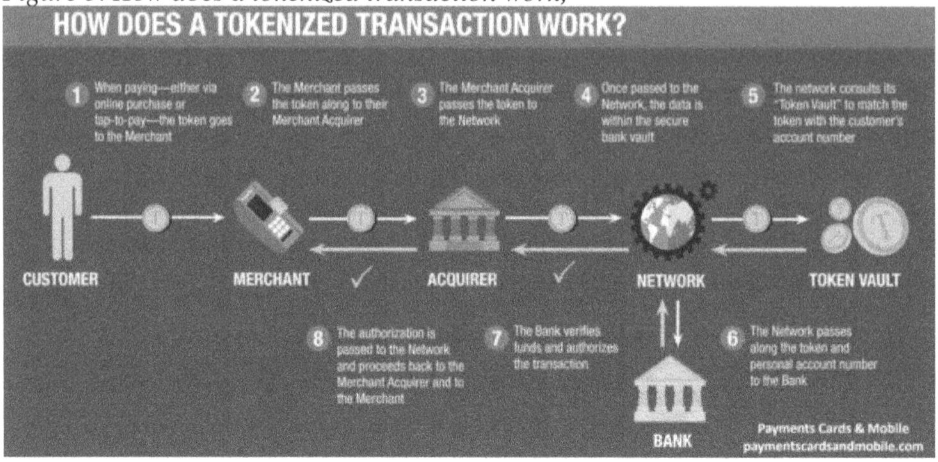

(Rolfe, 2015)

Enterprise Blockchain and Blockchain in Industry

While public Blockchains Products like Bitcoin and Ethereum are the most well-known, enterprise Blockchains are also gaining traction. These private Blockchains are used by companies to streamline processes, enhance transparency, and reduce costs. Companies like IBM and Microsoft are investing heavily in Blockchain technology for use cases such as supply chain management, identity verification, and financial services. One of the most notable examples of enterprise Blockchain is IBM Food Trust, a Blockchain-based system designed to improve transparency and traceability in the food supply chain. (IBM Team, 2023; IBM Team 2024).

IDENTIFYING RESEARCH GAPS

One of the most significant challenges facing the cryptocurrency market is regulatory ambiguity. Cryptocurrencies are not regulated in the same way as traditional financial assets, and this lack of oversight has contributed to market volatility and uncertainty. The absence of clear regulations makes it difficult for institutional investors to enter the market with confidence, and it leaves individual investors vulnerable to fraud and manipulation.

Countries around the world have adopted various approaches to cryptocurrency regulation. Some, like Japan and Switzerland, have implemented favorable frameworks that encourage innovation and adoption, while others, like China, have taken a more hostile stance, banning cryptocurrency mining and trading activities. The lack of standardized regulations across countries creates an environment where investors face different levels of risk depending on their jurisdiction.

In recent years, however, there have been efforts to bring more clarity to the regulatory landscape. One notable example is the European Union's Markets in crypto-assets (MICA) regulation, which aims to create a comprehensive regulatory framework for cryptocurrencies. The MICA regulation seeks to address issues related to investor protection, anti-money laundering (AML), and market stability, and it is seen as a step toward creating a more transparent and secure environment for cryptocurrency investors. (European Parliament, 2022; European Commission, 2023; European Commission, 2024)

Despite these developments, regulatory uncertainty remains a significant challenge for the industry. Until more countries adopt clear and consistent regulations, the risks associated with investing in cryptocurrencies will likely persist. Investors must remain vigilant and stay informed about the evolving regulatory landscape in order to navigate the market effectively.

As noted above, the uncertainty surrounding cryptocurrency regulation stems from its evolving nature and the complexity of existing legal frameworks. This creates uncertainty for users, investors, and regulators, especially because cryptocurrencies like Bitcoin, Ethereum, and others do not fit neatly into traditional categories like commodities, securities, or currencies.

While existing literature addresses the benefits of cryptocurrencies, little attention has been paid to the regulatory frameworks and cybersecurity risks. This study seeks to shed light on these gaps by attempting to provide a holistic analysis on the topic.

General Methodology of Research and Findings

A qualitative approach was adopted, synthesizing online data from peer-reviewed journals, industry reports, and case studies to analyze the historical trajectory, technological foundations, and market dynamics of cryptocurrencies.

All Data were collected from peer-reviewed journals, industry reports, online crypto-exchange, mentioned web sites and case studies published between 2009 and currently 2025.

Purposive sampling was used to select sources relevant to the research objectives. Thematic analysis was conducted to identify patterns and insights related to the research questions.

Triangulation of data sources ensured reliability, while peer-reviewed materials enhanced validity.

PERSPECTIVES OF CRYPTOCURRENCIES AND BLOCKCHAIN

3.1. Current Status of Cryptocurrencies and Investment Transaction Methods

Traditionally, two scenarios can be posited about the future direction of cryptocurrencies: the view that the technology will somehow remain in its current form or something similar, and that the system will continue in that form, and the view that the whole thing will be recognized as a massive worldwide fraud.

Today, some countries officially allow transactions with cryptocurrencies. They are usually treated as a commodity or investment asset and are subject to relevant legislation from a tax perspective. In other respects, in some countries, cryptocurrencies are recognized as a payment unit, in other it is a legal tender subject to tax legislation as a unit of exchange.

3.2. The Potential of Cryptocurrencies

In addition to Cryptocurrencies, Block-Chain technologies have a lot of potential, the technology, which can be adapted for various sectors, has already won the admiration of many experts and the potential to create the technological infrastructure for many applications in the future has been realized.

In a large-scale report prepared by Switzerland-based Credit Suisse, it was stated that Block-Chain is used in many areas, not just for digital currencies or financial services.

Cryptocurrencies as Investment Assets

Cryptocurrencies have rapidly evolved from niche digital currencies into highly sought-after investment assets, attracting both individual investors and large institutional players. These assets have demonstrated substantial returns, which have helped bolster their position in global financial markets. For instance, Bitcoin, the first and most well-known cryptocurrency, has consistently outperformed traditional financial instruments, delivering average annual growth rates exceeding 200% over the last decade. Despite their volatility, cryptocurrencies offer exceptional returns that have been difficult for conventional asset classes to match. Over the course of the past decade, Bitcoin has experienced significant price appreciation, rising from a few hundred dollars in 2013 to a peak of nearly $69,000 in late 2021 before experiencing corrections and in the middle of January 2025, Bitcoin price value stepped over the historical record line of $108,000. As of 2025, Bitcoin continues to be a focal point of institutional investment and interest, reinforcing its value proposition as a high-growth asset. (CoinGecko Team, 2025)

Institutional investors, once hesitant due to regulatory uncertainties and technological barriers, have gradually embraced cryptocurrency as an asset class. In early 2021, Tesla, the electric vehicle giant, made headlines by investing $1.5 billion in Bitcoin, signaling the growing acceptance of cryptocurrencies by mainstream corporations. This move was not just an experiment in adopting cutting-edge technology, but also a strategic decision to hedge against inflation and currency devaluation in an era of unprecedented monetary expansion and low interest rates. Tesla's investment in Bitcoin sparked similar moves by other major companies such as MicroStrategy, Square, and even traditional hedge funds, which saw cryptocurrencies as viable instruments for diversification and risk management. These institutional players recognized the potential of cryptocurrencies to act as a store of value, similar to precious metals, especially in an environment characterized by rising inflation, economic uncertainty, and a global push towards digitalization. The move by Tesla also underscored the increasing comfort and sophistication of institutional investors with digital assets, which were once thought of as risky and speculative. (Tesla Investor Relations Team, 2021; The Guardian Team, 2021)

The acceptance of cryptocurrencies as investment vehicles has been further validated by the increasing number of financial products tailored to institutional investors. In 2021, the U.S. Securities and Exchange commission (SEC) approved the launch of Bitcoin exchange-traded funds (ETFs), allowing investors to gain exposure to Bitcoin without the complexities of owning and securing the cryptocurrency directly. This innovation has made it easier for traditional investors, such as pension funds and endowments, to access cryptocurrency markets within the confines of established regulatory frameworks. ETFs have become a critical tool for diversifying portfolios, providing exposure to Bitcoin's price movements while offering the safety and convenience of traditional financial products. Bitcoin ETFs, such as those offered by companies ProShares, have attracted billions of dollars in investments, further signaling institutional adoption. (The U.S. Securities and Exchange Commission, 2021; Chen at al., 2024).

1. Advantages of Cryptocurrency Use in Modern Finance

Cryptocurrencies offer numerous advantages over traditional financial systems, particularly in areas such as financial inclusion, security, transparency, and transaction costs. These benefits have helped to further solidify cryptocurrencies as a viable and attractive alternative for both individual and institutional investors.

Financial Inclusion: One of the most significant contributions of cryptocurrencies to the global financial system is their ability to provide financial services to the unbanked and underbanked populations around the world. According to the World bank, approximately 1.7 billion adults worldwide remain unbanked, lacking access to basic banking services. Cryptocurrencies offer a decentralized, accessible platform for individuals to participate in financial transactions, invest, and even borrow without requiring a traditional bank account. For people in regions where access to banking infrastructure is limited or unreliable, cryptocurrencies offer an inclusive solution to economic participation. The use of Bitcoin for remittances has reduced fees that typically range between 3% and 7%, resulting in significant savings for recipients and fostering greater financial independence. (The World Bank Group, 2021; The World Bank Team, 2025)

Figure 6. Global account ownership between 2011 and 2021,

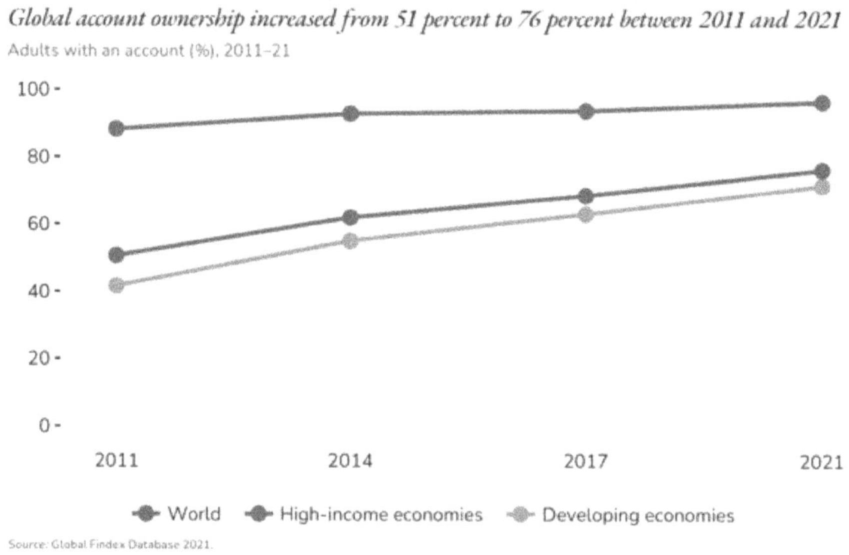

(The World Bank Group, 2021)

Transparency and Security: Blockchain technology, which underpins cryptocurrencies, provides an immutable, transparent ledger that enhances the security of transactions. This inherent transparency and immutability offer a level of trust that is often lacking in traditional financial systems, which are prone to errors, fraud, and manipulation. Blockchain's decentralized nature also reduces the risks associated with centralized data storage, where hacks and data breaches have led to massive financial losses.

IBM's Food Trust is a prominent example of how Blockchain can be used to ensure the transparency and traceability of products, from farm to home. (IBM Team, 2023)

This level of transparency extends beyond the food industry and can be applied to sectors such as pharmaceuticals, luxury goods, and electronics and many other sectors.

Lower Transaction Costs: Traditional banks and money transfer services often charge high fees for international remittances, with rates ranging from 3% to 7%. By contrast, cryptocurrencies offer relatively low transaction fees, with the average Bitcoin transaction costing around $1.50. This makes cryptocurrency an appealing option for businesses and individuals who need to send money across borders, particularly in emerging markets. (BitPesa, 2024; The World Bank Group, 2024)

Hedge Against Inflation: Unlike fiat currencies, which can be printed by central banks and subjected to inflationary pressures, cryptocurrencies capped supply ensures that it is not vulnerable to the same monetary policies that drive inflation. In countries facing hyperinflation, such as Venezuela, Argentina, and Turkey, Bitcoin has become a preferred alternative to traditional fiat currencies. Between 2021 and 2023, trading volumes in Bitcoin surged by over 300% in these countries, reflecting a growing demand for a more stable and inflation-resistant asset. (Chainalysis Team, 2023; Federal Reserve Economic Data, 2025)

2. Volatility and Speculation in Cryptocurrency Markets

Despite their numerous advantages, cryptocurrencies remain highly volatile, making them speculative assets prone to sharp price fluctuations. The extreme fluctuations in cryptocurrency prices are a result of a variety of factors, including speculative trading, market sentiment, and regulatory developments. The absence of centralized control means that cryptocurrencies are subject to the whims of the market, which can be influenced by anything from macroeconomic trends to public statements from influential figures.

Unlike traditional assets, cryptocurrencies lack centralized control, making them more susceptible to sudden market shifts. (Fidelity Digital Assets, 2024)

However, there is an emerging trend that suggests cryptocurrencies, particularly Bitcoin, are becoming increasingly correlated with traditional financial markets, especially during times of global economic stress. According to a study by Fidelity Digital assets, Bitcoin's correlation with the S&P 500 reached 0.44 in 2022, indicating that Bitcoin's price movements are becoming more closely aligned with broader market trends, and also in fact: in report of the 2024, compared to all S&P 500 constituents, Bitcoin has recently been exhibiting a lower historical annualized volatility figure than 33 of the approximately 500 companies in the S&P 500. Bitcoin was actually less volatile than 92 of the S&P 500 stocks in October of 2023 when using the 90-day realized historical volatility figures. Some of these names are also large-cap and mega-cap stocks. This could suggest that, over time, cryptocurrencies may no longer be completely decoupled from traditional assets, but instead be integrated into the larger financial ecosystem. (S&P Global, 2024; Erdaş abd Caglar 2018)

Increased institutional investment has contributed to this trend, as large financial institutions have begun to treat cryptocurrencies as part of their diversified portfolios. This shift represents a significant evolution in the way cryptocurrencies are viewed by traditional finance and has the potential to stabilize prices over the long term.

As seen, cryptocurrencies have cemented their place as a legitimate and growing asset class. Their advantages in terms of financial inclusion, transparency, lower transaction costs, and inflation hedging make them an attractive alternative to traditional financial assets. However, their volatility, regulatory uncertainty, and security risks continue to pose challenges. As the cryptocurrency market continues to mature, institutional involvement, regulatory clarity, and technological advancements will likely play key roles in determining the long-term stability and growth of digital assets. (European Parliament, 2022; European Commission, 2023; European Commission, 2024)

2.1. Jurisdictional Differences and Regulatory Ambiguity and Market Impact

One of the most significant challenges facing the cryptocurrency market is regulatory ambiguity. Cryptocurrencies are not regulated in the same way as traditional financial assets, and this lack of oversight has contributed to market volatility and uncertainty. The absence of clear regulations makes it difficult for institutional investors to enter the market with confidence, and it leaves individual investors vulnerable to fraud and manipulation.

Countries around the world have adopted various approaches to cryptocurrency regulation. Some, like Japan and Switzerland, have implemented favorable frameworks that encourage innovation and adoption, while others, like China, have taken a more hostile stance, banning cryptocurrency mining and trading activities. The lack of standardized regulations across countries creates an environment where investors face different levels of risk depending on their jurisdiction.

In recent years, however, there have been efforts to bring more clarity to the regulatory landscape. One notable example is the European Union's Markets in crypto-assets (MICA) regulation, which aims to create a comprehensive regulatory framework for cryptocurrencies. The MICA regulation seeks to address issues related to investor protection, anti-money laundering (AML), and market stability, and it is seen as a step toward creating a more transparent and secure environment for cryptocurrency investors. (European Parliament, 2022; European Commission, 2023; European Commission, 2024)

Despite these developments, regulatory uncertainty remains a significant challenge for the industry. Until more countries adopt clear and consistent regulations, the risks associated with investing in cryptocurrencies will likely persist. Investors must remain vigilant and stay informed about the evolving regulatory landscape in order to navigate the market effectively.

As noted above, the uncertainty surrounding cryptocurrency regulation stems from its evolving nature and the complexity of existing legal frameworks. This creates uncertainty for users, investors, and regulators, especially because cryptocurrencies like Bitcoin, Ethereum, and others do not fit neatly into traditional categories like commodities, securities, or currencies.

Different countries have different stances on cryptocurrency regulation. Some have embraced it, others have restricted or banned it.

2.2. Taxation Uncertainty

One of the biggest sources of regulatory ambiguity lies in how different countries treat crypto for tax purposes. This creates confusion for users, especially in countries with unclear guidelines. (Davenport, and Usrey, 2023)

A 2023 report from coinTracker found that over 50% of cryptocurrency holders in the U.S. are uncertain about their tax obligations, contributing to confusion during tax filing seasons. (Bhullar at al., 2025)

2.3. Regulatory Uncertainty for crypto Exchanges

The regulatory ambiguity surrounding crypto exchanges is a critical issue, especially since exchanges act as intermediaries between fiat currencies and cryptocurrencies.

e) Binance: Binance, one of the world's largest crypto exchanges, has faced regulatory scrutiny in multiple jurisdictions. In 2021, U.S. regulators began investigating Binance for compliance with anti-money laundering (AML) and know-your-customer (KYC) requirements. In 2023, Binance also faced legal actions from the SEC for allegedly listing unregistered securities. (The U.S. Department of Justice, 2023; The U.S. Department of the Treasury, 2023)
f) FTX collapse (2022): The downfall of FTX, once the third-largest crypto exchange, highlighted the dangers of regulatory gaps. It filed for bankruptcy, leaving investors with billions of dollars in losses. The collapse prompted U.S. lawmakers to call for clearer regulations for cryptocurrency exchanges. Following the FTX collapse, crypto exchange regulations in the U.S. gained more traction. The Biden administration proposed a new executive order in 2023 to regulate crypto more strictly, addressing risks posed by exchanges like FTX. (The U.S. Department of Justice, 2023; The U.S. Department of the Treasury, 2023)

2.4. Security Tokens and DeFi Regulation

DeFi (Decentralized Finance) and security tokens present another layer of ambiguity. While traditional financial systems have clear regulations regarding securities, cryptocurrencies blur the lines.

1. SEC's stance on DeFi: In 2023, the SEC issued warnings to several DeFi protocols for violating securities laws due to their offerings resembling traditional investment contracts without registration.
2. Security Token Offerings (STOs): STOs are regulated differently in various jurisdictions. In the U.S., an STO must comply with SEC regulations, which has led to a slowdown in their adoption. However, in countries like Switzerland, STOs are much more streamlined.
3. According to coinGecko, the DeFi market reached over $90 billion in total value locked (TVL) in 2023, underlining the importance of defining clear regulations for these financial products. (CoinGecko Team, 2023)

2.5. AML/KYC Compliance

Regulation around anti-Money Laundering (AML) and Know Your customer (KYC) policies is critical but often lacks global coordination.

- Global Disparity: cryptocurrencies are regulated very differently across the world. In the U.S., the SEC views some as securities, while other jurisdictions treat them as commodities or currencies.

- Tax confusion: Tax regulations are complex and varied. Some countries, like Japan, have ambiguous taxation systems, while others like the U.S. treat them as property for tax purposes.
- Exchange Scrutiny: High-profile exchanges like Binance and FTX highlight the need for clearer, more consistent regulations to protect investors.
- DeFi & Security Tokens: These emerging areas remain underregulated, with major concerns around investor protection and compliance. (The U.S. Government Accountability Office, 2023; The U.S. Department of Justice, 2023; The U.S. Department of the Treasury, 2023)

As the market continues to grow, the demand for clearer and more harmonized regulations is increasing. Governments are likely to continue adapting as the industry evolves.

2.6. Initial Coin Offerings (ICOs) and Token Sales

One of the key areas where regulation remains murky is Initial coin Offerings (ICOs). IcOs allow projects to raise capital by issuing new tokens to investors, often without clear regulatory oversight. This has led to a range of legal uncertainties about whether these offerings should be classified as securities, commodities, or something else entirely.

The case of Telegram's TON IcO (2018-2020): Telegram, a popular messaging app, attempted to raise funds through an IcO to launch its own Blockchain, the TON Blockchain. The U.S. SEC filed a lawsuit against Telegram in 2019, claiming that its IcO violated U.S. securities laws. Telegram eventually agreed to return $1.2 billion to investors and pay a $18.5 million fine. (The U.S. Securities and Exchange Commission, 2020)

2.7. Stablecoins and Central Bank Digital Currencies (CBDC's)

Stablecoins (Cryptocurrencies pegged to a stable asset like the U.S. dollar) introduce their own set of regulatory challenges. While stablecoins have grown rapidly, their regulation is still developing. Meanwhile, central bank Digital currencies (CBDCs) offer an alternative path that governments are exploring, but that too has raised questions on how to balance innovation with regulatory oversight.

Regulation of Stablecoins: Many countries have differing views on how to regulate stablecoins. The U.S. has proposed the Stablecoin TRUST act (2023), calling for more oversight, but the EU's MICA regulation is expected to be the first comprehensive framework for stablecoins across member states. (European Commission, 2024)

2.8. Global Coordination on Cryptocurrency Regulation

Because cryptocurrencies are inherently global, regulatory clarity is further complicated by the lack of international coordination. Cryptocurrencies move freely across borders, making it difficult for any one country to regulate them in isolation. Regulatory fragmentation can lead to regulatory arbitrage, where businesses move operations to jurisdictions with looser regulations, which can undermine the effectiveness of any single nation's rules.

Crypto Tax Havens: Jurisdictions like Malta, Switzerland, and Singapore have become popular destinations for cryptocurrency-related businesses due to their relatively lenient regulations. In contrast, countries like India and china have imposed outright bans on certain crypto activities.

Lack of General International coordination: There's a push from organizations like the Financial Action Task Force (FATF) to create global guidelines on cryptocurrency regulations, but nations vary widely in their willingness to cooperate. Without a global regulatory standard, enforcement becomes difficult, and businesses are left uncertain about where to operate.

(The U.S. Government Accountability Office, 2023; The U.S. Department of Justice, 2023; The U.S. Department of the Treasury, 2023; European Parliament, 2022; European Commission, 2023; European Commission, 2024)

2.9. DeFi (Decentralized Finance) Regulation

DeFi has surged in popularity, but its decentralized nature presents a unique regulatory challenge. DeFi platforms operate without centralized intermediaries, making it harder for regulators to impose controls over the system. This creates uncertainty about how to apply traditional financial regulations, such as securities laws, anti-money laundering (AML), and consumer protection laws.

The total value locked in DeFi decreased from $200 billion in April 2022 to $80 billion in July 2022, indicating that security in this area remained problematic. (Li at al., 2022)

As of 2025, the total value locked (TVL) in DeFi protocols was estimated at over $130 billion, showing the significance of this space. However, the lack of regulatory clarity on DeFi continues to hinder the development of clear legal frameworks. The decentralized nature of DeFi means there is no central authority to hold accountable. Regulatory bodies are still working out how to apply existing laws to a decentralized environment where there is no "company" or individual to regulate. (Atqa, 2025; DeFi Llama, 2025)

2.10. Cybersecurity Breaches and Hacks

One of the most significant risks in the cryptocurrency space is cybersecurity breaches and hacking incidents. The decentralized nature of Blockchain networks can make it more difficult to implement traditional security measures that are common in centralized financial systems. As cryptocurrencies become more valuable, they attract the attention of hackers seeking to exploit vulnerabilities in exchange platforms, decentralized finance (DeFi) protocols, and individual wallets.

In 2022 alone, losses from cryptocurrency-related hacks exceeded $3.8 billion, underscoring the growing problem of cybercrime within the sector. DeFi platforms, in particular, have become major targets for hackers, largely due to their reliance on smart contracts and the complexity of their ecosystems. One of the most high-profile hacks occurred in March 2022, when hackers breached the Ronin Network, resulting in the theft of over $625 million. The Ronin Network, which is tied to the popular play-to-earn game Axie Infinity, was targeted after hackers exploited a vulnerability. (Carpentier-Desjardins at al., 2023; Chainalysis Team, 2023; CNBC, 2021)

Figure 7. Total Value Stolen in Crypto Hacks and number of Hacks between 2016 and 2022,

(Chainalysis Team, 2025; Chainalysis Team, 2022)

2.11. Security and Hack Concerns: Exchange and Wallet Vulnerabilities

Centralized exchanges, which provide a platform for buying, selling, and trading cryptocurrencies, have become an essential part of the cryptocurrency ecosystem. However, they are also a primary target for cybercriminals due to their large user bases and substantial holdings of digital assets. While exchanges have implemented various security measures, they remain susceptible to breaches, often leading to significant financial losses for users. (Baldwin, 2021)

Figure 8. Global cryptocurrency market size

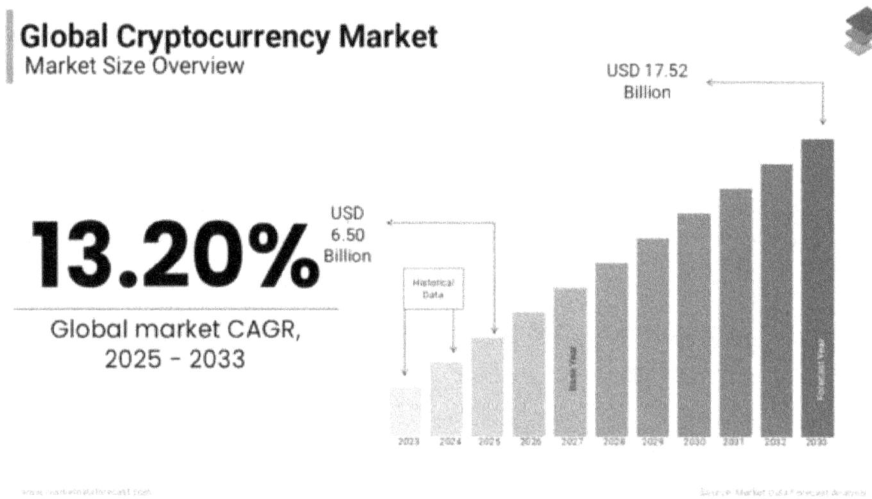

(Market Data Forecast Team, 2025)

Similarly, wallet providers, which offer storage solutions for cryptocurrencies, have also experienced breaches and attacks. Hardware wallets, while offering more secure storage options than software wallets, are not entirely immune to hacking attempts. In fact, phishing scams targeting users of popular wallet providers have become increasingly common.

The vulnerability of exchanges and wallet providers highlights the need for stronger security measures across the entire cryptocurrency ecosystem.

While Blockchain itself is secure, exchanges and wallets remain vulnerable to hacks. Notable incidents: (Sal and Moreland, 2025; George, 2024; Crystal Investigations Team, 2021)

Figure 9. The biggest exchange or network hack attacks,

№	Exchange or Network Name	Year	Amount of Attack
1	Cryptsy Hack	2013	$5 million
2	Mt. Gox Hack	2014	$300 million
3	Bitfinex Hack	2016	$65 million
4	ZAIF Hack	2018	$60 million
5	BANCOR Hack	2018	$23.5 million
6	Coincheck Hack	2018	$534 million
7	COINBENE Hack	2019	$105 million
8	Upbit Hack	2019	$45 million
9	BINANCE Hack	2019	$40 million
10	CRYPTOPIA Hack	2019	$15.5 million
11	KuCoin Hack	2020	$204 million
12	Poly Network Hack	2021	$600 million
13	Ronin Network Hack	2022	$600 million
14	DMM Bitcoin Hack	2024	$308 million
15	ByBit Hack	2025	$1.4 billion

(By Author, 2025; Sal and Moreland, 2025; George, 2024; Crystal Investigations Team, 2021)

Above, we've outlined some of the biggest security breaches in history of cryptocurrency market, the scale of these attacks is quite remarkable.

2.12. Ponzi Schemes and Rug Pulls

Ponzi schemes and rug pulls have become alarmingly common in the cryptocurrency space. These fraudulent schemes often involve project creators who attract investors with promises of high returns, only to disappear with investors' funds once they have accumulated enough capital. The unregulated nature of the cryptocurrency market makes it easier for bad actors to create projects that seem legitimate but are ultimately designed to defraud unsuspecting participants.

A particularly notable case occurred in 2022 with the "Frosties" NFT project. The creators of Frosties reportedly raised over $1.11 million from investors by selling non-fungible tokens (NFTs) that promised high returns and rewards. However, shortly after the sale, the project's creators disappeared, and the funds were stolen. This incident is a stark reminder of the risks inherent in investing in new and unproven cryptocurrency projects. Ponzi schemes and rug pulls exploit the lack of regulation in the cryptocurrency space, allowing creators to bypass the investor protections found in traditional finance. (Akartuna at al., 2022)

2.13. Phishing Attacks and Social Engineering

Phishing attacks and social engineering tactics are becoming increasingly prevalent in the cryptocurrency market. These attacks often involve hackers or fraudsters posing as legitimate companies or services in order to deceive users into revealing sensitive information, such as private keys or login credentials. Phishing scams targeting cryptocurrency holders can result in the loss of funds and personal information, making them a significant risk for investors.

2.14. Consumer Protection and Fraud Risk

The lack of clear regulations leaves consumers vulnerable to fraud, scams, and high levels of volatility. Cryptocurrencies have been linked to a high number of fraudulent schemes, from Ponzi schemes to pump-and-dump scams.

PlusToken Scam: One of the largest crypto Ponzi schemes, PlusToken, defrauded investors of approximately $2.9 billion in 2020. The perpetrators were able to exploit the regulatory gray area to target unsuspecting consumers, illustrating the risk of fraud in the crypto space. (Tsiattalou, 2021)

Generally

Chainalysis reported that $7.7 billion worth of cryptocurrency was stolen through hacks and fraud in 2022 alone, underscoring the significant consumer protection issues within the industry. Rapid growth in the use of decentralised finance (DeFi) services is creating a new soft spot for global financial systems, fostering new methods of cryptocrime for cybercriminals whose “rug pulls” and other attacks will, Cybersecurity Ventures predicts, cost the world $30 billion in 2025 alone. (Charoenwong, and Bernardi, 2021; Braue, 2022)

To truly foster a safe and innovative crypto environment, regulators need to create clear, consistent guidelines that can work internationally while balancing the need for consumer protection, financial stability, and innovation. With the growing role of digital assets in the global economy, regulatory clarity will be essential for ensuring the sustainable growth of the crypto ecosystem.

Figure 10. Hack attacks by types between 2015 and 2022

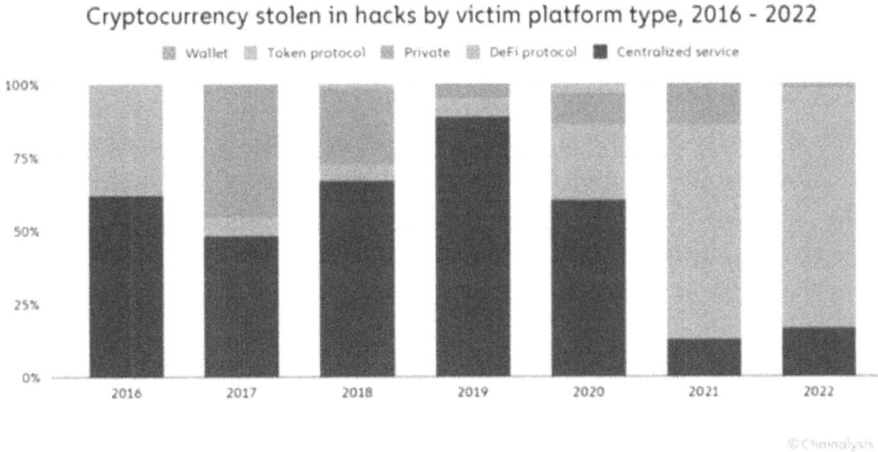

(Chainalysis, 2025)

3. MAJOR CRYPTOCURRENCIES: BEYOND BITCOIN AND ETHEREUM

While Bitcoin and Ethereum are the most widely recognized cryptocurrencies, the broader digital asset ecosystem is rich with numerous other cryptocurrencies that each serve unique purposes. As an investor, understanding the differences between these cryptocurrencies can be essential for creating a diversified portfolio and capitalizing on different use cases within the Blockchain ecosystem.

3.1. Bitcoin: The Pioneer and Digital Gold

Bitcoin, often referred to as "digital gold," remains the largest and most influential cryptocurrency by market capitalization. It was the first cryptocurrency, introduced in 2009, and continues to dominate the space both in terms of recognition and investment. Bitcoin's primary use case is as a decentralized store of value and medium of exchange. The cryptographic security of the Bitcoin network, coupled with its fixed supply cap of 21 million coins, has led many to compare Bitcoin to gold. This comparison is particularly relevant in the context of inflation hedging.

Bitcoin has demonstrated its potential as a store of value, particularly during periods of economic uncertainty. As central banks around the world have expanded monetary policy and printed large amounts of fiat currency, Bitcoin's deflationary nature has attracted increasing attention as a hedge against inflation. Unlike traditional fiat currencies that can be printed at will by central banks, Bitcoin's supply is capped, which means that its inflation rate decreases over time. As such, it is seen by many as a long-term store of value, similar to precious metals like gold.

However, Bitcoin's volatility remains a challenge. Its price is heavily influenced by market sentiment, technological advancements, regulatory news, and even public figures. This volatility has deterred some traditional investors who view Bitcoin as too risky. Nevertheless, Bitcoin's performance has outpaced traditional asset classes, and it is now considered by many to be an essential part of a diversified portfolio. (Nakamoto, 2008)

3.2. Ethereum: The Smart contract and DeFi Leader

Ethereum introduced a groundbreaking innovation: the ability to execute smart contracts. Smart contracts are self-executing contracts with the terms of the agreement directly written into lines of code. This capability allows for the creation of decentralized applications (DApps), which run on the Ethereum Blockchain without the need for central authority or intermediaries. For instance, smart contracts have been used to implement crowdfunding initiatives that raised a total of US$6.2 billion from January to June of 2018. (Buterin, 2013; Oliva, 2020)

Figure 11. CeFi vs DeFi and smart contracts scheme

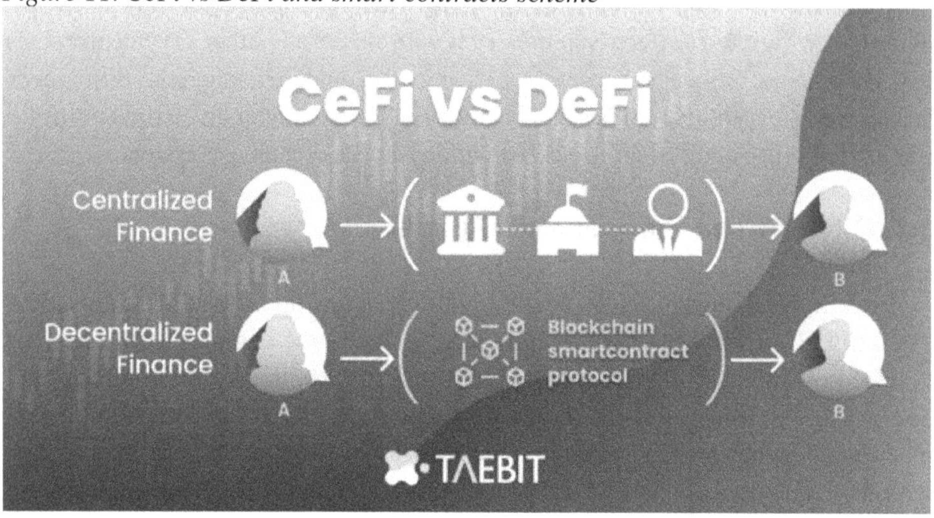

(Taebit DEX, 2022)

Ethereum has become a pivotal asset for DeFi protocols, which aim to offer financial services without traditional intermediaries like banks or brokers. The total value locked (TVL) in DeFi platforms was slightly exceeding $175 billion in 2022, with a large portion of those assets being held in Ethereum-based smart contracts. DeFi offers opportunities for investors to earn yield through activities such as staking, lending, and liquidity provision. Ethereum's role in this sector has cemented its place as one of the most important cryptocurrencies for both retail and institutional investors. (DeFi Llama, 2025; Li at al., 2022)

3.3. Other Prominent Cryptocurrencies

In addition to Bitcoin and Ethereum, there are several other cryptocurrencies that have gained significant traction in the market, each catering to different needs and use cases. Some of the most notable include:

- **Binance** coin (BNB): Initially launched as a utility token for discounted trading fees on Binance, one of the world's largest cryptocurrency exchanges. (Binance, 2023)
- **Cardano** (ADA): Cardano is a PoS Blockchain platform founded by Charles Hoskinson, co-founder of Ethereum. (Hoskinson, 2023)
- **Polkadot** (DOT): Polkadot is a multichain Blockchain platform designed to enable different Blockchains to interoperate. (Wood, 2023)
- **Solana** (SOL): Solana is a high-performance Blockchain designed for decentralized applications and crypto projects. (Solana Labs Team, 2023)

Each of these cryptocurrencies has unique features and capabilities that contribute to their growth and adoption. As an investor, diversifying across multiple cryptocurrencies can help mitigate the risks associated with individual assets while offering exposure to different Blockchain innovations.

CONCLUSION AND EVALUATION

Cryptocurrencies continue to reshape the financial landscape, their impact on global markets, investment strategies, and economic structures becomes increasingly evident. Despite inherent risks, including market volatility, regulatory ambiguity, and cybersecurity threats, digital assets have demonstrated their resilience, securing a place within the broader financial ecosystem. The adoption of Blockchain technology, which underpins cryptocurrencies, has also expanded beyond financial applications, influencing industries such as supply chain management, healthcare, and real estate, energy.

AND as a result, cryptocurrencies have mostly lost their monetary characteristics (as a means of payment) and are considered more as investment instruments. As such, their investment mobility and all other relevant aspects have not yet been fully analyzed, and the research literature in comparison with current investment instruments has not yet reached sufficient depth. Studies on its orientation have not reached a level to clearly reveal the general structure.

However, it can be said that it will firmly maintain its presence in today's economy. Cryptocurrencies, unfortunately, do not fully qualify as money in their current use, which can be attributed to their extremely unstable market value, which can constantly rise and fall. In addition, it can be characterized as a high-risk investment instrument as it contains various artificial speculation bubbles that can be encountered in the investment market.

On the other hand, despite all the mentioned risks, a considerable number of countries in the world economy are publicizing their national Cryptocurrencies projects, so it can be said that new regulations and rules will be brought to the markets in the near future.

This technology has not yet reached the limit of its potential and prevalence, and according to some experts, it is still in the development stage, but if we make an assessment in the light of the data considered, it can be said that positive development potentials are more dominant than negative developments.

Regulatory clarity remains a pivotal factor in shaping the future of cryptocurrencies. In the United States, the Securities and Exchange commission (SEC) has increased its scrutiny of cryptocurrency firms, emphasizing the need for investor protection while acknowledging the innovation potential of Blockchain technology. These regulatory developments indicate a maturing industry, where compliance and investor security will play a crucial role in determining long-term adoption.

The market capitalization of cryptocurrencies has also witnessed significant fluctuations, reflecting both the speculative nature of digital assets and their increasing mainstream acceptance. Bitcoin's dominance remains strong, accounting for approximately 50% of the total cryptocurrency market, which surpassed $2.5 trillion in late 2024. Ethereum, the second-largest cryptocurrency, continues to evolve through its transition to a Proof-of-Stake consensus mechanism, reducing energy consumption by over 99% compared to its previous proof-of-work model. The rise of layer-2 scaling solutions, such as the Lightning Network and Optimistic Rollups, further enhances transaction efficiency and lowers fees, making cryptocurrencies more viable for everyday use.

One of the most promising aspects of cryptocurrencies is their role as a hedge against inflation. In countries experiencing currency devaluation, such as Argentina and Turkey, Bitcoin adoption has surged, with local trading volumes increasing by over 300% in the past two years. Additionally, central bank digital currencies (CBDCs) have gained traction, with China leading the way through its digital yuan initiative, while the European central bank and the Federal Reserve explore potential implementations of digital versions of the euro and the US dollar. These developments illustrate how Blockchain technology is influencing monetary policies on a global scale.

Despite these advancements, risks remain. The collapse of major cryptocurrency firms, such as FTX in 2022, highlighted vulnerabilities in centralized exchanges, leading to calls for greater transparency and the adoption of decentralized finance (DeFi) solutions.

As of February 1, 2025, the cryptocurrency market has experienced remarkable growth, solidifying its position as a significant component of the global financial ecosystem. The total market capitalization of cryptocurrencies has surged to approximately $5 trillion, reflecting increased adoption and investment across various sectors. This expansion underscores the transformative impact of digital assets on modern finance.

Bitcoin (BTC), the pioneering cryptocurrency, continues to lead the market with the current approximate price of $90,000, despite a slight decrease from the 2025 middle of January historical record price. Its market capitalization now exceeds $2 trillion, accounting for a substantial portion of the overall crypto market.

The increasing involvement of institutional investors has been a pivotal factor in the market's expansion. companies such as MicroStrategy Inc. (MSTR) have made significant investments in Bitcoin, with MicroStrategy's holdings exceeding 150,000 BTC, currently valued at over $16 billion.

Projections for the cryptocurrency market generally remain optimistic views between holders and financial specialists. A report by Bernstein predicts that the total crypto market capitalization could reach $7.5 trillion by the end of 2025, driven by unprecedented institutional adoption and the maturation of Blockchain technologies. The report forecasts Bitcoin's market cap to rise to $3 trillion, Ethereum's to $1.8 trillion, and other leading Blockchains, such as Solana and avalanche, to collectively reach $1.4 trillion.

Looking ahead, the future of cryptocurrencies is poised for continued evolution. The integration of artificial intelligence (AI) with Blockchain, the development of tokenized real-world assets, and advancements in privacy-preserving technologies, such as zero-knowledge proofs, indicate a broader application of digital assets beyond speculative trading. As institutional adoption increases and regulatory frameworks mature, the volatility associated with cryptocurrencies may stabilize, fostering greater confidence in their use as long-term investment instruments and payment solutions.

While cryptocurrencies and the crypto market remain a dynamic and sometimes unpredictable young asset class, their fundamental advantages – decentralization, transparency, and borderless transactions – continue to drive their adoption around the world. As technological advances, regulatory clarity, and institutional participation shape the future of digital finance, cryptocurrencies have the potential to become integral components of the global economic system, redefining traditional financial structures and offering innovative solutions for the digital age.

It is also worth noting that cryptocurrencies have firmly established themselves as a transformative force in modern financial markets. Significant growth in market capitalization, increased institutional participation, and evolving regulatory frameworks all point to a future in which digital crypto assets will play an integral role in the global economy.

However, professionals and investors should remain vigilant in this industry, recognizing the hiper volatility inherent in this rapidly evolving market and the uncertainty surrounding current regulation in general.

REFERENCES

Akartuna, E. A., Nadini, M., DePow, C., & Annison, T. (2022). *NFTs and Financial Crime*.https://www.elliptic.co/hubfs/NFT%20Report%202022%20-full.pdf

Arner, D., Auer, R., & Frost, J. (2020). Stablecoins: Risks, potential and regulation (BIS Working Papers No. 905). Monetary and Economic Department. https://www.bis.org/publ/work905.pdf

Atqa, A. (2025). DeFi TVL Falls Below $95 Billion, Down 30% From December Levels. https://coinmarketcap.com/academy/article/defi-tvl-falls-below-dollar95-billion-down-30percent-from-december-levels

Awati, R. *(2024). What is Consensus Algorithm*.https://www.techtarget.com/whatis/definition/consensus-algorithm

Baldwin, C. (2021). Bitcoin worth $72 million stolen from Bitfinex exchange in Hong Kong. https://www.reuters.com/article/us-bitfinex-hacked-hongkong-idUSKCN10E0KP

Berwick, A. (2025). Bitcoin Price Rises Toward $80,000 in Crypto Rebound, The Wall Street Journal. https://www.wsj.com/livecoverage/stock-market-trump-tariffs-trade-war-04-08-25/card/bitcoin-price-rises-toward-80-000-in-crypto-rebound-NZ4M4z1zV52w5TalvNfl?utm_source=chatgpt.com

Bhullar, P., Joshi, M., Sharma, S., Phan, D., & Nguyen, A. (2025). Global taxation issues in cryptocurrencies – a synthesis of literature and future research agenda. *Journal of Accounting Literature*. Advance online publication. DOI: 10.1108/JAL-10-2024-0314

Binance, (2023). About Binance, BNB Whitepaper. https://www.binance.com/en/about

BitPesa. (2024). *Low-Cost Cross-Border Payments in Africa*. https://www.bitpesa.co/

Bitsgap Team. (2024). Consensus Technologies in Blockchain: PoW, PoS, PoA, DPoS, PoC, PoB, and Others. https://www.bitsgap.com/blog/consensus-technologies-in-blockchain-pow-pos-poa-dpos-poc-pob-and-others/

Braue, D. (2022). Cryptocrime To Cost The World $30 Billion Annually By 2025. https://cybersecurityventures.com/cryptocrime-to-cost-the-world-30-billion-annually-by-2025

Brown, A. (2021). Beeple NFT Sells For $69.3 Million, Becoming Most-Expensive Ever, Forbes. https://www.forbes.com/sites/abrambrown/2021/03/11/beeple-art-sells-for-693-million-becoming-most-expensive-nft-ever/

Buterin, V. (2013). Ethereum Whitepaper, Ethereum Foundation. https://ethereum.org/en/whitepap

Carpentier-Desjardins, C., Paquet-Clouston, M., Kitzler, S., & Haslhofer, B. (2023). Mapping the DeFi Crime Landscape: An Evidence-based Picture. ArXiv, abs/2310.04356. https://doi.org/.DOI: 10.1093/cybsec/tyae029

Carter, S. (2025, February 21). Latest on the Bybit record breaking 1.4 billion dollar crypto hack. Forbes. https://www.forbes.com/sites/digital-assets/2025/02/21/latest-on-the-bybit-record-breaking-14-billion-dollar-crypto-hack/

Chainalysis Team. *(2021). Poly Network Attacker Returning Funds After Pulling Off Biggest DeFi Theft Ever.*https://www.chainalysis.com/blog/poly-network-hack-august-2021/

Chainalysis Team. *(2022). Crypto Crime Report.*https://go.chainalysis.com/rs/503-FAP-074/images/Crypto-Crime-Report-2022.pdf

Chainalysis Team. (2023). 2022 Biggest Year Ever For Crypto Hacking with $3.8 Billion Stolen. https://www.chainalysis.com/blog/2022-biggest-year-ever-for-crypto-hacking/

Chainalysis Team. *(2023). Cryptocurrency Adoption in Emerging Markets.*https://www.chainalysis.com/blog/2024-global-crypto-adoption-index/

Chainalysis Team. (2025). 2025 Crypto Crime Trends. https://www.chainalysis.com/blog/2025-crypto-crime-report-introduction/

Chaleenutthawut, Y., Davydov, V., Evdokimov, M., Kasemsuk, S., Kruglik, S., Melnikov, G., & Yanovich, Y. (2024). Loan Portfolio Dataset From MakerDAO Blockchain Project. *IEEE Access : Practical Innovations, Open Solutions, 12*, 24843–24854. DOI: 10.1109/ACCESS.2024.3363225

Charoenwong, B., & Bernardi, M. *(2021).A Decade of Cryptocurrency 'Hacks': 2011 –2021. PSN: Exchange Rates & Currency (International) (Topic). https://doi.org/*DOI: 10.2139/ssrn.3944435

Chen, Y., Xu, K., & Yang, J. (2024). Market impact of the bitcoin ETF introduction on bitcoin futures. *International Review of Financial Analysis*. Advance online publication. DOI: 10.1016/j.irfa.2024.103810

CNBC. (2021). Cryptocurrency theft: Hackers steal $600 million in Poly Network hack. https://www.cnbc.com/2021/08/11/cryptocurrency-theft-hackers-steal-600-million-in-poly-network-hack.html

CoinGecko Team. (2023). DeFi Market Report 2023. https://www.coingecko.com/research/publications/2023-annual-crypto-report

CoinGecko Team. (2025). 2024 Annual Crypto Industry Report. https://coingecko.com/research/publications/2024-annual-crypto-report?utm_campaign=nil&utm_source=nil&utm_medium=nil

CoinGecko Team. *(2025). Bitcoin historical price data. CoinGecko.*https://www.coingecko.com/en/bitcoin

CoinMarketCap. *(2021). Cryptocurrency market capitalizations.*https://coinmarketcap.com

CoinMarketCap. *(2021). Exchanges ranking.*https://coinmarketcap.com/rankings/exchanges/

CoinMarketCap Team. (2021). Ethereum: Top 10 biggest projects. CoinMarketCap. https://coinmarketcap.com/academy/article/ethereum-top-10-biggest-projects

Cointelegraph Team. (2023). DeFi TVL reaches $100B as Bitcoin pumps sentiment. https://cointelegraph.com/news/defi-tvl-reaches-100b-bitcoin-pumps-sentiment

Cole, J. (2024). Staking in crypto: How PoS reduces energy consumption. BlockApps. https://blockapps.net/blog/staking-in-crypto-how-pos-reduces-energy-consumption/

Comparitech. (2021). The biggest cryptocurrency heists of all time. https://www.comparitech.com/crypto/biggest-cryptocurrency-heists/

Cong, L., Hui, X., Tucker, C., & Zhou, L. (2023). Scaling Smart Contracts via Layer-2 Technologies: Some Empirical Evidence. *Management Science*, *69*(12), 7306–7316. DOI: 10.1287/mnsc.2023.00281

Criddle, C. (2021). Bitcoin consumes 'more electricity than Argentina. https://www.bbc.com/news/technology-56012952

Crypto, R. B. C. *(2021). The largest hacks of crypto exchanges. RBC.*https://www.rbc.ru/crypto/news/611292c69a79472763747b32

Crystal Investigations Team. (2021). The 10 Biggest Crypto Hacks in History, Investigations. https://crystalintelligence.com/investigations/the-10-biggest-crypto-hacks-in-history/

Davenport, S., & Usrey, S. (2023). Crypto Assets: Examining Possible Tax Classifications. *Journal of Emerging Technologies in Accounting*, *20*(2), 55–70. Advance online publication. DOI: 10.2308/JETA-2022-033

De Best, R. *(2025). Ethereum price history. Statista.*https://www.statista.com/statistics/806453/price-of-ethereum/

DeFi Llama Team. *(2025). Total Value Locked in DeFi.*https://defillama.com/

Demidova, A. (2021). Paradise Stones: 1,500-Year-Old Bitcoin from Yap Island. https://bits.media/kamni-rai-bitkoin-s-ostrova-yap-vozrastom-v-1-500-let

Erdaş, M., & Caglar, A. (2018). Analysis of the relationships between Bitcoin and exchange rate, commodities and global indexes by asymmetric causality test. Eastern Journal of European Studies, *27-45.*

European Commission. (2023). MiCA Regulation Framework, EU Official Journal. https://cur-lcx.europa.eu/search.html?

European Commission. (2024). *MiCA Regulation Text.*https://www.eur-lex.europa.eu

European Parliament. (2022). Markets in crypto-assets (MiCA), EPRS, 2022, https://www.europarl.europa.eu/RegData/etudes/BRIE/2022/739221/EPRS_BRI(2022)739221_EN.pdf

Federal Reserve Economic Data (FRED). (2025). Global Monetary Policies and Inflation Trends, St. Louis Fed. https://fred.stlouisfed.org/

Fidelity Digital Assets. *(2024). Bitcoin's correlation with traditional markets. Fidelity.*https://www.fidelity.com/

Fitzpatrick, S. (2001). Archaeological investigation of Omis Cave: A Yapese stone money quarry in Palau. *Archaeology in Oceania*, *36*(3), 153–162. DOI: 10.1002/j.1834-4453.2001.tb00489.x

George, K. (2024). Investopedia: The Largest Cryptocurrency Hacks So Far. https://www.investopedia.com/news/largest-cryptocurrency-hacks-so-far-year/

Gokhan, E. (2025). Bitcoin hits historic high of $109,000. https://aa.com.tr/en/economy/bitcoin-hits-historic-high-of-109-000/3455945

Grachev, G. *(2021). Cryptocurrency investment options. In Equite SpQ Investment Expert, 79–88. Moscow Chamber of Commerce and Industry.*https://mostpp.ru/news/tsifrovaya-ekonomika/kriptovalyuty-neizbezhnoe-budushchee

Grasic, A., & Vidnjevic, M. (2024). Evolution Of Cryptocurrencies And Their Utilization In The Digital Economy. *MEST Journal. https://doi.org/.*DOI: 10.12709/mest.12.12.02.05

Hayes, A., & Khartit, K. I. (2022). investopedia, What Happens to Bitcoin After All 21 Million Are Mined? https://www.investopedia.com/tech/what-happens-bitcoin-after-21-million-mined/

Hoskinson, C. (2023). Cardano Whitepaper, Cardano Foundation, 2023, https://cardano.org/

Kharpal, A., & Browne, R. (2021, August 11). Hackers return nearly half of the $600 million they stole in one of the biggest crypto heists. CNBC. https://www.cnbc.com/2021/08/11/cryptocurrency-theft-hackers-steal-600-million-in-poly-network-hack.html

Knight, O., & Reback, S. (2023, December 5). DeFi market rebounds to $50B as speculators hunt for yield. CoinDesk. https://www.coindesk.com/business/2023/12/05/defi-market-rebounds-to-50b-as-speculators-hunt-for-yield

Kraken Learn Team. (2025). Bitcoin price history: Timeline of its evolution [2009–2025]. Kraken. https://www.kraken.com/learn/bitcoin-price-history

Kubinec, J. (2023). 2023's top 5 DeFi protocols by revenue: TVL has begun to fall out of favor, so which metric can be used to rank this year's top DeFi projects? Blockworks. https://blockworks.co/news/top-defi-protocols-by-revenue?utm_source=chatgpt.com

Levy, A. *(2023).* Bitcoin's 2023 rally drove some of the stock market's biggest gains this year.https://cnbc.com/2023/12/27/bitcoin-2023-rally-pumped-up-marathon-coinbase-microstrategy-gbtc.html

Lewis, M. (2024). Asset Tokenization: Unlocking New Possibilities for the Enterprise. https://www.kaleido.io/blockchain-blog/guide-to-tokenization

Li, W., Bu, J., Li, X., Peng, H., Niu, Y., & Chen, X. (2022). A Survey of DeFi Security: Challenges and Opportunities. ArXiv, abs/2206.11821. https://doi.org/.DOI: 10.1016/j.jksuci.2022.10.028

Lim, T. *(2022).* Predictive Crypto-Asset Automated Market Making Architecture for Decentralized Finance using Deep Reinforcement Learning. ArXiv, abs/2211.01346. https://doi.org//arXiv.*2211.01346.*DOI: 10.48550

Maiello, M. *(2021).* Researchers turn stones to find ancient origin of bitcoin. University of Oregon. https://around.uoregon.edu/content/researchers-turn-stones-find-ancient-origin-bitcoin

Makarov, I., & Schoar, A. (2022). Cryptocurrencies and Decentralized Finance (DeFi). *Brookings Papers on Economic Activity*, *2022*(1), 141–215. DOI: 10.1353/eca.2022.0014

Market Data Forecast Team. (2025). Global Cryptocurrency Market Size, Share, Trends & Growth Forecast Report Segmented By Component (Hardware and Software), Process, Type, and End Use, and Region (North America, Europe, Asia Pacific, Latin America, and Middle East & Africa), Industry Analysis From 2024 to 2033. https://www.marketdataforecast.com/market-reports/cryptocurrency-market

Myfin Team. *(2025). Bitcoin price chart throughout history.*https://myfin.by/crypto-rates/chart-bitcoin

Nakamoto, S. (2008). Bitcoin: A Peer-to-Peer Electronic Cash System, Bitcoin Whitepaper. https://bitcoin.org/bitcoin.pdf

Narayanan, A. (2016). *Bitcoin and cryptocurrency technologies*. Princeton University Press., https://www.lightspark.com/learn/lightning/how-does-the-bitcoin-lightning-network-work

News, T. R. T. (2021). Thodex founder investigated for $2 billion haul. https://www.trthaber.com/haber/ekonomi/thodexin kurucusu-2-milyar-dolarlik-vurgun-sorusturmasi-574973.html

Oku, B. *(2025). Why is Scalability a Big Debate in Crypto?*https://blog.ueex.com/scalability-in-crypto/

Oliva, G., Hassan, A., & Jiang, Z. (2020). An exploratory study of smart contracts in the Ethereum blockchain platform. *Empirical Software Engineering*, *25*(3), 1864–1904. DOI: 10.1007/s10664-019-09796-5

Pogosyan, A. M. (2021). Cryptocurrencies – the inevitable future. Moscow Chamber of Commerce and Industry. https://mostpp.ru/news/tsifrovaya-ekonomika/kriptovalyuty-neizbezhnoe-budushchee/

Poon, J., & Buterin, V. (2017). Plasma: Scalable autonomous smart contracts. Ethereum Foundation. https://plasma.io/plasma.pdf

Riley, C. (2024). Ethereum 2.0 Upgrades: Boosting Scalability, Security, and Sustainability, https://www.openmarketcap.com/ethereum-20-upgrades/?utm_source=chatgpt.com

Riley, C. (2024). Ethereum Smart Contracts: Revolutionizing Industries with Automation and Transparency. https://www.openmarketcap.com/ethereum-smart-contracts-2/

Rolfe, A. *(2015). Payment Card Tokenization.* https://www.paymentscardsandmobile.com/payment-card-tokenization/

Royal, J., & Barba, M. (2024). Cryptocurrency statistics 2024: Investing in crypto. Bankrate. https://www.bankrate.com/investing/cryptocurrency-statistics/

Sal, M., & Moreland, K. *(2025). Biggest Crypto Hacks in History.* https://www.webopedia.com/crypto/learn/biggest-crypto-hacks/

Sharma, R., Chavarria, A., & Ma, J. *(2024). What is decentralized finance? Investopedia.* https://www.investopedia.com/decentralized-finance-defi-5113835?utm_source=chatgpt.com

Sigalos, M. (2021, December 5). BitMart says it will compensate victims of $196 million hack and restore trading by Tuesday. CNBC. https://www.cnbc.com/2021/12/05/hackers-take-196-million-from-crypto-exchange-bitmart-in-large-breach.html

Solana Labs Team. (2023). Solana Whitepaper, Solana. https://solana.com/

SoluLab Team. *(2021). Ethereum and smart contracts: A symbiotic relationship.* https://www.solulab.com/ethereum-and-smart-contracts/

Songur, M. *(2019). Bitcoin market bubbles: A generalized SADF test. Anemon Muş Alparslan Üniversitesi Sosyal Bilimler Dergisi, 7(2), 459–466.* https://dergipark.org.tr/tr/pub/anemon/issue/50616/506466

S&P Global. (2024). Cryptocurrency Volatility vs. S&P 500 Stocks, S&P Dow Jones Indices. https://www.spglobal.com/spdji/

Taebit, D. E. X. *(2022). DeFi vs. CeFi: Which financial service is more suitable for you?* https://medium.com/coinmonks/defi-vs-cefi-which-financial-service-is-more-suitable-for-you-ca38af7609d9

Tapscott, D., & Tapscott, A. (2016). Blockchain Revolution: How the Technology Behind Bitcoin is Changing Money, Business and the World. *Penguin Random House, LLC*(1), 22–27.

Team, I. B. M. (2023). IBM Food Trust: Blockchain for Food Supply Chain Transparency. https://www.ibm.com/blockchain/solutions/food-trust

Team, I. B. M. (2024). IBM Food Trust overview, IBM. https://www.ibm.com/docs/en/food-trust?topic=overview

Tesla Investor Relations Team. (2021). *Tesla's bitcoin investment announcement.* https://ir.tesla.com/

The Cambridge Centre for Alternative Finance. (2025). *Bitcoin network power demand.*https://ccaf.io/cbnsi/cbeci?utm_source=chatgpt.com

The Guardian. (2021, February 8). Tesla buys $1.5bn in bitcoin, pushing price to new high. https://www.theguardian.com/technology/2021/feb/08/tesla-bitcoin-price-new-high-elon-musk-dogecoin

The Investopedia Team. (2023). *What was Mt. Gox? Definition, history, collapse, and future (Reviewed by E. Rasure, Fact checked by P. Rathburn). Investopedia.* https://www.investopedia.com/terms/m/mt-gox.asp

The U.S. Department of Justice. (2023). *Binance Settlement.*https://justice.gov

The U.S. Department of the Treasury. (2023). U.S. Treasury Announces Largest Settlements in History with World's Largest Virtual Currency Exchange Binance for Violations of U.S. Anti-Money Laundering and Sanctions Laws. https://home.treasury.gov/news/press-releases/jy1925

The U.S. Government Accountability Office. *(2023). Crypto Tax Compliance Challenges.*https://gao.gov

The U.S. Securities and Exchange Commission. (2020). Telegram to Return $1.2 Billion to Investors and Pay $18.5 Million Penalty to Settle SEC Charges. https://www.sec.gov/newsroom/press-releases/2020-146

The U.S. Securities and Exchange Commission. *(2021). SEC Approval of Bitcoin ETFs.*https://www.sec.gov/

The World Bank Group. *(2021). Global Findex Database 2021 survey headline findings on account ownership.*https://www.worldbank.org/en/publication/globalfindex/brief/the-global-findex-database-2021-chapter-1-ownership-of-accounts

The World Bank Group. (2021). The Global Findex Database 2021. https://www.worldbank.org/en/publication/globalfindex

The World Bank Group. *(2024). Remittance Prices Worldwide.*https://remittanceprices.worldbank.org/sites/default/files/rpw_main_report_and_annex_q324.pdf

The World Bank Team. (2025). Financial Inclusion, Financial inclusion is a key enabler to reducing poverty and boosting prosperity. https://worldbank.org/en/topic/financialinclusion/overview

Tselischev, P. B. *(2017). SWOT analysis of cryptocurrency as an innovative financial instrument. Nauka i obrazovanie.*http://naukovedenie.ru/PDF/34EVN417.pdf

Tsiattalou, B. *(2021). Bambos Tsiattalou examines whether cryptocurrency regulation goes far enough to mitigate white-collar crime.*https://www.stokoepartnership.com/bambos-tsiattalou-examines-whether-cryptocurrency-regulation-goes-far-enough-to-mitigate-white-collar-crime/

Tsihitas, T. *(2021). Worldwide cryptocurrency heists. Comparitech.*https://www.comparitech.com/crypto/biggest-cryptocurrency-heists/

Ullah, F., & Al-Turjman, F. (2021). A conceptual framework for blockchain smart contract adoption to manage real estate deals in smart cities. *Neural Computing & Applications*, *35*(7), 5033–5054. DOI: 10.1007/s00521-021-05800-6

Urusov, A., Berezovskiy, R., & Yanovich, Y. (2024). Backtesting Framework for Concentrated Liquidity Market Makers on Uniswap V3 Decentralized Exchange. Blockchain: Research and Applications. https://doi.org/.DOI: 10.1016/j.bcra.2024.100256

Vardai, Z. (2024). Ledger users targeted by malicious 'clear signing' phishing email. Cointelegraph. https://cointelegraph.com/news/ledger-users-targeted-malicious-clear-signing-phishing-email

Velasquez, F. (2023, February 27). Crypto crime hit all-time high of $20.6B in 2022: Chainalysis. CoinDesk. https://www.coindesk.com/business/2023/02/27/crypto-crime-hit-all-time-high-of-206b-in-2022-chainalysis

Wainwright, Z. (2024). A closer look at bitcoin's volatility. Fidelity Digital Assets. https://www.fidelitydigitalassets.com/research-and-insights/closer-look-bitcoins-volatility

Wegrzyn, K. E., & Wang, E. (2021). Types of Blockchain: Public, Private, or Something in Between. https://www.foley.com/insights/publications/2021/08/types-of-blockchain-public-private-between/

Wendl, M., Doan, M., & Sassen, R. (2022). The environmental impact of cryptocurrencies using proof of work and proof of stake consensus algorithms: A systematic review. Journal of environmental management, 326 Pt A, 116530. https://doi.org/.DOI: 10.1016/j.jenvman.2022.116530

Whiffin, A. (2024). Faster, Cheaper, Safer: How Tokenisation can change investing, Financial Tİmes. https://www.ft.com/content/dd020b5d-d031-4122-a488-ba7241b7f70d?utm_source=chatgpt.com

Wood, G. (2023). Polkadot Whitepaper, Web3 Foundation. https://polkadot.network

Z/Yen Group & University of Cambridge. (2021). 3rd global cryptoasset benchmarking study. Cambridge Centre for Alternative Finance. https://www.jbs.cam.ac.uk/wp-content/uploads/2021/01/2021-ccaf-3rd-global-cryptoasset-benchmarking-study.pdf

Chapter 6
Bank Challenges Facing the Emergence of FinTech:
From Competition to Collaboration

Bilel Jarraya
https://orcid.org/0000-0003-0167-6144
Department of Accounting, Qassim University, Buraidah, Saudi Arabia & Faculty of Economics and Management, Sfax University, Tunisia

Adel Alresheedi
Department of Accounting, Qassim University, Buraidah, Saudi Arabia

ABSTRACT

The fast acceleration of digital innovations has disrupted the banking system, putting conventional financial institutions under pressure. This study examines the fluctuating relationship between emerging Fintech startups and traditional banks in terms of opportunities for collaboration, regulatory challenges, and emerging trends. Initially competitive, this relationship has shifted to one of collaboration, as banks harness Fintech innovations to enhance digital efficiency. The research will also consider new banking models, including digital-first approaches and "Banking-as-a-Service," and how these changes influence customer demands for continuous and personalized services. Ultimately, the study will highlight the profound impact of Fintech and traditional bank integration on the modern financial ecosystem, particularly regarding cybersecurity and risk management.

DOI: 10.4018/979-8-3373-0355-0.ch006

1. INTRODUCTION

Within the past decade, the financial service industry has been facing transformational turmoil, catalyzed mainly by the exponential growth of fintech companies. It is one of the most energetic, fast-evolving areas, with its inceptive application of technology to refine and develop several aspects of financial services delivery. Fintech has, however, transformed the contours of banking as it has always been perceived to be the most steadfast institution of economic stability and one of the pillars of consumer confidence. Now, the game for traditional banking has altered into one in which they need to adjust to an unparalleled pace, adaptability, and an even higher emphasis on solutions for customer needs.

The rise of fintech has changed how financial transactions happen and has reworked what customers expect. Today's consumers tend to be inclined toward seamless experiences, customized interactions, and service models based on immediacy and convenience. These shifts in dynamics have forced traditional banks to go back to the drawing board, often ripping out their legacy systems and outdated ways, as they struggle to cope with the sophisticated demands of technology-savvy clients.

Initially, the interaction between fintech startups and established banking institutions was marked by rivalry and apprehension. Fintech companies with agile structures and a digital-first philosophy entered the market by offering diverse services such as mobile payment solutions, peer-to-peer lending platforms, wealth management tools, and insur-tech products. These products appealed to the technologically savvy consumer and significantly threatened the traditional banking model. The competitive landscape was set for a clash as fintechs sought significant market share to disrupt the established financial order.

However, as the years have passed in the fintech industry and the realities of a digital economy have set in a significant and strategic shift has emerged: the realization that collaboration-not outright competition-serves as the linchpin for sustainable success. Traditional banks have increasingly appreciated the worth of fintech innovations, especially in enhancing operational efficiencies, streamlining processes, and enriching customer experiences. The recognition has created a mutually beneficial relationship, moving away from the initial antagonism toward cooperation.

This transition from rivalry to partnership is not just a tactical adjustment to competitive pressures. In fact, it reflects a deep-seated evolution in the strategic outlook of both traditional banks and fintech firms. Banks can take advantage of strategic alliances to tap into fintech's technological capabilities and innovative prowess. Conversely, fintech companies can leverage banks' established trust, regulatory know-how, and comprehensive customer bases. Such collaborations open up new avenues for growth and more innovation in the financial space.

This paper navigates through the complex dynamics of this transformational partnership and outlines the rich tapestry of collaboration opportunities between fintechs and traditional banks. The discussion will examine the ways in which these partnerships shape the financial services ecosystem through emerging business models, such as "Banking-as-a-Service" (BaaS). Additionally, we will tackle both sectors' regulatory challenges, especially in critical areas such as anti-money laundering (AML) compliance and data privacy concerns. As the regulatory environment adapts to rapid technological advancements, successful collaboration between banks and fintechs can yield effective frameworks that foster innovation while ensuring that consumer protection remains paramount.

This paper will also highlight the emerging trends that are likely to shape the future of the financial industry, with a special emphasis on the critical role of cybersecurity and sound risk management practices in an increasingly interconnected and digitalized environment. As financial institutions navigate the many complexities of this new landscape, the ability to anticipate and manage risks is crucial for maintaining consumer confidence and meeting regulatory requirements.

This study aims to provide a comprehensive examination of the ongoing transformation within the banking sector, accentuating the urgent necessity for traditional financial institutions to embrace a collaborative mindset. Adaptability and innovation are indispensable in an era characterized by continuous technological disruption. By fostering a culture of collaboration, banks can secure their competitive standing in the evolving market and play a pivotal role in redefining the future of financial markets.

2. THE RELATIONSHIP BETWEEN FINTECH AND BANKS: FROM COMPETITION TO COLLABORATION

The collaboration patterns among FinTech firms and conventional banks show considerable differences in various global regions, shaped primarily by the differences in regulatory structures, the level of economic development, and the state of technological preparedness. In the other world- the most developed economies in Europe and America- the open banking spirit is followed actively with regulatory sandboxes. This spirit combines with the open settings to foster a rather dynamic platform for performance to create innovative and collaborative development away from conventional modern financial solutions. They encourage innovation, collaboration, and the development of advanced financial solutions to create a more dynamic environment. The most striking example could be considered as the FCA sandbox in the United Kingdom, which, since its introduction, has been highly efficient in enabling FinTech firms to experiment with their innovative ideas in a controlled environment and with exemption from most of the regulatory requirements (Haddad & Hornuf, 2019).

On the contrary, as it is recently developing areas such as Africa or Southeast Asia, they nearly entirely build their collaboration on the pressing need for financial inclusion. A collective effort from FinTechs and banks is mainly to fill the gaps in the underserved population segment, thus preventing them from garnishing accessible essential banking services and credit facilities beyond their reach. The most relevant example is perhaps in Kenya, where mobile banking facilities like M-Pesa have revolutionized the landscape of accessibility to financial services and, therefore, have enabled scores of people to conduct transactions and save money through their mobile phones.

It will encapsulate a holistic comparative analysis among these widely distant regions, which will not only show the different strategies adopted by these FinTech-bank partnerships but also throw light on how such partnerships are framed to meet specific economic conditions, and consumer needs that create the future of finance in their context.

2.1. Historical Context: Banks' Resistance to Change and Innovation

Historically, traditional banks dominated the financial landscape, setting guidelines and norms that governed various aspects of financial transactions and services. They were so influential that they could dictate everything about the rules of engagement with little regard for the consumer's will and interest. All this dramatically changed when financial technology companies began changing the dynamics within the industry. Fintech has accelerated innovation and placed additional emphasis on customer-centricity, thus redefining the way financial services are delivered.

The initial reluctance of traditional banks to adopt FinTech innovations is best illustrated through high-profile instances that reveal a deep-seated aversion to disruptive change. JPMorgan Chase exemplified this resistance in the United States, particularly in its early stance on blockchain and cryptocurrencies. The bank's CEO, Jamie Dimon, publicly dismissed Bitcoin as a "fraud" in 2017, reflecting a broader skepticism toward decentralized financial technologies that lacked regulatory oversight and were viewed as volatile and insecure. Despite this strong opposition, JPMorgan reversed its position a few years later by launching its blockchain-based payment system, JPM Coin, in 2020. This platform was designed to enable real-time, secure interbank transactions, signaling a strategic shift toward adopting blockchain once its practical utility became evident. In Europe, similar patterns of resistance were observed. In Germany, banks traditionally favored cash and conventional card-based payments, contributing to a slow national adoption of digital payment methods. Deutsche Bank, one of the leading financial institutions in the country, hesitated to widely integrate mobile wallet services like Apple Pay, only doing so after prolonged consumer demand and competitive pressures compelled them to reconsider. Similarly, BNP Paribas initially resisted collaborating with FinTech firms and digital wallet platforms in France due to concerns over data privacy and cybersecurity vulnerabilities. However, as digital transformation became essential, the bank launched its innovation accelerator program and began forming partnerships with FinTech companies to stay competitive. These cases illustrate how entrenched operational models and risk aversion among legacy banks initially hindered FinTech adoption before increasing market pressures and technological necessity forced a strategic shift toward digital integration. Such initial opposition stemmed from concerns over data privacy, cybersecurity risks, compliance complexities, and fears of disintermediation. Over time, these barriers weakened as banks recognized that technological stagnation could jeopardize their relevance and market position.

This has thrown banks into increasing competition not only among their kind but also from fleet-footed, tech-savvy fintech firms that are all about user experience and accessibility. This has forced banks to renew their technological infrastructure, adding state-of-the-art digital solutions to smoothen operations and improve service delivery. Furthermore, they are challenged to reinvent their business models and strategies to meet the diverse expectations of a more connected and discerning consumer base.

That is fascinating as it shows an evolution of the financial landscape toward collaboration. In many respects, a partnership between incumbent banks and innovative fintech players is no longer an option but a crucial aspect of staying competitive. This, therefore, will foster the exchange of ideas and resources that keep pushing the needle for the industry's progress much better than anyone can imagine; such would bring in more value regarding new and innovative financial products and services, serving the modern consumer.

2.2. Early Competition: How Fintech Startups Initially Disrupted Traditional Banking Models

The emergence of FinTechs has disrupted the traditional banking landscape. These new companies use technology to offer more agile, accessible, and often less expensive financial services than traditional banks. Initially, FinTechs represent a threat to traditional banks. By providing more agile, cheaper, and accessible services via digital platforms, FinTechs have captured significant market share in many areas, such as payments and consumer lending (Boustani, 2020). Whereas regulatory challenges represent a significant barrier to innovation in the banking sector, with the emergence of FinTechs, regulators must adapt their frameworks to respond to the rapid evolution of financial technologies (Restoy, 2019). Regulation aims to protect consumers and ensure the stability of the economic system while fostering innovation (BIS, 2019)[1]. FinTechs, often perceived as less regulated than traditional banks, must navigate a complex environment where requirements can vary significantly from one jurisdiction to another. One of the main challenges is compliance with anti-money laundering (AML)[2] and know-your-customer (KYC)[3] regulations. Vijayagopal et al.'s study (2024) states that Fintech companies consider regulatory compliance a significant challenge for their growth. However, as these startups have gained influence, banks have realized that collaborating rather than directly competing would be more beneficial.

2.3. Evolutionary Dynamics: From Competition to Collaboration

In the early 2010s, FinTechs were seen as rivals to traditional banks, mainly because of their ability to offer fast and cost-effective digital financial services. However, the trend has shifted to collaboration rather than direct confrontation (Anand & Mantrala, 2019, Horváth, et al., 2022). Today, many traditional banks are partnering with Fintechs to enhance their digital offerings and meet consumer expectations for convenience and innovation (Haddad & Hornuf, 2019). These collaborations often take the form of joint ventures, acquisitions, or incubator programs for Fintech startups. For example, Goldman Sachs launched its digital platform, Marcus, integrating multiple Fintech technologies to offer online loans and savings services (Goldstein et al., 2019). As banks have begun integrating Fintech innovations to enhance their digital service offerings, strategic partnerships have emerged in wealth management, peer-to-peer lending, and instant payments. For example, giant banks such as JPMorgan Chase[4] and HSBC[5] have invested heavily in Fintech startups or launched internal initiatives (BIS, 2023). In addition, the open banking model, introduced by the PSD2 directive in Europe, has facilitated this collaboration by allowing FinTechs to access bank customer data, paving the way for a new era of integrated financial services (Haddad & Hornuf, 2023). Anand and Mantrala (2019) argue that traditional banks have been adapting to the rise of fintechs. Although fintechs were initially perceived as an existential threat, banks have leveraged their assets (customer loyalty, financial resources) to remain competitive. The dominant trend is moving towards collaboration (coopetition) rather than fierce competition between banks and fintechs, which shows a mutual interest in leveraging each other's strengths. To this end, the article proposes a strategic framework based on the threat perceived by a fintech in terms of market and technology disruption. Four strategic responses are recommended depending on the case (Hold, Make, Ally, Buy, or Exit). The paper reinforces the proposed ideas with concrete examples, particularly in India, showing how some banks have chosen these responses, such as HDFC with its hold strategy and Canara Bank with its fully digitalized branches to counter fintechs. Authors conclude that banks can choose appropriate strategies depending on the degree of disruption to their business model by fintech.

However, banks must ensure that their collaborations with Fintechs comply with all regulatory requirements, which can slow down innovation. Regulators, meanwhile, are adopting more collaborative approaches, creating regulatory "testing sands" to allow Fintechs to test their products in a controlled environment (EBA[6], 2023). This approach helps foster innovation while preserving the safety and stability of the financial system. Indeed, Banks and FinTech companies must work together to create a regulatory framework that encourages innovation while protecting consumers.

2.4. Opportunities for Collaboration Between FinTech and Banks

The rise of Fintechs has radically changed the way consumers behave about banking. The modern customer seeks speed, simplicity, and personalization in financial solutions. Fintechs are very aptly providing these through their user-friendly interfaces and simplification of processes (Harasim, 2021). For example, personal finance apps like Mint allow users to track their spending in real time and analyze their financial situation (Agarwal & Ann, 2024).

Tanuwijaya and Oktavia's study (2023) found that banks providing better digital collaboration affect customers' switching behavior toward digital banks. Customers are inclined to switch to digital banks if they find the digital solutions more user-friendly, collaborative, and beneficial. Consequently, traditional banking services must adapt to retain their customers in the face of the rise of Fintechs. Moreover, the COVID-19 pandemic has accelerated this trend, with more people using online and mobile banking services to avoid physical interactions (Bapat, 2022).

Younger generations, especially Millennials, are increasingly becoming users of Fintech services. They tend to be lured in by new and innovative propositions that let them take ownership of their own financial future. This turn towards more digital and autonomous behavior not only challenges banks in modernizing their offerings but also requires changes in the way they reach out and communicate with customers (Tanuwijaya & Oktavia, 2023). The opportunities for collaboration between Fintechs and banks are vast. One of the main avenues of cooperation lies in open banking, which allows Fintechs to access bank customer data via APIs, offering personalized and integrated financial services (Chesini et al., 2017). Moreover, banks use Fintechs' flexibility to experiment with digital solutions such as instant payments, risk management via artificial intelligence, and predictive customer data analysis. Notable examples include the partnership between BBVA and the personal finance management app Simple, which allows customers to manage their accounts more efficiently (Mbama & Ezepue, 2018). These collaborations also benefit from Fintechs' UX/UI design skills, where the focus is on improving the user experience of banking services.

The opportunities for collaboration between Fintechs and banks are immense. With the advent of API (Application Programming Interface) technologies, traditional banks can now offer banking services integrated with Fintechs' applications through the "Banking as a Service" (BaaS) model. These platforms allow Fintechs to use banking infrastructures to quickly launch innovative financial products without the regulatory constraints that banks are subject to (Chen, 2022). For example, Fintech and bank collaboration in digital payments has seen a significant leap, facilitating instant and cross-border payment solutions, such as TransferWise (now Wise) and Revolute, which have implemented low-cost money transfer solutions (Karim & Lucey, 2024).

Implementing clear strategic frameworks that balance the drive for innovation with the necessary risk management practices is essential for effective collaboration between banks and FinTech firms. A key framework that can guide this process is the "Hold, Make, Ally, Buy, or Exit" model introduced by Anand and Mantrala (2019).

This comprehensive framework provides banking institutions with a structured approach to evaluate and determine the most appropriate strategies for engaging with FinTech companies. Each of the alternatives offered in the framework has inherent advantages and considerations. Under the first strategy, "Hold," banks maintain their existing operations and refrain from engaging with FinTech firms. This approach allows institutions to concentrate on improving their core offerings and fine-tuning the internal machinery, unencumbered by the complexity or distractions of external partnerships. The second strategy is "Make." This option encourages banks to innovate internally by developing proprietary technologies and solutions that can be as good, if not better, than FinTech firms. By exerting control over innovation, banks can guarantee that newly developed products and services follow their pre-existing standards and operational protocols. The third strategy, the "Ally" strategy, highlights the importance of cultivating strategic alliances between banks and FinTech companies. This collaborative approach allows financial institutions to take advantage of the technological expertise and innovative capabilities of their FinTech partners while developing their service offerings without losing control over operations and minimizing the risks associated with direct financial investments. The fourth option is the acquisition strategy, which includes buying FinTech companies when a deeper integration of technology and services is required. This pathway allows financial institutions to implement advanced innovations and flexible business frameworks directly within their operational structures to improve customer experience and the range of services offered continuously. The final strategy is the "Exit" strategy, which allows banks to terminate any collaboration or initiative deemed unaligned with their strategic objectives or that does not deliver expected outcomes. Knowing the right moments to retreat is a key to effective risk management and ensures wise use of resources.

By weighing these strategic options carefully, financial institutions can design customized approaches that capture the disruptive potential of FinTech innovations while cautiously dealing with inherent risks linked to collaboration. This delicate balance is vital in fostering successful and sustainable partnerships in the rapidly evolving financial services environment.

2.5. Inter-Regional Comparison of FinTech-Bank Collaboration

The dynamics of FinTech-bank collaboration differ significantly across global regions, shaped by unique regulatory philosophies, economic priorities, and varying technological readiness levels. Europe has become a leader in promoting FinTech integration due to its proactive and coordinated regulatory framework. Implementing the Revised Payment Services Directive (PSD2) in 2018 mandated open banking, requiring traditional financial institutions to provide third-party providers secure access to customer data via APIs. This legislation has established a competitive and innovation-friendly environment, particularly benefiting digital-first challenger banks like Monzo and Starling Bank. Complementing PSD2, the UK's Financial Conduct Authority (FCA) introduced its Regulatory Sandbox in 2016, allowing FinTech firms to experiment with new technologies under temporary regulatory waivers. These policies have collectively accelerated collaboration and lowered barriers to market entry. In contrast, North America, particularly the United States, has adopted a more fragmented, state-driven approach. Regulatory oversight is divided among federal entities, such as the Office of the Comptroller of the Currency (OCC) and state-level regulators, leading to compliance complexities and inconsistent enforcement. While FinTech innovation has thrived in hubs like Silicon Valley, traditional institutions such as Wells Fargo and Bank of America initially hesitated to embrace open APIs or engage with blockchain technologies. However, recent collaborations—like establishing the Zelle payment network and partnerships with data aggregators such as Plaid—reflect a shift toward cautious engagement with the FinTech sector. In Africa, regulatory priorities are less focused on innovation and more directed toward financial inclusion. Countries like Kenya spearheaded this movement by launching M-Pesa, a mobile-based money transfer service that transformed financial accessibility for unbanked populations. Despite initial resistance from commercial banks fearing deposit flight, institutions like Equity Bank adapted by partnering with M-Pesa and developing their own mobile platforms. Supportive regulatory actions from the Central Bank of Kenya, such as guidelines for mobile money interoperability, further facilitated these developments.

Meanwhile, Asia, particularly Singapore, has embraced a forward-looking, innovation-centric regulatory model. The Monetary Authority of Singapore (MAS) launched its Regulatory Sandbox in 2016, followed by the Payment Services Act (PSA) in 2020, consolidating and modernizing various payment-related regulations. These measures have positioned Singapore as a FinTech hub, attracting global startups and enabling collaborations with major banks like DBS through accelerator programs and digital partnerships. Comparing these models, Europe's customer-centric and unified regulatory strategy has proven highly effective in fostering structured and scalable collaboration. Singapore's agile, innovation-friendly framework strikes an optimal balance between oversight and experimentation, promoting deep FinTech penetration. Africa's inclusion-oriented model has successfully expanded access to financial services but remains heterogeneous across countries. In contrast, while encouraging robust innovation, the U.S. approach continues to struggle with regulatory fragmentation, which may hinder comprehensive FinTech-bank integration in the long run.

3. REGULATORY CHALLENGES: BALANCING INNOVATION AND COMPLIANCE

Regulatory challenges are among the most important barriers to innovation and growth for both FinTechs and traditional banks. While FinTechs continue to redefine the financial space with their innovative solutions, regulators struggle to adjust their regulatory frameworks to adapt to these new realities. This section addresses the regulatory landscape by structuring the discussion into key thematic areas: AML and Compliance, Data Privacy and Consumer Protection, and Regulatory Sandboxes. It concludes with a comparative table of major global regulatory frameworks.

3.1. AML and Compliance

Igbinenikaro and Adewusi (2024) argue that the rapid growth in FinTech innovations requires comprehensive regulatory frameworks that balance consumer protection and foster technological advancement. It researches different global regulatory strategies on how to ensure consumer safety while promoting innovation, specifically looking into privacy safeguards, AML, and mechanisms for consumer complaint resolution.

The paper also introduces the concept of regulatory sandboxes, where FinTech companies can experiment under regulatory oversight and offers recommendations for policymakers on how to adapt to the changing landscape in FinTech. A comprehensive study conducted by Vijayagopal et al. (2024) highlights that many FinTech companies perceive regulatory compliance as one of the most significant hurdles impeding their growth and development in an increasingly competitive landscape. This difficulty arises from the complex and frequently onerous regulatory environments that these businesses must negotiate in order to preserve their credibility and function efficiently.

In this regard, traditional banks take center stage, since every collaboration with FinTech companies has to be done without any breach in the standard set for data security and privacy. An example is the General Data Protection Regulation of Europe. Organizations wrestling with GDPR have to cautiously handle personal data, given that the repercussions of a breach could be very costly. Therefore, FinTech companies have to work together with banks to understand the regulations and implement them to the fullest extent.

3.2. Privacy and Data Protection

On the other hand, some of these challenges have waked regulatory authorities to be more proactive in creating an enabling environment for innovation within the financial sector. For instance, some regulators have already rolled out regulatory sandboxes that enable FinTech firms to test their products and services in a controlled environment. These sandboxes provide the opportunity to test new technologies and business models without the immediate burden of conventional regulatory requirements, therefore fostering creativity and experimentation while safeguarding the stability and integrity of the broader financial system.

The regulatory sandboxes are essential features that maintain the balance of innovation in the FinTech environment as mandated by the regulatory frameworks. Such regulatory sandboxes form a protected space for FinTech firms, allowing them to develop, test, and improve their services and products while under the watchful eye of the regulators. It is the effort towards the balancing act between having innovations on the one hand and having a critical need to comply on the other, which would eventually result in safety and efficiency regarding the financial solutions.

3.3. Regulatory Sandboxes and Innovation Support

Take the case of Singapore's Monetary Authority Sandbox (MAS): It has been a leading platform for digital payment and lending solutions innovations. The sandbox provided by the MAS assists in experimenting with new technologies and gathering substantial data for guiding regulatory frameworks and strategies concerning consumer protection (Igbinenikaro & Adewusi, 2024). In the same light, the Financial Conduct Authority (FCA[7]) in the UK has so far succeeded in supporting over 150 firms through its regulatory sandbox program, enabling innovative FinTech financial products to enter the market while ensuring that consumers' interests remain paramount.

Indeed, regulatory sandboxes would improve the relationship between traditional banks and FinTech companies. First, they would enable banks to comply better while easing the regulatory uncertainty necessary to inspire collaboration with the banks because of the reduced perception of services from FinTech providers. Furthermore, these sandboxes would inspire banks with the innovation in financial services that their FinTech partners are using.

3.4. Comparative Frameworks Across Regions

Incorporating successful case studies from various regions, especially those in Europe and Asia, would add more context and validation regarding how regulatory sandboxes work. For instance, the European Union has started to apply similar regimes across its member states, which has made it easier to harmonize rules and promote cross-border FinTech developments. Other countries, like Australia and Canada, have also introduced their sandbox programs, proving that there is a trend toward creating an environment that supports FinTech development but with protection for consumers and the financial system's stability.

Generally speaking, regulatory sandboxes represent a progressive approach to regulation, enabling the emergence of a more flexible and innovative financial ecosystem. Ultimately, the success of both banks and FinTech in the future depends on how they navigate this complex regulatory environment in cooperation. This requires further communication and coordination among market players, including financial institutions, FinTech companies, and regulators, to make sure that changing regulations will match the pace of rapid technological changes. They can provide a more flexible regulatory framework that fosters innovation while ensuring consumer protection and maintaining the safety of the financial system.

Table 1. To improve clarity and facilitate cross-regional comparison, the following table summarizes the main regulatory frameworks relevant to FinTech-bank collaboration across key jurisdictions:

Region	Key Regulation(s)	Focus Areas	FinTech Implications
European Union	PSD2, GDPR	Open banking, data privacy	Encourages API integration, strict data protection; fosters innovation with oversight
United States	OCC charters, CCPA, State Sandboxes	Fragmented oversight, consumer rights	Regulatory uncertainty; limited open banking; innovation led by private sector
United Kingdom	FCA Sandbox, Open Banking Initiative	Innovation support, interoperability	High collaboration with regulators; supports FinTech growth
Singapore	MAS Sandbox, Payment Services Act (PSA)	Licensing, payments, risk management	Streamlined FinTech entry; balanced innovation and regulation
Kenya (Africa)	Central Bank Mobile Money Guidelines	Financial inclusion, interoperability	Emphasis on mobile payments (e.g., M-Pesa); supports inclusion with flexible rules

4. SUSTAINABILITY AND BANKING TRANSFORMATION BUSINESS MODELS

4.1. Banking Transformation Business Models

Traditional banks have always built their business models on the basis of wide physical infrastructures and face-to-face customer contacts. This approach emphasized the importance of local branches and personal relationships with bank advisors as critical elements in attracting and maintaining client loyalty. However, this has dramatically changed in the emergence of financial technology companies, normally referred to as Fintechs, and the accelerated digitization of financial services. With these changes, traditional banks are bound to reassess and develop new operation methodologies.

A significant development in this evolution is the adoption of "Banking-as-a-Service" (BaaS)[8] models, which enable Fintech startups to leverage the existing infrastructure of established banks through Application Programming Interfaces (APIs). This approach allows these startups to develop and offer their financial services while utilizing the banking systems and functionalities that traditional banks have spent years establishing (Atabey, 2021). BaaS is not merely a technological transition but a fundamental shift in how banks view their role in the financial ecosystem (Hodges, 2022).

Besides, the digital transformation has also been involving several important activities, such as the automation of banking processes and the creation of completely digital banks, also known as neo-banks. These neo-banks, examples of which are N26 and Revolut, offer the whole range of financial services online, thus enabling customers to handle their finances without visiting a branch. By providing user-friendly interfaces, simple account opening procedures, and tools that will allow real-time tracking and management of financial transactions, their operating model focuses on improving the user experience (Bhatnagar, 2022).

The competition driven by the emergence of Fintechs has been detrimental to traditional banking business models. Banks are being encouraged to shift away from dependence on physical branch networks toward increasing digital outreach and service delivery. This shift corresponds to the changing expectations of modern consumers, who wish to have convenience, access, and efficiency in banking (Omarini, 2022).

Moreover, most of the banks are now moving toward advanced digital transformation strategies. They have been investing huge amounts in AI- and data analytics-based technologies to interpret and predict the needs of their customers in order to personalize their offerings accordingly. As mentioned by (Sheth, et al., 2022), such a transformation effort requires much more than technological upgrading; it actually calls for a cultural makeover of the institutions. Today, it is also encouraged that banks adopt agile and flexible mindsets, matching the innovative and rapid approaches that are typical for a Fintech startup. The ability to pivot quicker and faster and respond with what the market demands become critical to survival in this evolutionary financial landscape.

4.2. Sustainability in FinTech-Bank Partnerships

Sustainability and Environmental, Social, and Governance (ESG) considerations continuously become ingrained in partnerships with FinTech firms and banks. These make way for developing green financial products like sustainable loans and investment platforms. For example, FinTech solutions like CarbonChain use Artificial Intelligence and data analytics to track carbon emissions, enabling banks to offer ESG-compliant financing (Hodges, 2022). Moreover, such collaborations are likely to increase green digital banking initiatives, where financial institutions use FinTech to improve energy efficiency and reduce their carbon footprint (Omarini, 2022). Financial organizations worldwide are increasingly aligning their ambitions for sustainability with technology innovation to promote environmental responsibility and achieve their corporate objectives.

FinTechs play a crucial enabling role in this context by offering advanced digital infrastructure, automation tools, and real-time data analytics that support banks in implementing sustainable financial practices. From green lending algorithms to ESG investment scoring platforms, FinTech solutions are increasingly integrated into banks' sustainability strategies.

Here are some tangible instances where banks adopted sustainability as part of their strategy: First was Standard Chartered's [9]technology platform to align with the bank's business strategy and ESG goals. The system allows the bank to navigate risks and opportunities rightly, fast-tracking its path toward the ESG journey and improving sustainability ratings. Furthermore, Standard Chartered encourages its corporate clients to showcase their sustainability initiatives, creating a broader impact on its customer base. Second, another case in point is that, in partnership with Stockyard Beef[10], the Commonwealth Bank of Australia[11] (CBA) launched a Sustainability-Linked Loan (SLL) within the country's agribusiness sector. Sustainability performance metrics have turned the loan facility interest rates around to a point where they have entirely transformed Stockyard, not least in activities that deal with emissions reduction. The initiatives toward a solar portfolio in tandem with energy audits have enabled Stockyard to meet and exceed sustainability targets, adding a feather to its cap for international customers. In such examples, FinTech platforms are instrumental in embedding tracking, reporting, and verification systems into loan servicing mechanisms—making green financing not only possible but transparent and auditable. On the other hand, Commercial Bank of Ceylon[12] has taken a stance on sustainability through various technological developments. It has developed a Social and Environmental Risk Assessment System to evaluate the environmental impact of lending activities besides initiating paperless green banking for sustainability. Notably, it became the first bank in Sri Lanka to achieve 100% carbon neutrality, financing the country's first viable wind and large-scale solar power projects. Fourth, Banco Bilbao Vizcaya Argentaria[13]—BBVA is one of the founding members of the "Net-Zero Banking Alliance[14]" dedicated to achieving net-zero greenhouse gas emissions by 2050. BBVA has invested in venture capital funds specializing in climate change and has joined platforms like "Carbonplace[15]" to facilitate access to carbon credits. All these technology and financial innovations testify to BBVA's commitment to sustainability. Many of these innovations are the result of collaborations with FinTech startups focused on sustainability, ESG analytics, and climate-focused digital platforms. FinTech's agility allows traditional banks to experiment with decarbonization tools and test impact-driven financial products through digital channels without overhauling their core systems. Fifth, Firstmac[16] is an Australian non-bank lender leading the green lending landscape with its Green

Loans Program. Firstmac Puts Energy-Efficient Home Improvements and Electric Vehicle Purchases on Interest Rate Discounts through the Clean Energy Finance Corporation[17]. All projects are meant to support environmental sustainability and global climate objectives and fulfill clients' needs with a strong-growing demand for ESG practices. In sum, FinTech-bank collaborations are not only modernizing financial operations but are actively shaping the sustainability agenda in global banking. These partnerships are crucial in developing, scaling, and monitoring green financial instruments that align with emerging ESG standards and consumer expectations.

4.3. FinTech-Bank Partnerships Influence on SMEs

The increasing recognition of collaborations between the FinTech firms and banking institutions as key, and in most cases, the only support for small and medium enterprises has been particularly essential in developing economies where access to traditional banking services has otherwise proved difficult (Lu, 2018; Fasano & Cappa, 2022; Rehman et al., 2023). By being at the frontier of technology, these collaborations enable the innovative development and delivery of new financial products for SMEs and transform their operational space (Temelkov et al., 2018; Sheng, 2021). A significant characteristic of such collaboration is FinTech companies offering alternative lending platforms (Ivashchenko et al., 2018). Unlike traditional banks, which have very long, sometimes cumbersome application processes running into weeks or months, FinTechs provide the much-needed fresh air in the form of expedited and flexible options for credit (Yan, 2024). For example, consider the case of peer-to-peer (P2P) lending platforms like Funding Circle. These platforms directly match individual lenders with businesses raising capital; this way, a considerable amount of bureaucratic red tape is eliminated. By removing many of the barriers traditionally associated with obtaining loans, SMEs can access much-needed funds more quickly and with less administrative hassle, enabling them to seize growth opportunities as they arise (Temelkov et al., 2018; Fasano & Cappa, 2022).

Alternative lending transforms how SMEs manage their financial operations by directly integrating financial services into their workflows, eliminating the need for separate financial management tools. Also, a small business can use a software package that will be all-in-one for invoicing, payroll, and payment processing (Flögel & Beckamp, 2020). It provides convenience, enables small businesses to handle money better, increases cash flow, and gives small to medium enterprises better control over their financials. In addition, these systems have excellent real-time expense tracking and transaction management features (Sanga & Aziakpono, 2023). Beyond those benefits at the level of individual businesses, these collaborations by FinTechs and banks are instrumental in driving much broader economic growth. They drive innovation, growth, and jobs by concentrating on financing SMEs and helping these engines burn more vigorously in the local economy and overall economic dynamism. Financially inclusive partnerships target underserved communities and businesses that traditional financial systems have overlooked (Ebirim & Odonkor, 2024).

In the final analysis, the collaboration between FinTech companies and banks is not transactional; instead, it transforms how SMEs operate and succeed. Innovation and growth with enhanced financial inclusion can be nurtured in such environments by faster access to capital, such partnerships providing means of incorporating financial instruments into everyday operations for a much faster time in accessing such resources. It would make significant economic difference.

5. THE FINTECH IMPACT ON CUSTOMER BEHAVIOR

The emergence of Fintech companies has transformed the banking landscape in recent years, fundamentally shifting customer expectations of financial services. In today's digital age, consumers, especially those from younger generations, expect financial solutions that can be accessed 24/7 and come with a high degree of personalization and adaptability to meet their unique needs (Tanuwijaya & Oktavia, 2023).

Fintechs have been instrumental in setting up this new standard by offering seamless user experiences through innovative mobile applications and highly user-friendly digital platforms. The use of big data analytics is one of the recent developments in this field that allows Fintech to offer more detailed personalization of banking services. This also involves providing personalized financial product recommendations, personalized financial advisory, and special promotional packages based on detailed analyses of customers' financial histories and spending patterns (Benjamin et al., 2024). To this end, to understand consumer preference and increase consumer satisfaction, traditional banks are compelled to reassess their technology and strategy, and they are increasingly embedding artificial intelligence capabilities.

With Fintech services, very prominent changes have been introduced into consumer behavior in the banking sector. Presently, customers need banking solutions that are uncomplicated and transparent, accessible at any hour of the day, thereby forcing banks to change the way they engage with customers (Saputra, et al., 2023).

Fintech innovations, especially mobile applications, have made it easy for consumers to manage their finances, thus increasing the number of services that can meet the demands of today's tech-savvy users. Companies such as Revolut[18] and Robinhood[19] are exemplary cases of this shift; Revolut's multi-currency management features enable users to perform international transactions without traditional bank fees, while Robinhood has democratized stock market participation by eliminating trading commissions for its users (Rukiah, 2023).

The obvious and direct result of such an evolution in consumer expectations is an increased preference for mobile banking. Consumers are opting for the ease of their smartphones and other digital devices to conduct banking transactions rather than visiting physical branches of banks (Bapat, 2022). Some recent studies found that consumers prefer to interact with their banking services through mobile applications or via online platforms, up significantly.

In addition, Fintechs leverage data in powerful ways to provide personalized services that directly appeal to consumers' financial habits and preferences. All this builds a closer relationship between financial institutions and their customers and makes them more responsive to customer needs. Ultimately, such personalized services ensure significant optimization of the general user experience, making the customer feel wanted and heard (Soetan & Mogaji, 2024). This evolution in banking is a conscious movement toward a more consumer-driven model powered by technology and data analytics, which will likely define the future of financial services.

New rising FinTech innovations have completely revolutionized financial services' outlook, changing consumer behavior. In simple words, new technology advances in financial services make the services very friendly to consumers, personalized, and widely accessible to modern consumer demands. This, in turn, grants every individual unmatched control over their personal finances in very instinctive ways through applications and online platforms for managing and controlling one's private finances. In this overwhelming change, the rising emphasis on financial literacy is pivotal. Whereas the advantages of Fintech innovations are enumerable, they are subject to a caveat: Consumers must first know how to use these sophisticated tools efficiently to unlock their effectiveness fully. Recent studies show that many fintech platforms have contributed to increased financial literacy by providing users with real-time data, educational interaction, and personalized awareness about spending behavior (Benjamin et al., 2024). Such applications include Mint and "You Need a Budget" (YNAB). These tools help users track their spending, achieve their financial objectives, and enhance their understanding of personal finance management by making consumers informed and active. Apart from that, the emerging partnership FinTech is developing with traditional banking institutions has shown to be an important contributor to financial inclusion. This strategic alliance pragmatically helps reach out to those excluded or underserved by the services of a conventional banking system. This collaboration enables persons who are unable to avail themselves of certain financial products and services through innovative solutions and tailor-made offerings and guarantees greater participation of such persons in the economy itself.

Banks and FinTech companies today are all about digital financial literacy, thus growing their relationships with customers and building trust. However, consumers are becoming more knowledgeable participants in their financial journeys and are more likely to proactively engage with the financial services industry. This synergy will enrich individual consumer experiences while contributing to developing a financially stable and inclusive ecosystem for prosperity for all stakeholders involved.

6. EMERGING TRENDS IN FINTECH-BANK COLLABORATION

Banks are collaborating with FinTech through disruptive technologies like AI, blockchain, and big data. The advent of these technologies into banking operations can provide institutions with complex organizational changes, such as detecting fraudulent activities in real-time and very personalized financial services for customers. An excellent example is JPMorgan Chase, which uses sophisticated artificial intelligence systems to systematically review the pattern of transactions in a manner that allows for fast identification of possible fraud (Gigante & Zago, 2023). In the meantime, blockchain technology is developing significantly to offer transparency and security in financial systems with its decentralized network structure. This is epitomized by the groundbreaking cross-border payment solutions developed by Ripple, which has leveraged blockchain to make international transactions secure and efficient (Tapscott & Tapscott, 2016). Additionally, big data analytics can enable FinTech firms to gain deep insight into the behavior and preferences displayed by their clients. It may also allow the design and delivery of personalized financial products to meet individual users' specific needs, increasing engagement and overall satisfaction. The partnerships with FinTechs are expected to thrive soon as banks continue to embrace transformation technology. This will likely improve operational efficiency and provide an avenue for further innovation in the financial services space, thus leading to even more advanced services and solutions.

Several significant trends are shaping the collaboration landscape between Fintech companies and traditional banks. One of the most noteworthy is the rise of open banking, spurred by implementing the Revised Payment Services Directive (PSD2) in Europe. This directive mandates that banks provide third-party developers secure access to customer data through Application Programming Interfaces (APIs), creating new opportunities for Fintechs to develop innovative financial services and products. This collaborative ecosystem enhances personalization in banking services, as Fintechs can tailor offerings based on granular customer data, leading to a richer customer experience (Puschmann, 2017). In addition to open banking, the role of artificial intelligence (AI) in this collaboration cannot be overstated. AI technologies are revolutionizing banks' operations by automating various processes, such as risk management and fraud detection. For instance, leading banks like Wells Fargo and JPMorgan Chase have begun leveraging AI systems to analyze transaction patterns and detect anomalies efficiently (Gigante & Zago, 2023).

Furthermore, customer engagement has been enhanced through AI-driven chatbots and virtual advisors, which provide real-time customer assistance and support, drastically improving service quality and reducing operational costs (Arslanian & Fischer, 2019). Blockchain technology is also a transformative force in the financial sector and is increasingly being integrated into banking infrastructures. By facilitating decentralized finance (DeFi) solutions, block-chain provides a more transparent and secure manner for conducting transactions and managing assets. DeFi technologies empower users to engage in financial services without intermediary institutions, which can lead to reduced costs and increased accessibility for consumers. The potential of block-chain is profound, as it promises efficiency improvements and increases trust by allowing transactions to be immutable and publicly verifiable (Tapscott & Tapscott, 2016). The collaboration between banks and FinTechs has seen a significant uptick in partnerships as both sides recognize the strengths and innovations the other can bring. Many banks actively seek out Fintech partners to accelerate their innovation processes while FinTechs benefit from the established customer base, capital, and regulatory insights that banks offer. This symbiotic relationship illustrates how both entities work together to adapt to an increasingly competitive and fast-evolving financial services environment (Gomber et al., 2018). Looking ahead, these trends show no signs of abating. Experts predict that the collaboration between FinTechs and banks will continue to evolve toward even more integrated and sophisticated models, driven by technological advancements and changing consumer expectations. As these partnerships mature, they will likely usher in a new banking era characterized by heightened efficiency, security, and customer satisfaction, fundamentally reshaping the financial landscape (Tapscott & Tapscott, 2016; Puschmann, 2017).

7. CYBERSECURITY AND RISK MANAGEMENT IN THE FINTECH AGE

Cybersecurity risks are escalating alongside the growing digitalization of financial services. Cyberattacks that steal sensitive data or compromise financial systems have become increasingly sophisticated. To combat these threats, banks and FinTech companies must implement robust security strategies (Ng & Kwok 2017). The industry's primary threats include phishing attacks, ransomware, and large-scale data breaches.

The importance of cybersecurity in the financial sector has intensified, especially with the rise of FinTech companies that heavily rely on digital infrastructures. Emerging technologies such as blockchain and artificial intelligence (AI) are now employed to secure transactions and prevent fraudulent activities before they impact systems (Alenizi et al., 2024; Gafarov, 2024). Financial institutions are investing significantly in advanced cryptographic solutions to safeguard data and mitigate the risk of breaches (Olaiya et al., 2024).

For instance, in 2020, U.S.-based FinTech firm Robinhood suffered a data breach exposing personal details of over 7 million users due to social engineering tactics. Similarly, Capital One experienced a massive cybersecurity incident in 2019 when a hacker accessed the personal data of over 100 million customers by exploiting a cloud configuration vulnerability. These high-profile incidents underscore the critical need for constant vigilance and system testing.

Moreover, implementing RegTech[20] facilitates the automation of regulatory compliance processes, assisting banks and FinTechs in adhering to stringent security requirements while remaining agile (Adjasi et al., 2023). This technology allows for real-time monitoring of systems to identify potential vulnerabilities (Arner et al., 2022). As digital services expand, regulators strengthen their legislative frameworks to address emerging cyber threats (Lemieux, 2023).

As the financial services landscape evolves, the threat landscape concurrently shifts, necessitating a proactive stance on cybersecurity. The rise of interconnected systems has increased the attack surface for cybercriminals, making it imperative for organizations to adopt a comprehensive risk management approach. This includes not only technological solutions but also fostering a culture of security awareness among employees, as human error remains one of the most significant vulnerabilities (Georgiadou et al., 2022).

Moreover, collaboration between industry stakeholders is becoming crucial in the fight against cyber threats. Information-sharing initiatives among banks, FinTech firms, and governmental bodies can enhance collective defense mechanisms and streamline incident response. Financial institutions can better anticipate and mitigate cyber threats by pooling resources and intelligence, fostering a more resilient ecosystem (Hacioglu & Aksoy, 2021).

In response to these escalating threats, many institutions are also turning to cybersecurity insurance to transfer risk. This evolving market helps organizations manage potential financial losses from cyber incidents, complementing their internal security measures. However, the effectiveness of these insurance policies largely depends on establishing clear risk assessment frameworks and implementing adequate security controls (Elnagdy et al., 2016).

Rather than reiterating technological strategies discussed elsewhere, this section emphasizes that successful cybersecurity in the FinTech-bank environment relies on dynamic threat intelligence, cross-sector collaboration, and regulatory adaptability factors, which are often more decisive than any single tool or protocol.

Looking ahead, integrating machine learning and advanced analytics into cybersecurity protocols holds promise for enhancing threat detection and response capabilities. These technologies can analyze vast amounts of data to identify patterns indicative of unusual behavior or potential breaches. As cyberattacks become more complex and targeted, adapting and refining these technologies will be paramount to safeguarding the integrity of financial systems (Satav et al., 2023).

Ultimately, the financial sector must embrace a multi-layered approach to cybersecurity, combining technological advancements, regulatory compliance, and a robust culture of security awareness to navigate the challenges of an increasingly digital world. The future of financial services hinges on the ability to protect consumer data and maintain trust in the face of evolving cyber threats.

8. CONCLUSION

The rise of fintech has completely changed the dynamics of financial services and forced traditional banks to rethink their operational strategies, business models, and customer engagement methods. From what started as a competitive threat from fintech startups to established banks, as detailed in this paper, has graduated into collaborative partnerships leveraging the strengths of the two sectors. This will help improve service delivery and customer experience, thus keeping financial institutions relevant in this digital century of rapid technological changes and shifting consumer expectations. Our exploration shows that integrating fintech innovations has helped traditional banks renew their services, enhance operational efficiencies, and improve customer satisfaction. The emergence of BaaS models is a perfect example of how banks can use fintech capabilities to provide a frictionless and personalized banking experience. However, with these opportunities come some significant regulatory challenges that must be addressed collaboratively by both FinTech and banks. There needs to be a balanced regulatory framework that encourages innovation while protecting consumer interests.

Examining the effect of the collaboration of Fintech companies with traditional banks becomes an important aspect of measuring the efficacy and success of such alliances within the financial world. Therefore, the evaluation adopts qualitative and quantitative metrics to achieve a full assessment. On the quantitative side, key performance indicators (KPIs) such as cost reduction, customer acquisition rates, operational efficiency, and market share growth serve as significant benchmarks. For example, research indicates that implementing digital transformation initiatives can reduce operational costs by as much as 30% (Goldstein et al., 2019). However, such a remarkable decline is attributed to the strong potential to transform operations and the possibility of delaying expenses through direct digital collaborations with FinTechs. Banks then need metrics such as customer acquisition rates to help them understand how effectively those partnerships went into attracting more new clients. Besides all these measurements, the characteristic in measuring utility on the benchmark is also operational efficiency in the fintech bank value chain. Increased processing efficiency, faster transactions, and lower rates of errors can be quantified and expressed in value-related terms in the context of such partnership collaboration.

Additionally, expansion in market shares can easily capture the extent of changes caused by collaborative or co-branded new offerings and services introduced in partnership with FinTech firms. Qualitative measures are paramount to understanding the holistic benefits being gained from these partnerships. Customer satisfaction and loyalty indices create important footprints of sustainability success. Surveys, customer feedback, and net promoter scores (NPS) can be used to assess the clients regarding this partnership, including their sentiments and experiences. To further enhance these partnerships, banks need to become data-driven in tracking and measuring the economic value of consumer partnerships between FinTech and the internal operational benefits of the engagement. Through data analytics and performance tracking, banks can measure their successes in the past, predict future potential, and make adjustments to better meet their goals in a very changing financial world.

Theoretically, the paper enriches existing literature by providing a detailed understanding of the constantly changing dynamics between fintech and traditional banking institutions. It contributes to discussing competition or how competition and collaboration coexist within industries. By framing the relationship between fintech and banks as one of competition, the paper provides a new framework for analyzing strategic partnerships in other sectors that similarly face disruptive forces. The study further stresses that adaptability and innovation are the key survival competencies in a world marked by technological discontinuity.

Reflecting on the insights gained throughout this study, several key themes emerge. First, the initial tension between banks and FinTechs has evolved into a more cooperative relationship driven by shared goals and complementary capabilities. Rather than viewing each other as rivals, both sides increasingly recognize the benefits of strategic partnerships, particularly in enhancing customer experience, improving operational efficiency, and fostering product innovation. Second, this collaborative evolution is not uniform globally; regional regulatory environments, market maturity, and societal needs influence it. Jurisdictions such as Europe and Singapore have facilitated smoother collaboration through clear regulatory frameworks, whereas fragmentation and lack of cohesion continue to pose challenges in other regions. Third, regulatory compliance remains a significant pressure point, especially in areas like anti-money laundering and data privacy, yet innovations such as regulatory sandboxes and API-driven systems are helping to reduce these barriers. Fourth, FinTechs are becoming essential enablers of sustainable finance, providing banks with the tools and technologies to create more transparent, responsible, and climate-aligned financial services. Lastly, as digital systems become more integrated, cybersecurity and risk management must shift from reactive, siloed approaches to proactive, cooperative defense mechanisms shared between institutions and regulators.

From a policy perspective, there is a clear need for greater regulatory harmonization across jurisdictions. Governments and financial authorities should prioritize cross-border frameworks that support open banking, enforce consistent data privacy standards, and establish shared protocols for digital identity verification. Regulators should also expand the use of innovation hubs and sandboxes not only as testbeds for technology but as collaborative spaces for dialogue between banks, startups, and supervisory bodies.

Meanwhile, banks must commit to cultural transformation by investing in agile systems, enhancing cybersecurity training, and embedding sustainability and digital inclusion as strategic goals not merely compliance requirements. For future research, several promising directions arise. Scholars could conduct longitudinal studies to assess the long-term impact of FinTech-bank partnerships on customer retention, operational resilience, and financial inclusion, particularly in underserved regions. There is also potential for comparative case studies examining how different regulatory sandbox models influence innovation outcomes. Another area of exploration could be the role of FinTech in advancing climate finance, such as how real-time emissions tracking or tokenized green bonds are transforming ESG investing. Finally, deeper empirical work is necessary on the evolving nature of cyber threats, specifically how integrated defense ecosystems (e.g., shared threat intelligence networks) between FinTechs and banks are functioning in practice.

The findings of this chapter put into perspective the need to cultivate collaboration between banks and fintech to engender a more dynamic, responsive financial services ecosystem-one that is better poised to drive improved product offerings and enhance customer experiences while concurrently building operational efficiencies. The messages from this study can act as an excellent roadmap for traditional banks in the future in their journeys with fintech solutions. The guidance will help banks overcome internal resistance and introduce cultural changes favoring innovation. In this dynamically changing financial services environment, bank decision-makers should embrace collaboration through strategic partnerships with fintech companies to drive knowledge sharing and innovation in enhancing digital service offerings. Investment in state-of-the-art technology and skilled talent are required to drive this digital transformation and foster a culture of innovation and agility within the organization. Engaging proactively with regulators is necessary: banks and fintech must advocate for frameworks that allow innovation while ensuring compliance; participation in regulatory sandboxes has given valuable insights for balanced guidelines. Cybersecurity is of utmost priority, requiring advanced protocols, regular risk analysis, and security awareness among the employees.

Moreover, informed and changing customer expectations must be known and catered to through data analytics and feedback to facilitate solutions leading to superior user experiences and loyalty. Decision-makers can capitalize on new opportunities and be adequately positioned for disruptions with insight into emerging trends, including the increased focus on sustainability, artificial intelligence, and blockchain technology. Finally, this joint venture between fintech and traditional banks offers an unrivaled opportunity to shape the future of financial services for superior consumer value and a more robust financial ecosystem, thus stirring hope for what the future may bring.

The interaction of fintech and traditional banks offers different opportunities and challenges in various country types. Whereas developed nations can focus on enhancing personalization and innovation, developing countries can use fintech to advance financial inclusions for economic growth. This realization should constitute the core of policymakers' and decision-makers' awareness in both contexts to create effective collaboration between the sectors and progress in the global financial landscape. Thus, stakeholders may create sustainable financial ecosystems that benefit the consumers by addressing their respective markets' particular needs and capacities.

REFERENCES

Adjasi, C., Hamilton, C., & Lensink, R. (2023). Fintech and financial inclusion in developing countries. In *The Fintech disruption: How financial innovation is transforming the banking industry* (pp. 297–328). Springer International Publishing. DOI: 10.1007/978-3-031-23069-1_12

Agarwal, S., & Ann, T. C. (2024). Personal Finance Applications of Fintech. In *The Emerald Handbook of Fintech: Reshaping Finance* (pp. 235-248). Emerald Publishing Limited. DOI: 10.1108/978-1-83753-608-520241029

Alenizi, A., Mishra, S., & Baihan, A. (2024). Enhancing secure financial transactions through the synergy of blockchain and artificial intelligence. *Ain Shams Engineering Journal, 15*(6), 102733. DOI: 10.1016/j.asej.2024.102733

Anand, D., & Mantrala, M. (2019). Responding to disruptive business model innovations: The case of traditional banks facing fintech entrants. *Journal of Banking and Financial Technology*, *3*(1), 19–31. DOI: 10.1007/s42786-018-00004-4

Arner, D., Buckley, R., Charamba, K., Sergeev, A., & Zetzsche, D. (2022). Governing FinTech 4.0: BigTech, platform finance, and sustainable development. *Fordham J. Corp. & Fin. L.*, *27*, 1.

Arslanian, H., & Fischer, F. (2019). *The future of finance: The impact of FinTech, AI, and crypto on financial services*. Springer. DOI: 10.1007/978-3-030-14533-0

Atabey, A. (2021). Open banking & banking-as-a-service (BaaS): a delicate turnout for the banking sector. *Global privacy law review, 2*(1), 59-82.

Bapat, D. (2022). Exploring the relationship between lifestyle, digital financial element and digital financial services experience. *International Journal of Bank Marketing*, *40*(2), 297–320. DOI: 10.1108/IJBM-12-2020-0575

Benjamin, L. B., Amajuoyi, P., & Adeusi, K. B. (2024). Marketing, communication, banking, and Fintech: Personalization in Fintech marketing, enhancing customer communication for financial inclusion. *International Journal of Management & Entrepreneurship Research*, *6*(5), 1687–1701. DOI: 10.51594/ijmer.v6i5.1142

Bhatnagar, S. (2022). Technological evolution and emergence of new banking models. *IUP Journal of Bank Management*, *21*(2), 7–15.

BIS. (2019). Big tech in finance: Opportunities and risks (Annual Economic Report, pp. 55–79). Bank for International Settlements. https://www.bis.org/publ/arpdf/ar2019e3.htm

Boustani, N. M. (2020). Traditional banks and fintech: Survival, future and threats. In ICT for an Inclusive World: *Industry 4.0 – Towards the Smart Enterprise* (pp. 345–359).

Chen, D. (2024). Chinese Commercial Banks and Fintech-Competition and Collaboration. In *Commercial Banking in Transition: A Cross-Country Analysis* (pp. 333–353). Springer International Publishing. DOI: 10.1007/978-3-031-45289-5_16

Chesini, G., Giaretta, E., & Paltrinieri, A. (Eds.). (2017). *Financial markets, SME financing and emerging economies*. Springer. DOI: 10.1007/978-3-319-54891-3

EBA. (2023). Annual report 2023. European Banking Authority. https://www.eba.europa.eu

Ebirim, G. U., & Odonkor, B. (2024). Enhancing global economic inclusion with Fintech innovations and accessibility. *Finance & Accounting Research Journal*, *6*(4), 648–673. DOI: 10.51594/farj.v6i4.1067

Elnagdy, S. A., Qiu, M., & Gai, K. (2016, June). Understanding taxonomy of cyber risks for cybersecurity insurance of financial industry in cloud computing. In *2016 IEEE 3rd international conference on cyber security and cloud computing (CSCloud)* (pp. 295-300). IEEE. DOI: 10.1109/CSCloud.2016.46

Fasano, F., & Cappa, F. (2022). How do banking fintech services affect SME debt? *Journal of Economics and Business*, *121*, 106070. DOI: 10.1016/j.jeconbus.2022.106070

Flögel, F., & Beckamp, M. (2020). Will FinTech make regional banks superfluous for small firm finance? Observations from soft information-based lending in Germany. *Economic Notes*, *49*(2), e12159. DOI: 10.1111/ecno.12159

Gafarov, N. (2024). The Role of Blockchain in Securing Financial Transactions. *Next Generation Journal for The Young Researchers*, *8*(1), 105–105. DOI: 10.62802/2ad9q749

Georgiadou, A., Mouzakitis, S., Bounas, K., & Askounis, D. (2022). A cyber-security culture framework for assessing organization readiness. *Journal of Computer Information Systems*, *62*(3), 452–462. DOI: 10.1080/08874417.2020.1845583

Gigante, G., & Zago, A. (2023). DARQ technologies in the financial sector: Artificial intelligence applications in personalized banking. *Qualitative Research in Financial Markets*, *15*(1), 29–57. DOI: 10.1108/QRFM-02-2021-0025

Goldstein, I., Jiang, W., & Karolyi, G. A. (2019). To FinTech and beyond. *Review of Financial Studies*, *32*(5), 1647–1661. DOI: 10.1093/rfs/hhz025

Gomber, P., Kauffman, R. J., Parker, C., & Weber, B. W. (2018). On the fintech revolution: Interpreting the forces of innovation, disruption, and transformation in financial services. *Journal of Management Information Systems*, *35*(1), 220–265. DOI: 10.1080/07421222.2018.1440766

Hacioglu, U., & Aksoy, T. (Eds.). (2021). *Financial Ecosystem and Strategy in the Digital Era: Global Approaches and New Opportunities*. Springer Nature. DOI: 10.1007/978-3-030-72624-9

Haddad, C., & Hornuf, L. (2019). The emergence of the global fintech market: Economic and technological determinants. *Small Business Economics*, *53*(1), 81–105. DOI: 10.1007/s11187-018-9991-x

Haddad, C., & Hornuf, L. (2023). How do fintech start-ups affect financial institutions' performance and default risk? *European Journal of Finance*, *29*(15), 1761–1792. DOI: 10.1080/1351847X.2022.2151371

Harasim, J. (2021). FinTechs, BigTechs and banks—When cooperation and when competition? *Journal of Risk and Financial Management*, *14*(12), 614. DOI: 10.3390/jrfm14120614

Hodges, B. (2022). Banking-as-a-Service: Fintechs Walking the Regulatory Perimeter. *Brook. J. Corp. Fin. & Com. L.*, *17*, 127.

Horváth, D., Kerényi, Á., & Szabó, R. Z. (2022). Intended benefits and challenges of cooperation between FinTechs and commercial banks. *Acta Oeconomica*, *72*(3), 289–308. DOI: 10.1556/032.2022.00023

Igbinenikaro, E., & Adewusi, A. O. (2024). Financial law: policy frameworks for regulating fintech innovations: ensuring consumer protection while fostering innovation. *Finance & Accounting Research Journal*, *6*(4), 515–530. DOI: 10.51594/farj.v6i4.991

Ivashchenko, A., Britchenko, I., Dyba, M., Polishchuk, Y., Sybirianska, Y., & Vasylyshen, Y. (2018). Fintech platforms in SME's financing: EU experience and ways of their application in Ukraine. *Investment Management & Financial Innovations*, *15*(3), 83–96. DOI: 10.21511/imfi.15(3).2018.07

Karim, S., & Lucey, B. M. (2024). BigTech, FinTech, and banks: A tangle or unity? *Finance Research Letters*, *64*, 105490. DOI: 10.1016/j.frl.2024.105490

Lemieux, F. (2023). Digital transformation and artificial intelligence: opportunities and challenges. *Digital Strategies And Organizational Transformation*, 103-117.

Lu, L. (2018). Promoting SME finance in the context of the fintech revolution: A case study of the UK's practice and regulation. *Banking and Finance Law Review*, 317-343.

Mbama, C. I., & Ezepue, P. O. (2018). Digital banking, customer experience and bank financial performance: UK customers' perceptions. *International Journal of Bank Marketing*, *36*(2), 230–255. DOI: 10.1108/IJBM-11-2016-0181

Ng, A. W., & Kwok, B. K. (2017). Emergence of Fintech and cybersecurity in a global financial centre: Strategic approach by a regulator. *Journal of Financial Regulation and Compliance*, *25*(4), 422–434. DOI: 10.1108/JFRC-01-2017-0013

Olaiya, O. P., Adesoga, T. O., Adebayo, A. A., Sotomi, F. M., Adigun, O. A., & Ezeliora, P. M.Omolara Patricia OlaiyaTemitayo Oluwadamilola AdesogaAzeez Adekunle AdebayoFehintola Moyosore SotomiOluwaseun Aaron AdigunPaschal M Ezeliora. (2024). Encryption techniques for financial data security in fintech applications. *International Journal of Science and Research Archive*, *12*(1), 2942–2949. DOI: 10.30574/ijsra.2024.12.1.1210

Omarini, A. (2022). The changing landscape of retail banking and the future of digital banking. In *The Future of Financial Systems in the Digital Age: Perspectives from Europe and Japan* (pp. 133–158). Springer Singapore. DOI: 10.1007/978-981-16-7830-1_8

Puschmann, T. (2017). Fintech. *Business & Information Systems Engineering*, *59*(1), 69–76. DOI: 10.1007/s12599-017-0464-6

Rehman, S. U., Al-Shaikh, M., Washington, P. B., Lee, E., Song, Z., Abu-AlSondos, I. A., Shehadeh, M., & Allahham, M. (2023). FinTech adoption in SMEs and bank credit supplies: A study on manufacturing SMEs. *Economies*, *11*(8), 213. DOI: 10.3390/economies11080213

Restoy, F. (2019). Regulating fintech: What is going on, and where are the challenges? Bank for International Settlements. https://www.bis.org/speeches/sp191017a.htm

Rukiah, R. (2023). The Role of Fintech in the Evolution of Financial Management. *Riwayat: Educational Journal of History and Humanities*, *6*(4), 3239–3248.

Sanga, B., & Aziakpono, M. (2023). FinTech and SMEs financing: A systematic literature review and bibliometric analysis. *Digital Business*, 100067.

Saputra, M., Santosa, P. I., & Permanasari, A. E. (2023). Consumer Behaviour and Acceptance in Fintech Adoption: A Systematic Literature Review. *Acta Informatica Pragensia*, *12*(2), 468–489. DOI: 10.18267/j.aip.222

Satav, S. D., Patel, H. G., Walke, S., Megala, R., Patel, C. A., & Pavithra, G. (2023, December). Enhancing Network Security Algorithm Using Machine Learning. In *2023 10th IEEE Uttar Pradesh Section International Conference on Electrical, Electronics and Computer Engineering (UPCON)* (Vol. 10, pp. 1636-1643). IEEE. DOI: 10.1109/UPCON59197.2023.10434765

Sheng, T. (2021). The effect of fintech on banks' credit provision to SMEs: Evidence from China. *Finance Research Letters*, *39*, 101558. DOI: 10.1016/j.frl.2020.101558

Sheth, J. N., Jain, V., Roy, G., & Chakraborty, A. (2022). AI-driven banking services: The next frontier for a personalised experience in the emerging market. *International Journal of Bank Marketing*, *40*(6), 1248–1271. DOI: 10.1108/IJBM-09-2021-0449

Soetan, T. O., & Mogaji, E. (2024). Financial Services Consumers. In *Financial Services in Nigeria: The Path Towards Financial Inclusion, Economic Development and Sustainable Growth* (pp. 163–188). Springer Nature Switzerland. DOI: 10.1007/978-3-031-62340-0_6

Tanuwijaya, E., & Oktavia, T. (2023). Analysis of the Factors Influencing Customer Switching Behaviour of The Millennials in Digital Banks. *Journal of System and Management Sciences*, *13*(2), 122–133.

Tapscott, D., & Tapscott, A. (2016). *Blockchain revolution: how the technology behind bitcoin is changing money, business, and the world.* Penguin.

Temelkov, Z., & Gogova Samonikov, M. (2018). The need for fintech companies as non-bank financing alternatives for sme in developing economies. *International Journal of Information, Business and Management*, *10*(3), 25–33.

Vijayagopal, P., Jain, B., & Ayinippully Viswanathan, S. (2024). Regulations and Fintech: A Comparative Study of the Developed and Developing Countries. *Journal of Risk and Financial Management*, *17*(8).

Yan, J. (2025). How the use of alternative information in risk management fintech platforms influences SME lending: A qualitative case study. *Qualitative Research in Financial Markets*, *17*(1), 1–20. DOI: 10.1108/QRFM-08-2023-0198

KEY TERMS AND DEFINITIONS

FinTech (Financial Technology): Refers to a broad category involving the use of technology and innovation to deliver financial services more efficiently. FinTech firms create services such as mobile payments, peer-to-peer lending, and robo-advisors, which challenge traditional banking models.

Traditional Banks: Established financial institutions that offer a comprehensive range of banking services and primarily operate through physical branches. Historically, they have been slow to adopt new technologies because of regulatory constraints and outdated systems.

Banking-as-a-Service (BaaS): is a model that enables non-bank businesses, such as FinTechs, to provide financial services by leveraging the infrastructure and licenses of traditional banks via APIs. This approach fosters rapid innovation and seamless integration of digital services.

Regulatory Sandbox: A framework established by financial regulators that enables FinTech companies to test new products, services, or business models within a controlled environment, minimizing regulatory burdens to foster innovation.

Open Banking: This is a system in which banks grant third-party financial service providers access to consumer banking, transaction, and other financial data through APIs, thus promoting competition and innovation in financial services.

Anti-Money Laundering (AML): refers to a collection of laws and procedures aimed at preventing criminals from concealing illegally obtained funds as legitimate income. FinTechs and banks are required to ensure compliance with AML standards.

Know Your Customer: A regulatory requirement for banks and financial institutions to verify the identities of their clients to prevent fraud, money laundering, and other illicit financial activities.

Digital Transformation: The process by which traditional banks modernize their operations by incorporating digital technologies such as AI, mobile applications, and data analytics to better meet evolving customer expectations.

Neo-Banks: Fully digital banks that operate without physical branches. They provide banking services through apps and web platforms, focusing on customer experience, speed, and convenience.

Cybersecurity: Involves safeguarding digital financial infrastructure and data against cyberattacks. As banks and FinTechs increasingly embrace digital solutions, cybersecurity emerges as a vital concern to avert fraud and data breaches.

ESG (Environmental, Social, and Governance): Refers to a set of criteria used to evaluate a company's operations and impact regarding sustainability and ethical practices. Collaborations between FinTech and banks are increasingly focusing on ESG through sustainable finance initiatives.

Customer-Centricity: A business approach that puts customer needs and preferences first in all areas of service delivery, design, and communication, something that FinTechs do exceptionally well compared to traditional banks.

Coopetition: A combination of "cooperation" and "competition," this term refers to the evolving relationship between banks and FinTechs, where they both collaborate and compete within the financial ecosystem.

ENDNOTES

1 The Bank for International Settlements (BIS) is an international financial organization established in 1930 in the legal form of a public limited company whose shareholders are central banks. Its primary mission is cooperation between central banks, and it plays a decisive role in managing these institutions' foreign exchange reserves.

2 Anti-money laundering policies and practices are designed to ensure that financial institutions and other regulated entities prevent, detect, and report financial crime, particularly money laundering activities.

3 Know your customer is the name used to verify a company's customers' identities. The term is also used to refer to the banking regulations that govern these activities.

4 JPMorgan Chase, listed on the NYSE, is a financial holding company born from the merger between Chase Manhattan Bank and J.P. Morgan & Co. in January 2001. The group's headquarters are in New York, and the headquarters of the retail banking and commercial banking activities are in Chicago.

5 HSBC is a British international banking group with 60 million customers in eighty-four countries and territories. Its head office is in London's business district of Canary Wharf.

6 The European Banking Authority (EBA) was established by Regulation No 1093/2010 of 24 November 2010 to strengthen the European System of Financial Supervision. It has officially existed since 1 January 2011 and is the successor to the Committee of European Banking Supervisors.

7 Financial Conduct Authority (FCA): is a UK financial regulator, independent of government. It was established in March 2013, succeeding the Financial Services Authority. It oversees the application of UK securities law.

8 Banking-as-a-Service (BaaS): The term banking as a service (BaaS) appeared between 2014 and 2017, in the same trend as concepts such as software as a service (SaaS) or platform as a service (PaaS). These terms refer to solutions that allow the user to be billed only for the services they wish to use, or simply to choose the services (even free ones) they want to use.

9 Standard Chartered was founded in 1853 and has its head office in London. 1969, it expanded by merging two British overseas banks: Standard Bank of British South Africa and The Chartered Bank of India, Australia, and China. The bank is listed on the FTSE 100 index of the London Stock Exchange and

is among the 30 largest British stock market capitalizations. https://www.sc.com/en/

10 Stockyard Beef is an Australian company established in 1958. It is committed to being a leader in sustainable beef production.

11 Commonwealth Bank of Australia was Australia's central bank from 1920 to 1960 until the Reserve Bank Act of 1959 transferred these functions to the Reserve Bank of Australia on 14 January 1960. The bank floated on the Australian Stock Exchange in 1991 and was privatized by the government in 1996. https://www.commbank.com.au/

12 Commercial Bank of Ceylon: Founded in 1920 and originating from Eastern Bank, the Commercial Bank of Ceylon was incorporated as Commercial Bank Ceylon Limited on 25 June 1969, with Eastern Bank holding 40 percent of equity in it. The bank emphasizes its sustainability practice around three main areas including strategic business alignment, culture, and footprint along with responsibilities that reach further than the business of banking. https://www.combank.lk/

13 Banco Bilbao Vizcaya Argentaria is a Spanish multinational banking group based in Madrid and Bilbao. It is one of the largest financial institutions in the world, with a presence mainly in Spain, South and North America, and Turkey. https://www.bbva.com/en/

14 The Net-Zero Banking Alliance was established in 2021, comprising several international banking players, amounting to around 40% of total global banking assets. The Glasgow Financial Alliance for Net-Zero is the parent association for this forum, as made before COP 26. The motivation for forming this alliance is to help its members achieve net-zero emissions. Promerica Group, now two banks under the overlap of this NZBA, has signified their intention to use EcoAct to study their emissions and develop footprint reductions within a maximum temperature increase of 1.5°C. https://www.unepfi.org/net-zero-banking/

15 The carbonplace is an innovative technology, emerging from financial giants dedicated to large-scale climate action. It has established the essential elements of trust, transparency, and accessibility needed to transform the voluntary carbon market into a global platform and an acceleration mechanism for urgent climate actions worldwide. https://carbonplace.com/

16 Firstmac is a privately owned company that has been in business for over 44 years. It grew from a small family business to Australia's leading non-bank lender. https://www.firstmac.com.au/

17 The Clean Energy Finance Corporation (CEFC) is a government-backed green bank created in 2012. It invests in clean energy to propel Australia toward its objective of net zero emissions by 2050. The corporation's mission includes transforming the Australian energy grid and promoting sustainable housing initiatives alongside climate tech innovators. https://www.revolut.com/

18 Revolut is a British fintech company based in London, United Kingdom. Founded in 2014, it offers financial and banking services. Its mobile application provides various banking services, including debit cards, current accounts, investment products, foreign exchange and payments, insurance, consumer credit, and savings products. https://www.revolut.com/

19 Robinhood, founded in 2013, is an American financial services company based in Menlo Park, California. It builds high-frequency trading platforms for financial institutions in New York. https://www.robinhood.com/

20 The RegTech is the abbreviation for regulatory technology, which uses information technology to enhance regulatory and compliance processes. RegTech is most useful in regulated sectors such as financial services, healthcare, pharmaceuticals, energy, etc... It aims to enhance transparency and consistency, standardize regulatory processes, remove ambiguity from regulations, and provide higher-quality outcomes at a lower cost.

Chapter 7
FinTech and Financial Behavior Empirical Analysis of the Determinants of Adoption and Sustainability:
Key Drivers of FinTech Adoption and Sustainability in Digital Finance

Ismaili Safae
https://orcid.org/0009-0007-3971-8860
Private University of Fez, Morocco

ABSTRACT

This study examines the disruptive impact of FinTech technologies on financial behaviour, with a focus on transparency, personalisation and sustainability in digital financial services. It examines adoption trends and their impact on financial systems. Using a quantitative methodology, the analysis employs techniques such as PCA, MCA, ANOVA and logistic regressions, based on data collected through a survey of diverse consumers. The findings indicate that transparency and ease of use are key factors driving FinTech adoption, while sustainability criteria remain secondary for most users. This research enriches the literature by linking technology adoption theories with financial applications, and offers practical recommendations to en-

DOI: 10.4018/979-8-3373-0355-0.ch007

hance user trust and promote sustainable practices in digital financial ecosystems.

INTRODUCTION

The rapid advancements in financial technologies (FinTech) have had a profound impact on the global financial ecosystem, offering innovative solutions that enhance financial inclusion, streamline transaction processes, and cater to the personalised needs of consumers. Leveraging state-of-the-art technologies such as artificial intelligence (AI), blockchain, and big data, FinTech tools are reshaping the manner in which financial services are delivered and experienced. Notwithstanding the significant benefits that these technologies bring, including improved efficiency and accessibility, critical questions remain regarding their influence on financial behaviours, particularly in relation to transparency, simplicity, and sustainability. These aspects are becoming increasingly critical as consumer expectations evolve in a digitalised economy.

Existing literature provides valuable insights into the adoption of FinTech, highlighting frameworks such as the Technology Acceptance Model (TAM) and the Unified Theory of Acceptance and Use of Technology (UTAUT2), which emphasise the roles of perceived usefulness, trust, and ease of use. However, there are substantial research gaps, particularly in terms of exploring the interplay between FinTech innovations and user behaviours across diverse socio-economic and cultural contexts. Furthermore, while many studies examine factors that influence adoption, fewer focus on long-term user retention, the customisation of services, or the potential of FinTech tools to foster sustainable financial practices.

The primary objective of this study is to bridge these gaps by examining the impact of FinTech innovations on perceived transparency, personalisation, and sustainability in financial behaviours. Furthermore, the study seeks to evaluate the manner in which these technologies address consumer demands for simplicity and efficiency. Specifically, the study evaluates four hypotheses:

H1: The first hypothesis is that FinTech innovations enhance perceived financial transparency.

H2: The personalisation of financial services has a positive influence on consumer behaviour.

H3: Furthermore, it is hypothesized that FinTech tools promote sustainable financial practices.

H4: Furthermore, consumers have been shown to prioritise tools that integrate simplicity and efficiency for the purpose of improving their financial management (Smith, 2023).

The present study contributes to the existing body of knowledge by combining theoretical frameworks with empirical data from a large-scale survey involving 35,208 respondents. Employing advanced statistical techniques, such as Principal Component Analysis (PCA), Multiple Correspondence Analysis (MCA), and regression models, the study provides robust and actionable insights for stakeholders.

The article is structured as follows: The literature review synthesises existing studies and identifies research gaps in the adoption and sustainability of FinTech. The materials and methodology section outlines the rigorous design and analytical techniques used to collect and evaluate data. The results section presents key findings that explore relationships between FinTech innovations and financial behaviours. The discussion contextualises these findings within the broader literature, emphasizing theoretical and practical implications. Finally, the conclusion summarizes the study's contributions, offers strategic recommendations, acknowledges its limitations, and proposes directions for future research.

LITERATURE REVIEW

Financial technologies (FinTech) have transformed financial behavior, creating new opportunities and challenges for consumers and institutions. This literature review synthesizes previous research on FinTech adoption and sustainability, highlighting theoretical models such as the Technology Acceptance Model (TAM) and the Unified Theory of Acceptance and Use of Technology (UTAUT2). These frameworks emphasize critical adoption factors like perceived usefulness, simplicity, and trust, as illustrated by studies of mobile payment systems (Rahim et al., 2023) and machine learning applications for credit scoring (Carbo-Valverde, 2020).

Sociocultural influences, such as cultural norms and regional conditions, also play an important role, as illustrated by the adoption of bitcoin in South Africa (Mazambani, 2020) and buy now, pay later services in Southeast Asia (Koskelainen, 2023). In addition, government programs such as Saudi Arabia's Vision 2030 highlight the potential and challenges of public reforms in promoting digital financial inclusion.

Technological advances, including artificial intelligence and blockchain, have increased transparency and personalization in financial services (Pousttchi, 2018; Carbo-Valverde, 2020), but ethical concerns and privacy issues governed by frameworks like GDPR remain unresolved.

While FinTech has improved financial inclusion through models like M-Pesa in Africa, questions about scalability and regulation remain. Existing research largely focuses on adoption factors, but neglects long-term user retention. This study aims to fill this gap by examining the impact of service customization and user-centered approaches on the sustainability of FinTech platforms.

Building on previous research, this review identifies shortcomings in current studies and highlights the need for inclusive, sustainable FinTech models. Extensive references support this examination of behavioral, technological, and sociocultural dimensions, drawing clear distinctions from previous research and paving the way for future contributions.

MATERIALS AND METHODOLOGY

This study employs a robust methodological framework to analyze the adoption and impact of FinTech. The research is guided by four hypotheses: FinTech enhances the perception of financial transparency; the personalization of financial services positively influences consumer behavior; the use of FinTech fosters sustainable financial practices; and consumers prioritize tools that combine simplicity and efficiency in financial management.

The study's empirical underpinnings entail the meticulous collection of data over a substantial timespan of three years (2021–2023) through a comprehensive survey encompassing 35,208 respondents. The survey's design employs a non-probability convenience sampling method, meticulously stratified by age, education, professional status, and geolocation to ensure the representation of diverse perspectives. The study's comprehensive framework encompasses the examination of key variables, including security perceptions, service utilization, challenges encountered, and perceived benefits. These variables are meticulously categorized into nominal, ordinal, and continuous categories to enhance the analytical rigor and interpretability of the findings.

To analyze the data, advanced statistical techniques were employed. Principal Component Analysis (PCA) was utilized to identify latent dimensions and reduce complexity. Multiple Correspondence Analysis (MCA) was utilized to explore associations among categorical variables, while Confirmatory Factor Analysis (CFA) was employed to validate the identified latent structures. Logistic and Multiple Linear Regressions were utilized to assess relationships between predictive variables and behavioral outcomes, while Chi-Square Tests were employed to examine associations between categorical data.

The methodological rigor of the study was ensured through the implementation of internal consistency measures, including Cronbach's alpha values exceeding 0.70, a Kaiser-Meyer-Olkin (KMO) index of 0.853, and significant Bartlett's tests ($p < 0.001$). These measures confirmed the reliability and validity of the data and analytical models.

The methodology's strengths include the large sample size, providing high statistical power, and the use of multidimensional techniques, which uncovered complex patterns in FinTech adoption. However, the methodology is not without its limitations. These include potential selection bias from convenience sampling and limited generalizability to unrepresented geographic or cultural contexts. This comprehensive approach establishes a reliable foundation for understanding the behavioral, technological, and cultural dimensions influencing FinTech adoption and sustainability.

RESULTS

The study surveyed over 35,000 individuals across the national territory to ensure balanced representation of diverse demographic, socio-economic, and geographic groups. (Figure 1). The sample includes 55% women and 45% men, with a significant portion of respondents aged 25–34 (40%), categorized as "digital natives," followed by 35–44 (35%) and 45+ (25%). Geographically, 50% of participants resided in Casablanca-Settat, 20% in Rabat-Salé-Kénitra, 15% in Marrakech-Safi, 10% in Fez-Meknes, and 5% in rural areas. This distribution offers insights into disparities influenced by connectivity and infrastructure.

Regarding marital status, 65% of the participants are married, 25% are single, and 10% are divorced or widowed, reflecting a range of financial needs. Regarding educational attainment, 60% of the participants hold a Bac+5 degree or higher, 25% have a Bachelor's degree, and 15% have a Baccalaureate or lower degree. This indicates disparities in financial literacy. In terms of occupation, 50% of the sample are senior managers, 30% occupy intermediate roles, and 20% are self-employed, suggesting a variety of FinTech usage patterns.

This comprehensive sample provides a substantial foundation for analyzing research hypotheses. It identifies global trends and specific challenges across population segments, offering significant insights into the influence of FinTech on financial behaviors. The framework ensures rigorous empirical analysis, applicable across various contexts, and enables conclusions aligned with contemporary consumer realities.

Figure 1. Presentation of the sample

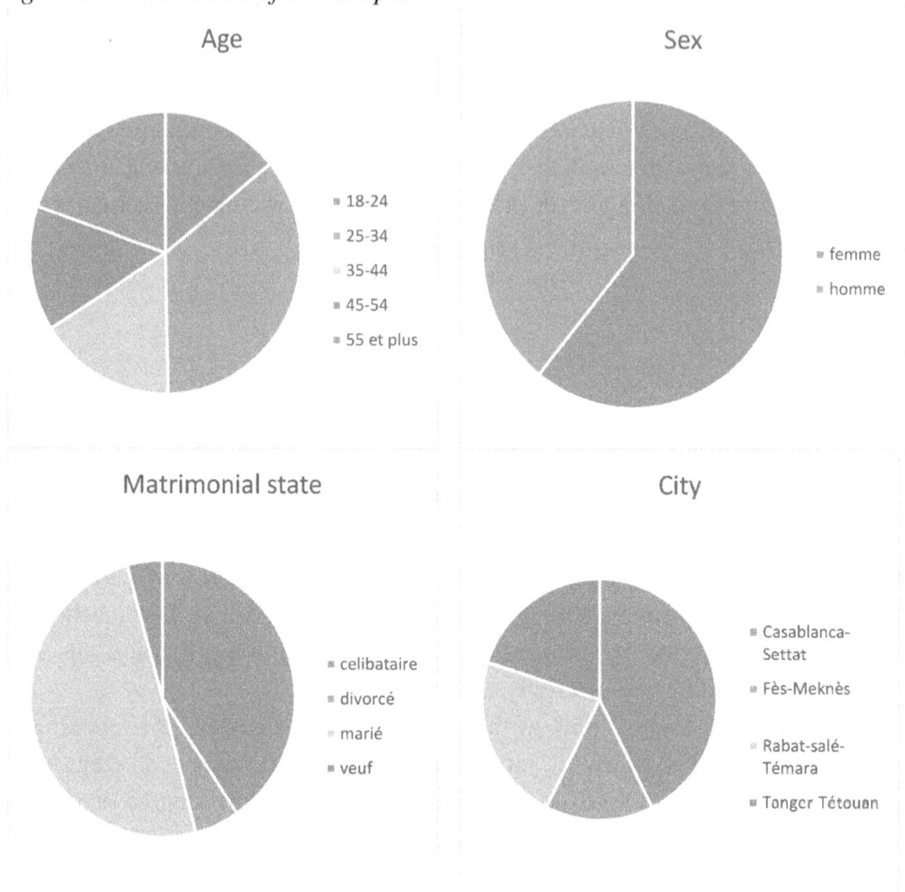

The reliability of the scales used in the study was assessed using Cronbach's alpha, which yielded a value of 0.760 for 26 items, thus aligning with established standards for exploratory social science research. This result is consistent with previous studies in technology adoption, such as the Technology Acceptance Model (TAM) and Unified Theory of Acceptance and Use of Technology (UTAUT), where reliability scores typically range between 0.7 and 0.8.

The study analysed 24 variables, categorised as nominal, ordinal, Likert-scale, and binary, from a sample of 35,208 participants, thus exceeding the statistical requirements for multivariate analyses. Minimal yet significant correlations were observed among variables, with no values exceeding 50%.Principal Component Analysis (PCA), supported by varimax rotation and Kaiser normalisation, was used to explain maximum variance with a minimal number of factors. The Kaiser-Meyer-Olkin (KMO) index of 0.853 indicated excellent sampling adequacy, and Bartlett's test of sphericity ($\chi^2 = 495{,}566.928$, $p < 0.001$) confirmed the robustness of variable relationships for dimensionality reduction. Eight principal components were identified, providing insights into key dimensions influencing consumer behaviour and perceptions of FinTech. These included security perception (e.g., online security: -0.781, app security: 0.735), demographic characteristics (e.g., age: 0.749, marital status: 0.860), socio-economic factors (e.g., occupational category: -0.839), service usage (0.606), and user satisfaction (e.g., perceived quality: 0.738, experience: 0.813).The scree plot (figure2) revealed that the first four components explained a significant portion of total variance, aligning with the eigenvalue criterion (≥1).

Figure 2. Confirmation PCA

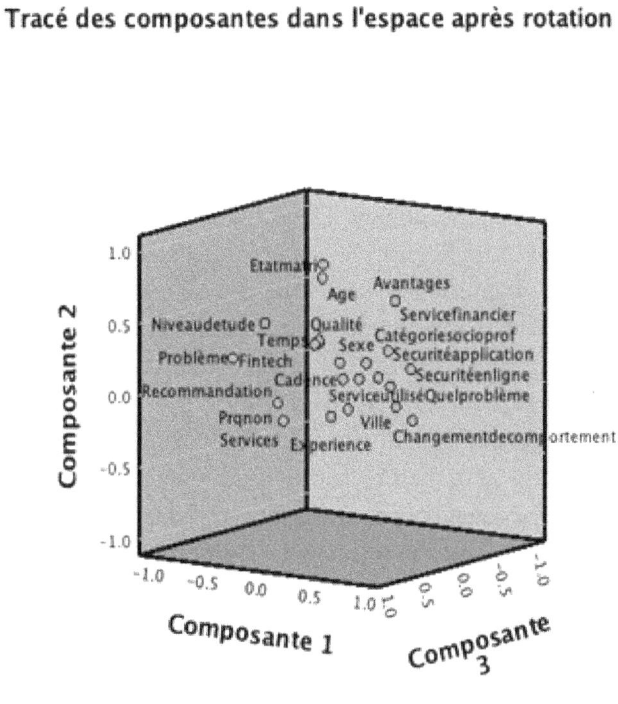

The findings demonstrate the effectiveness of exploratory factor analysis (EFA) in structuring data and identifying latent constructs. The results provide actionable insights for stakeholders in the FinTech sector, enabling them to enhance services, improve user satisfaction, and address diverse demographic needs effectively.

H1: Perceived Financial Transparency is Enhanced by FinTech Innovations

Is utilised to evaluate this hypothesis, with two principal dimensions being identified: global digital security and user-specific concerns. The first dimension combined online security (0.867) and app security (0.728), emphasising their contribution to overall digital security perceptions. Following varimax rotation, these correlations were further strengthened, with app security (0.859) and online security (0.772) consolidating their roles. The second dimension, user-specific concerns, was represented by “identified issues” (0.941), highlighting its relevance in understanding individual apprehensions related to security, accessibility, and trust.

The findings indicate that perceived transparency is significantly influenced by two pivotal factors: the overall safety of digital transactions and the resolution of specific user concerns. The scree plot confirmed the two dimensions' significant explanatory power, aligning with the eigenvalue criterion (≥ 1) and reinforcing their role as foundational constructs of perceived financial transparency.

A logistic regression analysis was conducted to quantify the impact of these dimensions on perceived transparency. The three variables — “Identified Issues,” “Online Security,” and “App Security” — were introduced incrementally.

In the initial step, “Identified Issues” emerged as a significant factor, demonstrating its importance in shaping transparency perceptions.

Subsequent steps revealed the impact of online and app security, highlighting their combined effect in reducing perceived risks.

The model demonstrated a predictive accuracy of 88% in the initial step and 81.8% in the concluding step, thereby substantiating its reliability and its capacity to elucidate transparency perceptions with efficacy. The findings underscore the significance of both dimensions in influencing the perceived transparency, thereby validating their role in promoting the adoption of FinTech.

The findings from both Principal Component Analysis (PCA) and Logistic Regression (LR) have consistently confirmed Hypothesis 1 (H1), thereby demonstrating that perceived financial transparency is significantly enhanced by FinTech innovations. This enhancement is primarily driven by strengthening digital security (online and app safety) and addressing specific user concerns. The results of this study provide a robust foundation for the development of strategies to enhance transparency and trust in FinTech services.

H2: The Personalization of Financial Services Positively Influences Consumer Behavior

The findings reveal considerable heterogeneity in the utilisation of financial services, as evidenced by the mean value of 9.26 and standard deviation of 6.85. This is accompanied by a noticeable variation in the perceived benefits, with a mean value of 8.22 and a standard deviation of 6.53. These observations are indicative of diverse consumer behaviours and perceptions. Conversely, user experience exhibited relative homogeneity (mean = 1.85; SD = 0.36), indicating comparable evaluations of service quality among respondents. Correlations between key variables reveal weak but significant relationships. Specifically, a negative correlation between service usage and user experience ($r = -0.126$, $p < 0.001$) suggests that increased usage may not necessarily lead to increased satisfaction. In a similar vein, a negative correlation between service usage and perceived benefits ($r = -0.086$, $p < 0.001$) suggests that increased usage does not necessarily enhance perceived benefits. The near-zero correlation between user experience and perceived benefits ($r = -0.017$, $p = 0.001$) highlights minimal direct connections between these factors.

Multiple linear regression analysis reveals limited influence of perceived benefits and user experience on financial service usage. While the model demonstrates statistical significance ($F = 422.693$, $p < 0.001$), it explains only a small proportion of the variance (Adjusted $R^2 = 0.024$). The regression coefficient for user experience is found to be significantly negative ($B = -2.427$, $p < 0.001$), suggesting that unfavourable experiences result in increased service usage, which may be indicative of exploratory behaviour among dissatisfied consumers. Furthermore, perceived benefits demonstrate a moderate negative effect ($B = -0.092$, $p < 0.001$), indicating that an increase in perceived benefits does not directly result in an increase in service usage.

Further diagnostic tests were conducted to ascertain the robustness of the model. The collinearity indicators (Tolerance = 1.000, VIF = 1.000) confirm the independence of predictors, while residual analysis shows a normal distribution and absence of patterns, validating the assumptions of normality, linearity, and homoscedasticity.

The findings of this study indicate that the impact of personalisation on consumer behaviour is indirect and influenced by other factors. Dissatisfied consumers may explore additional services, trust and security concerns could play significant roles, and aggregate results might mask differentiated effects across consumer profiles. While personalisation remains a strategic goal, its direct influence on financial service usage appears limited, necessitating further exploration of moderating factors to optimize its effectiveness.

H3: The Use of FinTech Tools Encourages Sustainable Financial Behaviour

The binary logistic regression model was employed to assess the relationship between the perceived benefits of FinTech tools, including convenience and sustainability, and the probability of recommendation. The regression coefficient for perceived benefits was -0.008 ($p < 0.001$), indicating a statistically significant but very weak relationship. The odds ratio (Exp(B) = 0.992) suggests that an increase in perceived benefits slightly decreases the likelihood of recommendation. The global significance of the model (χ^2 test, $p < 0.001$) was confirmed, thereby demonstrating that the incorporation of perceived benefits led to an enhancement in the model's performance. However, the pseudo-R^2 values (Cox and Snell: 0.000; Nagelkerke: 0.001) suggest that the variance in recommendation probability is only marginally explained by perceived benefits. The classification table demonstrated an overall accuracy of 86.5%, yet this was exclusively for positive recommendations, indicating a bias towards the dominant category. To provide further context to these findings, discriminant analysis was conducted to differentiate user groups based on their perceptions of sustainable behaviours associated with FinTech. The model achieved an overall accuracy of 73.3%, with superior predictive performance for non-users (76% accuracy, 88% recall) compared to active users (64% accuracy, 44% recall). This asymmetry underscores the model's limited capacity to capture the distinctive characteristics of active FinTech users. The findings reveal a multifaceted relationship between perceived benefits and sustainable financial behaviours. While the association between perceived benefits and the probability of recommendation is statistically significant, it is weak, and the model demonstrates limited explanatory and predictive capacity. Hypothesis H3 is partially validated: FinTech tools can promote sustainable financial behaviours, but perceived benefits alone are insufficient to fully explain these behaviours. The incorporation of additional variables, such as satisfaction, trust, and personal motivations, would serve to refine the analysis and provide a more profound understanding of the underlying mechanisms.

H4: Consumers are Looking for Tools That Combine Simplicity and Efficiency to Improve Their Financial Management

The analysis, which combines Multiple Correspondence Analysis (MCA) and Structural Equation Modeling (SEM), highlights the significance of perceived benefits such as simplicity and efficiency in shaping overall user satisfaction with

FinTech tools. While users value intuitive and user-friendly solutions, efficiency has the strongest influence on their satisfaction.

MCA identifies clear associations between perceived benefits and satisfaction levels. The association between simplicity (Advantage 1) and satisfaction levels is predominantly moderate, as evidenced by its placement in the "neutral" category. Conversely, efficiency (Advantage 18) has been found to correlate with higher satisfaction levels. The initial dimension of the analysis (accounting for 34.3% of the total inertia) juxtaposes these two categories of advantages with perceived quality. Users who prioritise efficiency tend to report higher levels of satisfaction, while those who prioritise simplicity remain neutral or only moderately satisfied.

The SEM further corroborates these findings by testing causal relationships between simplicity, efficiency, and overall satisfaction. The model fit indices (CFI = 0.92, TLI = 0.91, RMSEA = 0.048) indicate excellent model fit, thereby supporting the reliability of the results. The findings demonstrate that simplicity exerts a significant positive influence on satisfaction, as evidenced by the positive standardized coefficient of 0.42 and the p-value of less than 0.001. This suggests that users place a high value on intuitive tools. However, efficiency exerts an even stronger effect (standardized coefficient = 0.58, $p < 0.001$), emphasizing the importance of performance in meeting financial needs.

The strong correlation between simplicity and efficiency (correlation coefficient = 0.63, $p < 0.001$) indicates that these dimensions are perceived as complementary rather than independent. Users express a preference for tools that embody both characteristics, thereby optimising their financial management experience.

These findings support Hypothesis H6, which states that consumers seek FinTech tools that integrate simplicity and efficiency. While simplicity is regarded as a prerequisite, efficiency exerts the most significant influence on overall satisfaction. These findings underscore the importance for FinTech developers to design solutions that are both intuitive and effective, thereby addressing the growing expectations of users in financial management.

DISCUSSION

The study provides a comprehensive overview of the key insights into the adoption of FinTech, emphasising the pivotal roles of transparency, personalisation, sustainable behaviours, and the balance between simplicity and efficiency in shaping user satisfaction. The enhancement of financial transparency, driven by digital security and user-specific concerns, aligns with prior findings (e.g., Hofmann et al., 2020) but indicates room for improvement in addressing latent concerns.

The personalisation of services exhibited a modest direct impact on consumer behaviour, as dissatisfied users tend to explore alternatives, a finding consistent with Parasuraman et al. (2005), suggesting that trust and security are pivotal factors. The findings concur with those reported by Zhang et al. (2020) on the association between sustainable behaviours and perceived benefits, highlighting the necessity for a more comprehensive understanding of explanatory variables such as financial literacy and trust.

Simplicity and efficiency emerged as complementary factors, with efficiency exerting a stronger influence on satisfaction, validating the findings of Venkatesh et al. (2003) and Davis (1989). However, the interdependence of these factors emphasises the necessity for a joint optimisation approach.

While the findings provide actionable insights, the low R^2 values in the regression models suggest the influence of unexamined variables, such as cultural or institutional factors. The integration of broader frameworks, such as UTAUT, has the potential to enhance the robustness of future analyses. In conclusion, the study emphasises the necessity of refining FinTech solutions to address trust, personalisation, and performance for diverse consumer needs.

CONCLUSION

The present study provides a comprehensive examination of the adoption and impact of FinTech on financial behaviour, emphasising key dimensions such as transparency, personalisation, sustainable behaviours, and the interplay of simplicity and efficiency. The findings indicate that FinTech innovations have the potential to enhance perceived financial transparency through improvements in digital security and the resolution of user-specific concerns. The personalisation factor, on the other hand, exerts a modest direct influence on consumer behaviour, underscoring the pivotal role of trust and security in this regard. The study also highlights the potential of FinTech tools in promoting sustainable financial behaviours, although perceived benefits alone are insufficient to fully explain these patterns. Finally, the study identifies simplicity and efficiency as complementary factors, with efficiency exerting the strongest influence on user satisfaction.

The findings provide actionable insights for FinTech developers, policymakers, and financial institutions, and enhance our understanding of the importance of digital security and the addressing of specific user concerns in order to build trust and transparency. It is vital that FinTech solutions prioritise integrating simplicity and efficiency in order to meet the diverse needs of users. Furthermore, strategies which focus on user-centred design and personalisation, whilst addressing broader factors such as trust and financial literacy, could enhance both adoption and long-term user retention. Despite its robust methodology and large sample size, the study has limitations. The use of non-probability convenience sampling may introduce selection bias, limiting the generalisability of findings to unrepresented regions or cultural contexts. Additionally, the explanatory power of regression models was low, indicating that important variables, such as cultural norms, institutional factors, and psychological dimensions, were not included in the analysis.

Future studies could address these limitations by employing probability sampling methods and expanding the scope of research to include underrepresented regions. The incorporation of broader frameworks, such as UTAUT or psychological models, could enhance the understanding of the dynamics of FinTech adoption. Further research should explore the long-term impact of FinTech on user retention and sustainable financial practices, with a focus on trust-building mechanisms and the role of financial literacy. These directions would provide a deeper understanding of the evolving digital financial landscape and contribute to the development of more inclusive and effective FinTech solutions.

REFERENCES

Arias-Oliva, M., Pelegrín-Borondo, J., & Matías-Clavero, G. (2019). Variables influencing cryptocurrency use: A technology acceptance model in Spain. *Frontiers in Psychology*, *10*, 475. DOI: 10.3389/fpsyg.2019.00475 PMID: 30949085

Beldad, A., De Jong, M., & Steehouder, M. (2010). How shall I trust the faceless and the intangible? A literature review on the antecedents of online trust. *Computers in Human Behavior*, *26*(5), 857–869.

Carbo-Valverde, S. (2020). Artificial intelligence in financial services: The transformative potential and the regulatory challenges. *Journal of Banking & Finance*, *120*, 105948. DOI: 10.1016/j.jbankfin.2020.105948

Davis, F. D. (1989). Perceived usefulness, perceived ease of use, and user acceptance of information technology. *Management Information Systems Quarterly*, *13*(3), 319–340. DOI: 10.2307/249008

Hofmann, A., Räth, A., & Bühler, W. (2020). Security and trust in digital financial ecosystems: An empirical study. *Information Systems Journal*, *30*(4), 689–710. DOI: 10.1111/isj.12284

Koskelainen, T. (2023). The role of sociocultural factors in the adoption of Buy Now, Pay Later services: A Southeast Asian perspective. *Asia Pacific Journal of Marketing and Logistics*, *35*(1), 23–45. DOI: 10.1108/APJML-01-2022-0048

Mazambani, L., & Mutambara, E. (2020). Predicting FinTech adoption in South Africa: The role of trust and risk. *African Journal of Science, Technology, Innovation and Development*, *12*(1), 25–35. DOI: 10.1080/20421338.2019.1624008

Nunnally, J. C., & Bernstein, I. H. (1994). *Psychometric theory* (3rd ed.). McGraw-Hill.

Parasuraman, A., Zeithaml, V. A., & Malhotra, A. (2005). E-S-QUAL: A multiple-item scale for assessing electronic service quality. *Journal of Service Research*, *7*(3), 213–233. DOI: 10.1177/1094670504271156

Pousttchi, K. (2018). Digital transformation of financial services. *Electronic Markets*, *28*(3), 285–294. DOI: 10.1007/s12525-018-0309-3

Rahim, N. H. A., Ghani, E. K., & Said, J. (2023). Mobile payment adoption: Integrating security trust and technology acceptance model. *Journal of Financial Services Marketing*, *28*(2), 150–162. DOI: 10.1057/s41264-022-00136-5

Venkatesh, V., & Davis, F. D. (2000). A theoretical extension of the Technology Acceptance Model: Four longitudinal field studies. *Management Science*, *46*(2), 186–204. DOI: 10.1287/mnsc.46.2.186.11926

Venkatesh, V., Thong, J. Y., & Xu, X. (2012). Consumer acceptance and use of information technology: Extending the Unified Theory of Acceptance and Use of Technology (UTAUT2). *Management Information Systems Quarterly*, *36*(1), 157–178. DOI: 10.2307/41410412

Zhang, X., Zhang, J., & Zhao, K. (2020). A sustainable approach to financial inclusion: Understanding the impact of financial literacy and trust on FinTech adoption. *Sustainability*, *12*(1), 139. DOI: 10.3390/su12010139

ADDITIONAL READINGS

Gefen, D., Karahanna, E., & Straub, D. W. (2003). Trust and TAM in online shopping: An integrated model. *Management Information Systems Quarterly*, *27*(1), 51–90. DOI: 10.2307/30036519

Gomber, P., Koch, J. A., & Siering, M. (2017). Digital finance and FinTech: Current research and future research directions. *Journal of Business Economics*, *87*(5), 537–580. DOI: 10.1007/s11573-017-0852-x

Hair, J. F., Black, W. C., Babin, B. J., & Anderson, R. E. (2010). *Multivariate data analysis: A global perspective*. Pearson Education.

Zhou, T. (2012). Examining mobile banking user adoption from the perspectives of trust and flow experience. *Information Technology & People*, *25*(4), 405–422. DOI: 10.1108/09593841211278788

Zhu, K., & Kraemer, K. L. (2005). Post-adoption variations in usage and value of e-business by organizations: Cross-country evidence from the retail industry. *Information Systems Research*, *16*(1), 61–84. DOI: 10.1287/isre.1050.0045

Zwass, V. (2010). Co-creation: Toward a taxonomy and an integrated research perspective. *International Journal of Electronic Commerce*, *15*(1), 11–48. DOI: 10.2753/JEC1086-4415150101

Chapter 8
Assessing the Transformative Potential of Green Fintech Solutions in Mitigating Financed Emissions

Abdulkadri Toyin Alabi
https://orcid.org/0000-0002-3103-5367
Kwara State University, Nigeria

ABSTRACT

This study investigates the transformative potential of green fintech solutions in mitigating financed emissions, synthesizing insights through an integrative literature review approach. Green fintech which is powered by advanced technologies such as blockchain, artificial intelligence, big data analytics, and digital platforms, emerges as a critical driver of sustainability within financial systems. The study offers innovative solutions, including digital ESG analytics, green investment tools, crowdfunding platforms, and digital payment systems, to address the complex challenge of aligning financial activities with net-zero objectives. The study further identifies enabling factors which enhance transparency, efficiency, and accountability in green finance. However, these potentials are met with significant challenges, including limited data availability and standardization, high energy consumption of underlying technologies, regulatory and compliance uncertainties, greenwashing concerns, and unequal access to capital and technology.

DOI: 10.4018/979-8-3373-0355-0.ch008

1. INTRODUCTION

Financed emissions notably understood as the greenhouse gas emissions associated with the lending and investment activities of financial institutions, represent a significant challenge in achieving global climate goals (Nelson, 2022). These emissions account for the environmental impact of the entities that banks and other financial institutions support financially. For example, in 2023, Nordea (2024) reported that 99.9% of its total greenhouse gas footprint stemmed from financed emissions, underscoring the dominance of indirect emissions in the financial sector's climate footprint. The financial sector serves as a pivotal conduit in directing capital within society, influencing projects ranging from infrastructure development to innovation (Geddes et al., 2018). This is because when funds are being channelled into carbon-intensive sectors, financial institutions risk perpetuating emissions-heavy trajectories. Conversely, a deliberate shift in capital allocation toward sustainable activities could accelerate the transition to a low-carbon economy (Task Force on Climate-related Financial Disclosures, 2021). Despite these prospects, traditional financial mechanisms often lack the transparency, efficiency, and incentive alignment required to address the inherent complexities of financed emissions effectively.

In order to address these peculiar challenges of financed emissions, the transformative solution of Green Financial Technology (Green Fintech) has emerged which focuses on leveraging technology to align financial services with environmental sustainability (Addy et al., 2024; Jaiwant & Kureethara, 2023). Innovations such as blockchain-based carbon tracking, AI-driven sustainability assessments, and digital platforms for green investments can enhance transparency, reduce transaction costs, and enable more accurate emission accounting (Alabi, 2025; Bayram et al., 2022). Green FinTech offers financial institutions tools and strategies to proactively engage in climate action (Muganyi et al., 2021). The role of Green FinTech in this context can be understood through its capacity to enhance transparency, efficiency, and decision-making in sustainable finance (Kwong et al., 2023; Mishra, 2023).

Importantly, real-time data analytics is a cornerstone of Green FinTech, enabling financial institutions to monitor, analyze, and respond to climate-related risks dynamically. According to Addy et al. (2024), the incorporation of advanced analytics tools allows institutions to process large volumes of environmental, social, and governance (ESG) data with unprecedented speed and precision. For example, satellite data integrated with machine learning algorithms can provide granular insights into deforestation rates, carbon emissions, and water usage, helping institutions assess the environmental impact of their investments. This capacity not only aids in compliance with regulatory frameworks, such as the EU Taxonomy for Sustainable Activities, but also facilitates proactive risk management by identifying and mitigating climate risks before they materialize. Further, the availability of real-time data enhances transparency, a critical factor in building trust among stakeholders in sustainable finance. Financial institutions equipped with real-time analytics can provide more accurate and timely disclosures to investors, thereby encouraging greater participation in green financial products (Sharma et al., 2024). This aligns with the growing demand for actionable, data-driven insights in aligning portfolios with net-zero targets, as emphasized by the Task Force on Climate-Related Financial Disclosures (TCFD).

In terms of directing capital toward sustainable ventures, Green FinTech also plays a crucial role. Digital platforms and blockchain technologies have emerged as transformative tools, allowing financial institutions to streamline green bond issuance, crowdsource climate-friendly projects, and democratize access to sustainable investment opportunities (Anshari et al., 2021; Macchiavello & Siri, 2022). These innovations reduce transaction costs and improve the traceability of funds, ensuring that capital is allocated to projects with genuine environmental impact. Blockchain, in particular, enhances accountability by providing immutable records of transactions, which is crucial for mitigating issues such as greenwashing (Dorfleitner & Braun, 2019; Kabaklarlı, 2022). For instance, decentralized platforms can facilitate peer-to-peer energy trading, where renewable energy producers and consumers transact directly, bypassing intermediaries. Such platforms also enable microfinance solutions for climate-resilient projects in underserved regions, expanding the reach of sustainable finance. Additionally, artificial intelligence (AI)-powered investment platforms can analyse market trends and ESG metrics to identify profitable yet sustainable investment opportunities (Hui & Loang, 2023; Mohammed et al., 2024). These platforms empower investors to make informed decisions by aligning their financial objectives with sustainability goals. This integration of technology and sustainability not only attracts more green investments but also incentivizes traditional financial institutions to adopt greener practices.

Despite Green FinTech's representation of a paradigm shift in how financial institutions approach climate strategies, the adoption of green fintech faces barriers, including regulatory uncertainty, technological integration challenges, and varying levels of institutional commitment to sustainability (Bhattacharya & Sachdev, 2024; Singh & Kaunert, 2024). Addressing these issues requires a synergistic effort between policymakers, financial institutions, and technology innovators. This study therefore explores the transformative potential of green fintech in mitigating financed emissions, analysing its capacity to overcome existing obstacles and its implications for global climate strategies.

2. LITERATURE REVIEW

Various studies in the past have investigated how green fintech solutions leverage AI, IoT, and smart contracts to enhance ESG assessments in order to promote sustainable investments and facilitate green lending. These technologies can significantly mitigate financed emissions by embedding environmental considerations into financial decision-making, driving a transition to a sustainable economy. (Elias et al., 2024; Stern & Valero, 2021). As these innovations continue to evolve, they hold the potential to reshape traditional financial models, encouraging greater transparency and accountability in how capital is allocated towards environmentally responsible initiatives (Shahbaz et al., 2020). This shift not only benefits the planet but also appeals to a growing demographic of socially conscious investors who prioritize sustainability in their portfolios.

The integration of green fintech solutions into the financial sector demonstrates transformative potential in addressing financed emissions, as evidenced by the growing body of literature on this subject. Wang et al. (2024) underscore the role of fintech in mitigating corporate carbon emissions through multiple mechanisms, including alleviating financing constraints, promoting energy efficiency, and driving green innovation. Their findings position fintech as a critical tool in fostering a shift towards sustainable finance, catalyzing systemic changes in corporate strategies to align with environmental goals. This alignment promotes long-term resilience and growth, highlighting the need for a paradigm shift in financial practices to support global sustainability objectives. Zhang et al. (2024) extend this argument, emphasizing the regional effectiveness of fintech solutions in enhancing carbon emission efficiency, particularly in eastern regions and urban centres. They illustrate how fintech innovations in green finance and technological advancements contribute to low-carbon economic transformations, thus enabling the financial sector to become a proactive agent of environmental change. Similarly, Anwar et al. (2024) reveal that in Australia, supportive policies amplify the effectiveness of fintech-enabled green finance initiatives, leading to significant carbon reductions. This correlation between regulatory frameworks and fintech innovation underscores the necessity of policy integration to optimize the transformative potential of green financial technologies.

The transformative capabilities of green fintech are further elaborated by Osby and Anderson (2023) and Hui and Loang (2023), who identify its ability to facilitate renewable energy financing, support sustainable investments, and enhance climate risk assessment. These advancements not only empower businesses to align their practices with environmental sustainability but also enable consumers to make informed, eco-friendly financial decisions. Addy et al. (2024) reinforce this perspective by demonstrating how fintech leverages data analytics to assess sustainability risks, develop green investment strategies, and promote eco-friendly projects. Collectively, these insights emphasize the role of green fintech in transitioning towards a low-carbon economy. Adaramola et al. (2024) explore the specific applications of green fintech, such as sustainable investments and carbon credit trading, which enhance resource management in smart cities and mitigate financed emissions. Their findings stress the importance of transparency and environmentally friendly practices in fostering sustainable urban development. Complementing this, Sharma et al. (2024) highlight the role of green banking awareness in driving the adoption of fintech solutions, which subsequently enhances environmental, social, and governance (ESG) investments. These innovations promote sustainable banking practices and encourage eco-conscious economic behaviors, creating a positive feedback loop between technology, finance, and environmental responsibility.

Wang (2023) and Wu (2024) offer additional dimensions to the discussion by examining the broader impact of green financial products and industrial optimization. Wang (2023) highlights the optimization of investment structures to support corporate low-carbon initiatives, while Wu (2024) emphasizes the role of fintech in reducing carbon emissions intensity through enhanced energy efficiency and innovative financial services. Both studies underscore the necessity of integrating green finance and technology to drive sustainable economic development and foster resilience against climate change. Sengupta and Patil (2023) argue that the integration of fintech and green finance is pivotal in achieving near-zero emissions. Their findings highlight the importance of government support, innovation, and heightened awareness in maximizing green fintech's impact, underscoring its role as a critical enabler in the transition towards sustainable economies. Tao and Chao (2024) build on this foundation, focusing on the Chinese energy sector. They emphasize how fintech enhances corporate green investments by fostering data-driven decision-making, transparency, and sustainable strategies. This approach not only mitigates financed emissions but also aligns with broader goals of environmental protection and sustainable development. Similarly, Zhang et al. (2024) reveal fintech's effectiveness in managing pollution and reducing emissions in heavily polluting industries, showcasing its transformative impact through increased environmental investments and green innovation.

Regional studies, such as Zhang (2023) on South Asian economies, highlight fintech's role in CO2 emissions reduction, environmental protection, and green economic recovery. These insights illustrate how green fintech can tailor solutions to meet specific regional challenges, enhancing its global applicability. Zhou et al. (2024) contrast traditional finance with digital finance, emphasizing the latter's superior capability in reducing industrial waste gas emissions through support for green credit projects and energy conservation. This comparative analysis reinforces the criticality of transitioning to digital financial solutions. Shan et al. (2024) extend this discourse by examining fintech's role in BRICS countries, highlighting its ability to reduce CO2 emissions within the banking industry and promote green innovations. This demonstrates the scalability of green fintech solutions across diverse economic landscapes. Huang and Yang (2023) delve into the technological mechanisms underpinning fintech's impact, such as the use of big data for product circulation and green credit realization. Their study underscores fintech's role in fostering environmental transparency and supporting carbon neutrality goals.

The interconnectedness between fintech adoption and sustainable performance is further explored by Alsadoun and Alrobai (2024). Their findings suggest that green fintech effectively mitigates financed emissions by driving green innovation and promoting sustainability initiatives within the financial sector. Cai et al. (2024) complement this perspective by demonstrating the role of green finance in creating supportive financial environments that reduce carbon intensity, facilitating transitions to low-carbon economies. The mediating effect of green finance on climate change and bank stability is a notable focus in Hishamuddin et al. (2024). Their findings suggest that sustainable banking practices and environmentally friendly investments can simultaneously enhance financial stability and mitigate financed emissions. Similarly, Muganyi et al. (2021) highlight fintech's role in promoting green consumption and enhancing institutional capacity for green credit, particularly in reducing sulphur dioxide emissions.

In developing economies, Coffie et al. (2024) and Nenavath (2022) explore the practical implications of green fintech. Coffie et al. (2024) highlight its potential in mobile money economies within sub-Saharan Africa, emphasizing the need for green financial products to promote cleaner production. Nenavath (2022) focuses on India, showcasing how strategic funding mechanisms and innovative technologies enhance environmental protection and mitigate emissions. From a broader industry perspective, Puschmann et al. (2020) argue that startups are leading the charge in deploying green fintech solutions, significantly impacting the financial services value chain. This innovation-driven approach contrasts with the slower adaptation seen in traditional incumbents. Likewise, Macchiavello and Siri (2022) discuss green fintech's potential to address gaps in sustainable finance, such as improving ESG disclosures and access to retail financing, further reinforcing its role in fostering a sustainable financial ecosystem. The reviewed literature collectively underscores green fintech's transformative potential in mitigating financed emissions across diverse sectors and regions. It is therefore evident that green fintech facilitates the transition to low-carbon economies if organizations are able to leverage technology-driven solutions, which will enhance transparency, and promote sustainable investments.

3. METHODOLOGY

In an attempt to achieve the objective of this study, an integrative literature review approach was considered in consistency with the methodologies of Callahan (2010) and Torraco (2016). This approach is believed to be well-suited for synthesizing existing knowledge and identifying patterns, gaps, and trends in order to reach a conclusion regarding the relationship between Green FinTech and financed emissions. A thorough search of existing literature was conducted to locate pertinent articles, research papers, reports, and publications that explore the relationship between green fintech and net-zero solutions concerning financed emissions. This search was conducted on academic databases including Scopus, Web of Science, Google Scholar, PubMed, and IEEE Xplore.

Several articles were selected based on their relevance to Green Fintech, financed emissions and their connection to specific analytical threads. Further inclusion criteria involved focus on studies published within the last six years to incorporate the latest developments in the field. The chosen publications were further subjected to a quality assessment to ensure the reliability and reputation of the sources. The evaluation also considered factors such as authors credibility, the peer-review process, and the relevance and accuracy of the data and findings. This method thus allowed for a broad synthesis of sources, providing a comprehensive overview of the current state and progress of green fintech in mitigating financed emissions.

4 CONCEPTUALIZATION OF GREEN FINTECH

4.1 Green Fintech

This study contends that green fintech is regarded as the convergence of financial technology and environmental sustainability which aims to leverage technological innovations in order to promote green finance and support sustainable development goals. This encompasses the strategic use of fintech solutions to create positive environmental impacts by financing green projects, promoting sustainable consumer behaviour, and enhancing the efficiency of environmental markets. Table 1 presents the conceptual definitions from previous related studies.

Table 1. Green Fintech: Definition of concept

Authors	Conceptual Definition
Puschmann and Khmarskyi (2024)	Green Fintech involves the application of fintech innovations to enhance green finance, towards supporting sustainable development and environmental protection. This integration facilitates the development of financial services that promote environmental sustainability.
Aboalsamh et al., (2023)	Green Fintech encompasses technology-driven innovations applied to financial processes and products with the intentional goal of supporting Sustainable Development Goals or mitigating sustainability risks. This includes leveraging data analytics, blockchain, and other technologies to create financial solutions that address environmental challenges.
Kwong et al., (2023)	Green Fintech refers to the use of fintech solutions to promote environmental sustainability by providing financial services that support green projects, reduce carbon footprints, and encourage sustainable practices among consumers and businesses.
Macchiavello and Siri (2022)	Green Fintech is defined as the utilization of technological advancements in the financial sector to drive sustainable finance, enabling the development of financial products and services that contribute to environmental conservation and sustainability.
Aboalsamh et al., (2023)	Green Fintech involves the deployment of digital financial innovations aimed at addressing climate change by facilitating investments in renewable energy, energy efficiency, and other environmentally friendly initiatives. This includes the use of blockchain for transparent tracking of green investments.
Geetha and Biju (2024)	Green Fintech refers to digital platforms that offer financial services designed to support environmental sustainability, including green loans, eco-friendly investment options, and carbon footprint tracking tools for consumers and businesses.
Blakstad et al., (2018)	Green Fintech involves fintech innovations that drive green economic growth by providing financial solutions that facilitate the transition to a low-carbon economy, support renewable energy projects, and promote sustainable business practices.
Kabaklarlı (2022)	Green Fintech represents eco-innovative financial technologies that integrate environmental considerations into financial services, aiming to reduce ecological footprints, support sustainable resource management, and promote environmental stewardship through financial mechanisms.

4.2 Solutions Towards Financed Emissions

The transformative potential of green fintech solutions in mitigating financed emissions is vital, as evidenced by the reviewed various studies highlighting the integration of technology and finance to promote sustainability. Green fintech leverages innovations such as AI, IoT, and smart contracts to enhance environmental assessments, improve energy efficiency, and facilitate green investments, thereby embedding sustainability into financial decision-making. In line with Green Digital Finance Alliance (2024), this study breaks down green fintech solutions into six

(6) categories namely: green digital asset solutions, digital ESG-data and analytics solutions, green digital investment solutions, green digital deposit and lending solutions, green digital crowdfunding and syndication platforms and lastly green digital payment and account solutions. All these solutions and other related mechanism are presented in Figure 1.

4.2.1 Green Digital Asset Solutions

The growing focus on sustainability in finance has catalysed the emergence of green fintech solutions, with green digital asset solutions playing a pivotal role in mitigating financed emissions. Green digital asset solutions are defined by their use of blockchain to represent real or intangible green assets digitally, facilitating ownership, investment, and accountability in ways that traditional systems often fail to achieve. This typifies the integration of technological innovation and environmental stewardship, leveraging blockchain infrastructure to enable the tokenization of green assets. The transformative potential of these solutions lies in their ability to align financial markets with sustainability goals (Blakstad et al., 2018; Cen & He, 2018; Sergeeva et al., 2023). These innovative solutions are powered by smart contracts, which automate processes such as settlement, reconciliation, and distribution, ensuring efficiency and reliability. For instance, integration platforms like the green security token offering platforms involves the tokenization of green assets that has to do with the conversion of tangible or intangible environmental resources, such as carbon credits or biodiversity offsets, into digital tokens tradable on blockchain platforms. This process not only enhances transparency but also democratizes access to green investments. As noted by Chikov et al. (2023) and Omar et al. (2021), the automation capabilities of smart contracts significantly reduce transaction costs and enhance scalability, making green fintech solutions an attractive avenue for institutional and retail investors alike.

4.2.2 Digital ESG-Data and Analytics Solutions

A defining feature of digital ESG-data solutions is their ability to automate and standardize data collection and analysis across diverse sources. artificial intelligence (AI) and machine learning (ML) algorithms scrape unstructured data from reports, media, social platforms, and public databases, translating this information into actionable insights. This implies that digital environmental, social, and governance (ESG) data and analytics solutions represent a cornerstone of green fintech's transformative potential towards addressing financed emissions. These solutions empower financial stakeholders to make informed, sustainability-aligned decisions by automating the collection, analysis, and reporting of ESG data (Sarkar, 2024;

Saxena et al., 2022). For instance, automated ESG ratings and carbon accounting offer a real-time, data-driven understanding of a company's or asset's environmental footprint (Lee et al., 2024; Sokolov et al., 2021).

4.2.3 Green Digital Investment Solutions

Green digital investment solutions involve green robo-advisory platforms that are transforming the investment landscape by embedding sustainability directly into portfolio allocation and risk assessment. This represents a significant innovation in sustainable finance which offers automated algorithm-driven tools for managing investments with a focus on environmental, social, and governance (ESG) considerations (Javeed et al. 2024; Ye et al., 2023). A hallmark of green digital investment solutions is their use of sophisticated algorithms to provide automated investment advice and portfolio management. These platforms enable investors to align their financial goals with sustainability objectives through minimal manual intervention (Lei et al., 2024). For example, robo-advisors tailor investment strategies based on user-defined preferences, such as prioritizing renewable energy or circular economy initiatives. The automation inherent in these platforms ensures efficiency and scalability, allowing for greater democratization of sustainable investing. Alekseenko (2023) highlights how automation reduces barriers to entry for retail investors, thereby broadening participation in green finance.

4.2.4 Green Digital Deposit and Lending Solutions

Green digital deposit and lending solutions encompass a range of services, including loans for green purchases, transition finance, and mortgages tailored to support energy efficiency. These products leverage digital tools, such as IoT devices and automated monitoring, to assess and promote environmentally responsible behaviours (Suri et al., 2021). Green digital deposit and lending solutions exemplify the integration of digital innovation and sustainability in the financial sector. These solutions leverage technology to channel financial resources into environmentally beneficial projects, promoting green consumption and sustainable asset management. Digital savings and loan products incentivize green behaviours, as these tools have the potential to significantly reduce financed emissions (Cho & Lee, 2024). This is because this category of green fintech not only facilitates the transition to a low-carbon economy but also reshapes consumer and corporate financial practices. For instance, green digital loans and transition loans are tied to measurable environmental outcomes, enabling lenders to monitor compliance and incentivize borrowers to adopt sustainable practices. Such mechanisms ensure that financing aligns with green objectives, addressing a critical gap in traditional lending frameworks where environmental impacts are often overlooked (Emmert 2023; Kusumasari et al., 2024).

4.2.5 Green Digital Crowdfunding and Syndication Platforms

Green digital crowdfunding and syndication platforms are at the forefront of democratizing access to sustainable finance (Cumming et al., 2024; Testa et al., 2019). Green digital crowdfunding platforms leverage the power of collective action to fund sustainable projects. Energy efficiency projects, and renewable energy initiatives connect funders and users of funds through web-based platforms, these solutions enable the mobilization of capital for green ventures. Osby and Anderson (2023) highlights the transformative potential of these platforms which lies in their ability to decentralize financing, promote grassroots involvement in sustainability, and align investment flows with environmental goals. The degree of automation on these platforms enhances their scalability and operational efficiency because any green crowdfunding solutions incorporate features like automated due diligence, real-time tracking of investments, and integration with payment systems (O'Reilly 2024; Sreelekshmi & Biju, 2023).

4.2.6 Green Digital Payment and Account Solutions

Green digital payment and account solutions leverage payment data and open banking standards to provide users with insights into the environmental impact of their purchases. This represent a cutting-edge approach to fostering environmentally conscious consumer behaviour and facilitating carbon accountability (Shala & Berisha, 2024; Zhang 2023). Through innovative tools and gamified reward systems, these solutions encourage behavioural shifts toward sustainability while enabling automated carbon accounting and offsetting. Moreover, the core functionality of green payment and account solutions lies in their ability to analyze payment transactions and quantify their environmental impact (Sergeeva et al., 2023). Such platforms hypothetically translate purchase prices into corresponding carbon emissions and recommending offset measures, which consequently empower consumers to understand and address the environmental costs of their actions. This implies that green digital payment and account solutions can bridge the gap between individual consumption and collective environmental goals, fostering greater awareness and responsibility.

Figure 1. Green Fintech's potential and financed emissions

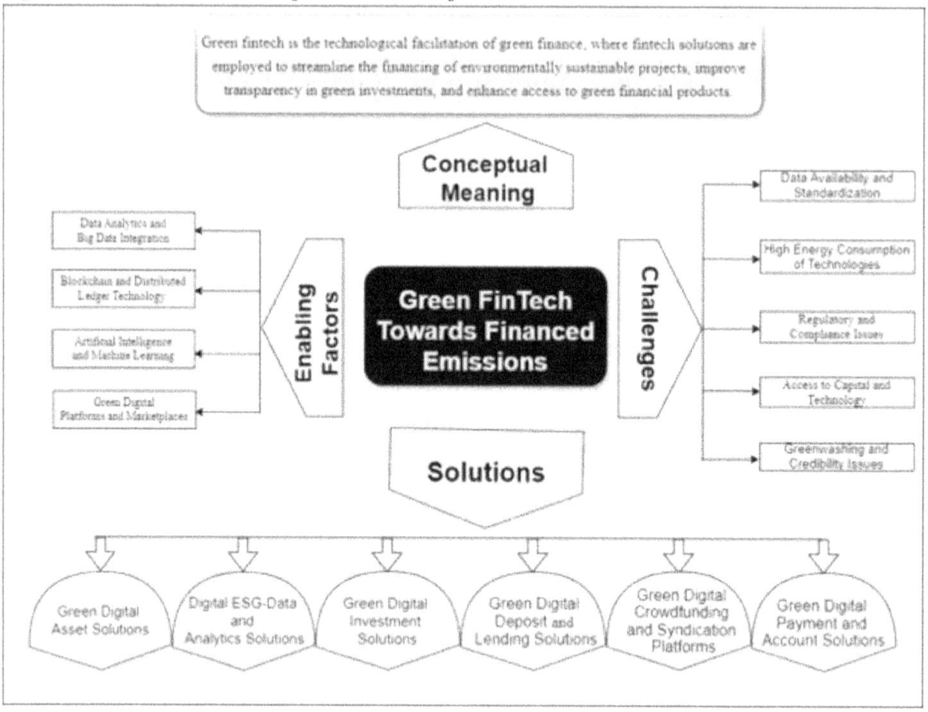

4.3 Enabling Factors of Green Fintech

Green fintech, enabled by data analytics, blockchain, AI, digital platforms, and supportive policy frameworks, is argued in this study to be critical in transforming financed emissions. However, the effective realization of its potential requires addressing various specific enablers such as data analytics and big data integration, blockchain and distributed ledger technology, artificial intelligence and machine learning, and green digital platforms and marketplaces (see Figure 1).

4.3.1. Data Analytics and Big Data Integration

Data analytics and big data integration form the cornerstone of green fintech's ability to mitigate financed emissions by enabling robust environmental assessments and decision-making. This is because green fintech utilizes advanced data analytics and big data to track, measure, and report financed emissions, enabling more informed decision-making (Addy et al., 2024). Sarkar (2024) argue that the integration of big data in fintech platforms significantly enhances environmental risk assessment capabilities due to the prowess of big data technologies allowing financial institutions to gather, process, and analyse vast amounts of ESG data. This prospect allows institutions to channel resources toward initiatives that deliver the highest environmental and financial returns, thus mitigating financed emissions on a broader scale

4.3.2. Blockchain and Distributed Ledger Technology

Blockchain ensures transparency and traceability in green finance transactions, providing immutable records of environmental impact which offers an unparalleled transparency and traceability in mitigating financed emissions (Dorfleitner & Braun, 2019). The immutable nature of blockchain records ensures that all transactions, whether related to carbon credits, green bonds, or sustainable investments, are transparent and verifiable. This feature addresses long-standing issues of fraud and double-counting in carbon trading markets, fostering trust among stakeholders (Bhattacharya & Sachdev, 2024; Hui & Loang, 2023). This capability thus helps bridge the trust gap in green finance, enabling the scaling of sustainability-focused investments. Moreover, transitioning to energy-efficient algorithms is crucial to align with green fintech objectives due to the high energy consumption of certain blockchain technologies, like Bitcoin, poses sustainability challenges (Singh & Kaunert, 2024). This implies that when supportive ecosystem for blockchain adoption is fostered, green fintech can drive significant progress in reducing financed emissions and advancing the transition to a low-carbon economy.

4.3.3. Artificial Intelligence and Machine Learning

Artificial Intelligence (AI) and Machine Learning (ML) enhance the predictive capabilities of green fintech systems, enabling proactive identification of emission-intensive activities through their ability to assess environmental impacts with high precision and scalability (Mohammed et al., 2024; Sokolov et al., 2021). AI tools accurately predict emissions, assess environmental risks, and guide sustainable investment decisions through processing vast amounts of environmental data. Another key role of AI and ML in green fintech is the automation of ESG data collection and analysis (Alekseenko 2023). Green fintech solutions leverage AI to scrape and analyze unstructured data from a range of sources, including company reports, social media, satellite imagery, and environmental data, to evaluate a company's sustainability performance. Machine learning models then aggregate this data to offer more accurate ESG ratings, predict long-term environmental risks, and assess a company's carbon emissions reduction potential (Cen & He, 2018; Sokolov et al., 2021). These AI-driven capabilities enable investors and financial institutions to assess their portfolios and align investments with low-carbon or environmentally friendly ventures, thus mitigating financed emissions.

4.3.4. Green Digital Platforms and Marketplaces

Digital platforms enable seamless interactions between stakeholders in green finance, due to their integral nature towards the transformative potential of green fintech, thereby providing the infrastructure for more transparent, efficient, and inclusive financial systems that mitigate financed emissions (Macchiavello & Siri, 2022; Zhang 2023). These platforms facilitate the aggregation and dissemination of critical sustainability data, enabling stakeholders to make informed decisions aligned with environmental goals (Dorfleitner & Braun, 2019). For instance, digital platforms like carbon trading marketplaces and climate-focused investment platforms connect investors, corporations, and consumers, promoting accountability and aligning financial flows with sustainability benchmarks. These platforms leverage blockchain technology to enhance traceability and trust in carbon credit transactions, ensuring that funds are directed toward verified emission reduction projects. Through transparent digital marketplaces, corporations can offset their financed emissions while gaining access to high-quality carbon credits. these platforms make participation in sustainability efforts more accessible, through increased efficiency and reducing the costs associated with traditional carbon markets, particularly for small and medium-sized enterprises (SMEs), which often face barriers in conventional finance mechanisms (Alabi, 2025).

4.4 Associated Challenges

Green fintech faces significant challenges in transforming financed emissions despite its potential to foster sustainability. This study posits several challenges to include data standardization, energy consumption, regulatory uncertainties, greenwashing, and resource constraints. However, robust frameworks are required to address these barriers in order to overcome these obstacles and unlock the full potential of green fintech in transforming financed emissions.

4.4.1. Data Availability and Standardization

One of the primary challenges in data availability is the fragmented nature of ESG reporting, of which green fintech relies heavily on accurate, reliable, and timely ESG data to drive decision-making, assess financed emissions, and implement sustainable financial solutions (Gade, 2023). However, the lack of standardized frameworks and inconsistent data availability across industries, regions, and institutions creates significant barriers to the effective deployment of green fintech solutions. Companies and financial institutions often use disparate methodologies to measure and disclose their carbon footprints and other sustainability metrics. This lack of uniformity makes it difficult for green fintech solutions to aggregate, analyse, and compare data effectively. For instance, variations in how companies report Scope 3 emissions, those from their supply chains and downstream activities can result in substantial gaps or inaccuracies, undermining the ability of fintech tools to offer precise carbon accounting or emission mitigation strategies. Without comprehensive and inclusive data, green fintech solutions cannot fully address emissions at a systemic level (Sarkar, 2024).

Standardization of data is equally problematic (Addy et al., 2024). Even when ESG data is available, the absence of universal standards leads to discrepancies in definitions, measurement criteria, and reporting formats. For example, different financial institutions may use varying benchmarks to assess the same ESG factor, such as energy efficiency or biodiversity impact. This lack of harmonization complicates the integration of ESG data into green fintech tools, hindering their ability to provide actionable insights. It also reduces investor confidence in the reliability of the data, potentially deterring participation in green financial markets. Addressing these challenges requires coordinated action across multiple dimensions. Initiatives such as the Partnership for Carbon Accounting Financials (PCAF) and the Task Force on Climate-related Financial Disclosures (TCFD) are crucial in promoting standardized reporting frameworks.

4.4.2. High Energy Consumption of Technologies

The high energy consumption of certain technologies poses a significant challenge for green fintech in its efforts to mitigate financed emissions and foster sustainability (Zhu et al., 2024). Technologies such as blockchain, AI, and large-scale data analytics, which underpin many green fintech solutions, often require substantial computational power, leading to increased energy demands. This paradoxically weakens the sustainability goals of green fintech by contributing to indirect emissions, particularly in regions where energy grids are still reliant on fossil fuels (Nenavath, 2022). Blockchain technology, widely celebrated for its potential to enhance transparency and traceability in carbon markets and green finance, exemplifies this challenge. Public blockchains like Bitcoin and Ethereum rely on energy-intensive consensus mechanisms, such as Proof of Work (PoW), which involve solving complex mathematical problems to validate transactions. Studies estimate that the annual energy consumption of Bitcoin alone rivals that of entire countries, raising concerns about its environmental impact. While newer blockchain models, such as Proof of Stake

(PoS), offer more energy-efficient alternatives, widespread adoption remains limited due to technological and institutional inertia.

Moreover, the challenge of high energy consumption is compounded by the global disparities in energy infrastructure. In developing regions, where energy grids are less stable or more carbon-intensive, the adoption of green fintech technologies can inadvertently exacerbate local environmental issues. This highlights the need for localized strategies that prioritize renewable energy integration and energy efficiency alongside technological deployment. Addressing the high energy consumption of these technologies requires coordinated efforts at multiple levels. Innovations in green computing, such as energy-efficient hardware and optimized algorithms, can play a crucial role in reducing the carbon footprint of green fintech operations. Policy measures that incentivize the use of renewable energy for powering fintech infrastructures, such as tax benefits or subsidies, can also accelerate the transition to more sustainable practices. Furthermore, collaboration among stakeholders to prioritize energy-efficient technologies, such as PoS blockchains, can enhance the alignment of green fintech solutions with their overarching sustainability objectives.

4.4.2. Regulatory and Compliance Issues

A key challenge lies in the fragmented nature of environmental and financial regulations across jurisdictions. Different countries and regions have varying standards for carbon accounting, ESG disclosures, and sustainable finance practices (Pollman, 2019). This regulatory patchwork creates complexities for green fintech firms operating across borders, making it difficult to ensure compliance while maintaining efficiency. For example, a green fintech platform offering carbon credit trading might face different rules for credit verification, taxation, and reporting in each country, increasing operational costs and risks. Moreover, regulatory frameworks for emerging technologies like blockchain, AI, and digital platforms which are foundational to green fintech are still evolving. Blockchain-based carbon marketplaces, for instance, often encounter legal uncertainties regarding the recognition of digital assets, data privacy, and cross-border transactions (Bhattacharya & Sachdev, 2024). Similarly, AI-driven ESG rating systems may face scrutiny over algorithmic transparency and accountability, raising concerns about bias and fairness. These regulatory uncertainties can deter investment and innovation, slowing the deployment of green fintech solutions. In order to address these challenges, collaborative efforts between regulators, policymakers, and industry stakeholders are essential. Regulatory sandboxes, for instance, can provide green fintech companies with a controlled environment to test their innovations while ensuring compliance with regulatory standards. Initiatives like the EU's Sustainable Finance Disclosure Regulation (SFDR) and the establishment of green taxonomies aim to create clearer and more consistent regulatory guidelines, which can support the growth of green fintech solutions.

4.4.4. Greenwashing and Credibility Issues

Greenwashing which could mean misrepresenting financial products as environmentally sustainable undermines trust in green fintech solutions. Greenwashing occurs when companies or financial products falsely present themselves as environmentally friendly, either by exaggerating their sustainability claims or by misrepresenting their environmental impact. These practices undermine the trust that is essential for green fintech solutions to function effectively, thereby stalling their transformative potential (Macchiavello & Siri, 2022; Mohammed et al., 2024). This credibility gap is particularly problematic for technologies like blockchain-based carbon marketplaces or AI-driven ESG scoring systems. While these innovations have the potential to enhance transparency, they are vulnerable to manipulation if the underlying data is inaccurate or unverifiable. For instance, if a blockchain-based platform allows unverifiable carbon credits to be traded, it perpetuates the problem of "phantom credits," where emissions reductions exist only on paper. Similarly, ESG ratings derived from biased or incomplete data can mislead investors, steering capital toward entities that fail to meaningfully address financed emissions.

The rise of greenwashing also puts green fintech companies at risk of reputational damage. If a green fintech solution is found to facilitate greenwashing intentionally or unintentionally, it can lose the trust of its users and stakeholders (Rousseau et al., 2021). This distrust can deter individuals and institutions from adopting green fintech platforms, slowing their ability to scale and create meaningful environmental impact. The resulting lack of accountability not only hampers efforts to mitigate financed emissions but also undermines the broader credibility of sustainable finance. However, technologies such as blockchain can help improve traceability and verification in green finance by providing immutable records of transactions and ensuring that funds are directed toward verified projects. For example, using smart contracts to release funds only upon meeting predefined sustainability milestones can enhance accountability.

4.4.5. Access to Capital and Technology

Green fintech start-ups and initiatives often struggle with securing adequate funding and technological resources, especially in developing regions due to the unequal access to financial and technological resources, which creates disparities that hinder the widespread adoption and impact of green fintech, particularly in emerging markets and among smaller enterprises (Bollaert et al., 2021; Giakoumelou et al., 2024). A primary barrier is the high upfront investment required to develop innovative green fintech solutions (Bollaert et al., 2021). For instance, building AI-driven platforms to assess ESG risks or blockchain-based systems for carbon trading involves substantial costs for research, infrastructure, and skilled labour. Many green fintech start-ups struggle to secure the necessary funding due to the perceived risks and uncertainties associated with their business models. Traditional investors may be hesitant to back these ventures, viewing them as less profitable or more speculative compared to conventional fintech or financial services.

Access to technology further compounds the issue, particularly in developing regions where digital infrastructure is limited. Green fintech solutions often rely on robust internet connectivity, cloud computing, and advanced data processing capabilities. In regions with underdeveloped technological infrastructure, these prerequisites are not readily available, creating a digital divide that prevents equitable participation in green fintech ecosystems. Additionally, smaller enterprises and start-ups frequently face challenges in adopting cutting-edge technologies due to their cost and complexity. Unlike large financial institutions, which can afford to invest in proprietary platforms and hire specialized talent, smaller players often lack the resources to integrate advanced green fintech solutions into their operations. This limits the reach of green fintech and its ability to address financed emissions comprehensively, as SMEs play a critical role in global supply chains and economic systems (Alabi, 2025). Importantly, targeted interventions are necessary to overcome these challenges through public and private funding mechanisms which can play a key role in bridging the capital gap for green fintech start-ups. On the technological front, partnerships between fintech companies, technology providers, and governments can help address infrastructure gaps. For instance, collaborations to deploy affordable cloud computing solutions or enhance rural internet connectivity can expand access to green fintech tools in underserved regions. Open-source platforms and shared technological resources can further reduce costs for smaller players, enabling them to participate in and contribute to the green fintech ecosystem.

5. CONCLUSIONS

This study explores the transformative potential of green fintech solutions in mitigating financed emissions, synthesizing insights from recent literature. Green fintech, characterized by its integration of advanced technologies like AI, blockchain, big data, and digital platforms, has emerged as a fundamental force in aligning financial systems with sustainability objectives. Various solutions ranging from digital ESG analytics and green investment tools to crowdfunding and payment systems, demonstrate the capability to address financed emissions through innovation, transparency, and stakeholder engagement. Enabling factors such as data analytics, blockchain, and AI underpin the success of green fintech, facilitating real-time decision-making, enhancing accountability, and democratizing access to green financial products. However, this potential is tempered by significant challenges, including data availability and standardization, regulatory uncertainties, and concerns about greenwashing and credibility. The study concludes that while green fintech holds immense promise, its success in addressing financed emissions depends on overcoming these systemic barriers and fostering collaborative efforts across sectors.

The study puts forward that financial institutions and green fintech start-ups should prioritize transparency and accountability to enhance stakeholder trust. Investing in energy-efficient technologies, embracing standardized ESG reporting frameworks, and mitigating the risks of greenwashing are essential steps for scaling impactful solutions. Further, regulatory bodies should create harmonized global standards for ESG data and green finance, addressing fragmentation and ensuring market integrity. Initiatives like regulatory sandboxes can encourage innovation while safeguarding consumer interests.

Despite the study's comprehensive approach, this study faces few limitations. The study mainly relies on secondary data from existing literature, limiting its ability to provide direct empirical evidence on the effectiveness of specific green fintech solutions in mitigating financed emissions. In addition, the study draws from globally available literature but may not fully capture region-specific dynamics, particularly in emerging markets where green fintech adoption faces unique challenges. Addressing these limitations calls for more region-specific, longitudinal, and empirical studies that analyse the real-world implementation of green fintech solutions. Given these limitations, the study's deductions are still relevant and pivotal with respect to green fintech's role in mitigating financed emissions and promoting long-term sustainability in the financial sector.

REFERENCES

Aboalsamh, H. M., Khrais, L. T., & Albahussain, S. A. (2023). Pioneering perception of green fintech in promoting sustainable digital services application within smart cities. *Sustainability (Basel)*, *15*(14), 11440. DOI: 10.3390/su151411440

Adaramola, T. S., Omole, O. M., Wada, I., Nwariaku, H., Arowolo, M. E., & Adigun, O. A. (2024). Internet of thing integration in green fintech for enhanced resource management in smart cities. *World Journal of Advanced Research and Reviews*, *23*(2), 1317–1327. DOI: 10.30574/wjarr.2024.23.2.2448

Addy, W. A., Ofodile, O. C., Adeoye, O. B., Oyewole, A. T., Okoye, C. C., Odeyemi, O., & Ololade, Y. J. (2024). Data-driven sustainability: How fintech innovations are supporting green finance. *Engineering Science & Technology Journal*, *5*(3), 760–773. DOI: 10.51594/estj.v5i3.871

Alabi, A. T. (2025). *A Critical Exploration of Industry 5.0 in Sustainable Business Performance of SMEs: A Framework for i5.0 Integration. In the Future of Small Business in Industry 5.0*. IGI Global Scientific Publishing.

Alekseenko, A. (2023). Rights of investors in the context of algorithmic Artificial Intelligence technologies and automatization. Brazilian Journal of Law. *Technology and Innovation*, *1*(2), 42–62.

Alsadoun, M., & Alrobai, F. S. (2024). Influence of Fintech Adoption on Sustainable Performance via mediation role of Green Finance and Green Innovation. [AJBSP]. *American Journal of Business Science Philosophy*, *1*(1), 61–72. DOI: 10.70122/ajbsp.v1i1.13

Anshari, M., Almunawar, M. N., Masri, M., Hamdan, M., Fithriyah, M., & Fitri, A. (2021, November). Digital wallet in supporting green FinTech sustainability. In *2021 Third International Sustainability and Resilience Conference: Climate Change* (pp. 352-357). IEEE. DOI: 10.1109/IEEECONF53624.2021.9667957

Anwar, H., Waheed, R., & Aziz, G. (2024). Importance of FinTech and green finance to achieve the carbon neutrality targets: A study of australian perspective. *Environmental Research Communications*, *6*(11), 115007. DOI: 10.1088/2515-7620/ad853d

Bayram, O., Talay, I., & Feridun, M. (2022). Can FinTech promote sustainable finance? Policy lessons from the case of Turkey. *Sustainability (Basel)*, *14*(19), 12414. DOI: 10.3390/su141912414

Bhattacharya, S., & Sachdev, B. K. (2024). Regulatory Challenges and Opportunities for Future Fintech and Green Finance. Blockchain's Transformative Potential of Financial Technology for Sustainable Futures, 103-122.

Blakstad, S., Allen, R., Blakstad, S., & Allen, R. (2018). Green Fintech. FinTech Revolution: Universal Inclusion in the New Financial Ecosystem, 183-199.

Bollaert, H., Lopez-de-Silanes, F., & Schwienbacher, A. (2021). Fintech and access to finance. *Journal of Corporate Finance*, *68*, 101941. DOI: 10.1016/j.jcorpfin.2021.101941

Cai, Q., Chen, W., Wang, M., & Di, K. (2024). How Does Green Finance Influence Carbon Emission Intensity? A Non-Linear fsQCA-ANN Approach. *Polish Journal of Environmental Studies*. Advance online publication. DOI: 10.15244/pjoes/190658

Callahan, J. L. (2010). Constructing a manuscript: Distinguishing integrative literature reviews and conceptual and theory articles. *Human Resource Development Review*, *9*(3), 300–304. DOI: 10.1177/1534484310371492

Cen, T., & He, R. (2018, December). Fintech, green finance and sustainable development. In 2018 international conference on management, economics, education, arts and humanities (MEEAH 2018) (pp. 222-225). Atlantis Press.

Chikov, I. A., Koliadenko, S. V., Supryhan, V. A., Tabenska, O. I., Nitsenko, V. S., & Holinko, O. V. (2023). Smart contracts and business process automation: The technical aspect. *Naukovyi Visnyk Natsionalnoho Hirnychoho Universytetu.*, *2023*(5), 186–192. DOI: 10.33271/nvngu/2023-5/186

Cho, I., & Lee, B. G. (2024). Consumer Behavior in Achieving the Goals of ESG Banking Products: Focusing on environmental awareness and saving behavior. *Journal of Service Research and Studies*, *14*(2), 117–137.

Coffie, C. P. K., Yeboah, F. K., Emuron, A. S. O., & Ahiabenu, K. (2024). FinTech and CO2 emission: evidence from (top 7) mobile money economies in Africa. Journal of Financial Regulation and Compliance.

Cumming, D., Meoli, M., Rossi, A., & Vismara, S. (2024). ESG and crowdfunding platforms. *Journal of Business Venturing*, *39*(1), 106362. DOI: 10.1016/j.jbusvent.2023.106362

Dorfleitner, G., & Braun, D. (2019). Fintech, digitalization and blockchain: possible applications for green finance. The rise of green finance in Europe: opportunities and challenges for issuers, investors and marketplaces, 207-237.

Elias, O., Awotunde, O. J., Oladepo, O. I., Azuikpe, P. F., Samson, O. A., Oladele, O. R., & Ogunruku, O. O. (2024). The evolution of green fintech: Leveraging AI and IoT for sustainable financial services and smart contract implementation. *World Journal of Advanced Research and Reviews, 23*(1), 2710–2723. DOI: 10.30574/wjarr.2024.23.1.2272

Emmert, F. (2023). Smart Money for the People: Using Financial Innovation and Technology to Promote ESG. *Duke Law & Technology Review, 22*(1), 21.

Gade, K. R. (2023). The Role of Data Modeling in Enhancing Data Quality and Security in Fintech Companies. *CIT. Journal of Computing and Information Technology, 3*(1).

Geddes, A., Schmidt, T. S., & Steffen, B. (2018). The multiple roles of state investment banks in low-carbon energy finance: An analysis of Australia, the UK and Germany. *Energy Policy, 115*, 158–170. DOI: 10.1016/j.enpol.2018.01.009

Geetha, S., & Biju, A. V. N. (2024). Is green FinTech reshaping the finance sphere? Unravelling through a systematic literature review. *Environmental Science and Pollution Research International, 31*(2), 1790–1810. DOI: 10.1007/s11356-023-31382-y PMID: 38057679

Giakoumelou, A., Salvi, A., Bekiros, S., & Onorato, G. (2024). ESG and FinTech funding in the EU. *Research in International Business and Finance, 69*, 102233. DOI: 10.1016/j.ribaf.2024.102233

Green Digital Finance Alliance. (2024). Interim Report: A Green Fintech Taxonomy and Data Landscaping https://www.greenfinanceplatform.org/sites/default/files/downloads/resource/GreenFintechTaxonomyDataLandscaping-v5%20.pdf

Hishamuddin, M. R. A., Bujang, I., & Rimin, F. (2024). *Transformative Impacts Of Green Finance On Co2 Emissions And Bank Stability*. European Proceedings of Social and Behavioural Sciences., DOI: 10.15405/epsbs.2024.05.30

Huang, H., & Yang, Q. (2023). FinTech Promotes the Development of Green Finance. In International Conference on Economic Management and Green Development (pp. 885-892). Singapore: Springer Nature Singapore.

Hui, X., & Loang, O. K. (2023). Fintech solutions for green finance: Leveraging blockchain and AI in financing renewable energy projects. International Journal of Accounting [IJAFB]. *Finance and Business, 8*(50), 337–348.

Jaiwant, S. V., & Kureethara, J. V. (2023). Green finance and Fintech: Toward a more sustainable financial system. In *Green Finance Instruments, FinTech, and Investment Strategies: Sustainable Portfolio Management in the Post-COVID Era* (pp. 283–300). Springer International Publishing. DOI: 10.1007/978-3-031-29031-2_12

Javeed, S. A., Latief, R., Cai, X., & San Ong, T. (2024). Digital finance and corporate green investment: A perspective from institutional investors and environmental regulations. *Journal of Cleaner Production*, *446*, 141367. DOI: 10.1016/j.jclepro.2024.141367

Kabaklarlı, E. (2022). Green FinTech: Sustainability of bitcoin. *Digital Finance*, *4*(4), 265–273. DOI: 10.1007/s42521-022-00053-x PMID: 35698578

Kusumasari, L., Anggoro, R. W., & Wulandari, R. (2024). Enhancing Member Welfare in Savings and Loan Cooperatives: The Roles of Digitalization, Product Quality, and Financial Performance. *Journal of System and Management Sciences*, *14*(3), 203–225.

Kwong, R., Kwok, M. L. J., & Wong, H. S. (2023). Green FinTech innovation as a future research direction: A bibliometric analysis on green finance and FinTech. *Sustainability (Basel)*, *15*(20), 14683. DOI: 10.3390/su152014683

Lee, H., Lee, S. H., Park, H., Kim, J. H., & Jung, H. S. (2024). ESG2PreEM: Automated ESG grade assessment framework using pre-trained ensemble models. *Heliyon*, *10*(4), e26404. DOI: 10.1016/j.heliyon.2024.e26404 PMID: 38404885

Lei, Y., Bao, Y., & Zhao, B. (2024). Measuring the performance of green investment portfolios for zero-carbon environment: A comparative analysis of digital finance and asset-backed securities. *Environmental Science and Pollution Research International*, *31*(1), 357–370. DOI: 10.1007/s11356-023-31009-2 PMID: 38012491

Macchiavello, E., & Siri, M. (2022). Sustainable finance and fintech: Can technology contribute to achieving environmental goals? A preliminary assessment of 'green fintech'and 'sustainable digital finance'. *European Company and Financial Law Review*, *19*(1), 128–174. DOI: 10.1515/ecfr-2022-0005

Mishra, N. (2023). Green Fintech Revolution: Assessing the Impact of Sustainable Finance Platforms on Environmental and Social Outcomes. *Journal of Management & Entrepreneurship*, *17*(3), 103–111.

Mohammed, K. S., Serret, V., Jabeur, S. B., & Nobanee, H. (2024). The role of artificial intelligence and fintech in promoting eco-friendly investments and non-greenwashing practices in the US market. *Journal of Environmental Management*, *359*, 120977. DOI: 10.1016/j.jenvman.2024.120977 PMID: 38678903

Muganyi, T., Yan, L., & Sun, H. P. (2021). Green finance, fintech and environmental protection: Evidence from China. *Environmental Science and Ecotechnology*, *7*, 100107. DOI: 10.1016/j.ese.2021.100107 PMID: 36160697

Nelson, C. S. (2022). *Financed Emissions–Methodologies and Implication for Global Financial Institutions. Nicholas Sch. Environ.* Duke University.

Nenavath, S. (2022). Impact of fintech and green finance on environmental quality protection in India: By applying the semi-parametric difference-in-differences (SDID). *Renewable Energy*, *193*, 913–919. DOI: 10.1016/j.renene.2022.05.020

Nenavath, S., & Mishra, S. (2023). Impact of green finance and fintech on sustainable economic growth: Empirical evidence from India. *Heliyon*, *9*(5), e16301. DOI: 10.1016/j.heliyon.2023.e16301 PMID: 37234625

Nordea. (2024). What are financed emissions? Retrieved December 31, 2024, from https://www.nordea.com/en/news/what-are-financed-emissions

O'Reilly, S. (2024). Equity Crowdfunding as a Source of Finance for Early-Stage Cleantech Firms: Exploring the Role of Patents. *International Review of Entrepreneurship*, *22*(1).

Omar, I. A., Jayaraman, R., Debe, M. S., Salah, K., Yaqoob, I., & Omar, M. (2021). Automating procurement contracts in the healthcare supply chain using blockchain smart contracts. *IEEE Access : Practical Innovations, Open Solutions*, *9*, 37397–37409. DOI: 10.1109/ACCESS.2021.3062471

Osby, K., & Anderson, J. (2023). Unleashing the Green: Why Green Fintech Should be the Focus in Climate Change Mitigation. *Journal of Student Research*, *12*(4). Advance online publication. DOI: 10.47611/jsrhs.v12i4.5811

Pollman, E. (2019). Corporate social responsibility, ESG, and compliance. Forthcoming, Cambridge Handbook of Compliance (D. Daniel Sokol & Benjamin van Rooij eds.), Loyola Law School, Los Angeles Legal Studies Research Paper, (2019-35).

Puschmann, T., Hoffmann, C. H., & Khmarskyi, V. (2020). How green FinTech can alleviate the impact of climate change—The case of Switzerland. *Sustainability (Basel)*, *12*(24), 10691. DOI: 10.3390/su122410691

Puschmann, T., & Khmarskyi, V. (2024). Green fintech: Developing a research agenda. *Corporate Social Responsibility and Environmental Management*, *31*(4), 2823–2837. DOI: 10.1002/csr.2675

Rousseau, S., Gendron, E., Morales, M., & Payette, D. (2021). ESG Tech: Attractions and Challenges for Fintechs in the Age of COVID-19. *Banking & Finance Law Review*, *37*(1), 57–96.

Sarkar, R. (2024). Advancements in ESG data management: Harnessing cloud technologies for sustainability. [IJCET]. *International Journal of Computer Engineering and Technology*, *15*(6), 337–356.

Saxena, A., Singh, R., Gehlot, A., Akram, S. V., Twala, B., Singh, A., Montero, E. C., & Priyadarshi, N. (2022). Technologies empowered environmental, social, and governance (ESG): An industry 4.0 landscape. *Sustainability (Basel)*, *15*(1), 309. DOI: 10.3390/su15010309

Sengupta, R., & Patil, A. A. (2023). Green Finance Products and Investments in the Changing Business World. In Handbook of Research on Sustainable Consumption and Production for Greener Economies (pp. 344-357). IGI Global. DOI: 10.4018/978-1-6684-8969-7.ch020

Sergeeva, S., Gladilina, I., Bulochnikova, N., Shichkin, I., Kochetkov, E., & Filonova, A. (2023). Impact of green fintech solutions on achieving esg principles. *Revista Gestão & Tecnologia*, *23*(1), 335–345.

Shahbaz, M., Nasir, M. A., Hille, E., & Mahalik, M. K. (2020). UK's net-zero carbon emissions target: Investigating the potential role of economic growth, financial development, and R&D expenditures based on historical data (1870–2017). *Technological Forecasting and Social Change*, *161*, 120255. DOI: 10.1016/j.techfore.2020.120255 PMID: 32904903

Shala, A., & Berisha, V. (2024). The Impact of Fintech on the Achievement of Environmental, Social, and Governance (ESG) Goals. In Sustainable Development Goals: The Impact of Sustainability Measures on Wellbeing (pp. 55-77). Emerald Publishing Limited.

Shan, H., Wong, W. K., Hu, H., Al Shraah, A., Alromaihi, A., Cong, P. T., & Uyen, P. T. M. (2024). Fintech innovation for sustainable environment: Understanding the role of natural resources and human capital in BRICS using MMQR. *Resources Policy*, *88*, 104468. DOI: 10.1016/j.resourpol.2023.104468

Sharma, R., Vasishta, P., & Singla, A. (2024). Impact of green banking awareness on green FinTech adoption: A way towards profitable and sustainable practices. *Managerial Finance*.

Sharma, R., Vasishta, P., & Singla, A. (2024). Impact of green banking awareness on green FinTech adoption: A way towards profitable and sustainable practices. *Managerial Finance*. Advance online publication. DOI: 10.1108/MF-04-2024-0272

Singh, B., & Kaunert, C. (2024). Revealing Green Finance Mobilization: Harnessing FinTech and Blockchain Innovations to Surmount Barriers and Foster New Investment Avenues. In Harnessing Blockchain-Digital Twin Fusion for Sustainable Investments (pp. 265-286). IGI Global.

Sokolov, A., Mostovoy, J., Ding, J., & Seco, L. (2021). Building machine learning systems for automated ESG scoring. *The Journal of Impact and ESG Investing*, *1*(3), 39–50. DOI: 10.3905/jesg.2021.1.010

Sreelekshmi, G., & Biju, A. V. (2023). Leveraging of Climate Fintech Model in India: Scoping through a Qualitative Approach using a Sentiments Analysis.

Stern, N., & Valero, A. (2021). Innovation, growth and the transition to net-zero emissions. *Research Policy*, *50*(9), 104293. DOI: 10.1016/j.respol.2021.104293 PMID: 36540358

Suri, T., Bharadwaj, P., & Jack, W. (2021). Fintech and household resilience to shocks: Evidence from digital loans in Kenya. *Journal of Development Economics*, *153*, 102697. DOI: 10.1016/j.jdeveco.2021.102697

Tao, Z., & Chao, J. (2024). Investing in green, sustaining the planet: The role of fintech in promoting corporate green investment in the Chinese energy industry. *Journal of Environmental Management*, *370*, 122990. DOI: 10.1016/j.jenvman.2024.122990 PMID: 39461146

Task Force on Climate-related Financial Disclosures. (2021). Implementing the recommendations of the Task Force on Climate-related Financial Disclosures. Retrieved from https://www.fsb-tcfd.org/publications/

Testa, S., Nielsen, K. R., Bogers, M., & Cincotti, S. (2019). The role of crowdfunding in moving towards a sustainable society. *Technological Forecasting and Social Change*, *141*, 66–73. DOI: 10.1016/j.techfore.2018.12.011

Torraco, R. J. (2016). Writing integrative literature reviews: Using the past and present to explore the future. *Human Resource Development Review*, *15*(4), 404–428. DOI: 10.1177/1534484316671606

Wang, C. A., Wang, L., Zhao, S., Yang, C., & Albitar, K. (2024). The impact of Fintech on corporate carbon emissions: Towards green and sustainable development. *Business Strategy and the Environment*, *33*(6), 5776–5796. DOI: 10.1002/bse.3778

Wu, Q. (2024). From bits to emissions: How FinTech benefits climate resilience? *Empirical Economics*, *67*(5), 1–29. DOI: 10.1007/s00181-024-02609-9

Ye, F., Ouyang, Y., & Li, Y. (2023). Digital investment and environmental performance: The mediating roles of production efficiency and green innovation. *International Journal of Production Economics*, *259*, 108822. DOI: 10.1016/j.ijpe.2023.108822

Zhang, D. (2023). Can digital finance empowerment reduce extreme ESG hypocrisy resistance to improve green innovation? *Energy Economics*, *125*, 106756. DOI: 10.1016/j.eneco.2023.106756

Zhang, X., Zhao, Y., & Liu, X. (2024). Can fintech promote environmental investment in heavily polluting enterprises? evidence from China. *Frontiers in Environmental Science*, *12*, 1373685. DOI: 10.3389/fenvs.2024.1373685

Zhang, Y. (2023). Impact of green finance and environmental protection on green economic recovery in South Asian economies: Mediating role of FinTech. *Economic Change and Restructuring*, *56*(3), 2069–2086. DOI: 10.1007/s10644-023-09500-0

Zhang, Y., Chen, M., Zhong, S., & Liu, M. (2024). Fintech's role in carbon emission efficiency: Dynamic spatial analysis. *Scientific Reports*, *14*(1), 23941. DOI: 10.1038/s41598-024-74834-2 PMID: 39397038

Zhou, C., Zhou, J., Chen, G., & Xu, C. (2024). Digital finance and greener emissions: Evidence from China. *Applied Economics*, *56*(49), 5830–5844. DOI: 10.1080/00036846.2023.2266599

Zhu, Y., Lin, Y., Tan, Y., Liu, B., & Wang, H. (2024). The potential nexus between fintech and energy consumption: A new perspective on natural resource consumption. *Resources Policy*, *89*, 104589. DOI: 10.1016/j.resourpol.2023.104589

ADDITIONAL READINGS

De Haas, R., & Popov, A. A. (2019). Finance and carbon emissions. Available at *SSRN* 3459987.

Chapter 9
Decentralized Financial Assets:
An Attraction Beyond Stock Market Investments

Anand Patil
https://orcid.org/0000-0003-4736-6571
Christ University, India

V. K. R. Akilesh
Christ University, India

R. B. Lakshmi
https://orcid.org/0009-0004-1613-4473
Christ University, India

ABSTRACT

The Chapter revolves around the attraction that the blockchain-backed decentralized financial products have created in the investment market. These new-gen decentralized finances would be Defi, Metaverse coins, Stablecoins, Cryptocurrency, smart contracts, and privacy coins. The influence of social media and high internet penetration levels have made the retail and partial investors very well aware of many alternate financial investment products. Though there is an increase seen in the Stock Market investments as well a drastic increase in the alternate market is seen. In recent times major stock markets have created indexes separately. It is the icing on the cake with the highest attention of the globe. The regulations have put the Defi Assets under unpredictable volatility but the attraction towards cryptocurrency and other digital asset gain still exists. The study aims to identify the relationship between the increase in investment in the stock market and the alternative.

DOI: 10.4018/979-8-3373-0355-0.ch009

1. INTRODUCTION

The financial models of traditional days have been shaken to their very foundations by the offensive of cryptocurrencies and decentralized finance (DeFi) that led to a realignment in the way investors monitor and analyze market trends. Because of the increasing popularity of digital assets the new indexes based on cryptocurrencies have taken birth, but long-established traditional indexes like the S&P 500, DJIA, and Bloomberg U.S. Aggregate Bond Index have been utilized for decades as benchmarks for the stock and bonds market (Smith & Jones, 2019; Wilson, 2020). Organized by financial institutions such as Bloomberg, the crypto indices help us to track the performance of major digital currencies, such as Bitcoin and Ethereum, according to market capitalization (Davis & Thompson, 2021).

Despite the growth in Defi, not much research has been undertaken by scholars to know how these new cryptocurrency indexes influence investor strategies and interact with existing market standards. The literature that is currently available tends to oversimplify the impacts of blockchain technology and cryptocurrencies without fully addressing how they can be incorporated into index-based investing for portfolio management. In addition, from the viewpoint of diversified investment approaches, the volatile and speculative character of cryptocurrency markets, often compounded by leveraged products such as spread bets and CFDs, presents particular risks that warrant additional scrutiny (Fisher & Martin, 2021; Brown et al., 522).

This research scrutinizes how the interaction between cryptocurrency indices and international stock market indices is growing in an attempt to bridge this research gap. Specifically in developing nations such as India, where traditional benchmarks such as the Nifty index are predominant in investment culture, it particularly aims to measure how these indexes interact to influence investor behavior (Kumar & Patel, 2023). This supports in presenting a broader view of how financial decision-making is evolving within today's investing environment as a result of digital disruption.

2. LITERATURE REVIEW

Index Changes and Market Reactions

Research on how index changes affect stock prices and trading volumes constitutes a fundamental area of study in emerging markets. Rahman and Rajib (2014) researched how changes in the Nifty Index affected the market from 1998 through 2011. The study results support the Price Pressure Hypothesis (PPH), demonstrating temporary increases in both price and volume when stocks undergo index status changes. The research establishes market sentiment and short-term imbalances as important factors in the Indian stock market. Their research analysis remains centered on historical patterns while failing to examine the modern impacts of digital and algorithmic trading methods.

Investor Behavior in Times of Market Volatility

Investor psychology in times of crises has also been a subject of academic interest. Chittineni (2020) studied investor reactions during the COVID-19 pandemic and observed that Indian investors, unlike worldwide panic sentiments, were quite immune to short-term volatility. Rather, they made choices based on fundamental signals, viewing falls in the market as opportunities to buy. This change in behavior is an indication of the more mature investor pool within emerging markets, even though this report has not included newer classes of investors transacting with decentralized finance (DeFi) and digital assets.

Geopolitical Events and Market Dynamics

Building upon the subject of volatility, Chittineni (2020) examined further how geopolitics affects market behavior through the India VIX (IVIX). The research indicated that improved understanding of such events could improve market predictions. However, it provided scant analysis of the ways these dynamics engage with newer asset classes such as cryptocurrencies, which are similarly affected by geopolitics.

Sectoral Interdependence and Index Correlations

Amuthan (2021) examined the performance and interdependence of different NSE indices in the post-2019 recovery. His research indicated strong correlations between the Nifty 50, Nifty IT, and Nifty Consumer Durables indices, while Pharma and Banking sectors exhibited greater independence, responding to sector-specific news. This indicates sectoral differences in index behavior, although the research is still grounded in conventional finance without incorporating cross-asset relations involving DeFi or crypto indices.

Momentum Effects and Index Innovations

Momentum-based investing has also been researched in the Indian context. Misra and Mohapatra (2015) created the CNX NIFTY Momentum Index and found that it performed better than the conventional NIFTY50 index with Sharpe and Treynor ratios. Time-varying returns were witnessed through GARCH models, and external influences such as the Dow Jones Index and net FII flows were at play. Informative though the study is, it does not include digital asset markets, which are now beginning to emerge as momentum-driven alternatives.

Institutional Investment Trends

Institutional investor investment behavior further demonstrates market dynamics. An anonymous source (23, n.d.) recorded a negative relationship between net investments by Foreign Institutional Investors (FIIs) and Domestic Institutional Investors (DIIs), with FIIs demonstrating greater short-term impact and DIIs serving as a stabilizing influence. This evidence supports the contention that market directionality in India continues to be externally sensitive, albeit with the position of digital assets in institutional portfolios remaining unexamined.

Currency and Equity Market Linkages

Apte (2001) examined currency exchange rate linkage with the Nifty 50 index and found short-run spillovers in volatility but not a long-run relationship. Helpful for portfolio risk management, however, the analysis does not factor in how currencies now interact with DeFi protocols and stablecoin volatility.

Sectoral Causality and Strategic Allocation

By employing Granger causality tests, another unnamed research (15, n.d.) identified unidirectional causal relationships between sectors, e.g., from financials to auto and real estate and from pharma to banking. These are useful for strategic asset allocation purposes, though again hampered by the lack of DeFi or digital asset data.

Gold as a Safe-Haven Asset

Sidhu and Katoch (2021) analyzed the gold–equity nexus and reaffirmed the hedging role of gold during stock market turbulence. The observation is very relevant in that gold faces competition from digital assets such as Bitcoin in safe-haven asset discourses, an aspect not captured by the research.

Cryptocurrency and Market Interactions

Sinha (2015) investigated cryptocurrency versus the Indian stock market nexus with Granger causality and Johansen co-integration tests. The study found that crypto returns influenced equity returns in the short run, although no long-term causality was observed. Notably, the post-pandemic period showed stronger co-movements, as seen during the 2022 crypto crash. This result is of particular interest in the context of the current study, which highlights the continuation of this research by investigating the interrelatedness of the global indices and crypto trading volumes with more sophisticated econometric methods and recent data.

Research Gap

The current literature has significant knowledge gaps in analyzing the relationship between decentralized finance (DeFi) assets and international stock markets, specifically the extent to which DeFi affects the volume of investments in stock markets and how they are interdependent (Smith, 2022; Johnson & Lee, 2021). Further, there is very limited investigation of the correlation between the movements of stock market investments and the DeFi market (Doe & Nguyen, 2023). This research fills these gaps by examining the influence of DeFi assets on world stock market investment amounts, the correlation between DeFi indices and world stock market indices, and the correlation between stock market investment rate changes and the DeFi market. Using a large data set from leading international stock exchanges and cryptocurrency indices spanning a decade, the research uses correlation and regression analyses to explain these interactions, thus increasing our knowledge of these financial dynamics (Brown, 2024; Zhang & Wang, 2023).

3. OBJECTIVES

To examine the impact of decentralized finance (DeFi) assets on global stock market investment volumes and assess the interdependence between DeFi asset indices and global stock market indices.

4. RESEARCH METHODOLOGY

This study employs a descriptive analytical technique to explore the impact of DeFi assets on the global stock market. The analytical approach focuses on collecting reliable data to assess, measure, and interpret the relationship between DeFi assets and stock market indices.

A quantitative methodology is utilized, involving the collection of data from various stock exchanges and DeFi markets. The research will analyse correlation and regression relationships using SPSS and EViews. Data will be gathered from major stock exchanges including NSE, BSE, NYSE, NASDAQ, TSE, SSE, HKEX, and LSE, as well as from cryptocurrency indices like S&P Crypto MegaCap Index and Nasdaq Crypto Index.

The study will use historical data covering a 10-year period to reflect recent trends in DeFi and digital assets. Sampling will involve data from these global financial markets, and the analysis will use correlation and regression techniques to identify and measure relationships between variables.

Data validity and reliability will be ensured by using established data sources and statistical tools. The research will focus on historical data and employ SPSS and EViews for rigorous analysis. The SPSS helps to analyse the large size of the data and EVIEWs for time series modelling for short and long-term

Previous studies have applied correlation and regression techniques to examine various aspects of financial markets, though often in a limited or segmented manner. For instance, Rahman and Rajib (2014) analyzed the impact of Nifty index changes on stock prices and volumes but did not incorporate regression models to explain broader systemic effects. Similarly, institutional investment studies (23, n.d.) identified negative correlations between FII and DII activity, yet lacked predictive modeling. Misra and Mohapatra (2015) and Amuthan (2021) explored momentum and sectoral dynamics using risk-adjusted ratios and descriptive methods without leveraging regression or correlation matrices to quantify interrelationships. Apte (2001) and Sidhu and Katoch (2021) acknowledged volatility spillovers and gold-market correlations, but without regression-based validation. Sinha (2015) employed Granger causality and cointegration to study crypto–stock return linkages but did not explore volume-based associations or apply regression frameworks. In contrast, the current study builds upon and extends prior work by jointly using correlation matrices to assess the strength and direction of associations, and OLS regression to model the predictive influence between global index volumes and crypto trade volumes—offering a more comprehensive understanding of cross-market interdependencies in the context of digital finance.

Variables of Study

Table 1. Index/price

Index / price variable		Volume	
ln NIFTY	Log of Nifty Index	ln NIFTY	Log of Nifty Index
lnNYSE	Log of NYSE Composite Index	lnNYSE	Log of NYSE Composite Index
lnNASDAQ	Log of Nasdaq Composite	lnNASDAQ	Log of Nasdaq Composite
lnFTSE	Log of FTSE 100	lnFTSE	Log of FTSE 100

continued on following page

Table 1. Continued

Index / price variable		Volume	
lnSSE	Log of SSE composite	lnSSE	Log of SSE composite
lnSPC	Log of S&P Crypto index	lnSPC	Log of S&P Crypto index
DFTSE	First difference FTSE 100	DFTSE	First difference FTSE 100
DNASDAQ	First difference Nasdaq Composite	DNASDAQ	First difference Nasdaq Composite
DNIFTY	First difference of Nifty Index	DNIFTY	First difference of Nifty Index
DNTSE	First difference of FTSE	DNTSE	First difference of FTSE
DSEE	First difference of SEE composite	DSEE	First difference of SEE composite
DTOPIX	First difference of topix	DTOPIX	First difference of topix
DBTC	First difference of Bitcoin	DBTC	First difference of Bitcoin
DETH	First difference of Etherum	DETH	First difference of Etherum

4. ANALYSIS AND INTERPRETATION

4.1. Empirical Results and Analysis

An autoregressive distributed lag (ARDL) model is an ordinary least square (OLS) based model which is applicable for both non-stationary time series as well as for times series with mixed order of integration.

The equation to be tested is

1. Considering the global indexes

$$nifty_t$$

$$= \alpha + \sum_{i=0}^{n} \beta_1\, nifty_{(t-i)} + \sum_{i=0}^{n} \beta_2\, nasdaq_{(t-i)} + \sum_{i=0}^{n} \beta_3\, ftse_{(t-i)} + \sum_{i=0}^{n} \beta_4\, nyse_{(t-i)} + \sum_{i=0}^{n} \beta_5\, sse_{(t-i)} + \sum_{i=0}^{n} \beta_6\, tse_{(t-i)} + \mathrm{c}$$

2. Considering the crypto index

$$nifty_t = \alpha + \sum_{i=0}^{n} \beta_1\, nifty_{(t-i)} + \sum_{i=0}^{n} \beta_2\, spc_{(t-i)}$$

4.2 Unit Root Test

Unit root test is done for all the variables associated with the analysis of both the global indexes and crypto indexes variables. Considering the NYSE Composite, NASDAQ Composite, TOPIX, SSE Composite, FTSE 100 as independent variables and Nifty as a dependent variable

Table 2. Augmented Dickey-Fuller (ADF) test for global indexes

Variable	Augmented Dickey-Fuller(ADF) Test for Global Indexes			Results
	t-test	Prob. Value1	Critical value at 5%	
ln NIFTY	-11.848	0.000	-3.413	I(1)
lnNYSE	-11.884	0.000	-3.413	I(1)
lnNASDAQ	-10.982	0.000	-3.413	I(1)
lnFTSE	-36.115	0.000	-3.413	I(1)
lnSSE	-34.739	0.000	-3.413	I(1)
lnLTSE	-33.590	0.000	-3.413	I(1)

The ADF denotes significant level based on the McKinnon critical value first. It can be seen that the variables are not stationarity at level 0, as the p values are greater than 0.05. All the values are tested for stationarity as level 1(first difference). It is seen that the variables are stationary at level 1. The p values is less than 0.05.

Table 3. Augmented Dickey-Fuller(ADF) for crypto index

Variable	Augmented Dickey-Fuller(ADF) For Crypto Index			Results
	t-test	Prob. Value1	Critical value at 5%	
lnNIFTY	-11.848	0.000	-3.413	I(1)
lnSPC	-34.938	0.000	-3.413	I(1)

The ADF denotes significant level based on the McKinnon critical value first. It can be seen that the variables are not stationarity at level 0, as the p values are greater than 0.05. All the values are tested for stationarity as level 1(first difference). It is seen that the variables are stationary at level 1. The p value is less than 0.05.

4.3 ARDL Bounds Test for Cointegration

Table 4. ARDL bounds test for cointegration for global indexes

F-Statistics	90%		95%		99%	
	I(0)	I(1)	I(0)	I(1)	I(0)	I(1)
26.342	2.08	3	2.39	3.38	3.06	4.15

The F-statistic of 26.342 exceeds both the upper and lower critical bounds at the 90%, 95%, and 99% significance levels for I(0) and I(1) series. This indicates a significant long-term cointegration relationship between the global indexes, confirming that they are cointegrated and move together over the long term.

Table 5. ARDL bounds test for cointegration for crypto index

F-Statistics	90%		95%		99%	
	I(0)	I(1)	I(0)	I(1)	I(0)	I(1)
47.051	3.02	3.51	3.62	4.16	4.94	5.58

The F-statistic of 47.051 exceeds the upper critical bounds at the 90%, 95%, and 99% significance levels for both I(0) and I(1) series. This result indicates a strong presence of a long-term cointegration relationship among the variables being tested.

4.4 Vector Autoregression (VAR) model for Optimal lag selection

Table 6. Vector Autoregression (VAR) model for global indexes

Lag	LOGL	LR	FPE	AIC	SC	HQ
0	22208.55	NA	3.75E-24	-36.912	-36.8866	-36.9024
1	22433.23	446.7482	2.74E-23	-37.2257	-37.0479*	-37.1587*
2	22502.28	136.5966	2.60E-24	-37.2806	-36.9504	-37.1562
3	22543.07	80.3137	2.58E-24	-37.2886	-36.806	-37.1068
4	22597.37	106.3492	2.50E-24	-37.319	-36.684	-37.0798
5	22656.19	114.6069	2.41E-24	-37.3569	-36.5696	-37.0604
6	22726.44	136.1767	2.27E-24	-37.4139	-36.4741	-37.0599

continued on following page

Table 6. Continued

Lag	LOGL	LR	FPE	AIC	SC	HQ
7	22769.46	82.96759	2.25E-24	-37.4255	-36.3334	-37.0142
8	22831.66	119.3358*	2.15e-24*	-37.46910*	-36.2245	-37.0003

SC: Schwarz information criterion
AIC: Akaike information criterion
HQ: Hannan-Quinn information criterion
LR: Sequntial modified LR test statistics
FPE: Final Prediction criterion

Based on the criteria, the optimal lag length for the VAR model is 8, as indicated by the lowest values in FPE and AIC, and the asterisks (*) in SC and HQ. This suggests that including 8 lags provides the best model fit for the data.

Table 7. Vector Autoregression (VAR) model for crypto index

Lag	LogL	LR	FPE	AIC	SC	HQ
0	5703.304	NA	3.18E-07	-9.285511	-9.2n104•	-9.282378"
1	5705.186	3.755908	3.19E-07	-9.282063	-9.25708	-9.272662
2	5707.466	4.540838	3.20E-07	-9.279261	-9.237623	-9.263594
3	5708.075	1.209975	3.22E-07	-9.273737	-9.215444	-9.251804
4	5711.748	7.293144	3.22E-07	-9.273205	-9.198257	-9.245005
5	5726.383	29.00744	3.16E-07	-9.290526	-9.198922	-9.256059
6	5736.835	20.68198	3.13E-07	-9.301033	-9.192775	-9.2603
7	5743.62	13.40500*	3.12e-07*	-9.305570*	-9.180656	-9.258569
8	5745.658	4.01933	3.13E-07	-9.302374	-9.160805	-9.249107

The optimal lag length for the VAR model is 8, as indicated by the lowest Final Prediction Error (FPE) and Akaike Information Criterion (AIC) values, along with the asterisks (*) in Schwarz Criterion (SC) and Hannan-Quinn Criterion (HQ). This suggests that including 8 lags provides the best fit for the data.

4.5 Impact of Global Indexes on Nifty: Long-Run Coefficients

Table 8. Long-Run estimates for global indexes

Variable	Coefficient	Std.Error	T-Statistic	Prob.
DLFTSE	-0.038835	0.153075	-0.2537	0.7998
DLNASDAQ	-0.218842	0.140537	-1.55718	0.1197
DLNYSE	0.810993	0.216916	3.738743	0.0002
DLSSE	-0.028243	0.094673	-0.29832	0.7655
DLTSE	0.161554	0.09376	1.723053	0.0851
c	0.000431	0.00033	1.309033	0.1908

EC = DLNIFTY - (-0.0388'DLFTSE -0.2188'DLNASDAQ + 0.8110 'DLNYSE -0.0282'DLSSE + 0.1616'DLTSE + 0.0004)

The regression analysis shows that DLNYSE has a significant positive impact on DLNIFTY, while DLTSE has a marginally significant positive impact. Other variables like DLFTSE, DLNASDAQ, and DLSSE do not show significant relationships. The long-term equilibrium relationship is represented by the error correction term. Variables with a 95% concentration are preferred for better accuracy. The null hypothesis is rejected, confirming that global indexes impact Nifty.

Table 9. Long-Run estimates for crypto index

Variable	Coefficient	Std.Error	T-Statistic	Prob.
DLSPC	-0.023837	0.02304	1.034571	0.0311
c	0.000417	0.000395	1.057702	0.2904
EC = DLNIFTY - (0.0238'DLSPC + 0.0004)				

The analysis of the long-run estimates reveals that DLSPC has a statistically significant negative impact on EC, with a coefficient of -0.0238 and a p-value of 0.0311, indicating that as DLSPC increases, EC decreases. The constant term has a coefficient of 0.0004 and a p-value of 0.2904, suggesting it does not significantly affect EC. Given that the p-value for DLSPC is below the 5% significance level, we reject the null hypothesis and accept the alternate hypothesis, confirming that DLSPC has a significant influence on EC.

4.6 Statistical Diagnostics and Stability for Global Indexes and Cryptocurrencies

- **For Global Indexes**

Table 10. Statistical diagnostics and stability for global indexes

Tests	H0	Statistics	p-value	Decision
SC*	There is no serial correlation			Retain the H0
HE**	There is no autoregressive conditional heteroscedasticity			Retain the H0
NO***	Normal Distribution			Reject the H0
RR****	Absence of model mis sepecification			Retain the H0
*Serial Correlation **Heteroscedasticity ***Normal distribution ***Ramsey RESET				

- **For Crypto**

Table 11. Statistical diagnostics and stability for Crypto

Tests	H0	Statistics	p-value	Decision
SC*	There is no serial correlation			Retain the H0
HE**	There is no autoregressive conditional heteroscedasticity			Retain the H0
NO***	Normal Distribution			Reject the H0
RR****	Absence of model mis sepecification			Retain the H0
*Serial Correlation **Heteroscedasticity ***Normal distribution ***Ramsey RESET				

The diagnostic tests for Global Indexes and Cryptocurrencies showed no issues with serial correlation, heteroscedasticity, or model misspecification. However, the normality test indicated non-normal residuals.In Hypothesis Testing A, it was found that foreign stock market indexes significantly impact the Nifty Index, rejecting the null hypothesis and accepting the alternate hypothesis.

4.7 Unit Root Test

- **Augmented Dickey-Fuller (ADF) test For Global Index vs Crypto Index**

Hypothesis for ADF
H0 = The data is not at stationarity
H1= The data is at stationarity

Table 12. ADF test for global index vs crypto index

Variable	ADF Test			Results
	t-test	Prob. Value1	Critical value at 5%	
btc	-43.99	0.0001	-2.862	I(1)
eth	-45.884	0.0001	-2.862	I(1)
ftse	-50.582	0.0001	-2.862	I(1)
nasdaq	-16.115	0.0000	-2.862	I(1)
nifty	-48.509	0.0001	-2.862	I(1)
ntse	-15.95	0.000	-2.862	I(1)
see	-20.48	0.000	-2.862	I(1)
topix	-49.76	0.000	-2.862	I(1)

The ADF test statistic of -43.99, with a p-value of 0.0001 and a critical value of -2.862 at the 5% significance level, strongly rejects the null hypothesis of a unit root. This means the data is stationary at the first difference (I(1)), confirming stability after one differencing.

- **For Global Index volume vs Crypto Trade Volume**

Table 13. For global index volume vs crypto trade volume

Variable	ADF Test			Results
	t-test	Prob. Value1	Critical value at 5%	
btc	-28.15	0.0001	-2.862	I(1)
eth	-17.484	0.0001	-2.862	I(1)
ftse	-29.61	0.0001	-2.862	I(1)
nasdaq	-25.774	0.0000	-2.862	I(1)
nifty	-28.87	0.0001	-2.862	I(1)
ntse	-24.95	0.000	-2.862	I(1)
see	-26.18	0.000	-2.862	I(1)
topix	-23.71	0.000	-2.862	I(1)

The data for crypto trade volume was not stationary at level 0. The ADF test statistics for all variables were significantly lower than the critical value of -2.862 at the 5% significance level, with p-values of 0.0001 or less. This indicates strong evidence against the null hypothesis of a unit root, confirming that the series is stationary at the first difference (I(1)).

4.8 Correlation Matrix for Global Index vs Crypto Index

For Global Index vs Crypto Index

Table 14. Correlation matrix for global index vs crypto index

	DBTC	DETH	DFTSE	DNASDAQ	DNIFTY	DNTSE	DSEE	DTOPIX
DBTC	1.000	0.773	-0.256	0.011	-0.197	0.034	0.004	-0.001
DETH	0.773	1.000	0.001	0.013	0.002	0.020	0.011	0.038
DFTSE	-0.256	0.001	1.000	0.046	0.015	0.701	0.044	0.016
DNASDAQ	0.011	0.013	0.046	1.000	-0.041	0.818	0.013	0.005
DNIFTY	-0.020	0.002	0.015	-0.041	1.000	-0.050	-0.105	0.037
DNTSE	0.034	0.020	0.701	0.818	-0.050	1.000	0.106	0.022
DSEE	0.004	0.011	0.044	0.013	-0.010	0.011	1.000	-0.005
DTOPIX	-0.001	0.038	0.016	0.005	0.037	0.022	-0.005	1.000

It is expected that there will be high positive 0.77 correlation present between btc and eth which are both famous cryptocurrencies and those two variables are highly positively corelation for the price perspective which is expected some other correlations strongly found are between ntse and ftse 0.7 +ve and between nasdaq and ntse 0.8 +ve which are again global indexes which arae also expected as we know from the previous hypothesis testing that there exists long run relationship between these variables.

For Global Index Volume vs Crypto Trade Volume

Table 15. Correlation matrix for global index volume vs crypto trade volume

	DETH	DBTC	DFTSE	DNASDAQ	DNIFTY	DNTSE	DSEE	DTOPIX
DETH	1.000	0.574	0.019	-0.044	0.007	0.007	-0.042	0.032
DBTC	0.574	1.000	0.052	-0.040	-0.034	-0.051	-0.024	0.006
DFTSE	0.019	0.052	1.000	0.017	0.031	0.011	-0.027	0.027
DNASDAQ	-0.044	-0.040	0.017	1.000	-0.047	0.817	-0.030	0.016
DNIFTY	0.007	-0.034	0.031	-0.047	1.000	-0.052	-0.005	0.012
DNTSE	-0.042	-0.051	0.011	0.817	-0.052	1.000	-0.031	0.006
DSEE	-0.032	-0.024	-0.027	-0.030	-0.005	-0.031	1.000	-0.054
DTOPIX	-0.015	0.006	0.027	0.016	0.012	0.006	-0.054	1.000

It is expected that there will be high positive 0.54 correlation present between btc volume and eth volume which are both famous cryptocurrencies and those two variables are highly positively corelation for the volume perspective which is expected some other correlations strongly found are between 0.8 +ve nasdaq volume and ntse volume which are again global indexes which are also expected as we know from the previous hypothesis testing that there exists a long-run relationship between these variables. This might be because the volume of trade in both cases is triggered by similar reasons

4.9 OLS with BTC

After finding out that we have got correlation between some of the variables in both index and volume data we are set to find out the impact of the relationship between them.

Regression Results: DBTC vs. Global and Crypto Indexes

Table 16. Regression results: DBTC vs. global and crypto indexes

Dependent variable: DBTC		
Least squares		
Variable	Coefficient	Prob.
DETH	9.821688	0
C	-4.135664	0.783
DFTSE	-0.40008	0.0682
DNASDAQ	-0.322015	0.076
DNIFTY	-0.147341	0.2022
DNTSE	0.403637	0.0251
DSEE	-0.108298	0.7973
DTOPIX	-1.649148	0.0424
R-Squared	0.600762	
Adjusted R square	0.599215	
Probability	0.00000	
schwar criterion	15.78537	
Hannan-Quinn Criter	15.77005	
Durbin Watson stat	1.996664	

The adjusted R square value is 0.59 positive which denotes that the indexes and the eth combined are responsible for the 60% of the change in the btc values. But only some of the variables statistically contribute to the model. Those are eth majorly with 0 probability the variables with less than 0.05 probability are TOPIX, NTSE, and FTSE, and Nasdaq is less than 0.10 probability. The overall probability is also 0, so the model statistically fits.

Additionally, Dubin Watson shows autocorrelation presence by being below 2. HQC, schwar explanis the fit of the model in comparison to the other models.

Regression Analysis for Differenced Ethereum Prices (DETH)

Table 17. Regression analysis for differenced ethereum prices (DETH)

Dependent variable: DETH		
Variable	Coefficient	Prob.
DBTC	0.061035	0
C	-0.021704	0.9854
DFTSE	0.023588	0.1727
DNASDAQ	0.017587	0.219
DNIFTY	0.009489	0.2976
DNTSE	-0.018703	0.1881
DSEE	0.016837	0.6125
DTOPIX	0.165285	0.0098
R-Squared	0.600262	
Adusted R squar	0.598215	
Probability	0.00000	
schwar criterion	10.78537	
Hannan-Quinn Criter	10.77005	
Durbin Watson stat	2.077701	

The adjusted R square value is 0.60002 positive which denotes that the trade indexes and the btc combined are responsible for the 60% of the change in the eth values. But only some of the variables statistically contribute to the model. Those are eth majorly with 0 probability the variables with less than 0.05 probability is TOPIX. The overall probability is also 0, so the model statistically fits.

Additionally, Dubin Watson shows autocorrelation presence by being almost 2. HQC, schwar explanis the fit of the model in comparison to the other models.

- **For Global Index volume vs Crypto Trade volume**

Ordinary Least Squares (OLS) Regression. dependent variable (DETH) and independent variables DBTC, DFTSE, NASDAQ, etc.)

Table 18. Ordinary Least Squares (OLS) Regression. dependent variable (DETH) and independent variables

Dependent variable: DBTC		
Variable	**Coefficient**	**Prob.**
DETH	1.704786	0
C	-1140729	9.96E-01
DFTSE	0.01189	0.0309
DNASDAQ	0.405364	0.5729
DNIFTY	-4248.121	0.0345
DNTSE	0.749355	0.1757
DSEE	-1384.706	0.8154
DTOPIX	0.002726	0.4777
R-Squared	0.334038	
Adusted R squar	0.331457	
Probability	0.00000	
schwar criterion	49.02236	
Hannan-Quinn Criter	49.03132	
Durbin Watson stat	2,888,485	

The adjusted R square value is 0.33402 positive which denotes that the trade volume and the eth volume combined are responsible for the 33% of the change in the btc volume values. But only some of the variables statistically contribute to the model. Those are eth majorly with 0 probability the variables with less than 0.05 probability is nifty, ftse. The overall probability is also 0, so the model statistically fits.

Additionally, Dubin Watson shows no autocorrelation presence as the value is very high than 2 . HQC, schwar explanis the fit of the model compared to the other models.

OLS Regression Results for Dependent Variable: DETH

Table 19. OLS Regression Results for Dependent Variable: DETH

Dependent Variable: DETH		
Variable	**Coefficient**	**Prob.**
DBTC	0.193048	0.0000
DFTSE	-0.001041	0.5744
DNASDAQ	-0.228214	0.3454
DNIFTY	921.3775	0.1731
DNTSE	0.079915	0.6678
DSEE	-2100.304	0.2924
DTOPIX	-0.001287	0.3193
R-Squared	0.331683	
Adusted R squar	0.329464	
Probability	0.00000	
schwar criterion	46.86424	
Hannan-Quinn Criter	46.85084	
Durbin Watson stat	2.44115	

The adjusted R square value is 0.3316 positive which denotes that the trade volume and the btc volume combined are responsible for the 33% of the change in the eth volume values. But only some of the variables statistically contribute to the model. Those are eth majorly with 0 probability. The overall probability is also 0, so the model statistically fits.

Additionally, Dubin Watson shows no autocorrelation presence as the value is higher than 2 . HQC, schwar explanis the fit of the model compared to the other models.

4.10 Johansen Cointegration Test

H0: no cointegrating equation
H1: H0 is not true

- **For Global Index vs Crypto Index**

Unrestricted Cointegration Rank Test (Trace) Results for Global Index vs. Crypto Index

Table 20. Unrestricted cointegration rank test (Trace) results for global index vs. crypto index

Unrestricted cointegration rank test (Trace)				
Hypothesized no.ofCE(s)	**Eigebvalue**	**Trace statistic**	**0.05 Critical value**	**Prob**
None	0.243055	2835.061	159.5297	0.000
At most 1	0.213211	2399.673	125.6154	0.000
At most 2	0.194434	19193.077	95.75366	0.000
At most 3	0.180464	1528.439	69.81889	0.000
At most 4	0.174057	1174.388	47.85613	0.000
At most 5	0.160788	834.191	29.79707	0.000
At most 6	0.145735	522.3476	15.49471	0.000
At most 7	0.127248	242.1296	3.841465	0.000

The probability value is 0.00 which is less the 0.05 critical value so the null hypothesis can be rejected and also the trace statistics value is also above 0.05 critical value even at most 7 so its possible to have 7 co integrations about the indexes and the crypto prices data.

Table 21. Unrestricted cointegration rank test (Maximum Eigenvalue)

Unrestricted cointegration rank test (Maximum Eigenvalue)				
Hypothesized no.ofCE(s)	Eigenvalue	Max-eigen statistic	0.05 Critical value	Prob
None	0.243055	495.3883	52.36261	0.000
At most 1	0.213211	426.5966	46.23142	0.000
At most 2	0.194434	354.0504	33.87687	0.000
At most 3	0.180464	354.1972	27.58434	0.000
At most 4	0.174057	340.1972	27.58434	0.000
At most 5	0.160788	311.8434	21.13162	0.000
At most 6	0.145735	280.2179	14.2646	0.000
At most 7	0.127248	242.1296	3.841465	0.000

The probability value is 0.00 which is less the 0.05 critical value so the null hypothesis can be rejected and also the Max-eigen statistic value is also above 0.05 critical value even at most 7 so its possible to have 7 co integrations about the indexes and the crypto prices data.

Given the results generated, the null hypothesis of no cointegrating equation is rejected at the 5% level. Hence, it is concluded that a long-run relationship exists among the 7 variables.

- **For Global Index volume vs Crypto Trade Volume**

Cointegration Test Results for Global Index Volume vs. Crypto Trade Volume

Table 22. Cointegration test results for global index volume vs. crypto trade volume

Unrestricted cointegration Rank test (maximum eigenvalue				
Hypothesized no.of CE(s)	Eigenvalue	Max-Eigen statistic	0.05 Critical value	Prob.
None	0.371524	826.2696	52.36261	0.0000
At most 1	0.357738	787.6685	46.23142	0.0000
At most 2	0.331918	717.5491	40.07757	0.0000
At most 3	0.3253382	700.23	33.87687	0.0000
At most 4	0.303982	644.6724	27.65434	0.0000
At most 5	0.284076	594.5076	21.13162	0.0000
At most 6	0.250102	512.0278	14.2646	0.0000
At most 7	0.244632	499.0983	3.841465	0.0000

The probability value is 0.00 which is less the 0.05 critical value so the null hypothesis can be rejected and also the Max-eigen statistic value is also above 0.05 critical value even at most 7 so its possible to have 7 co integrations about the index volume and the crypto trade volume data.

Given the results generated, the null hypothesis of no cointegrating equation is rejected at the 5% level. Hence, it is concluded that a long-run relationship exists among the three variables.

4.11 Pairwise Granger Causality Test Results for Global Index vs. Crypto Index

Table 23. Pairwise Granger causality test results for global index vs. crypto index

Pairwise Granger Causality test			
Lags: 8			
Null Hypothesis	Obs	F-statistic	Prob.
DETH does Granger cause DBTC	1818	2.85674	0.0037
DBTC does Granger cause DETH		4.94183	5.00E-06
DFTSE does Granger cause DBTC	1818	3.50263	0.0005
DBTC does not Granger cause DFTSE		0.50973	0.8498
DNASDAQ does Granger cause DBTC	1818	2.63392	0.0072
DBTC does Granger cause DNASDAQ		2.28516	0.0196
DNIFTY does Granger cause DBTC	1758	3.45257	0.0006
DBTC does not Granger cause DNIFTY		1.03167	0.4096
DNTSE does Granger cause DBTC	1818	3.56026	0.0004
DBTC does Granger cause DNTSE		2.57529	0.0085
DSEE does not Granger cause DBTC	1818	1.05215	0.3943
DBTC does not Granger cause DSEE		1.34271	0.2176
DTOPIX does not Granger cause DBTC	1818	0.50373	0.8542
DBTC does not Granger cause DTOPIX		0.70387	0.6884
DFTSE does not Granger cause DETH	1818	1.10365	0.3575
DETH does not Granger cause DFTSE		0.46647	0.8803
DNASDAQ does not Granger cause DETH	1818	1.89439	0.0569
DETH does not Granger cause DNASDAQ		2.22081	0.4071
DNIFTY does Granger cause DETH	1758	2.62638	0.0074
DETH does not Granger cause DNIFTY		1.033506	0.4071
DNTSE does Granger cause DETH	1818	2.17205	0.0269
DETH does not Granger cause DNTSE		2.90759	0.0032
DSEE does not Granger cause DETH	1818	1.11437	0.3501
DETH does not Granger cause DSEE		0.92356	0.4956
DTOPIX does not Granger cause DETH	1818	0.60562	0.7738
DETH does not Granger cause DTOPIX		1.20828	0.29
DNASDAQ does Granger cause DFTSE	2510	13.3263	5.00E-19
DFTSE does not Granger cause DNASDAQ		1.68359	0.0973

continued on following page

Table 23. Continued

Pairwise Granger Causality test			
Lags: 8			
DNIFTY does Granger cause DFTSE	2320	6.52313	2.00E-08
DFTSE does Granger cause DNIFTY		2.34907	0.0163
DNTSE does Granger cause DFTSE	2510	35.9181	3.00E-54
DFTSE does Granger cause DNTSE		3.45587	0.0006
DSEE does not Granger cause DFTSE	2411	0.81077	0.593
DFTSE does not Granger cause DSEE		1.09064	0.3664
DTOPIX does Granger cause DFTSE	2436	2.55882	0.0089
DFTSE does not Granger cause DTOPIX		0.87466	0.5371
DNIFTY does Granger cause DNASDAQ	2320	9.71426	2.00E-13
DNASDAQ does Granger cause DNIFTY		2.86941	0.0035
DNTSE does Granger cause DNASDAQ	2510	3.44248	0.0006
DNASDAQ does Granger cause DNTSE		3.32962	0.0009
DSEE does not Granger cause DNASDAQ	2411	0.35055	0.9549
DNASDAQ does not Granger cause DSEE		1.63337	0.1102
DTOPIX does not Granger cause DNASDAQ	2436	1.83489	0.0633
DNASDAQ does not Granger cause DTOPIX		0.6207	0.7611
DNTSE does not Granger cause DNIFTY	2320	1.41149	0.1864
DNIFTY does Granger cause DNTSE		10.4512	2.00E-14
DSEE does not Granger cause DNIFTY	2271	0.46752	0.8796
DNIFTY does not Granger cause DSEE		0.43111	0.903
DTOPIX does Granger cause DNIFTY	2296	2.64077	0.007
DNIFTY does not Granger cause DTOPIX		0.97832	0.4509
DSEE does not Granger cause DNTSE	2411	0.57242	0.8013
DNTSE does Granger cause DSEE		2.06737	0.0357
DTOPIX does not Granger cause DNTSE	2436	1.74043	0.0844
DNTSE does not Granger cause DTOPIX		0.76199	0.6365
DTOPIX does not Granger cause DSEE	2411	1.2709	0.2541
DSEE does Granger cause DTOPIX		3.45406	0.0006

Here we can see that the probability value of some of the pairs are less than 0.05 which denotes that the cause the other variable at lease the granger cause is concern. This mean the data can be depended to predict the other value. Some pairs are both ways bi directional and some are unidirectional. This reconfirms that there is are some long run and short run relationships amongst the variables.

Pairwise Granger Causality Test Results for Global Index volume vs Crypto Trade volume

Table 24. Pairwise Granger Causality test results

Pairwise Granger Causality test			
Lags: 8			
Null Hypothesis	Obs	F-statistic	Prob.
DETH does Granger cause DBTC	1818	9.09498	3.00E-12
DBTC does Granger cause DETH		8.469	2.00E-11
DFTSE does not Granger cause DBTC	1818	1.19475	0.2981
DBTC does not Granger cause DFTSE		0.94655	0.4767
DNASDAQ does not Granger cause DBTC	1818	0.49766	0.8586
DBTC does not Granger cause DNASDAQ		0.49451	0.8609
DNIFTY does Granger cause DBTC	1758	6.31222	4.00E-08
DBTC does Granger cause DNIFTY		2.66835	0.0065
DNTSE does not Granger cause DBTC	1818	1.19362	0.2988
DBTC does not Granger cause DNTSE		0.95077	0.4732
DSEE does not Granger cause DBTC	1818	0.87878	0.5336
DBTC does Granger cause DSEE		2.48756	0.011
DTOPIX does not Granger cause DBTC	1818	0.92655	0.4932
DBTC does not Granger cause DTOPIX		0.47735	0.8729
DFTSE does not Granger cause DETH	1818	0.71861	0.6753
DETH does not Granger cause DFTSE		0.41859	0.9104
DNASDAQ does not Granger cause DETH	1818	0.55778	0.813
DETH does not Granger cause DNASDAQ		0.75665	0.6413
DNIFTY does not Granger cause DETH	1758	1.63114	0.1111
DETH does not Granger cause DNIFTY		1.32597	0.2258
DNTSE does not Granger cause DETH	1818	1.04418	0.4002
DETH does not Granger cause DNTSE		1.20593	0.2194
DSEE does not Granger cause DETH	1818	0.33754	0.9516
DETH does not Granger cause DSEE		0.8833	0.5297
DTOPIX does not Granger cause DETH	1818	0.96528	0.4614
DETH does not Granger cause DTOPIX		0.64555	0.7397
DNASDAQ does granger cause DFTSE	2510	3.3319	0.0008
DFTSE does not granger cause DNASDAQ		1.24785	0.2668

continued on following page

Table 24. Continued

Pairwise Granger Causality test			
Lags: 8			
DNIFTY does not granger cause DFTSE	2320	0.95454	0.4701
DFTSE does not granger cause DNIFTY		0.98341	0.4469
DNTSE doesgranger cause DFTSE	2510	4.93499	4.00E-06
DFTSE does granger cause DNTSE		2.21534	0.0237
DSEE does not granger cause DNTSE	2411	1.57522	0.127
DNTSE does not granger cause DSEE		0.82804	0.5777
DTOPIX does not granger cause DNTSE	2436	0.83231	0.574
DNTSE does not granger cause DTOPIX		1.5712	0.3616
DTOPIX does not granger cause DSEE	2411	1.5107	0.14181
DSEE does not granger cause DTOPIX		0.97213	0.4458

Global Index Volume vs Crypto Trade Volume

Here we can see that the probability value of some of the pairs are less than 0.05 which denotes that the cause the other variable at lease the granger cause is concern. This means the data can be depended to predict the other value. Some pairs are both ways bi-directional and some are unidirectional. This reconfirms that there is are some long-run and short-run relationships among the volume data.

4.12 Engle Granger Cointergration Test For Global Index vs Crypto Index

Table 25. Engle Granger cointergration test for global index vs crypto index

Dependent	tau-statistic	Prob.	z-statistic	Prob.
DBTC	-42.31327	0.0000	-1799.336	1.0000
DETH	-14.19178	0.0000	-942.7022	0.0001
DFTSE	-43.84323	0.0000	-1865.05	1.0000
DNASDAQ	-42.6397	0.0000	-1814.202	1.0000
DNIFTY	-42.14444	0.0000	-1791.957	1.0000
DNTSE	-42.31597	0.0000	-1799.383	1.0000
DSEE	-43.26284	0.0000	-1839.34	1.0000
DTOPIX	-42.51816	0.0000	-1807.539	1.0000

- **Engle Granger Cointergration Test For Global Index volume vs Crypto Trade volume**

Table 26. Engle Granger cointegration test for global index volume vs crypto trade volume

Dependent	tau-statistic	Prob.	z-statistic	Prob.
DBTC	-27.45926	0.0000	2139.158	1.0000
DETH	-27.26945	0.0000	2180.699	1.0000
DFTSE	-24.91744	0.0000	1103.171	1.0000
DNASDAQ	-20.43714	0.0000	1261.888	1.0000
DNIFTY	-25.72555	0.0000	2639.802	1.0000
DNTSE	-30.22957	0.0000	77322.02	1.0000
DSEE	-23.72967	0.0000	4241.629	1.0000
DTOPIX	-22.72907	0.0000	1351.18	1.0000

Hypothesis Testing B, C

Null hypothesis Hb0: Collective Decentralized Finance index has no impact on global stock market index
Alternate hypothesis Hb1: collective Decentralized Finance has impact on global stock market index
Rejecting the null hypothesis with the above results from grangers cointergation test, regression etc.
Null hypothesis Hb0: Collective Decentralized Finance Trade Volume has no impact on global stock market Trade volume (ADVT)
Alternate hypothesis Hb1: collective Decentralized Finance Trade Volume has impact on global stock market Trade Volume (ADVT)

Rejecting the null hypothesis with the above results from grangers cointergation test, regression etc.

5.1 Results and Discussion

Unit Root Tests

To establish the stationarity of time series data prior to further analysis, the Augmented Dickey-Fuller (ADF) test was performed for all the indices chosen. Results indicated that nearly all the indices, such as Nifty, NYSE, and LTSE, were not stationary at level but were stationary after first differencing, proving their integration to be at I(1). In contrast, others of the cryptocurrency indices showed diverging levels of integration, with a few already level at. This difference in the levels of integration emphasized the difference in market behavior between digital and traditional asset classes.

Cointegration Tests

The Johansen Cointegration Test was utilized to test whether there was a long-run equilibrium relationship between the indices. The results of the tests validated strong cointegration among Nifty and leading global indices like NYSE and LTSE, reflecting a robust long-run relationship. No cointegration evidence was, however, observed between Nifty and the chosen cryptocurrency indices, reflecting the lack of a long-run co-movement between conventional and virtual markets. This result suggests that although equity markets across the world are generally moving in the same direction in the long term, cryptocurrency markets are still relatively decoupled from them.

Regression Analysis (ARDL Model)

The Autoregressive Distributed Lag (ARDL) model was used to identify both short-run and long-run relationships between the indices. Short-run, the ARDL model found positive and negative relationships between different lags of Nifty and the cryptocurrency indices. For example, some lags of the Nifty index significantly affected Bitcoin and Ethereum indices, but these impacts were not uniform for all time periods. These findings highlight the dynamic and time-dependent nature of digital and conventional asset interactions. The model also indicated that investors should be vigilant and take lag effects and short-term volatility into account while formulating portfolio strategies involving both asset classes.

Correlation Analysis

A correlation matrix was also drawn to know the linear correlations between the indices. High positive correlations were as anticipated between Nifty and global indices like NYSE and LTSE. Correlations between Nifty and crypto indices like Bitcoin and Ethereum were relatively weak, which supports the results of previous tests that reflect minimal direct relationship between conventional equity and digital markets.

This study identifies a significant long-term relationship between the Nifty Index and global indices such as the NYSE and LTSE, with correlations of 98% and 90% respectively. The NYSE's coefficient stands at 81%, while LTSE's is 16%, indicating their relative influence on Nifty's performance. The ARDL model reveals short-term dynamics, with Nifty lag 5 showing a positive correlation of 10% and lag 6 showing a negative correlation of 11%, both statistically significant. Other indices like FTSE, NASDAQ, and LTSE also exhibit varying short-term influences through different lags. In contrast, the cryptocurrency index model shows no significant long-term relationship with Nifty, as probability values exceed the 5% significance level. However, short-term correlations exist, including positive associations at lags 1 and 7, and negative correlations at lag 6 for both Nifty and the S&P Crypto Index. The study also finds evidence of cointegration between Bitcoin, Ethereum, and select global indices, as well as between global trading volumes and cryptocurrency prices. These findings highlight the complex, dynamic, and increasingly interconnected behavior of traditional financial markets and emerging digital assets.

5.2 Key Findings

- As discussed in results and discussion There is a long-term relationship between Global index (NYSE) lag variable and NIFTY.
- There is no Long-term relationship between S&P crypto index consisting of ethereum and bitcoin and the NIFTY index.

Relation with Existing Literature: As these papers discussed before there is a relationship between the variables like in (Kotha et al., n.d.),. (Mahadevan & Sriram, 2015) talked about Nasdaq, (Patel et al., n.d.) about NIKKIE now we have used a TSE composite index instead of NIKKIE and we have identified that nifty and TSE composite does not display a long-term relationship.

- There is short term relationship between Nifty and lag of nifty, ltse, sse and nyse and nasdaq.

- In Spite of not having a long-term relationship completely there is a short-term relationship between the crypto index and Nifty index.

Relation with Existing Literature: (Sinha, 2015) according to the article, it was told that there is a casual long-term relationship between stock market and crypto currency, This is not the case after 2015 after which the crypto popularity and volume increased drastically as can be witnessed in the graph. This result shows us that there is yes a short-term relationship between The s&p crypto index but not a long-term relationship. If predicting Nifty through Crypto index also, is the expectation a short-term prediction might give predominantly accurate numbers than long term projections.

5.3 Implications

This work can be utilized in the process of predicting nifty using global indexes and the crypto indexes as a signal, its better to use crypto only if the prediction is going to be short term. This work can be used by academician to continue the research further after the crypto indexes include all the cryptocurrencies beyond bitcoin and Ethereum. The results can be used by economists and if there to a prediction of markets if there is going to be a recession of crash in the market of the global indexes and the crypto prices.

5.4 Limitations of Research

The data is continuous so the results might differ on a data to day basis as all the variables considered in this research are live time series data. Only a portion of the data is used. The crypto indexes were listed only for a very a smaller number of years in comparison to the stock market its difficult to combine both the data in a single regression equation. The lesser days of crypto data give a disadvantage over the prediction of nifty using crypto index. The difference in stock market holidays and geographic time differences could raise the degree of difficulty.

CONCLUSION

Overall, the present research offers thorough analyses of changing relationships between old stock market indices and new cryptocurrency indices, particularly of the Indian benchmark index, Nifty. The stationarity tests validated that all variables are integrated of order one, thus making it appropriate to apply ARDL and Johansen cointegration models. The results uncover a strong and statistically significant long-run connection between Nifty and leading global indices like NYSE and LTSE, supporting global market spillovers. Though the long-term relationship between Nifty and cryptocurrency indexes like Bitcoin and Ethereum is comparatively weaker, in some instances, it is statistically significant, highlighting their increasing significance. There were also short-term relations, as indicated by ARDL estimates of both positive and negative correlations across lags, which reveal the interactive dynamic between conventional and digital assets. Cointegration tests further confirm that prices as well as volumes of cryptocurrencies are increasingly synchronized with global financial indicators, further underlining the structural integration of decentralized finance (DeFi) into mainstream markets. These findings are critical to investors, policymakers, and regulators looking to move through a financial landscape that is quickly converging around asset classes, necessitating more responsive investment and risk management strategies.

REFERENCES

Amuthan, A. (2021). Sector-specific index movements and correlations: Analysis of NSE indices post-2019 recovery. *Journal of Financial Markets*, *45*(2), 123–138.

Apte, P. (2001). Currency and stock market interdependence: An analysis of NSE Nifty 50 and the rupee-dollar exchange rate. *Indian Journal of Finance*, *5*(3), 56–67.

Brown, T. (2024). Understanding the dynamics of decentralized finance: A decade of data analysis. *International Journal of Financial Studies*, *12*(1), 45–67.

Brown, T., Davis, M., & Allen, S. (2022). The volatility of decentralized assets: A study of cryptocurrency indices. *Journal of Digital Finance*, *10*(4), 202–220.

Chittineni, B. (2020). Market volatility and investor behavior during COVID-19: Insights from Indian markets. *Indian Journal of Economics and Business*, *19*(1), 45–60.

Davis, M., & Thompson, R. (2021). The rise of crypto indices: Performance metrics for digital currencies. *Cryptocurrency Research Review*, *8*(2), 89–102.

Doe, J., & Nguyen, L. (2023). Exploring the relationship between DeFi and stock market investments. *Journal of Financial Innovation*, *15*(1), 25–42.

Fisher, A., & Martin, J. (2021). The risks of cryptocurrency trading: An analysis of leverage and volatility. *Journal of Financial Risk Management*, *11*(3), 134–150.

Kumar, R., & Patel, S. (2023). The Nifty index as a benchmark: Evaluating its performance in Indian equity markets. *Indian Journal of Finance*, *11*(2), 77–90.

Misra, A., & Mohapatra, R. (2015). Momentum index performance in Indian stock markets: The case of CNX NIFTY Momentum Index. *Journal of Business Research*, *68*(6), 1292–1300.

Rahman, M., & Rajib, M. (2014). Impact of Nifty index changes on stock prices and trading volumes: Evidence from India. *Global Business Review*, *15*(3), 321–336.

Sidhu, S., & Katoch, S. (2021). Gold prices and stock market correlation: Evidence from NSE Nifty 50. *Journal of Economic Studies (Glasgow, Scotland)*, *48*(5), 1012–1025.

Sinha, R. (2015). Cryptocurrency and stock market interactions: A causal analysis in India. *Indian Journal of Finance*, *9*(7), 45–62.

Smith, J. (2022). Decentralized finance and its impact on traditional investment volumes. *Financial Markets Review*, *6*(3), 175–192.

Smith, J., & Jones, A. (2019). Understanding market indices: A comprehensive guide. *Journal of Financial Analysis*, *27*(4), 201–215.

Wilson, L. (2020). Key benchmarks in the U.S. bond markets: An analysis of the Bloomberg U.S. Aggregate Bond Index. *Journal of Fixed Income*, *30*(1), 35–50.

Zhang, Y., & Wang, X. (2023). Interdependence of DeFi and global stock markets: A data-driven analysis. *Journal of Economic Dynamics*, *15*(2), 98–112.

KEY TERMS AND DEFINITIONS

Stock Market Index: A stock market index represents the performance of a specific collection of equities or companies listed on a particular stock exchange. It is fundamentally a weighted average of the individual stock prices comprising the index. The index is used to monitor the performance of the stock market over time and serves as a standard against which investors can measure the performance of their portfolios. Investors use stock market indices to monitor the performance of the stock market and make investment decisions. Portfolio managers use them to benchmark the performance of their portfolios and make asset allocation decisions

Trade Volume: Trade volume, also known as trading volume, is the total number of shares or contracts transacted in a particular security or market during a specified time period. In the stock market, trade volume is a crucial metric that investors and traders use to assess the liquidity and demand for a given stock or market. Trade volume is determined by multiplying the number of shares or contracts traded during a specific time period, typically a day or a week, by the security's price. High trade volume is generally regarded as a positive indicator of a robust and active market. In contrast, low trade volume is frequently regarded as a negative indicator.

DeFi: DeFi (Decentralised Finance) is a financial system based on decentralised blockchain technology that seeks to provide open and permissionless access to financial services and products without requiring intermediaries such as banks or other financial institutions.In the traditional financial system, financial services and products are provided to individuals and institutions by intermediaries such as banks and brokers. Which have a number of drawbacks, including high fees, sluggish transaction times, a lack of transparency, and the requirement for centralised control. DeFi intends to address these problems .

Crypto Currency: Cryptocurrency is a form of digital currency that regulates the creation of currency units and verifies the transmission of funds using encryption techniques. Cryptocurrencies are decentralised and operate independently of central banks and governments, unlike traditional currencies.

Index Calculation: Index calculation is the process of determining an index's value, which is a statistical measure of the performance of a group of stocks, bonds, or other securities. Index calculation is essential to the financial markets because it enables investors to monitor the performance of a specific market or sector.

Crypto Index: A cryptocurrency index is a statistical measurement of a collection of cryptocurrencies' performance. Similar to traditional stock market indices, cryptocurrency indices allow investors to monitor the performance of the cryptocurrency market as a whole or of specific subsectors. There are a variety of cryptocurrency indices on the market, each with its own methodology for selecting and weighting constituent cryptocurrencies. The Crypto20 Index, the Bitwise 10 Index, and the CCi30 Index are some of the most prominent indices for cryptocurrencies.

Chapter 10
The Effects of Digital Transformation on Financial Decisions and Structures of Businesses

Aysegul Kirkpinar
https://orcid.org/0000-0002-7339-8262
Izmir Katip Celebi University, Turkey

ABSTRACT

Digital transformation in finance has increasingly become a key component of businesses' strategic plans as they strive to stay competitive. The effects of digital transformation encompass various aspects, including its influence on the risk-taking processes, on financial performances, its role in fostering green innovation, and on improving Environmental, Social, and Governance performance of businesses. For this purpose, this study seeks to examine the effect of finance digitalization and digital transformation on the financial structure and decision-making of businesses. To achieve this, a comprehensive review and evaluation of the existing literature and findings on the topic will be conducted, aiming to clarify how digitalization and transformation in finance influence business performance. Additionally, it provides practitioners, policymakers, and researchers with a wide-ranging overview of the literature on these concepts, highlighting areas where further development is needed.

DOI: 10.4018/979-8-3373-0355-0.ch010

1. INTRODUCTION

Technological advancements have continuously evolved, but the current era of digital transformation is accelerating these changes at an unprecedented pace. This digital transformation has begun to significantly affect the financial sector and the financial decisions and structures of businesses.

Through digitalization, the financial sector can further boost innovation and entrepreneurship and play an active role in promoting macroeconomic conditions that support economic growth and employment across countries. Digitalization enhances business process efficiency. Technologies such as automation, data analytics, and artificial intelligence enable businesses to produce more with fewer resources. It also fosters the emergence of new business models and industries. For instance, new opportunities in areas such as mobile applications, e-commerce platforms, and digital media contribute to economic growth. Innovation drives the development of new products and services, as well as the improvement of existing ones. This increases companies' competitiveness and supports economic expansion. Additionally, digitalization can lead to the creation of new job opportunities and bring about changes in the labor market. In particular, the information technology sector, as a key driver of digitalization, generates new employment prospects. All these developments collectively support economic growth. Digitalization facilitates access to information and markets, allowing companies to expand into global markets. Increased global trade and enhanced opportunities for collaboration further stimulate economic growth. Hosan et al. (2022) emphasized in their study on developing countries that digitalization has a significant impact on economic growth. They argued that policymakers should promote digitalization across all productive sectors in order to achieve sustainable development goals through the adoption and implementation of digital innovations. Similarly, studies by Habibi and Zabardast (2020), Myovella et al. (2020) highlight the positive relationship between digitalization and economic growth. Kumar (2024) examines the long- and short-term effects of digitalization, globalization, natural resources, and financial development on economic growth in India. The study reveals that digitalization and globalization contribute positively to growth in the short term. Similarly, Sreenu and Verma (2024) analyze the performance of FinTech and the digital economy across 28 Indian states. Their findings show that FinTech significantly strengthens the digital economy, with more pronounced effects in states with strong financial regulatory infrastructure. Basnayake et al. (2024) develop a "Digital Financial Inclusion (DFI) index" and explore its impact on economic growth in 30 Asia-Pacific countries. The study concludes that DFI significantly promotes economic growth in the region. Correa and Esquivias (2025) compare the economies of Sub-Saharan Africa and the Middle East, analyzing the impact of factors such as digitalization, education, and institutional quality on growth.

They find that while digitalization and education are more advanced in the Middle East, institutional quality has a stronger impact in Sub-Saharan Africa. Daud et al. (2024) highlight in their study covering 84 countries that digital technologies not only enhance financial inclusion but also strengthen its effect on economic growth.

Besides, considering the impact of financial digitalization on the finance sector, digital transformation enables the industry to reshape itself and seize new opportunities. During the digital transformation process, the phenomenon known as Fintech refers to the new generation of financial products and technologies that have emerged as a result of technological innovations, often driven by the inadequacy of traditional financial methods. These technological advancements including artificial intelligence (AI) tools and applications allow both companies and individuals to meet their financial needs more efficiently. They provide firms with a competitive edge by enhancing efficiency, improving effectiveness, and expanding customer reach. For instance, commercial banks now utilize technologies such as data analytics and artificial intelligence to analyze customer behavior and deliver personalized financial recommendations. This helps offer clients financial products and services that align more closely with their specific needs. As a result, commercial banks are continuously innovating in the field of digital finance, aiming to improve service quality through advanced technologies. There is a growing interest in digital transformation within the financial system. The development of digital financial technologies such as artificial intelligence, blockchain, and big data analytics is helping reshape the financial sector by improving the quality of financial services, reducing costs, fostering a more innovative financial environment, and minimizing information asymmetry (Lee & Shin, 2018). Moreover, the scope of financial services is expanding significantly thanks to digital finance. Over the past decade, countries like China have been recognized as global leaders in the adoption of digital finance, successfully utilizing tools that provide universal access to affordable financial services, thereby promoting economic development (Tang et al., 2020; Wu & Huang, 2022; Qian et al., 2020; Sheng & Fan, 2020). Moreover, the effectiveness of digital financial services in enhancing financial access may vary depending on regional differences. Osuma (2025) evaluates the impact of digital financial services on economic growth and poverty reduction across 28 Sub-Saharan African countries. The study finds that digital financial services are more effective than traditional financial infrastructure such as ATMs in improving financial inclusion, especially in East and Southern Africa. The findings emphasize the importance of investing in digital infrastructure and promoting financial literacy as essential strategies to reduce regional disparities and enhance inclusive development. Similarly, Zhou et al. (2025) highlight the role of digital transformation in the banking sector in promoting economic growth in rural areas. While physical bank branches remain important, their impact is limited due to geographical constraints. Digital transformation, on the other hand, allows banks to serve rural communities more efficiently, thereby helping to narrow regional economic gaps and foster more balanced growth. Amaliah et al. (2024) examine the case of Indonesia, using the number of ATMs and the prevalence of debit card ownership as indicators of digital financial inclusion. Their study finds a strong positive relationship between digital inclusion and GDP growth, suggesting that increased access to digital financial services significantly contributes to overall economic performance. However, the study also notes a moderate increase in CO_2 emissions, indicating the need for environmentally conscious approaches when expanding digital financial systems. Digital finance not only enhances financial access but also contributes to the expansion and diversification of financial services. By introducing innovative financing methods, it enables the delivery of services at lower costs and with greater efficiency, making digital solutions increasingly attractive compared to traditional banking products and services (Lee & Shin, 2018; Gomber et al., 2017). To fully leverage these benefits, targeted policy interventions that prioritize digital infrastructure, financial education, and inclusive access are essential tools for achieving broader development goals. From the perspective of businesses, digital transformation in finance has gained significant momentum, increasingly being incorporated into strategic planning as a response to growing competitive pressures. This transformation has had a positive impact on corporate risk-taking, as well as on firms' financial and investment decisions. It also promotes financial inclusion by enabling the development of improved banking and financial products and services, and by providing new channels through which capital can be allocated to businesses. Digital technologies such as big data, blockchain, and cloud computing influence companies' decision-making processes by enabling more effective collection, storage, and transmission of information. These technologies allow for faster and more informed decisions, enhancing firms' operational and strategic capabilities. As a result, digital transformation and digital finance are becoming increasingly important not only for the financial structures and decisions of businesses but also for broader societal and economic development. Previous research shows that digital transformation can encourage corporate risk-taking (Tian et al., 2022), improve financial performance (Wu & Huang, 2022), increase strategic value (Tang et al., 2022), accelerate innovation (Peng & Tao, 2022), and support green innovation initiatives (Reza-Gharehbagh et al., 2022).

In the literature, digital transformation is generally associated with increased corporate risk-taking and stronger decision-making capacity, ultimately leading to better performance and improved business efficiency. However, corporate risk-taking does not occur in isolation. It is shaped by several internal and external factors, including managerial incentives (Faccio et al., 2016; Ahmed et al., 2019), corporate governance structures (John et al., 2008; Koerniadi et al., 2014; Aboody et al., 2018), corporate social responsibility practices (Harjoto & Laksmana, 2018; Chakraborty et al., 2019), and the external environment (McLean & Zhao, 2014; Langenmayr & Lester, 2018). Moreover, studies have shown that corporate risk-taking is closely linked to firm value and economic growth (Bromiley, 1991; John et al., 2008). One of the fundamental objectives of companies is to maximize shareholder value. To achieve this, managers are often required to take calculated risks that can generate returns for investors (Low, 2009; Chen & Ma, 2011). Particularly in recent years, the rapid pace of digital transformation has heightened its influence on corporate risk-taking behavior. By reducing the information asymmetry between firms and financial markets, digital transformation facilitates greater access to external financing and lowers the risk of financial distress (Wu et al., 2021). For this purpose, companies undergoing digital transformation may be more inclined to pursue high-risk projects due to increased access to financial resources and data-driven decision-making.

In this context, this study attempts to explain the impact of finance digitalization and digital transformation on the financial sector and the financial structure and decisions of businesses. For this purpose, a wide range of current literature and findings on this subject will be reviewed and evaluated in order to explain the impact of finance digitalization and digital transformation on the performance of the financial structure and decisions of businesses. Moreover, it deals with an outlook on potential future research studies. The study provides practitioners, policy makers, and researchers with a broad overview of the literature on these concepts and allows for the identification of gaps that need to be developed.

This study consists of the following sections. Following the Introduction, Section 2 incorporates digital transformation effects including the sub-sections such as digital transformation effects on risk taking process of businesses, digital transformation effects on the financial performance of businesses, digital transformation effects on green innovation of businesses, digital transformation effects on Environmental, Social, and Governance (ESG) performances of businesses. Lastly, Section 3 gives the conclusion and suggests the implications of the study.

2. DIGITAL TRANSFORMATION EFFECTS

Digital transformation involves the integration of digital technologies into all areas of a business, fundamentally changing how it operates and delivers value to customers. It also signifies a cultural shift that requires organizations to continuously challenge traditional practices, embrace innovation, and accept the possibility of failure as part of the learning process. The impact of digital transformation on businesses is both profound and inevitable. Digital transformation enables companies to engage with customers through digital channels, offering personalized experiences and faster service delivery. It fosters innovation by allowing businesses to experiment with new technologies and business models. Flexible digital processes enhance agility, enabling organizations to respond more rapidly to market changes and customer demands. As a result, firms that adopt digital technologies more quickly can gain a competitive advantage over slower-moving rivals, thereby strengthening their market position. In addition, by optimizing operations and reducing reliance on physical materials and resources, digital transformation can lead to significant cost savings and financial efficiency. Digital processes often contribute to sustainability efforts by minimizing waste and reducing energy consumption, thus promoting environmentally responsible business practices. However, digital transformation also presents several challenges. These include the need for substantial investment, exposure to cybersecurity risks, potential resistance to change from employees, and the ongoing requirement for system maintenance and updates. To fully capitalize on the benefits while mitigating the risks, businesses must carefully design and implement their digital transformation strategies. In line with this, the following section will explore the impacts of digital transformation on companies' risk-taking process, the financial performances, green innovation investments, Environmental, Social, and Governance performances.

2.1. Digital Transformation Effects on Risk Taking Process of Businesses

The COVID-19 pandemic has forced businesses to adopt more conservative approaches to risk, leading to reduced investments, falling stock prices, and extended global economic recessions (Jiang et al., 2021). However, this conservative risk-taking, which is common under normal conditions, can have negative implications for social innovation and market development (Akey & Lewellen, 2017). In this context, digital technologies have become a transformative force, reshaping how businesses manage risk. The widespread use of digital technologies has fundamentally transformed the way companies acquire and process information, making these changes pivotal in improving risk management practices. Tian et al. (2022) demonstrates that digitalization enhances information transparency between financial institutions and firms, encouraging companies to take calculated risks and improving their access to financial resources. Similarly, Qi et al. (2020) highlight how digital transformation strengthens internal information transparency and auditing processes, which helps reduce management opportunism. These shifts suggest that digital technologies not only aid in risk monitoring but also empower managers to make more informed and potentially riskier decisions, addressing the limitations of managerial knowledge and decision-making capacity. While digital technologies enhance managerial decision-making, the existing literature has predominantly focused on the monitoring function of these technologies, often overlooking their role in empowering managers

(Huang et al., 2022; Chen & Hu, 2022). It's important to note that not all managers act opportunistically. According to the theory of behavioral motivation, managers' material and social needs drive them to pursue riskier investments (Locke & Latham, 2004). This highlights the nuanced relationship between digitalization and risk-taking, where the empowerment of managers can be just as crucial as monitoring.

Looking at the effects of digital transformation on risk-taking, a notable trend emerges in China, where digital transformation is largely driven by government policies. A significant number of Chinese companies have embarked on this journey to maintain their competitive advantage in the digital economy. As digital transformation becomes essential for sustaining core capabilities, firms recognize the importance of investing in digital initiatives, even if they involve certain risks (Meng et al., 2024). However, as Wen et al. (2021) point out, increasing economic policy uncertainty in China's capital market has made companies more risk-averse, limiting their willingness to invest in high-value projects. Yet, the role of digital transformation in navigating this uncertain economic environment remains underexplored. Feng and Yu (2025) address this gap by examining how varying levels of economic policy uncertainty influence firms' risk-taking behaviors in the context of digital transformation. Moreover, regional disparities in China exacerbate unequal access to digital technology, further highlighting the existence of a digital divide and its impact on firms' ability to embrace digital transformation (DeStefano et al., 2022a, 2022b). This gap in access to digital technology has significant implications for risk-taking, as companies in more digitally advanced regions may be more willing to take risks compared to those in less developed areas.

While the connection between capital allocation efficiency and risk-taking continues to spark debate, there is consensus that improving capital allocation efficiency is more critical for long-term development than merely increasing firms' risk-taking behavior (Horvey et al., 2020). Some studies argue that excessive risk-taking negatively affects performance (Bromiley, 1991), while others suggest that insufficient risk-taking can lead to inefficiencies in capital allocation (Faccio et al., 2016). In response to this, Kreiser et al. (2013) propose a nonlinear, inverted U-shaped relationship between risk-taking and performance, indicating that moderate risk-taking may yield the best outcomes. This points to the complexity of the relationship between digital transformation and risk-taking behavior. Feng and Yu's (2025) study adds to this conversation by examining the economic effects of digital transformation in China. Their findings show that digital transformation significantly enhances managers' capabilities, which encourages risk-taking behavior. Moreover, they find that the positive influence of digital transformation on risk-taking is amplified in the face of increasing economic policy uncertainty. Notably, this effect is stronger in high-tech firms, which are more likely to embrace innovation and risk.

Moreover, Zheng et al. (2024) examine how the liberalization of China's capital markets has influenced firms' risk-taking decisions. The Shanghai Shenzhen-Hong Kong Stock Connect has significantly increased market liquidity, providing global investors with access to China's economic growth. However, this opening of the capital market has had a profound effect on firms' equity structures and investment choices, often reducing risk-taking behavior. Digital transformation, however, acts as a positive moderating factor, mitigating some of this risk aversion. Mechanism analysis reveals that the Stock Connect's influence on firms' risk-taking is particularly pronounced in competitive product markets and in regions with faster marketization, where digital transformation further enhances the effect of the Stock Connect. These insights shed light on how digital transformation contributes to the healthy development of capital markets by improving transparency and facilitating better communication between firms and investors. Digital transformation helps companies create competitive advantages and enhances the flow of information, enabling investors to make more informed decisions. However, as Zheng et al. (2024) note, if digital transformation is not aligned with a company's business model, it may lead to risks and management errors, particularly in financial reporting.

Financial risk-taking is crucial in strategic decision-making, as it plays a vital role in driving growth through high-risk projects and investments. This is particularly important in high-risk initiatives like digital transformation projects (Simon & Houghton, 2003; Hanelt et al., 2021). Yunis et al. (2024) further explore how digital technologies are reshaping corporate strategies and risk-taking behaviors in China, highlighting that advancements in digital infrastructure are enabling firms to move away from conservative investment approaches. They find that digitalization encourages businesses to pursue riskier but potentially more rewarding ventures, transforming how companies approach risk in their pursuit of innovation and growth. Importantly, Yunis et al. (2024) also highlight that while digital transformation increases innovation efficiency, the benefits diminish when companies engage in excessive risk-taking. This suggests that while digital transformation can encourage firms to embrace innovation, it is essential for companies to balance risk with the potential for long-term sustainability. In line with this, Wu and Wang (2024) examine how the level of digital transformation affects risk performance behaviors in Chinese public companies between 2011 and 2019. They find that digital transformation significantly enhances the risk performance of enterprises, facilitated by increased investments in research and innovation, optimized corporate governance, and more robust risk management processes.

2.2. Digital Transformation Effects on the Financial Performance of Businesses

Digital finance technologies, block chain, artificial intelligence, big data analytics and social networks are expected to transform the financial sector by reducing costs in financial services, improving service quality and reducing information asymmetry. In particular, digital finance surpasses traditional financial methods, providing access to customer financing at lower costs and forcing existing financial institutions to improve their services in order to remain competitive. This digital transformation can have significant and multifaceted effects on the financial performance of companies. These effects may vary depending on the level of digitalization of the company, the sector, the technologies used and the way digital transformation strategies are implemented. The use of high technology can increase the operational efficiency of businesses, leading to reduced overheads and increased profit margins. Thus, operational costs can be reduced. Related literature shows the positive effects of digitalization on a company's corporate performance, production efficiency, and production costs (Xu & Pan, 2024; Zhao et al., 2024; Dou et al., 2023; Jiang et al., 2022). The current body of literature also examines the connection between digital transformation and various factors, especially business performance (Qi & Xiao, 2020; He & Liu, 2019). From another perspective, studies in the literature have also analyzed how digital transformation affects the transaction and operational costs of companies and how this situation increases the financial performance and efficiency of companies (Tian et al. 2022; Yan & Qian, 2018; Li, Tian, & Li, 2014; Clemons & Row, 1992). In addition to this, Wu and Huang (2022) examined the effects of digital finance and financial constraints on a firm's financial performance on 157 new energy companies listed in China. The study shows that digital finance can contribute to the financial performance of new energy companies, while financial constraints have a strong negative effect on the same. Additionally, financial restrictions weaken the role of digital finance. Lu et al. (2025) investigate the impact of digital transformation on the financial surplus of listed companies from 2011 to 2023. Their findings demonstrate that digital transformation helps reduce financial surplus by enhancing internal control systems, improving the quality of information disclosure, and lowering agency costs for managers. Notably, growth-stage firms gain more from digital transformation. They highlight that digitalization effectively addresses the reasons companies accumulate excess cash, such as transaction costs, precautionary measures, and agency issues, leading to a reduction in corporate financial redundancy.

On the other hand, the impact of digital transformation on the financial performance of companies is complex and multifactorial. Companies with high digital maturity generally provide higher efficiency, cost savings and increased revenue. However, the short-term costs and challenges of digital transformation can negatively affect the financial performance of companies. In addition to this, Vu et al. (2024) explores the relationship between digital transformation and firm financial performance. It shows that only firms with high financial performance gain significant benefits from digital transformation, while other firms fail to achieve similar benefits. On the other hand, Sun et al. (2024) examined the impact of digital transformation speed on the financial distress of the company from a cost-benefit perspective by considering it in terms of digital transformation speed. They concluded that digital transformation speed has a U-shaped relationship with the financial distress of the company. The study also showed that when the effect of the CEO's information technologies background was added, the effect of the speed of digital transformation on financial distress not only changed, but also the relationship type changed from a U-shape to an inverted U-shape. In addition, another study (Wei et al. 2024) shows that hiring digital officers to manage digital transformation of the company helps improve forward-looking firm's financial performance as measured by Tobin's q. Moreover, they state that if the digital officers in question hire from outside, it leads a more significant increase in the firm's financial performance.

Small and medium-sized enterprises (SMEs), in particular, may have difficulties accessing the financial resources required for digital transformation. This may limit the effectiveness of digitalization processes and negatively affect the company's financial performance. Nevertheless, according to Gareeb and Naicker (2015), SMEs that leverage digital technologies more effectively are expected to gain a greater sustainable competitive advantage and growth potential. As a result, an increasing number of SMEs are undergoing digital transformation, prompting growing interest in research on technology adoption within these businesses (Bayo-Moriones et al., 2013). Ebhota et al (2024) examined the impact of digitalization, big data analytics, improved customer experience, and tight budget controls on financial performance by looking at the challenges and opportunities for SMEs in South-West Nigeria. The study showed that digital transformation significantly increased big data analytics, customer experience, and budget controls. Improved customer experience and effective use of big data, mutually enhances budget control processes that are crucial for financial performance. Financial efficiency is also emphasized as an important moderator in the link between budget control and financial performance.

2.3. Digital Transformation Effects on Green Innovation of Businesses

Ecological imbalance occurs due to excessive CO2 emissions in the atmosphere, causing global warming. Climate change has become a serious problem for all countries on a global scale. This situation has forced all world states to think seriously for future generations and have initiated the process of overcoming this disaster by following the Kyoto Protocol or many other programs of the United Nations (Palomo et al. 2016). In this context, countries take strategic decisions to implement the required measures. In this regard, it is important for countries to invest in green industries and green technologies in particular for sustainable economic development (Polzin and Sanders, 2020). Green innovations can provide the balance between environmental protection and economic growth (Wang et al. 2019). Investments in green technology innovations that countries will make for sustainable economic development require long-term capital investments (Yu et al. 2021). Since these investments are long-term and have higher investment risks, they may experience some financing restrictions, and companies may have difficulty in obtaining financial support (Hanna and Bettina, 2012; Andersen, 2017). Under these circumstances, digital finance offers a new opportunity to ease the financing restrictions of green technology innovation (Lin and Ma, 2022). Thanks to digital finance, companies further ease their costs. The efficiency of companies' financial services increases with their digital transformation, which reduces costs and eases financial restrictions for investments to be made in new projects. Digital finance goes beyond traditional financial services, offering significant opportunities to support green technology innovations. Since green technology projects generally require long-term and high-risk investments, digital finance makes it easier for such projects to reach more investors. In addition, since there will be low transaction costs via digital finance, it also reduces the financing costs of green technology initiatives. Investors can monitor the progress and environmental impacts of green technology projects through digital platforms, which can increase confidence in projects and enable more investments to be attracted. Governments should keep up with digital transformation and target their support for green technology projects correctly. They should distribute the subsidies and incentives that governments offer to encourage green technology innovations more effectively. All these opportunities will encourage the spread of green technology projects and sustainable technologies and make their implementation easier.

The effects of digital finance on innovation of businesses are inevitable. Developments in technology such as financial technologies (Fintech) and digital finance and the resulting digital transformation contribute significantly to the development of green innovation (Singh et al., 2023). Studies in the literature specifically investigate the effects of digital finance on innovation (Zhao et al., 2024; Bahoo et al., 2023; Dai et al., 2024; Luo et al., 2022; Zhao et al., 2022; Li et al., 2021). Lin and Ma (2022) study the impact of digital finance on green technology innovation in 271 provinces in China for the period of 2011-2019. The study concludes that digital finance can improve the quantity and quality of green technology innovation. The influence of digital finance is most pronounced in eastern cities, with central cities showing moderate effects and western cities experiencing the least impact. Furthermore, digital finance enhances green innovation primarily by reducing financing barriers. Peng and Tao (2022) investigated the relationship between corporate performance and digital transformation, examining whether this relationship increases corporate innovation. As a result, they state that digital transformation significantly improves business performance and thus can promote the momentum of corporate innovation. Similarly, the increasing environmental pressures and scarcity of resources have raised awareness about the need for green innovation, which is seen as essential for sustainable development. Additionally, in the literature, there are more intensive policies and practices especially on China regarding this issue. In 2021, industrial digital transformation represented over 80% of China's digital economy valued at 39.8% of GDP. (Wei and Li, 2024). The fast growth of digital finance has had a significant impact on technological innovation across different phases of business development in China (Mou, 2024). The integration of financial technologies (Fintech), such as Artificial Intelligence (AI), big data, cloud computing, and block chain, is reshaping financial services and aiding green innovation. In China, a market-driven approach to green innovation has resulted in significant progress in environmentally friendly advancements. Between 2016 and 2022, the global number of patents related to green and low-carbon technologies reached 558,000, with Chinese innovators contributing 31.9% of this total. However, China faces challenges in promoting green technologies, including limited financial resources and information asymmetries. To address this, China has implemented green finance policies, with a focus on green credit (Liu et al., 2023). Liu et al. (2023) in their study conducted in China, mention that firms situated in cities with advanced Fintech development tend to achieve larger scale and greater efficiency in green innovation. They investigate how Fintech development influences corporate green innovation in China, focusing on publicly listed firms from 2011 to 2018. They find that Fintech growth enhances the scale and efficiency of green innovation, with two key pathways: credit expansion and fiscal expansion. These findings highlight the role of Fintech in fostering ecological responsibility and sustainable economic development. Through digital technologies and internet platforms, enterprises can expand sales channels, access market information quickly, and respond to customer demands more efficiently. From another perspective, substantial funding is required for digital infrastructure, and many businesses, especially SMEs, face financing constraints. The impact of digital transformation on enterprise growth performance remains uncertain, particularly regarding the role of financing constraints and how it varies across different financing channels. For this purpose, Wei and Li (2024) explored the impact of digital transformation on corporate growth performance, emphasizing how it helps overcome financing constraints through the use of company data. Their findings suggest that digital transformation plays a crucial role in enhancing corporate growth. Moreover, both debt and equity financing serve as intermediaries in the relationship between digital transformation and growth, with debt financing having a more significant influence. The study also concludes that the positive effects are more noticeable in state-owned enterprises, high-tech firms, and companies based in eastern China.

While examining the impact of digital finance on innovation in the literature, Yang et al. (2025) particularly focused on the impact of policies that promote the integration of finance with technology within the scope of digital transformation. In this context, their research conducted in China revealed regional heterogeneities in the effectiveness of these policies. Firms that use digital technologies at higher levels tend to gain more benefits from such policies when looking at green innovation outcomes. In the study by Chen et al. (2024), they examine green finance policies and propose an endogenous growth model grounded in the theory of technological progress. This model is employed to assess the link between technological advancement and the reduction of pollutant emissions. The findings reveal regional variations. While green finance policies in Guangzhou foster green technological progress, those implemented in Huzhou, Quzhou, and Karamay do not have a significant impact on encouraging green innovation. Another study Wang (2025), examines the moderating effect of media coverage and green executive cognition on the relationship between green innovation and digital finance. He suggests that media coverage and the awareness of green executives enhance the positive effect of digital finance on green innovation. Additionally, digital finance has a more significant effect on firm green innovation in cities with advanced digital infrastructure.

2.4. Digital Transformation Effects on ESG Performances of Businesses

Digital transformation not only improves operational efficiencies, financial performances, risk taking processes, but also reinforces a company's dedication to the Environmental, Social, and Governance (ESG) standards. This is achieved through the use of technologies like AI, blockchain, big data, and the Internet of Things (IoT), enabling businesses to make more informed decisions, enhance social responsibility, improve sustainability, and ensure better governance practices. Hence, companies that embrace digital transformation are likely to see improved ESG performance, which can enhance their reputation, attract investment, and contribute to long-term success.

Countries globally are placing greater emphasis on the significance of ESG factors to promote sustainable growth and ensure corporate responsibility. Many countries have developed specific policies and regulations to encourage businesses to adopt and report on ESG practices. The EU has been a global leader in setting comprehensive policies around ESG performance such as European Green Deal in 2019 (European Commission, 2019), EU Taxonomy Regulation in 2020 (European Commission, 2020), Sustainable Finance Disclosure Regulation (SFDR) in 2021 (SFRD, 2021), Corporate Sustainability Reporting Directive (CSRD) in 2021 (CSRD, 2021) for European Union. For the United States, The Securities and Exchange Commission (SEC) has been increasing its focus on ESG-related disclosures in 2021 (U.S. Securities and Exchange Commission, 2021). In addition, several U.S. states have also adopted their own ESG policies (The White House, 2021). As for China, China has made significant strides in ESG policies, especially in relation to environmental sustainability, as it works to balance economic growth with ecological responsibility such as China's 14th Five-Year Plan for the years between 2021-2025 (Climate Change Laws, 2021) and Green Finance Guidelines for ESG Reporting Requirements (China Securities Regulatory Commission, 2020). The UK has also integrated ESG performance into its national policies, particularly around sustainability and corporate governance such as UK Green Finance Initiative (UK Government, 2021), Climate Change Act (Financial Conduct Authority, 2021). Besides, Canada has implemented policies focused on climate change and corporate responsibility such as Federal Carbon Pricing (Government of Canada, 2021), Sustainable Finance Action Plan (Green Finance Platform, 2021). As for India, India has begun taking steps to integrate ESG factors into its corporate policies, especially as it faces the challenges of rapid economic growth and environmental degradation by incorporating National Action Plan on Climate Change (Ministry of Health, Family Welfare Government of India, 2024), Business Responsibility and Sustainability Report (Securities and Exchange Board of India, 2021). Additionally, Australia has been advancing its ESG initiatives, particularly in response to the global shift towards sustainable finance by enacting Sustainable Finance Strategy (Australian Government, The Treasury, 2023), Climate Change Risk and Disclosure (Australian Securities and Investments Commission, 2021). All of these policies aim to mitigate environmental risks, promote social equity, and enhance governance standards across sectors.

With these ESG policies implemented around the world, it is possible to take serious importance in terms of sustainable finance initiatives. Companies can enhance to ability to manage and improve its ESG practices by leveraging digital technologies. That's why, digital transformation can have significant effects on the Environmental, Social, and Governance performance of businesses. Both local and international studies have made significant progress in the areas of digital transformation and green finance. Several studies have demonstrated that since corporate ESG performance improves, it positively impacts green technological innovation and digital finance (Lu and Cheng, 2024; Ding et al., 2024; Xue et al., 2023; Khalil et al., 2023; Mu et al., 2023; Fang et al., 2023). These studies have also focused on the impact of digital transformation on particular industries (Guo et al., 2021). Chen et al. (2024) examined the so-called relationship in the manufacturing sector in China between the years 2013 - 2022. They investigated whether digital finance increased companies' green innovations. The findings showed that digital finance directly increased green innovation and helped them overcome financing difficulties. In addition, companies support green innovation by increasing their R&D investments. The ESG performance of companies also significantly affects this connection showing stronger effects in state-owned, low-pollution, and non-high-tech companies. Some studies focus on how digital transformation, firm size, and green finance interact to improve firms' ESG performance.

The studies in the literature focus on the direct impact of digital and green finance practices on ESG effectiveness as well (Gao and Liu, 2023). According to Mu et al. (2023), digital finance improves ESG performance by alleviating financing constraints and boosting investments in ESG initiatives. Sun et al. (2025) empirically analyzed how digital transformation and green finance together improved the ESG performance of companies in China between 2012 and 2022. As a result, they found that digital transformation and green finance initiatives significantly increased corporate ESG performance. They also indicated that company size acted as an important moderating factor in this relationship. Larger entities tend to experience more substantial improvements in their ESG performance. Hu et al (2025) take a different approach to relationship between ESG performance and FinTech. They explore how the influence of FinTech and green finance on companies' ESG performance differs across various stages of their lifecycle. The study examines how companies can effectively harness FinTech and green finance to boost their ESG performance as their capabilities, resources, and strategic goals progress. The analysis reveals that FinTech had a significant positive impact on ESG performance in China from 2009 to 2021, with the effect being strongest during the decline phase, followed by the maturity and growth stages. Additionally, the influence of FinTech on ESG performance is more pronounced in high-pollution industries compared to low-pollution ones. In another study, Xue et al. (2023) conducted on firms in China using the China Digital Financial Inclusion Index between 2010-2019. It shows that digital finance significantly and positively affects the ESG performance of firms. It shows that digital finance improves ESG performance by reducing agency costs, enhancing the reputation of the firm, and supporting green innovation. On the other hand, they suggest that political connections have a negative moderating effect on the positive relationship between digital finance and ESG performance. Similarly, Ren et al. (2023) state that digital finance can greatly enhance corporate ESG performance, particularly in terms of more environmental and social outcomes than the social outcomes. Furthermore, it influences ESG performance through green innovation and external oversight. Additionally, its impact is more noticeable in companies with lower levels of digitalization and profitability, as well as in firms operating in regulated sectors and industries with high carbon emissions.

3. CONCLUSION

This study discusses the rapid advancements in digital technology and its significant impact on the financial sector, business financial decisions, and overall economic growth. Digital finance is highlighted as a key driver of innovation, efficiency, and new business models, such as mobile apps, e-commerce platforms, and digital media, which contribute to economic development. In the financial sector, the rise of FinTech, driven by advancements in artificial intelligence, blockchain, and big data analytics, has reshaped traditional finance by improving service quality, reducing costs, and enabling more inclusive financial services. China, in particular, is noted for its success in utilizing digital finance to promote economic development. From a business perspective, digital transformation has a positive effect on corporate risk-taking, decision-making, and financial performance. Digital technologies also influence businesses' investment strategies, corporate governance, and innovation capabilities. By reducing information asymmetry, digital finance allows firms to access external financing more easily, which can encourage risk-taking and new projects.

The study highlights the importance of digital transformation for businesses, helping them navigate competition, improve financial inclusion, and make more informed decisions. As a result, the importance of digital finance in reshaping financial structures and decisions is inevitable. This study provides insight to practitioners, policy makers and researchers by providing a comprehensive overview of how financial digitalization affects financial decisions and performance of businesses. The effects of digital transformation on businesses, especially risk-taking process, the financial performances, green innovation investments, ESG performances, are of great importance.

Digital finance technologies such as big data analytics, artificial intelligence, blockchain, and social networks are expected to transform the financial sector by reducing costs, improving service quality, and addressing information asymmetry. Digital finance surpasses traditional financial methods by offering customer financing at lower costs and pushing existing financial institutions to enhance their services to remain competitive. This digital transformation can have significant effects on companies' financial performance, though the impact may vary based on the company's level of digitalization, sector, technologies used, and implementation strategies. Studies show that digitalization improves operational efficiency, reduces costs, and increases profit margins, positively affecting corporate performance, production efficiency, and production costs. Digital finance also helps alleviate financial constraints and improve financial performance. However, the impact of digital transformation is complex; companies with higher digital maturity generally experience greater benefits, while the short-term costs and challenges of transformation may negatively affect performance. Some studies highlight that only firms with high financial performance gain significant benefits from digital transformation. For SMEs, accessing financial resources for digital transformation can be challenging, potentially limiting the effectiveness of digitalization. However, when SMEs effectively adopt digital technologies, they can achieve a sustainable competitive advantage and growth potential.

Digital finance has also significant impact on green technology innovation. Studies show that it improves both the quantity and quality of green innovations, especially in areas with advanced digital infrastructure. In China, digital finance has boosted technological innovation, with financial technologies like AI, big data, and blockchain aiding green projects. The government's green finance policies, such as green credit, have also supported this development, although challenges remain, such as limited resources and information asymmetry. Companies in cities with advanced Fintech development have shown greater efficiency in green innovation. Additionally, digital transformation has a broader impact on business growth by easing financing constraints, as evidenced by studies in China. Policies promoting the integration of finance and technology have proven effective, particularly for firms with higher digital technology usage. Some studies also highlight regional differences in the effectiveness of green finance policies, with some cities encouraging innovation more than others.

Digital transformation plays a crucial role in improving not only business efficiencies and financial performance but also in strengthening a company's dedication to the principles of ESG. By utilizing technologies like AI, blockchain, big data, and IoT, businesses can make better decisions, enhance social responsibility, improve sustainability, and foster better governance. This, in turn, boosts a company's ESG performance, improving its reputation, attracting investments, and ensuring long-term success. Many countries have recognized the importance of ESG factors and developed policies to encourage businesses to adopt them. For this purpose, most countries have developed certain policies related to these issues. Digital transformation significantly impacts ESG performance by enabling better management of ESG practices through technology. Several studies have shown that digital finance and green innovation contribute positively to ESG performance. Research in China, for example, demonstrated that digital finance boosts green innovation and helps companies overcome financial barriers, leading to improved ESG outcomes. Some studies also highlight how firm size influences the effectiveness of digital transformation and green finance on ESG performance. In summary, digital transformation and green finance have been shown to positively affect ESG performance, with larger companies experiencing more substantial improvements. The role of FinTech, in particular, has been crucial in boosting ESG outcomes, especially in high-pollution sectors. However, factors such as political connections and industry regulations can moderate these effects.

As a result, financial digitalization has become a critical factor that not only accelerates economic growth but also fundamentally transforms businesses' decision-making and risk management processes. With the widespread adoption of digital financial tools, firms are now able to access information more quickly, improve their integration into financial markets, and make bolder, data-driven decisions. At the same time, financial digitalization contributes to the deepening of capital markets by facilitating access to external financing, thereby supporting economic growth. In this context, several implications can be drawn for policymakers and practitioners to ensure that the digital transformation process is effectively managed at both economic and institutional levels. In particular, in developing countries, comprehensive internet infrastructure should be expanded nationwide to reduce digital inequality. Improving access to digital services in rural areas is essential to ensure that businesses in these regions can also benefit from digital financial tools. Incentives such as tax benefits, funding support, and incubation programs should be introduced to foster the growth of Fintech startups. Regulations should be designed to be flexible enough to promote innovation while maintaining a balanced approach that ensures security. National and international data protection policies must be developed to ensure the ethical and secure use of artificial intelligence and big data tools by businesses. Strengthening cybersecurity infrastructure is also essential to make the financial system resilient against digital threats. Given the higher levels of uncertainty and volatility in the digitized business environment, firms should be supported in developing and implementing digital risk monitoring and early warning systems tailored to their needs. In conclusion, while policymakers should prioritize infrastructure, regulation, and inclusivity, practitioners should focus on data-driven strategies, technology integration, and effective risk management. In doing so, financial digitalization not only promotes economic growth but also enables businesses to make more courageous, informed, and innovative decisions. The opportunities offered by financial digitalization do not merely improve macroeconomic indicators, but also reshape strategic decision-making processes at the micro level. For this transformation to contribute to sustainable economic growth, it is crucial that the digitalization process is directed in a planned, inclusive, and secure manner. In this regard, comprehensive policy frameworks will contribute to the creation of competitive, resilient, and adaptive economic structures that are aligned with the demands of the digital age. This study aims to provide a theoretical contribution to the existing body of literature through an extensive review of relevant scholarly works. Future research may build upon these theoretical insights by employing empirical methods and appropriate datasets to validate the proposed relationships.

REFERENCES

Aboody, D., Levi, S., & Weiss, D. (2018). Managerial incentives, options, and cost-structure choices. *Review of Accounting Studies, 23*(2), 422–451. DOI: 10.1007/s11142-017-9432-0

Ahmed, S., Sihvonen, J., & Vähämaa, S. (2019). CEO facial masculinity and bank risk-taking. *Personality and Individual Differences, 138*, 133–139. DOI: 10.1016/j.paid.2018.09.029

Akey, P., & Lewellen, S. (2017). Policy uncertainty, political capital, and firm risk-taking. *Econometric Modeling: Corporate Finance & Governance eJournal*. https://ssrn.com/abstract=2758395

Amaliah, I., Firmansyah, I., & Kurniasih, H. (2024). Digital financial inclusion and sustainable economic growth: Evidence from Indonesia. *Journal of Sustainable Finance & Investment, 14*(1), 45–62. DOI: 10.1080/20430795.2024.000000

Andersen, D. C. (2017). Do credit constraints favor dirty production? Theory and plant-level evidence. *Journal of Environmental Economics and Management, 84*, 189–208. DOI: 10.1016/j.jeem.2017.04.002

Australian Government. The Treasury. (2023). *Sustainable Finance Strategy*. Retrieved January 23, 2025, from https://treasury.gov.au/sites/default/files/2023-11/c2023-456756.pdf

Australian Securities and Investments Commission. (2021). *Climate Change Risk and Disclosure*. Retrieved January 23, 2025, from https://asic.gov.au/about-asic/news-centre/speeches/corporate-governance-update-climate-change-risk-and-disclosure/

Bahoo, S., Cucculelli, M., & Qamar, D. (2023). Artificial intelligence and corporate innovation: A review and research agenda. *Technological Forecasting and Social Change, 188*, 122264. DOI: 10.1016/j.techfore.2022.122264

Basnayake, B. M. H. S., Li, Y., & Zaman, Q. (2024). Digital financial inclusion and economic growth: Evidence from the Asia-Pacific region. *Economic Modelling, 132*, 105986. DOI: 10.1016/j.econmod.2024.105986

Bayo-Moriones, A., Billon, M., & Lera-Lopez, F. (2013). Perceived performance effects of ICT in manufacturing SMEs. *Industrial Management & Data Systems, 113*(1), 117–135. DOI: 10.1108/02635571311289700

Bromiley, P. (1991). Testing a causal model of corporate risk taking and performance. *Academy of Management Journal, 34*(1), 37–59. DOI: 10.2307/256301

Bromiley, P. (1991). Testing a causal model of corporate risk taking and performance. *Academy of Management Journal, 34*(1), 37–59. DOI: 10.2307/256301

Chakraborty, A., Silva Gao, L., & Sheikh, S. (2019). Managerial risk taking incentives, corporate social responsibility, and firm risk. *Journal of Economics and Business, 101*, 58–72. DOI: 10.1016/j.jeconbus.2018.07.004

Chen, D. Q., & Hu, Q. (2022). Corporate governance research in the digital economy: New paradigms and frontiers of practice. *Journal of Management World, 38*(6), 213–240. DOI: 10.53935/jomw.v2023i2.245

Chen, W., Arn, G., Song, H., & Xie, Y. (2024). The influences of digital finance on green technological innovation in China's manufacturing sector: The threshold effects of ESG performance. *Journal of Cleaner Production, 467*, 142953. DOI: 10.1016/j.jclepro.2024.142953

Chen, X., Mao, Y., Cheng, J., Wei, P., & Li, X. (2024). Green financial policy, technological advancement reversal, assessment of emission reduction effects. *Energy Economics. Energy Economics, 107678*, 107678. Advance online publication. DOI: 10.1016/j.eneco.2024.107678

Chen, Y.-R., & Ma, Y. (2011). Revisiting the risk-taking effect of executive stock options on firm performance. *Journal of Business Research, 64*(6), 640–648. DOI: 10.1016/j.jbusres.2010.02.012

China Securities Regulatory Commission. (2020). *Guidelines for ESG Reporting*. Retrieved January 23, 2025, from http://www.csrc.gov.cn/csrc_en/c102063/c1606114/1606114/files/CSRC%20Annual%20Report%202020.pdf

Clemons, E. K., & Row, M. C. (1992). Information technology and industrial cooperation: The changing economics of coordination and ownership. *Journal of Management Information Systems, 9*(1), 9–28. DOI: 10.1080/07421222.1992.11517956

Climate Change Laws. (2021). *14th Five-Year Plan for Ecological Environmental Protection.* Retrieved January 23, 2025, from https://climate-laws.org/document/14th-five-year-plan_0496

Correa, R., & Esquivias, M. A. (2025). Digitalization, education and institutional quality as drivers of growth: A comparative analysis of Sub-Saharan Africa and the Middle East. *Applied Economics, 57*(4), 321–340. DOI: 10.1080/00036846.2025.000000

Dai, J., Farooq, U., & Alam, M. M. (2024). Navigating energy policy uncertainty: Effects on fossil fuel and renewable energy consumption in G7 economies. *International Journal of Green Energy*, •••, 1–14. DOI: 10.1080/15435075.2024.XXXXXXX

Daud, S. N. M., Yusof, Z., & Aziz, M. R. A. (2024). Digital technology, financial inclusion and economic growth: A cross-country analysis. *Journal of Economic Studies (Glasgow, Scotland)*, *51*(2), 210–227. DOI: 10.1108/JES-10-2023-0598

DeStefano, T., Kneller, R., & Timmis, J. (2022). The (fuzzy) digital divide: The effect of universal broadband on firm performance. *Journal of Economic Geography*, *23*(1), 139–177. DOI: 10.1093/jeg/lbac006
DeStefano, T., Kneller, R., & Timmis, J. (2022). The (fuzzy) digital divide: The effect of universal broadband on firm performance. *Journal of Economic Geography*, *23*(1), 139–177. DOI: 10.1093/jeg/lbac006
Ding, J., Li, L., & Zhao, J. (2024). How does fintech prompt corporations toward ESG sustainable development? Evidence from China. *Energy Economics*, *107387*, 107387. Advance online publication. DOI: 10.1016/j.eneco.2024.107387

Dou, B., Guo, S., Chang, X., & Wang, Y. (2023). Corporate digital transformation and labor structure upgrading. *International Review of Financial Analysis*, *90*, 102904. DOI: 10.1016/j.irfa.2023.102904

Ebhota, O. S., Hongxing, Y., & Sampene, A. K. (2024). Investigating the influence of digital transformation, budgeting and budgetary control on the financial performance of SMEs. *Scientific African*, *26*, e02429. DOI: 10.1016/j.sciaf.2024.e02429

European Commission. (2019). *The European Green Deal.* Retrieved January 23, 2025, from https://ec.europa.eu/

European Commission. (2020). *EU Taxonomy Regulation (2020/852).* Retrieved January 23, 2025, from https://ec.europa.eu/info/business-economy-euro/banking-and-finance/sustainable-finance_en

European Commission. (2020). *European Green Deal.* European Commission. Retrieved January 23, 2025, from https://ec.europa.eu/

European Commission. (2021). *Proposal for a directive of the European Parliament and of the Council on corporate sustainability reporting*. Retrieved January 23, 2025, from https://ec.europa.eu/info/publications/210421-sustainable-finance-communication_en

Faccio, M., Marchica, M., & Mura, R. (2016). CEO gender, corporate risk-taking, and the efficiency of capital allocation. *Journal of Corporate Finance*, *39*, 193–209. DOI: 10.1016/j.jcorpfin.2016.02.008

Faccio, M., Marchica, M.-T., & Mura, R. (2016). CEO gender, corporate risk-taking, and the efficiency of capital allocation. *Journal of Corporate Finance*, *39*, 193–209. DOI: 10.1016/j.jcorpfin.2016.02.008

Fang, M., Nie, H., & Shen, X. (2023). Can enterprise digitization improve ESG performance? *Economic Modelling*, *118*, 106101. DOI: 10.1016/j.econmod.2022.106101

Gao, W., & Liu, Z. (2023). Green credit and corporate ESG performance: Evidence from China. *Finance Research Letters*, *55*, 103940. DOI: 10.1016/j.frl.2023.103940

Gareeb, P. P., & Naicker, V. (2015). Determinants for South African SMEs to adopt broadband Internet technologies. *The Electronic Journal on Information Systems in Developing Countries*, *68*(1), 1–24. DOI: 10.1002/j.1681-4835.2015.tb00491.x

Gomber, P., Koch, J. A., & Siering, M. (2017). Digital finance and FinTech: Current research and future research directions. *Journal of Business Economics*, *87*(5), 537–580. DOI: 10.1007/s11573-017-0852-x

Government of Canada. (2021). *Canada's Federal Carbon Pricing*. Retrieved January 23, 2025, from https://publications.gc.ca/site/archivee-archived.html?url=https://publications.gc.ca/collections/collection_2021/eccc/En4-423-1-2021-eng.pdf

Green Finance Platform. (2021). *Sustainable Finance Action Plan*. Retrieved January 23, 2025, from https://www.greenfinanceplatform.org/policies-and-regulations/sustainable-finance-action-councilIndia

Guo, Y., Li, P., & Li, A. (2021). Tail risk contagion between international financial markets during the COVID-19 pandemic. *International Review of Financial Analysis*, *73*, 101649. DOI: 10.1016/j.irfa.2020.101649

Habibi, F., & Zabardast, M. A. (2020). Digitalization, education and economic growth: A comparative analysis of Middle East and OECD countries. *Technology in Society*, *63*, 101370.

Hanelt, A., Bohnsack, R., Marz, D., & Antunes Marante, C. (2021). A systematic review of the literature on digital transformation: Insights and implications for strategy and organizational change. *Journal of Management Studies*, *58*(6), 1159–1197. DOI: 10.1111/joms.12639

Hanna, H., & Bettina, P. (2012). Innovative capability and financing constraints for innovation: More money, more innovation? *The Review of Economics and Statistics*, *94*(6), 1126–1142. DOI: 10.1162/REST_a_00209

Harjoto, M., & Laksmana, I. (2018). The impact of corporate social responsibility on risk taking and firm value. *Journal of Business Ethics*, *151*(2), 353–373. DOI: 10.1007/s10551-016-3202-y

He, F., & Liu, H. X. (2019). Evaluation of performance improvement effect of digital transformation of entity enterprises from the perspective of digital economy. *Reform*, *302*, 137–148.

Horvey, S. S., & Ankamah, J. (2020). Enterprise risk management and firm performance: Empirical evidence from Ghana equity market. *Cogent Economics & Finance*, *8*(1), 1840102. Advance online publication. DOI: 10.1080/23322039.2020.1840102

Hu, H., Jia, Z., & Yang, S. (2025). Exploring FinTech, green finance, and ESG performance across corporate life-cycles. *International Review of Financial Analysis*, *97*, 103871. DOI: 10.1016/j.irfa.2024.103871

Huang, D. Y., Xie, H. B., Zou, M. T., & Meng, X. Y. (2022). The impact of enterprise digital transformation on risk taking: The mechanism and path. *Keji Jinbu Yu Duice*, *39*, 1–10. DOI: 10.1016/j.ribaf.2024.102285

Jiang, J., Hou, J., Wang, C., & Liu, H. (2021). COVID-19 impact on firm investment—Evidence from Chinese publicly listed firms. *Journal of Asian Economics*, *75*, 101320. Advance online publication. DOI: 10.1016/j.asieco.2021.101320 PMID: 36567767

Jiang, K., Du, X., & Chen, Z. (2022). Firms' digitalization and stock price crash risk. *International Review of Financial Analysis*, *82*, 102196. DOI: 10.1016/j.irfa.2022.102196

John, K., Litov, L., & Yeung, B. (2008). Corporate governance and risk-taking. *The Journal of Finance*, *63*(4), 1679–1728. DOI: 10.1111/j.1540-6261.2008.01372.x

Khalil, R. G., Damrah, S., Bajaher, M., & Shawtari, F. A. (2023). Unveiling the relationship of ESG, fintech, green finance, innovation and sustainability: Case of gulf countries. *Environmental Science and Pollution Research International*, *30*(54), 116299–116312. DOI: 10.1007/s11356-023-30584-8 PMID: 37910364

Koerniadi, H., Krishnamurti, C., & Tourani-Rad, A. (2014). Corporate governance and risk-taking in New Zealand. *Australian Journal of Management*, *39*(2), 227–245. DOI: 10.1177/0312896213478332

Kreiser, M. D., Marino, L. D., Kuratko, D. F., & Weaver, K. M. (2013). Disaggregating entrepreneurial orientation: The non-linear impact of innovativeness, proactiveness, and risk-taking on SME performance. *Small Business Economics*, *40*(2), 273–291. DOI: 10.1007/s11187-012-9460-x

Kumar, R. (2024). Natural resources, globalization, financial development, and digitalization in India's economic growth. *Resources Policy*, *88*, 104110. DOI: 10.1016/j.resourpol.2024.104110

Langenmayr, D., & Lester, R. (2018). Taxation and corporate risk-taking. *The Accounting Review*, *93*(3), 237–266. DOI: 10.2308/accr-51872

Lee, I., & Shin, Y. J. (2018). Fintech: Ecosystem, business models, investment decisions, and challenges. *Business Horizons*, *61*(1), 35–46. DOI: 10.1016/j.bushor.2017.09.003

Li, H. J., Tian, Y. X., & Li, W. J. (2014). Internet thinking and traditional enterprise reengineering. *China Industrial Economy*, *10*, 135–146.

Li, J., Li, N., & Cheng, X. (2021). The impact of fintech on corporate technology innovation based on driving effects, mechanism identification, and heterogeneity analysis. *Discrete Dynamics in Nature and Society*, *7825120*, 1–12. Advance online publication. DOI: 10.1155/2021/7825120

Lin, B., & Ma, R. (2022). How does digital finance influence green technology innovation in China? Evidence from the financing constraints perspective. *Journal of Environmental Management*, *320*, 115833. DOI: 10.1016/j.jenvman.2022.115833 PMID: 35940011

Lin, B., & Ma, R. (2022). How does digital finance influence green technology innovation in China? Evidence from the financing constraints perspective. *Journal of Environmental Management*, *320*, 115833. DOI: 10.1016/j.jenvman.2022.115833 PMID: 35940011

Liu, J., Zhang, Y., & Kuang, J. (2023). Fintech development and green innovation: Evidence from China. *Energy Policy*, *183*, 113827. DOI: 10.1016/j.enpol.2023.113827

Locke, E. A., & Latham, G. P. (2004). What should we do about motivation theory? Six recommendations for the twenty-first century. *Academy of Management Review*, *29*(3), 388–403. DOI: 10.2307/20159050

Low, A. (2009). Managerial risk-taking behavior and equity-based compensation. *Journal of Financial Economics*, *92*(3), 470–490. DOI: 10.1016/j.jfineco.2008.05.004

Lu, H., & Cheng, Z. (2024). Digital inclusive finance and corporate ESG performance: The moderating role of executives with financial backgrounds. *Finance Research Letters*, *60*, 104858. DOI: 10.1016/j.frl.2023.104858

Luo, S., Sun, Y., Yang, F., & Zhou, G. (2022). Does fintech innovation promote enterprise transformation? Evidence from China. *Technology in Society*, *68*, 101821. DOI: 10.1016/j.techsoc.2021.101821

McLean, R. D., & Zhao, M. (2014). The business cycle, investor sentiment, and costly external finance. *The Journal of Finance*, *69*(3), 1377–1409. DOI: 10.1111/jofi.12047

Meng, M., Fan, S., Li, X., & Lei, J. (2024). Digital transformation and strategic risk taking dataset for China's public-listed companies. *Data in Brief*, *54*, 110511. DOI: 10.1016/j.dib.2024.110511 PMID: 38868384

Ministry of Health, Family Welfare Government of India. (2024). National Action Plan on Climate Change (NAPCC). Retrieved January 23, 2025, from https://ncdc.mohfw.gov.in/wp-content/uploads/2024/04/27505481411548674558.pdf

Mou, Y. (2024). The impact of digital finance on technological innovation across enterprise life cycles in China. *Heliyon*, *10*(14), e33965. DOI: 10.1016/j.heliyon.2024.e33965 PMID: 39104497

Mu, W., Liu, K., Tao, Y., & Ye, Y. (2023). Digital finance and corporate ESG. *Finance Research Letters*, *51*, 103426. DOI: 10.1016/j.frl.2022.103426

Myovella, G., Karacuka, M., & Haucap, J. (2020). Digitalization and economic growth: A comparative analysis of Sub-Saharan Africa and OECD economies. *Telecommunications Policy*, *44*(2), 101856. DOI: 10.1016/j.telpol.2019.101856

Osuma, G. O. (2025). Digital financial services, economic growth, and poverty reduction in Sub-Saharan Africa. *African Development Review*, *37*(1), 65–82. DOI: 10.1111/1467-8268.12689

Palomo, I., Felipe-Lucia, M. R., Bennett, E. M., Martín-López, B., & Pascual, U. (2016). Disentangling the Pathways and Effects of Ecosystem Service Co-Production. [). Academic Press.]. *Advances in Ecological Research*, *54*, 245–283. DOI: 10.1016/bs.aecr.2015.09.003

Peng, Y., & Tao, C. (2022). Can digital transformation promote enterprise performance?—From the perspective of public policy and innovation. *Journal of Innovation & Knowledge*, *7*(3), 100198. DOI: 10.1016/j.jik.2022.100198

Peng, Y., & Tao, C. (2022). Can digital transformation promote enterprise performance?—From the perspective of public policy and innovation. *Journal of Innovation & Knowledge*, *7*(3), 100198. DOI: 10.1016/j.jik.2022.100198

Polzin, F., & Sanders, M. (2020). How to finance the transition to low-carbon energy in Europe? *Energy Policy*, *147*, 111933. DOI: 10.1016/j.enpol.2020.111863

Qi, H. J., Cao, X. Q., & Liu, Y. X. (2020). The influence of digital economy on corporate governance: Analyzed from information asymmetry and irrational behavior perspective. *Reform*, *4*, 50–64.

Qi, Y. D., & Xiao, X. (2020). Enterprise management reform in the digital economy era. *Management World*, *36*, 135–152.

Qian, Haizhang, Tao, Y., Cao, S., & Cao, Y. (2020). Theoretical and empirical analysis on the development of digital finance and economic growth in China. *Shuliang Jingji Jishu Jingji Yanjiu*, *37*(06), 26–46.

Raziyeh, R.-G., Hafezalkotob, A., Makuib, A., & Sayadic, M. K. (2022). Financing green technology development and role of digital platforms: *Insourcing vs. outsourcing.Technology in Society*, *69*, 101967. DOI: 10.1016/j.techsoc.2022.101967

Ren, X., Zeng, G., & Zhao, Y. (2023). Digital finance and corporate ESG performance: Empirical evidence from listed companies in China. *Pacific-Basin Finance Journal*, *79*, 102019. DOI: 10.1016/j.pacfin.2023.102019

Securities and Exchange Board of India. (2021). *Business Responsibility and Sustainability Report (BRSR)*. Retrieved January 23, 2025, from https://www.sebi.gov.in/sebi_data/commondocs/may-2021/Business%20responsibility%20and%20sustainability%20reporting%20by%20listed%20entitiesAnnexure1_p.PDF

Shahadat, H., Karmaker, S. C., Rahman, M. M., Chapman, A. J., & Saha, B. B. (2022). Dynamic links among the demographic dividend, digitalization, energy intensity and sustainable economic growth: Empirical evidence from emerging economies. *Journal of Cleaner Production*, *330*, 1. DOI: 10.1016/j.jclepro.2021.129858

Sheng, T. X., & Fan, C. C. (2020). FinTech, optimal banking market structure and small and micro enterprise credit supply. *Financial Research*, *6*, 114–132.

Simon, M., & Houghton, S. M. (2003). The relationship between overconfidence and the introduction of risky products: Evidence from a field study. *Academy of Management Journal*, *46*(2), 139–149. DOI: 10.2307/30040610

Singh, A. K., Raza, S. A., Nakonieczny, J., & Shahzad, U. (2023). Role of financial inclusion, green innovation, and energy efficiency for environmental performance? Evidence from developed and emerging economies in the lens of sustainable development. *Structural Change and Economic Dynamics*, *64*, 213–224. DOI: 10.1016/j.strueco.2022.12.008

Sreenu, R., & Verma, S. S. (2024). FinTech and the digital economy in India: A state-wise CRITIC analysis. *Technological Forecasting and Social Change, 201*, 122097. DOI: 10.1016/j.techfore.2024.122097

Sun, T., Liu, S., & Guo, M. (2025). Effects of green finance and digital transformation on enhancing corporate ESG performance. *Finance Research Letters, 74*, 106774. DOI: 10.1016/j.frl.2025.106774

Tang, S., Wu, X., & Zhu, J. (2020). Digital finance and Enterprise technology innovation: Structure feature, mechanism identification and effect difference under financial supervision. *Management World, 36*(05), 52–67.

The White House. (2021). F*act Sheet: The Biden-Harris Administration's Actions to Tackle the Climate Crisis.* Retrieved January 24, 2025, from https://www.energy.gov/sites/default/files/2024-06/130.%20FACT%20SHEET%20Biden-Harris%20Administration%20New%20Actions%20to%20Reduce%20Greenhouse%20gas%20Emissions%20%28Sept.%2021%2C%202023%29.pdf

Tian, G., Li, B., & Cheng, Y. (2022). Does digital transformation matter for corporate risk-taking? *Finance Research Letters, 49*, 103107. DOI: 10.1016/j.frl.2022.103107

Tian, G., Li, B., & Cheng, Y. (2022). Does digital transformation matter for corporate risk-taking? *Finance Research Letters, 49*, 103–107. DOI: 10.1016/j.frl.2022.103107

Tian, G., Li, B., & Cheng, Y. (2022). Does digital transformation matter for corporate risk-taking? *Finance Research Letters, 49*, 103107. Advance online publication. DOI: 10.1016/j.frl.2022.103107

UK Government. (2021). *UK Green Finance Initiative.* Retrieved January 23, 2025, from https://www.gov.uk/government/publications/uk-government-green-financing

U.S. Securities and Exchange Commission. (2021). *Climate and ESG Task Force.* Retrieved January 24, 2025, from https://www.sec.gov/newsroom/press-releases/2021-42

Vu, D. A., Nguyen, T. V., Nhu, Q. M., & Tran, T. Q. (2024). Does increased digital transformation promote a firm's financial performance? New insights from the quantile approach. *Finance Research Letters, 64*, 105430. DOI: 10.1016/j.frl.2024.105430

Wang, Q., Qu, J., Wang, B., Wang, P., & Yang, T. (2019). Green technology innovation development in China in 1990-2015. *The Science of the Total Environment, 696*, 134008. DOI: 10.1016/j.scitotenv.2019.134008 PMID: 31465930

Wang, W. (2025). Digital finance and firm green innovation: The role of media and executives. *Finance Research Letters, 74*, 106794. DOI: 10.1016/j.frl.2025.106794

Wei, L., & Li, M. (2024). Digital transformation, financing constraints and firm growth performance–From the perspective of financing channels. *Finance Research Letters, 63*, 105272. DOI: 10.1016/j.frl.2024.105272

Wei, Y., Zhu, M., Lam, H. K. S., Liu, F., Yeung, A. C. L., Cheng, T. C. E., & Zhou, H. (2024). Protagonists in digital transformation: The impact of chief digital officers on firms' financial performance. *Journal of Industrial Information Integration, 41*, 100678. DOI: 10.1016/j.jii.2024.100678

Wen, F., Li, C., Sha, H., & Shao, L. (2021). How does economic policy uncertainty affect corporate risk-taking? Evidence from China. *Finance Research Letters, 41*, 101840. Advance online publication. DOI: 10.1016/j.frl.2020.101840

Wu F., H. Hu, H. Lin, X. Ren (2021). Enterprise digital transformation and capital market performance: empirical evidence from stock liquidity, *Management World,* 37, pp. 130-144+110, DOI: 10.19744/j.cnki.11-1235/f.2021.0097

Wu, H., & Wang, Y. (2024). Digital transformation and corporate risk taking: Evidence from China. *Global Finance Journal, 62*, 101012. DOI: 10.1016/j.gfj.2024.101012

Wu, Y., & Huang, S. (2022). The effects of digital finance and financial constraint on financial performance: Firm-level evidence from China's new energy enterprises. *Energy Economics, 112*, 106158. DOI: 10.1016/j.eneco.2022.106158

Xu, G., & Pan, X. (2024). Can digital transformation reduce the cost of debt capital? Empirical analysis based on the perspective of financial situation. *Applied Economics Letters*, •••, 1–6. DOI: 10.1080/13504851.2024.2351592

Xue, L., Dong, J., & Zha, Y. (2023). How does digital finance affect firm environmental, social and governance (ESG) performance? — Evidence from Chinese listed firms. *Heliyon, 9*(10), e20800. DOI: 10.1016/j.heliyon.2023.e20800 PMID: 37867793

Yan, R. S., & Qian, X. Y. (2018). Strategic analysis of digital transformation of Chinese operators in the era of digital economy. *China Soft Science, 328*, 172–182.

Yunis, M., Mirza, N., Safi, A., & Umar, M. (2024). Impact of audit quality and digital transformation on innovation efficiency: Role of financial risk-taking. *Global Finance Journal, 62*, 101026. DOI: 10.1016/j.gfj.2024.101026

Zhao, G., Xin, Z., & Wang, Y. (2024). Effect of the sci-tech finance pilot policy on corporate environmental information disclosure—Moderating role of green credit. *Finance Research Letters, 105177*, 105177. Advance online publication. DOI: 10.1016/j.frl.2024.105177

Zhao, S., Zhang, L., Peng, L., Zhou, H., & Hu, F. (2024). Enterprise pollution reduction through digital transformation? Evidence from Chinese manufacturing enterprises. *Technology in Society*, *77*, 102520. DOI: 10.1016/j.techsoc.2024.102520

Zhao, X., Sun, X., Zhao, L., & Xing, Y. (2022). Can the digital transformation of manufacturing enterprises promote enterprise innovation? *Business Process Management Journal*, *28*(4), 960–982. DOI: 10.1108/BPMJ-01-2022-0018

Zheng, F., Jiang, J., Ma, J., & Liou, J. (2024). Capital market opening, digital transformation, and corporate risk-taking: Empirical evidence from Shanghai Shenzhen-Hong Kong stock connect. *International Review of Economics & Finance, 96*(B), 103680. https://doi.org/DOI: 10.1016/j.iref.2024.103680

Zhou, Y., Chen, X., & Lin, W. (2025). Bank digital transformation and rural economic growth: Evidence from China. *Journal of Rural Studies*, *103*, 187–200. DOI: 10.1016/j.jrurstud.2025.000000

KEY TERMS AND DEFINITIONS

Digital Transformation: It is the integration of digital technologies into all areas of a business to improve operations and deliver greater value to customers.

Green Innovation: It refers to the development of new products, services, or processes that reduce environmental impact and promote sustainability.

Financial Performance: It is a measure of how well a company uses its assets to generate revenue and profit over a specific period.

Risk-Taking Process: It involves identifying, assessing, and deciding to pursue opportunities that carry potential uncertainties or losses in order to achieve desired outcomes or gains.

ESG Performance: It refers to how well a company manages its environmental, social, and governance responsibilities to create sustainable long-term value.

Digital Finance: It refers to the use of digital technologies to deliver financial services, such as payments, lending, and investing, more efficiently and accessibly.

Fin Tech: short for financial technology, refers to the use of innovative technology to improve, automate, and disrupt traditional financial services and processes.

Chapter 11

FinTech for Sustainability: A Treatise

Batani Raghavendra Rao
https://orcid.org/0000-0003-0663-2394
Jain University, India

Anusha Batni
https://orcid.org/0009-0003-8584-7904
Deloitte, Ireland

Raghavendra Navaratna
Princeton University, USA

ABSTRACT

Sustainability is the way forward. In order to achieve the sustainable growth, technology is a pivot. Leveraging Artificial Intelligence (AI), Blockchain, Big Data, Cloud Computing and related tech tools there is an applied tech lever, known as FinTech. FinTech is a tech revolution which has disrupted the business landscape. It has transformed the economy by enabling the institutions, firms, consumers and society at large. The chapter dwells on the role of FinTech in creating sustainable development. Further, the chapter covers aspects like FinTech architecture and ecosystem, impact of FinTech in realising Sustainable Development Goals (SDGs) and the significance of FinTech as an enabler of ESG theme and philosophy. Use cases of FinTech in promoting sustainability are documented. The chapter outlines the challenges of FinTech in terms of risks and ethical issues. Also, the chapter establishes the need for a sustainable FinTech ecosystem in the wake of its impact, potential, associated risks and ethical challenges.

DOI: 10.4018/979-8-3373-0355-0.ch011

1.0 INTRODUCTION TO FINTECH AND SUSTAINABILITY

Essentially FinTech is an application of technology and data science for developing, reaching out, serving the needs of human in a sustainable manner. United Nations has brought out 17 Sustainable Development Goals (SDGs) for the overall development of the mankind (United Nations, 2015). The 17 SDGs can be grouped into 3 pillars such as Environmental, Social and Governance (ESG). FinTech companies invariably address at least one SDG and this can be found by examining the mission and target markets of FinTech companies (Carè et al., 2023). FinTech is omni present and one can appreciate its application in almost all sectors, be it agriculture, banking, insurance, micro finance, audit, risk management and modelling, manufacturing, supply chain, governance, climate modelling etc. FinTech plays a crucial role in transforming agriculture and agri business (Hinson et al., 2019). The dawn of FinTech has brought in digital transformation of the banking's "bricks-and-mortar" and spectacularly changed the way financial services are rendered (Agarwal & Zhang, 2020).

In the words of "Reserve Bank of India", FinTech is a "A linguistic blend of two individual terms 'Finance' & 'Technology', FinTech is being used to denote the wide array of technological innovations that have a bearing on financial services" (RBI, 2024).

Bank of England defines FinTech as, "Combining the words 'financial' and 'technology', fintech is technology-enabled financial innovation, which is changing the way financial institutions provide – and consumers and businesses use – financial services" (Bank of England, 2024).

Fostering the sustainability requires the protagonist role of technology among other resources. In fact, all other resources do require technology as an enabler. Sustainability is the final target key performance indicator (KPI) for the overall well-being of the human race. To achieve this KPI, advent of FinTech is a boon. FinTech along with its related advanced tech tools can transform the sustainability agenda. ABCD of FinTech is the set of other tech tools. The acronym ABCD refers to Artificial Intelligence (AI), Blockchain, Cloud computing and Data analytics. While discussing the use cases, the application of these ABCD tech tools would come to the fore. Similarly, the 5 Ds (namely, Digitization, Disruption, Democratization, Decentralization, and Data) are critical in developing and applying FinTech solution and these 5 Ds are drivers of FinTech innovation. The study covers the FinTech space development over the years in the main geographies like India, USA, Europe and Asia Pacific. A more detailed attention is given to FinTech in India as India is a growing tech economy with a high adoption of digitisation. Also, India is acclaimed as a strong FinTech hub. India is often portrayed as a global pioneer in spearheading the technological innovations in the world of finance (Jeff Kearns and Ashlin Mathew, 2022). As far as most FinTech unicorns in the world, India bags the third place for 2023 and US is in first place (Livemint, 2023).

2.0 FINTECH ECOSYSTEM

The ecosystem of Fintech comprises the regulatory framework, FinTech firms and the other stakeholders. The apex or the central bank of the country will normally provide the FinTech regulatory framework under the state patronage. The self-regulatory bodies join the apex bank in the regulation and governance of FinTech. The core of the ecosystem is made up of the FinTech firms, made up of start-ups, small, big and bank / financial institution led FinTech firms. The other stakeholders are Government, tech providers, investors, consumers, society and public at large.

2.1 FinTech Regulatory and Governance Framework

There is a strong nexus between Fintech governance and performance. Also, studies show that robust Fintech governance and financial stability are strongly correlated (Greta Benedetta Ferilli et al., 2024). FinTech firms vary in size and scope and hence, the regulatory framework should factor entity-based requirements in terms of competition and operational resilience so that a level playing field is ensured and thereby competitive distortions can be mitigated (Restoy, 2021). The policy framework is critical for the development of FinTech and in turn for the development of financial market (Muganyi et al., 2022).

FinTech Regulatory and governance framework in various countries may be summarised as follows.

2.1.1 India

Indian FinTech ecosystem is robust and worth emulating (Migozzi et al., 2023). RBI i.e., "Reserve Bank of India" is the chief regulator of FinTech ecosystem. For regulating and orderly growth of FinTech ecosystem in India, Reserve Bank of India started Sandbox on 13th August 2019. Sandbox is the outcome of a detailed and extensive consultations with stakeholders. The FinTech entities can live test their novel FinTech products or services in RBI's Sandbox in a controlled environment. Reserve Bank Innovation Hub, a subsidiary of RBI has a mission, named, "FinTech and Startup Acceleration" (FAST). Its objective is to make India a FinTech nation.

During May 2024, Reserve Bank of India, RBI, came out with a framework for self-regulation of FintTechs (Reserve Bank of India - Reports, 2024).The framework outlines the concept of Self-Regulatory Organisations (SRO) for FinTech (FT), and it is known as SRO-FT. Further, the framework contains the need for SRO-FT, characteristics, operations, eligibility for becoming SRO-FT, membership criteria, governance of board, key managerial personnel, functions and responsibilities, grievance redressal etc. RBI recognised FACE as self-regulatory organisation in FinTech sector (Gonsalves, 2024). FACE stands for Fintech Association for Consumer Empowerment. It is the first SRO-FT.

Indian government has provided a solid fillip to the growth of FinTech by instituting award for the innovative ideas and thereby recognising the startups in FinTech space (Startup India, 2023). The "Centre for Advanced Financial Research and Learning" opines in its first "India Finance Report" (IFR)' that close to half of startups are fintech and the report says that the projected lending by FinTech firms to overtake banks by 2030 (Fintech Must Wait for Spring, 2024).

2.1.2 US

SEC ("Securities and Exchange Commission") has established the "Office of Strategic Hub for Innovation and Financial Technology (FinHub)". FinHub is the Strategic Hub for Innovation in FinTech. It aids the SEC's nimble engagement with various stakeholders of FinTech such as developers, innovators and entrepreneurs of FinTech. It coordinates the support systems for distributed ledger technology, AI (artificial intelligence) and ML (machine learning).

2.1.3 Australia

Chief Regulator is ASIC i.e., "Australian Securities and Investments Commission". APRA i.e., "Australian Prudential Regulation Authority" and Reserve Bank of Australia assist in FinTech monitoring. To facilitate financial innovation, Australian Securities and Investment Commission has a robust regulatory sandbox.

2.1.4 UK

Bank of England, Financial Conduct Authority (FCA), Prudential Regulation Authority and Payment Systems Regulator are the main regulators. Artificial Intelligence Consortium provides a platform for public-private engagement. This consortium gathers inputs from stakeholders regarding capabilities, development, deployment and leveraging of AI. The AI Public-Private Forum (AIPPF) was established by Bank of England and FCA to facilitate and further the dialogue on AI innovation and the adoption of AI safely within financial services. FCA urges the significance of ethical AI in the practice and application of FinTech solutions.

2.1.5 Singapore

MAS i.e., "Monetary Authority of Singapore" is the chief regulator of FinTech. MAS has enabled "FinTech Regulatory Sandbox framework" for FinTech players to experiment their innovative products and services. Singapore hosts every year FinTech Fest, which is considered to be the biggest in the world. From 134 countries, 65,000 participants attended in the recently concluded FinTech fest (8th November 2024).

2.1.6 China

FinTech journey began in December 2004 with Alipay first came online. FinTech is known as "Internet Finance" in China. Real boost and momentum did begin in 2013, when Ant Financial launched its online money market fund called as "Yu'ebao". Three key drivers for FinTech growth in China: (i) Shortage of supply in formal financial market (ii) strong support from government giving thrust to financial inclusion leveraging digital technology (iii) More "tolerant" regulatory environment.

2.2 World Bank and FinTech Regulation

The World Bank's note on FinTech regulation, (World Bank, 2020a) brings out a template for apprising the benefits versus risks of various approaches to FinTech regulation. The summary or abridged version is presented in Table 1.

Table 1. Examining benefits versus risks of various regulatory approaches

Regulatory Approaches	**Benefits**	**Risks**
1. Wait and See	(a) Regulators have time to respond (b) Determine the timing and type of policy intervention	Could lead to eruption of risks
2. Test and Learn	(a) Regulates have active learning environment (b) Calibrated approach	Cannot be used indefinitely
3. Innovation Facilitators: Innovation Hubs:	(a) Supports the Fintech ecosystem, appreciation of FinTech trends and fosters an open dialogue with industry (b) Potential issues drive policy formulation	Requires some dedicated resources
4. Innovation Facilitators: Regulatory Sandboxes:	(a) Allows pilot study before going big (b) Risks are better controlled (c) Dynamic, evidence based and suitable for matured FinTech market	Resource demanding

continued on following page

Table 1. Continued

Regulatory Approaches	Benefits	Risks
5. Innovation Facilitators: Regulatory Accelerators or Regtech Labs:	(a) Increased collaboration between the regulators and stakeholders (b) effective and efficient way of regulation	Substantial and dedicated resources are required
6. New regulatory reform	(a) Level playing field is created (b) Transformative in nature	Time-consuming process

Source: Compiled from World Bank, 2020a

2.3 FinTech Segments

FinTech has many segments or variations such as Regtechs, Suptech, Wealthtech, SaaS (Software as a Service), Insuretech, Investech, Challenger bank, Cryptocurrecy, BNPL (Buy Now and Pay Later), Openbanking etc. Some of these segments of FinTech worth a closer look and attention.

2.3.1 Regtech

According to Moody's, "Regtech or regulatory technology refers to a range of SaaS (software as a service) solutions designed to automate regulatory compliance processes for businesses across different sectors" (Moody's Analytics, 2022). It comes handy in know your customer (KYC). For mitigating and allaying anti-money laundering (AML), Regtech is of great help. Complex regulatory requirement and huge penalty for non-compliance are the chief drivers of Regtech.

Banks are increasingly integrating AI across operations. The regulators are closely watching and making right intervention to ensure transparency and to take note of ethical issues. Therefore, going forward, banks and other financial institutions need to balance between innovation and regulatory compliance (ETBFSI Research & ET BFSI, 2024). Regtech has end use is maximum in BFSI. It is applied in other industry segments like manufacturing, IT, Telecom, Government and Healthcare.

2.3.2 Suptech

SupTech furnishes regulators with advanced AI tools to keep a check and monitor financial institutions on real time basis (marketing@asiaverify.com, 2024).

Reserve Bank of India (RBI), the Apex Bank of India launched DAKSH platform on October 6, 2022. It is a FinTech application for enabling a competent and efficient supervisory process (ET Online, 2022). This variant of FinTech is known as

Suptech. This platform is a web-based application. Identification of the challenges, issues and risks at an early stage is a key. Further, the financial institutions should address them adequately. In this backdrop, supervisory technology (suptech) comes in handy. The authorities of regulatory bodies must leverage on their past experience in developing robust suptech platforms (Jermy Prenio, 2024).

2.3.3 Wealthtech

The complexity in investment management calls for applying welathtech tools. Welathtech essentially applies technology for managing investment and wealth. Robo advisors, tools for managing portfolio, digital trading platforms and data analytics form the core of wealthtech. Wealthtech is imperative in the light of trends like pressure on profitability and return on investment, changing client needs and demographics, asset allocation shifts, regulatory changes and paradigm shift in investment advice space (EY, 2025). Wealthtech assumes a critical role for expanding the financial inclusion and taking the investment management to masses.

2.3.4 Challenger Banks

These are tech-led neobanks and are popular in Europe. They are also present in US, Asia including India. They usually target millennials by offering superior and delighted customer experience (Challenger Banks and the Future of Digital Banking, 2021). They are known for hyper-personalised services and offerings. With the advent of technology and digital solutions, Challenger Banks are novel as they are (i) unpacking the banking value chain (ii) redefining the boundaries of banking industry (iii) prioritise growth over profit in the initial years (iv) compelling value proposition and customer experience (iv) targets digitally native, dissatisfied, unbanked or underbanked customers and (v) thrive on emotive branding (Cowell & Borkenhagen, 2020).

Challenger Banks are making their way. These Challenger Banks are tech driven and pose a formidable competition to the old-fashioned traditional banks (FT Partners | Financial Technology Investment Bank San Francisco, 2024). With the advent of new generation technology brought up by a new generation of FinTech companies, the traditional banks can stand-off or face the competition of the Challenger Banks (Coleman, 2024).

Global FinTech companies across the major countries are presented in Table 2.

Table 2. Top global FinTech companies

Country	Names and (Segment) of FinTech Firms
US	(a) Visa (Paytech) (b) Mastercard (Paytech) (c) Intuitt (Accounting) (d) Fiserv (Open banking) (e) Paypal (Paytech) (f) Coinbase (Cryptocurrency) (g) Acrisure (Insuretech) (g) Robinhood (Wealthtech) (h) Chime (Challenger Bank) (i) Ally Financial (Challenger Bank)
Australia	(a) Afterpay (Fintech) (b) Immutable (Blockchain) (c) Airwallex (Challenger Bank) (d) Judo Bank (Challenger Bank)
India	(a) Policybazaar (Insuretech (b) Cred, Razorpay, PhonePe, Pine Labs, Oxyzo (Payech) (c) Digit Insurance (Insuretech) (d) Polygon (Blockchain) (e) Groww (Wealthtech) (f) BillDesk, PineLabs (FinTech) (f) CredAvenue, Zeta, Onecard (Challenger Bank), (g) Upstox (SaaS) (h) Coinswitch Kuber, CoinDCX (Cryptocurency)
UK	(a) Wise (TransferWise), Lendable, Copper, eToro, (FinTech) (b) Revolut, Oaknorth, Monzo, Starling Bank (Challenger Bank) (c) Ziltch (BNPL) (d) Quantexa (SaaS) (e) FNZ (Investech) (f) Checkout.com, ZEPZ, Remitly, Zepz (WorldRemit) (Paytech)
Singapore	(a) Amber Group, Crypto.com, Matrixport (Cryptocurrency) (b) Singlife, Bolttech (Insurtech) (c) Coda Payments, Nium (Paytech) (d) Atome (FinTech)
China	(a) Ant Financial (Open Banking) (b) Tencent (Fintech business) (Regtech) (c) WeBank, Jiedaibao, China UnionPay, FirstP2P, AI Bank (Baixain) (FinTech) (d) Lufax, cgtz (Wealthtech) (e) Du Xiaoman Financial (BNPL) (f) Waterdrop (Insurtech) (g) WeLab (Challenger Bank) (h) PingPong, Lakala, Xtransfer (Paytech) (i) HashKey (Cryptocurrency) (j) Fenbeitong (SaaS)

Source: Author's Compilation from published sources + https://courses.cfte.education/ranking-of-largest-fintech-companies/

2.4 Global Spread of FinTech Firms

FinTech firms may be grouped or segmented as per the business models as (a) Digital Payments (b) Digital Lending (c) Insurtech (d) Digital Capital Raising and (e) Digital Banking and Services. The World Economic Forum conducted a survey of 227 FinTech firms from across the globe (The Future of Global Fintech: Towards Resilient and Inclusive Growth, 2024). The spread of FinTech segments as per this survey is captured in Table 3.

Table 3. Spread of FinTech segments across the globe

Segments	Percentage
Digital Payments	30
Digital Lending	27
Insurtech	17
Digital Capital Raising	14
Digital Banking and Savings	12

Source: (The Future of Global Fintech: Towards Resilient and Inclusive Growth, 2024)

2.5 FinTech Trends

The recent trends of FinTech (Lazarow, 2024) may be recollected in terms of (a) Crypto is getting renewed interest due to factors like rise in institutional interest, recently concluded the 2024 US election, approval by U.S. regulators in select Bitcoin ETFs, bitcoin portfolio recommendation by BlackRock (b) Evinced interest in digital currency by the Central Banks of 134 countries (in fact, countries like India and Brazil are preparing for early-stage implementation of Central Bank Digital Currency) (c) FinTech startups getting into capital market through Initial Public Offerings (d) Leveraging generative AI for developing innovative FinTech solutions in spaces like operational automation, detection of fraud, personalized financial services, automated underwriting etc.

3.0 DRIVERS OF FINTECH

There is a continuous disruption in financial services space due to regulations and the technology advancement. Further this space is going through a flux, thanks to new generation customers who demand for innovative products and services. These developments have led to a phenomenal growth in FinTech *(PwC, Mapping the FinTech Innovation Landscape in India, 2024)*. Both supply side and demand side growth drivers have contributed for the surge in FinTech. The drivers of FinTech may be examined with the help of the following figure 1.

Figure 1. Drivers of FinTech

Country	Use Cases
Brazil	Social transfer program, "Bolsa Familia" resulted in decrease in transaction costs from 15% in 2001 to below 3% in 2005 and this contributed to drop in poverty by 12%–18%
BURKINA FASO	"Mobile money" users save 3 times more for emergency and health risks than non-users of "Mobile money" = Saving for rainy days
NIGER	"Delivering safety net programs" resulted in time and cost savings for beneficiaries more or less equal to 20 kilograms of grain per person
INDONESIA	"Card based vouchers" for subsidized rice distributed by the Government of Indonesia to 1.4 million for poor people resulting in more and higher quality food for 9 out of 10
UGANDA	(a) "Digitized payments" helping 12,000 Ugandan farmers in terms of cost cutting by 27%. Also, it eliminated inefficient and risky process of transporting cash. This led to indirect benefits like higher business productivity and access to funds (b) Flexible education loan app, "payas-you-go", helped parents to service school fees payment on time and facilitated to school education for children
BANGLADESH	"MAMA Bangladesh" by paying incentives with mobile money impacted over one million new mothers through maternal mHealth program and created history in health and wellness of women from pregnancy to infancy
NEPAL	"Digital technology" led to increased spending on education for women
LIBERIA	"Paytech" led to savings in teachers' time and effort for getting salary and led to increase in classroom time
AFRICA	"Mobile payments enabled pay-as-you-go e-learning" empowered disadvantaged or remote populations to access affordable education digitally. Case in point is Eneza Foundation's mobile education platform. It has has more than three million users across Africa (70% are from rural areas)
KENYA	(a) "Mobile money" during 2008 to 2014 lifted roughly one million people (around 2 percent of population) from extreme poverty (b) "One Acre Fund" enabled digitisation of agricultural loan repayments reduced effort and cost of servicing loan by 80%. Also, it freed up farmers' resources
TANZANIA	Farmers took microinsurance through "Mobile money" benefitted 16% more than their uninsured peers
EAST AFRICA	Farmers who accessed "Mobile Agricultural microinsurance" for farmers through mobile reduced financial losses and farmers earned 16% more than their uninsured peers

Source: UN. 2018, https://sdgs.un.org/sites/default/files/publications/2655SDG_Compendium_Digital_Financial_Inclusion_September_2018.pdf

Source: Author's Compilation

GDP, Lerner Index (The Lerner index, "LERNER", was formalised by Abba Lerner, the British Economist. It measures the firm's market power and is usually applied to banks for ascertaining the level of bank's market power.), bank branches, regulation, ease of doing business are the drivers of FinTech (Cornelli et al., 2022).

3.1 Covid - Disruptor and Driver of FinTech

With the outbreak of Covid in 2020, the world was grappled with many challenges which were unheard and unknown. The governments of the countries of the world came up with set of varied measures. The aggregate effect of these measures resulted in leveraging technology (AI and related tools) to kick start and stimulate the economy. The pandemic, Covid 19, gave a big push and thrust to the FinTech's evolution, innovation and revolution. Covid was a huge disruptor and arguably a significant driver of FinTech.

4.0 FINTECH'S USE CASES

The role of FinTech is fuelled by several levers or drivers like advancement in technology, digitisation, favourable regulatory framework and rise in consumer expectations. The protagonist role pf FinTech is envisaged in government budget allocations for capex spending. The private players (corporates) also leveraging FinTech across the value chain of business to achieve efficiency, speed, reach, experience, inclusion and scale. There are many use cases of FinTech in achieving SDGs and three pillars of ESG. ESG and SDGs can be integrated for examining sustainability aspect and to strategies the action plan (Markopoulos & Maria Barbara Ramonda, 2022). SDGs can be mapped to ESG and this is presented in Appendix 1.

As FinTech is resource effective, it will be a great boon for achieving sustainability. There are several segments in FinTech. The prominent ones are Regtech, Insurtech, BaaS (Banking as a Service), Paymentstech, Wealthtech, Lendtech, Climate FinTech, Blockchain etc. FinTech is reshaping the future of finance (World Bank, 2022). FinTech revolutionised the banking and insurance space (Thakor, 2020).Through green finance, FinTech contributes to climate issues as green finance supports environmentally friendly investments and thus contributing for low-carbon technologies projects (Omeragic et al., 2024). The study pertaining to the patent filings from 2003 to 2017, revealed the role of FinTech in financial sector. The study confirms the role of internet of things (IoT), robo-advising, and blockchain as the valuable innovation tools (Chen et al., 2019). For financial inclusion, FinTech plays a critical role. The central banks leverage FinTech for promoting financial inclusion. For a country like India, financial inclusion is the key driver for achieving many SDGs.

United Nation's task force on digital finance and SDGs prescribes that the FinTech firms must innovate products and services which can meet consumer demand and to channel finance to sustainable development goals. FinTech firms should be committed to philosophy and principles of SDG-aligned digital financing by developing corporate governance mechanisms to ensure they operationalize them (UN Secretary General, 2018).

FinTech empowers the MSMEs (Micro, Small and Medium Enterprises) by addressing the barriers that they face. In this context, two organisations, (1) "Nacional Financiera (NAFIN) development bank system in Mexico" and (2) the Indian "Trade Receivables Discounting System (TReDS) system", allow for strengthening relationships between SME suppliers and anchor buyers (World Bank, 2022).

FinTech lending platforms pave the way for access to finance and capital to new entrepreneurs (Bollaert et al., 2021). Fintech serves as a mediator to mitigate the risks of the banks and thereby helps the banks to show up their ESG performance (Samir Srairi, 2024). While the world is shifting towards sustainable energy, FinTech plays a vital role for the smooth transition (Polat et al., 2024).

For promoting green finance, FinTech's contribution can be examined in many broad areas; (i) application of blockchain for sustainable development (ii) use-cases of blockchain for renewable energy, (iii) FinTech led green bonds, carbon credits, and climate finance and decentralized electricity market (Nassiry, 2019).

The top sustainable FinTech trends are green loans and sustainable finance, Open banking, digital payments and signatures, Carbon tracking, green investments, Mobile payments etc are contributing for the sustainability (Thompsett, 2024). Many studies have found that FinTech underpins sustainable financing and green bonds issuance. FinTech lifts the green bond issuance by increasing ESG awareness and empowers the intermediary institutions (Huang et al., 2024). FinTech and renewable energy financing are correlated and promote the green bonds (Sreenu, 2024).

Financial super app is a great innovative tool and it has revolutionised the trajectory of FinTech space (Team, 2024). Fintech solutions are the way forward to the compounded, ever-evolving financial crime and cybersecurity problems and issues (Risk Insights & Advisory Fintech Spotlight | Darwinium, 2024).

Using blockchain technology, FinTech firms are mitigating supply chain issues by offering decentralized network connects buyers, suppliers, and funders directly. Fintech is a catalyst for Industry 4.0 as it can help smart manufacturing by tracking raw materials, spare parts and products at every stage.

Insurance companies all over the world are factoring ESG considerations and are aiming for net zero commitment and mitigate climate risks (Pricewaterhouse-Coopers, 2022). In this regard, Insurtech plays a significant role.

Bank for International Settlement (BIS) is all set to leverage fintech for sustainable finance by adding two green finance projects, one in Hong Kong's innovation hub and another in Singapore's innovation hub (Walker, 2024). The strategic partnerships between FinTech companies and the banks could expand the scope of customer base by targeting near-prime, and low-prime consumers (Elliehausen & Hannon, 2023). As per Mastercard, the tech advancement and disruption will push greater inclusivity, efficiency, and security and create a revolution in the FinTech space (ETBFSI Research & ET BFSI, 2024).

The use case of Regtech, an important segment of FinTech is presented in Table 4.

Table 4. Regtech use case: ESG and sustainability assessment

Publisher	**Hong Kong Monetary Authority (HKMA)**
Tech Solution Provider	Seneca Technologies Private Limited
Tech Platform	Seneca ESG's ZENO platform
Target Beneficiary and End Use	Banks for assessing loan applications from ESG perspective. Banks can evaluate and monitor corporate barrowers performance vis-a-vis ESG philosophy (Green banking products and services)
How	Banks are assisted by ZENO ESG platform in (i) creating assessment questionnaires (ii) Data collection and analysis of responses (iii) real time monitoring borrowers' profiles
Outcomes	(i) Streamlined assessment (ii) Improved sustainability profiles for borrowers (iii) Improved loan terms (iv) Impact Financing

Source: Bulletin on Regtech Use Case ESG and Sustainability Assessment, 2024, https://www.hkma.gov.hk/media/eng/doc/key-functions/banking-stability/Bulletin_on_Regtech_Use_Case_Seneca.pdf

Insurance is an important financial product from sustainability view point. It provides a great leverage for masses. There is a need for sustainable underwriting. Insurtech, a segment of FinTech comes handy. Leveraging technology, Insurtech can offer dynamic pricing post considering the insured's context and thereby insurance service can be at a compelling price. Insurance companies are helping the clients in promoting sustainability and taking head on the climate challenges through Insurtech (Team. 2024). The use cases of Insurtech are summarised in Table 5.

Table 5. Role of Insurtech in promoting sustainability

Technology Tool	**Application or Outcome**
Satellite imagery	assess environmental risks and leveraging on geospatial data and analytics
Cloud-based solutions	Insurance companies to share resources through Cloud computing. It cuts down on the energy needed to power data centres, resulting in energy saving and cost efficiency, which can be passed on to clients
Leveraging AI and ML	Real time risk monitoring and hence enhanced risk management
Promoting the use of renewable energy	Providing insurance cover for equipment used for reviewable energy production, for example, solar panel

Source: (Team, 2024). https://www.bobsguide.com/5-ways-insurtech-drives-environmental-sustainability-in-business/

5.0 FINTECH AND SDGs

FinTech significantly contributes to accomplishment of SDGs through (i) financial inclusion enabling under-banked populations (ii) empowering sustainable investments (iii) facilitating green finance initiatives to take on climate challenges (iv) transparency and resilience are achieved by blockchain and Insurtech (v) encouraging responsible production and (vi) empowering overall inclusive economic growth (Jain et al., 2024).

FinTech can further and foster the cause of Sustainable Finance by providing the underpinning support. By default, the FinTech solutions are innovative and path breaking. This leads to sustainable growth. FinTech contributes for transparency and authenticity of ESG data, and thus preventing companies from manipulating ESG data. This reduces the risk of ESG greenwashing (Liu & Li, 2024). Financial inclusion, ease of access to capital markets, reducing cost of service, risk mitigation are some of the use cases of FinTech in sustainable and green finance (PricewaterhouseCoopers, 2021). There is a direct nexus between mobile payment and financial inclusion (Ahmad et al., 2020). Financial inclusion is driven by accessibility. FinTech is a huge contributor of financial stability and as per the study, spatial dependence determines the financial stability in the globe. Further the financial stability has a positive spillover effect to adjacent countries (Koranteng & You, 2024). Although Fintech is an invaluable tool for enabling the unbanked people to access and avail financial services, the intentions of the user will largely depend on social, cultural, economic and technological factors (Hoque et al., 2023).

Paytech is a boon to financial inclusion. It provides cost effective transfer of money and reaches out to remote areas. Paytech's digital payment solution reduces poverty (UN, 2018). Paytech promotes accessibility and thus contributes for sustainable goal (Hinton, 2024). The United Nations's compendium on the impact of Paytech with use cases across the globe is worth reading (UN, 2018). The summary of the UN's compendium is provided in Table 6.

Table 6. Paytech's digital finance impact or outcome

Country	Use Cases
Brazil	Social transfer program, "Bolsa Família" resulted in decrease in transaction costs from 15% in 2001 to below 3% in 2005 and this contributed to drop in poverty by 12%–18%
BURKINA FASO	"Mobile money" users save 3 times more for emergency and health risks than non-users of "Mobile money" = Saving for rainy days
NIGER	"Delivering safety net programs" resulted in time and cost savings for beneficiaries more or less equal to 20 kilograms of grain per person

continued on following page

Table 6. Continued

Country	Use Cases
INDONESIA	"Card based vouchers" for subsidized rice distributed by the Government of Indonesia to 1.4 million for poor people resulting in more and higher quality food for 9 out of 10
UGANDA	(a) "Digitized payments" helping 12,000 Ugandan farmers in terms of cost cutting by 27%. Also, it eliminated inefficient and risky process of transporting cash. This led to indirect benefits like higher business productivity and access to funds (b) Flexible education loan app, "payas-you-go", helped parents to service school fees payment on time and facilitated to school education for children
BANGLADESH	"MAMA Bangladesh" by paying incentives with mobile money impacted over one million new mothers through maternal mHealth program and created history in health and wellness of women from pregnancy to infancy
NEPAL	"Digital technology" led to increased spending on education for women
LIBERIA	"Paytech" led to savings in teachers' time and effort for getting salary and led to increase in classroom time
AFRICA	"Mobile payments enabled pay-as-you-go e-learning" empowered disadvantaged or remote populations to access affordable education digitally. Case in point is Eneza Foundation's mobile education platform. It has has more than three million users across Africa (70% are from rural areas)
KENYA	(a) "Mobile money" during 2008 to 2014 lifted roughly one million people (around 2 percent of population) from extreme poverty (b) "One Acre Fund" enabled digitisation of agricultural loan repayments reduced effort and cost of servicing loan by 80%. Also, it freed up farmers' resources
TANZANIA	Farmers took microinsurance through "Mobile money" benefitted 16% more than their uninsured peers
EAST AFRICA	Farmers who accessed "Mobile Agricultural microinsurance" for farmers through mobile reduced financial losses and farmers earned 16% more than their uninsured peers

Source: UN. 2018, https://sdgs.un.org/sites/default/files/publications/2655SDG_Compendium_Digital_Financial_Inclusion_September_2018.pdf

Agri-FinTech enables farmers in terms of market access, transparency in pricing and risk management and thus driving financial inclusion and updating rural economy Gloria Mathias. (2024). Climate FinTech essentially provides collaborative support to take on climate challenges by coordinating between digital technologies, climate challenges and financial ecosystem (Fintech Applications for Boosting Climate Finance, 2024).

The surge and growth in FinTech are driven by the potential of FinTech platforms and tools for impacting the society at a huge scale (Bulger & Rouen, 2022). FinTech has a huge impact on UN SDGs as a consequence of enablers like inclusive finance, economic growth, and environmental perspectives and dimensions (Hasan et al., 2024). In forwarding, financial inclusion agenda (which is a strong requirement for sustainability), technology plays a crucial role. Fintech lenders bring out more practicality in lending to consumers in terms of value proposition than the traditional banks (Buchak et al., 2017).

The studies have shown that (i) FinTech is an excellent accelerator of sustainable economic growth, (ii) FinTech's technological innovations contribute for sustainable finance (iii) banks can contribute to sustainability better by collaborating or associating with FinTech companies (Galeone et al., 2024).

Agentic AI is a more developed and sophisticated version of AI. Agentic AI, being an advanced AI tool, can foster the FinTech revolution (The Gartner Top Strategic Technology Trends for 2025, 2024). It helps in developing a sustainable FinTech solution. Solution with Agentic AI can help in achieving SDG4 (Quality and inclusive education goal). UNESCO has partnered with Capgemini in developing transformative AI tool for achieving inclusive, effective and equitable education for all (Enabling Evidence-Based Education Decision-Making with Agentic AI, 2024).

6.0 BANKS AND FINTECH

The volume of credit offered and lent by FinTech companies negatively influences the bank market power and stability (Cuadros-Solas et al., 2024). There is a strong case for the collaboration between traditional banks and FinTech firms. This is both crucial and critical for competitive differentiation. Several partnership models are available. To take on technology disruption, many banks leverage or outsource the services from specialised FinTech firms, where, both bank and FinTech firm can focus on their respective core competencies. For the effective collaboration or partnership between banks and FinTech firms, due diligence is the key. Also, integration areas must be identified. The common integration areas are:

(i) Digital and mobile transactions (both payments and deposits)
(ii) User experience and interface: Creating better user interface via technology
(iii) Extending portfolio of services i.e., offering investment and wealth management
(iv) Customer onboarding and KYC
(v) Credit underwriting
(vi) Data integrity and security

6.1 Ways of Collaboration / Integration between Banks and FinTech firms

FinTech firms turning into bank or banks where the DNA is FinTech i.e., banks can have their own inhouse FinTech R&D (Research and Development) or innovation lab. The best example in this case is Challenger bank or Neo bank. This is the one end of the spectrum. The other end is community banks. In between these extremes, for both banks and FinTech firms, a lot of permutations and combinations are possible.

7.0 TECH FOR FINTECH

The technology platforms underpin the FinTech development. Big data and AI play a critical role in the development of FinTech (KPMG China Fintech 50 Report, 2023). KPMG's report brings out the distribution of technology platforms in the development of FinTech and this is captured in Figure 2.

Figure 2. Tech for FinTech

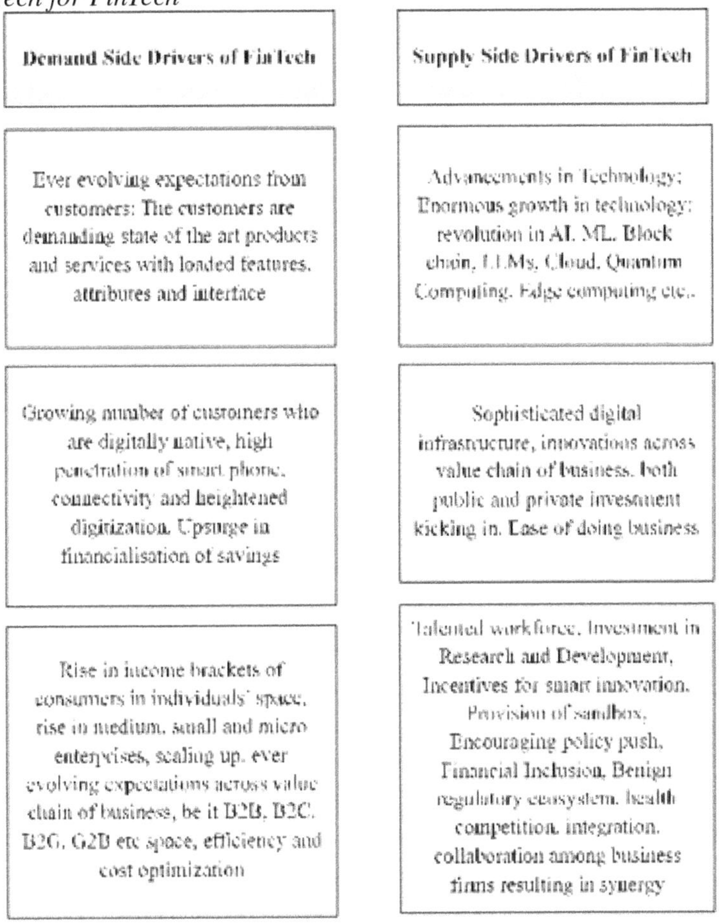

Source: KPMG China Fintech 50 Report, 2023

Some of the major tech tools, as presented the Figure 2, do need additional note.

(a) Big data plays a key role in the development of FinTech, as data is the oil of the disruptive world. Big data is certainly is the way to progress towards SDGs as it provides invaluable insight for informed decision making (United Nations, 2022). Several studies have shown that the transformative power of big data in FinTech must be underscored in a march towards inclusive financial ecosystem (Mhlanga, 2024). A detailed note on how big data helps to achieve each SDG is presented in Appendix 2.
(b) AI is the staple of FinTech. AI enables the exploitation of the big data for developing innovative FinTech solution. Invariably, all FinTech firms make use of AI. As far as achieving SDGs, AI driven solutions are critical (UNESCO, 2019).
(c) Knowledge graph organises data from multiple sources and thereby big data will be deployed gainfully. "Clausematch", a Regtech platform leverages knowledge graph to drive digitalisation of regulation.
(d) Deep learning is an extension or a sub set of AI. It enables to get insights from big data as it has interpretation and decoding capabilities. Deep learning models are extensively applied in the key finance and banking domains (Huang et al., 2020).
(e) Natural Language Processing (NLP) makes sense out of language (written or spoken). It draws sentiment, patterns, emotions and intent by analysing the words and non-verbal communication. In simple words, the outcome is text mining i.e., converting unstructured data into structured data. NLP is extensively used in Welathtech, Insurtech, developing credit scoring and appraisal models for banks, fraud detection, cloud-based chat, CRM optimisation etc. Integrating NLP in developing FinTech solution for social good is the emerging trend and as a consequence of it, we have Financial Natural Language Processing (FinNLP) (Ghosh & Naskar, 2023).
(f) TCS, a bellwether IT company form India, came out with a white paper on "Computer Vision: A Game Changer for Banking, Financial Services and Insurance". This throws a detailed light on the application and use case of compute vision (Tata Consultancy Services, 2020). The study shows that the applications of computer vision is far reaching in the areas of Know Your Customer (KYC), customer on boarding, Data extraction from trade documents, Data extraction from insurance documents, damage assessment in insurance claims and the list goes on.
(g) Cloud computing helps to access computing resources over internet i.e., cloud. The resources such as storage, servers, and databases may be accessed over cloud. This tech tool offers scalability and flexibility. Microsoft Azure is a case

in point. Customer centric solutions can be provided by FinTech using Cloud computing. Data security and cost effectiveness are the chief benefits of cloud computing.

(h) Blockchain provides additional and much needed security to FinTech applications. The smart contracts are possible due to blockchain. Features like superb traceability and accountability, data integrity (immutable ledger), partner interoperability (decentralised), courtesy of its tamper-proof records etc., enable developing robust FinTech applications. Thus, blockchain is a significant tech tool in fraud prevention and and reduces the risk of fraudulent activities.

(a) Internet of Things (IoT) is a foundation for FinTech application, especially in case of Insurtech. Combining sensor device and drone in claim assessment is a classic example of use case of IoT.

8.0 FINTECH RISKS AND ETHICAL CHALLENGES

Every opportunity comes with risks. FinTech is no exception. This section documents challenges in developing FinTech innovative solutions, implementing FinTech. One should find a fine balance between innovation and prudence. Unregulated FinTech firms (Alibaba of Jack Ma) in China created a havoc and balance in banking was lost as Fintech firms started dominating over regulated banks. Stringent regulations and compliances, fraud, data privacy, poor infrastructure like inadequate digital and tech penetration, bad mobile network, platform downtimes are some of the challenges bothering FinTech revolution.

FinTech is fraught with many risks and challenges. The Table 7 provides a glimpse of risks associated with FinTech.

Table 7. Risks and ethical challenges of FinTech

Sl	Source Literature	Risks and Ethical Challenges
1	Aldboush & Ferdous, 2023	(a) Privacy risks bother more millenials than seniors (b) Ethical issues like data privacy, transparency and security concerns arise while leveraging big data and AI (c) Firms with Corporate Digital Responsibility (CDR) enjoy more benefits from FinTech
2	Tóth & Blut, 2024	Ethical AI could be a reality and possibility with CDR as CDR takes into account the ethical dilemma, responsible use of technology, social impact and prevention of potential side effects and cyber security risks
3	Crisanto et al., 2024	Risks surface due to the deployment of third party AI Services
4	AI Risks in Fintech: 10 AI Challenges Finechs Still Struggle With, 2023	Machine learning biases come in the way of achieving financial inclusion of FinTech firms
5	Agarwal et al., 2024	Security threats and third party risks

continued on following page

Table 7. Continued

Sl	Source Literature	Risks and Ethical Challenges
6	Kovas, 2018	Data Privacy related ethical issues
7	Consumer Risks in Fintech, 2021, April	Algorithmic risk i.e., Proprietary nature of algorithms leading to failure on the part of regulators to evaluate algorithmic systems due to lack of technical expertise
8	Bank, 2022 (World Bank)	Financial exclusion, data privacy, security risks, Cyber risks
9	Reserve Bank of India - Reports, 2024	internal governance, financial system integrity, data privacy, cyber security, customer protection, grievance handling,
10	PwC, Beyond the Cloud: Navigating FinTech Cyber Threats and Fortifying Defences, 2024	Phishing, deepfakes and tech-based fraud, Open-source S/W vulnerabilities and supply chain risks, Insider threats (stemming from gig economy), Complex third-party integrations and APIs, Vulnerable data pipelines, Major cyberthreats for cloud-native FinTech, Multi-cloud threat complexity, GenAI threats
11	Ghosh, 2024	(a) Possibility of bias (b) Issues related to the ethical use of data

Source: Author's Compilation

8.1 Managing and Mitigation of Risks and Ethical Challenges

FinTech firms must bear in mind the fact that security is of paramount importance from the consumer's point of view and also security should be an integral part of Fintech solutions. It is the responsibility of the FinTech firms and the onus for which lies with them. Going forward, security and data privacy are critical to win the consumer confidence *(Security Challenges in the Evolving Fintech Landscape, 2016).*

The policy response is the key to circumvent some of the challenges in regulating FinTech. The regulatory policy framework should factor in elements like; (i) adjust prudential and conduct requirements (ii) clarity and clarification of supervisory expectations especially in areas like financial soundness, consumer protection, operational resilience and Anti-Money Laundering and Combating the Financing of Terrorism (AML/CFT). There should be review and reassessment of regulatory framework supported by international policy cooperation (Barakova et al., 2024).

Critical areas that need a further regulatory attention are; (i) governance (ii) expertise and skills (iii) model risk management (iv) data governance (v) non-traditional players in the financial sector (vi) new business models and (vii) third-party AI service providers (Crisanto et al., 2024). When the FinTech firms partner with the banks for FinTech application, it gives rise to opportunities and certain issues for the banks and supervisors. This is especially true when the FinTech firm is a big tech firm (Barakova et al., 2024).

Initiatives like, SRO-FT of RBI promotes innovation while ensuring regulatory compliances and risk mitigation.

The report of Price Waters Coopers has brought to the fore certain recommendations for managing and taking on risk (PwC, Beyond the Cloud: Navigating FinTech Cyber Threats and Fortifying Defences, 2024). The recommendations are:

(a) Cloud-native security
(b) API security and secure DevOps i.e., integration of security mechanisms
(c) Identity and access management
(d) Implementation of ML and AI solutions
(e) Continuous monitoring and audit, security testing and compliance
(f) Incident response plan
(g) Risk analysis of GenAI and large language models (LLMs)
(h) Training and awareness for employees
(i) Strengthen risk culture
(j) Regulatory compliance

8.2 Tokenisation as a Risk Containment Tool

Current financial paradigm has caught up with issues such as breaching of trust and exclusivity. To remedy these risks or challenges, Tokenization has been innovated. Tokensiation in FinTech circumvents these complexities (Vardari & Qabrati, 2024).

Tokenisation is a big innovation in the FinTech world. Tokenisation helps in securing the data as it replaces the original data by a blockchain token. Trust Commerce was the first company to create and introduce the concept of tokenisation in 2001. Tokenization refers to "the process of constructing digital representations (crypto tokens) for non-crypto assets (reference assets)" (Carapella et al., 2023).

Tokenization does protect sensitive data while still permitting it to be utilised for transaction or commercial purposes and thus tokenization is different from encryption. Encryption does not permit or allow data being used for business objectives (JisaSoftech, 2022).

Tokenization created a revolution in the financial space as it has many enablers like democratisation of access to capital, inclusivity and easy access to finance, fuelling liquidity in illiquid assets by allowing fractional ownership, prooting cross-border investment opportunities and the list goes on *(FinTech Magazine Article, 2023)*. Tokenisation brings in a significant impact on how trades are finalized. The other benefits of it are risk reduction, spur in liquidity and creating new ways for investors to make gains and collateral management (Yuval Rooz, 2024).

The global regulatory bodies like "Hong Kong Monetary Authority" (HKMA) and "The Reserve Bank of India" (RBI) have promoted tokenisation and "The Monetary Authority of Singapore" (MAS) is planning to introduce and promote tokenisation any time soon.

One of the key outcomes of FinTech is achieving financial inclusion. In achieving this outcome, banks and other FinTech firms have to fraught with many risks. To mitigate these risks, the World Bank's generic framework on risk assessment is interesting (Celik, 2019) and worth noting. This is captured in Appendix 3. The Appendix 3 clearly illustrates as to what are the dimensions and sources of risks across the various FinTech segments.

9.0 CONCLUSION

FinTech is creating new opportunities and challenges in the world of finance. The stakeholders like consumers, financial institutions, regulators and new entrants must exploit the opportunities and mindful of challenges (World Bank, 2020). FinTech is a passport to achieve ESG drive. The environmental risks, social challenges (stigma) and governance matters are well handled by FinTech solution. Achieving the UN's seventeen SDGs are the pressing need of the mankind. The future of the global economy is contingent upon the sustainability metric. In this demanding situation, FinTech being the portfolio of tech tools is the way forward. FinTech reduces and eases the financing constraints and directly enhances ESG outcomes for new energy automobile firms (Huang et al., 2025).

The environment for innovation, supported by a thoughtful policy framework, is needed to propel the growth of FinTech in achieving sustainable growth. An extensive study in China reveals that FinTech is a key driver of sustainable development and to foster FinTech innovation, a robust regulatory sandbox for FinTech needs to be promoted (Deng et al., 2019). FinTech is the key as far as sustainability agenda is concerned as it makes financial services accessible, affordable and contribute for customer delight in terms of user interface and experience (Fuelling Innovation: Closing Fintech Funding Gaps, 2024).

FinTech is pervasive of Generative AI, blockchain, tokenisation and such other novel trends. The novel means of finance like crowdfunding is the outcome of FinTech revelation. FinTech innovation is forging ahead, but caught up with challenges. FinTech is fraught with many challenges such as regulatory uncertainties, illegal transactions and data abuse (Hua & Huang, 2020). There is an acute need for taking cognisance of FinTech related risks and challenges so as to reap the full potential of FinTech. The success and sustained growth of FinTech would largely depend on the solid risk management controls in place (Deloitte, 2024).

REFERENCES

Agarwal, R., Kremer, A., Kristensen, I., & Luget, A. (2024, March 1). How generative AI can help banks manage risk and compliance | McKinsey. Www.mckinsey.com. https://www.mckinsey.com/capabilities/risk-and-resilience/our-insights/how-generative-ai-can-help-banks-manage-risk-and-compliance

Agarwal, S., & Zhang, J. (2020). FinTech, Lending and Payment Innovation: A Review. *Asia-Pacific Journal of Financial Studies*, *49*(3), 353–367. DOI: 10.1111/ajfs.12294

Ahmad, A. H., Green, C., & Jiang, F. (2020). Mobile money, financial inclusion and development: A review with reference to African experience. *Journal of Economic Surveys*, *34*(4), 753–792. DOI: 10.1111/joes.12372

Aldboush, H. H. H., & Ferdous, M. (2023). Building Trust in Fintech: an Analysis of Ethical and Privacy Considerations in the Intersection of Big Data, AI, and Customer Trust. International Journal of Financial Studies, 11(3), 90. MDPI. DOI: 10.3390/ijfs11030090

Bank, W. (2022). Finance, Competitiveness, and Innovation Global Practice Regulation and Supervision of Fintech: Considerations for EMDE Policymakers. https://documents1.worldbank.org/curated/en/099735204212215248/pdf/P173006033b45702d09522066cbc8338dcb.pdf

Bank of England. (2024). Fintech. Www.bankofengland.co.uk. https://www.bankofengland.co.uk/research/Fintech

Barakova, I., Ehrentraud, J., & Leposke, L. (2024). *A two-sided affair: banks and tech firms in banking*. Bank for International Settlements, Financial Stability Institute.

Bollaert, H., Lopez-de-Silanes, F., & Schwienbacher, A. (2021). Fintech and Access to Finance. *Journal of Corporate Finance*, *68*, 101941. DOI: 10.1016/j.jcorpfin.2021.101941

Buchak, G., Matvos, G., Piskorski, T., Seru, A., Agarwal, S., Carey, M., Cembalest, M., Claessens, S., Cochrane, J., Duffie, D., Diamond, D., Fuster, A., Gopalan, R., Mueller, H., Palmer, C., Philippon, T., Rajan, R., Schnabl, P., Sedunov, J., & Shin, H. (2017). Nber Working Paper Series Fintech, Regulatory Arbitrage, And The Rise Of Shadow Banks. https://www.nber.org/system/files/working_papers/w23288/w23288.pdf

Bulger, Z., & Rouen, E. (2022, July 15). How Fintech Can Deliver on Its Social Impact Promises. *Harvard Business Review*. https://hbr.org/2022/07/how-fintech-can-deliver-on-its-social-impact-promises

Bulletin on Regtech Use Case ESG and Sustainability Assessment. (2024). https://www.hkma.gov.hk/media/eng/doc/key-functions/banking-stability/Bulletin_on_Regtech_Use_Case_Seneca.pdf

Carapella, F., Chuan, G., Gerszten, J., Hunter, C., & Swem, N. (2023). Tokenization: Overview and Financial Stability Implications. Www.federalreserve.gov.https://www.federalreserve.gov/econres/feds/tokenization-overview-and-financial-stability-implications.htm

Carè, R., Boitan, I. A., & Fatima, R. (2023). How do FinTech companies contribute to the achievement of SDGs? Insights from case studies. *Research in International Business and Finance*, *66*, 102072. DOI: 10.1016/j.ribaf.2023.102072

Celik, K. (2019). FinTech and ML/TF Risks Insert footer from top menu. https://thedocs.worldbank.org/en/doc/187851572546448957-0130022019/original/Day3KuntayCelikWBFinTECHRisks002.pdf

Challenger banks and the future of digital banking. (2021). https://www.pwc.in/assets/pdfs/consulting/financial-services/fintech/publications/challenger-banks-and-the-future-of-digital-banking.pdf

Chen, M. A., Wu, Q., & Yang, B. (2019). How Valuable Is FinTech Innovation? *Review of Financial Studies*, *32*(5), 2062–2106. DOI: 10.1093/rfs/hhy130

China Fintech, K. P. M. G. 50 Report. (2023). https://assets.kpmg.com/content/dam/kpmg/cn/pdf/en/2024/01/china-fintech-50-2023.pdf

Coleman, A. (2024, July 31). How Fintechs Are Helping Traditional Banks Catch Up With Challenger Banks. Forbes. https://www.forbes.com/sites/alisoncoleman/2024/07/31/how-fintechs-are-helping-traditional--banks-catch-up-with-challenger-banks/

Consumer Risks in Fintech. (2021, April). https://documents1.worldbank.org/curated/en/515771621921739154/pdf/Consumer-Risks-in-Fintech-New-Manifestations-of-Consumer-Risks-and-Emerging-Regulatory-Approaches-Policy-Research-Paper.pdf

Cornelli, G., Frost, J., Gambacorta, L., Rau, P. R., Wardrop, R., & Ziegler, T. (2022). Fintech and big tech credit: Drivers of the growth of digital lending. *Journal of Banking & Finance*, *148*, 106742. DOI: 10.1016/j.jbankfin.2022.106742

Cowell, N., & Borkenhagen, M. (2020). The DNA of Digital Challenger Banks 2020. https://www2.deloitte.com/content/dam/Deloitte/us/Documents/financial-services/us-dna-of-digital-challenger-banks.pdf

Crisanto, J. C., & Leuterio, C. B. Jermy Prenio, & Yong, J. (2024, December 12). *Regulating AI in the financial sector: recent developments and main challenges*. Bis.org. https://www.bis.org/fsi/publ/insights63.htm

Cuadros-Solas, P. J., Cubillas, E., Salvador, C., & Suárez, N. (2024). Digital disruptors at the gate. Does FinTech lending affect bank market power and stability? *Journal of International Financial Markets, Institutions and Money*, *92*, 101964–101964. DOI: 10.1016/j.intfin.2024.101964

Deloitte. (2024). The Future of Fintechs: Risk and Regulatory Compliance. Deloitte United States. https://www2.deloitte.com/us/en/pages/regulatory/articles/future-of-fintechs-risk-and-regulatory-compliance.html

Deng, X., Huang, Z., & Cheng, X. (2019). FinTech and Sustainable Development: Evidence from China based on P2P Data. *Sustainability (Basel)*, *11*(22), 6434. DOI: 10.3390/su11226434

Elliehausen, G., & Hannon, S. M. (2023, October 4). FinTech and Banks: Strategic Partnerships That Circumvent State Usury Laws. Federalreserve.gov. https://www.federalreserve.gov/econres/feds/fintech-and-banks-strategic-partnerships-that-circumvent-state-usury-laws.htm

Capgemini. (2024). Enabling evidence-based education decision-making with agentic AI. Capgemini USA. https://www.capgemini.com/us-en/news/client-stories/enabling-evidence-based-education-decision-making-with-agentic-ai/

ETBFSI Research, & ET BFSI. (2024, December 14). Mastercard identifies key payment trends for 2025: AI, Blockchain, and Tokenisation. ETBFSI.com; ETBFSI. https://bfsi.economictimes.indiatimes.com/news/fintech/mastercard-identifies-key-payment-trends-for-2025-ai-blockchain-and-tokenisation/116301292?utm_source=Mailer&utm_medium=newsletter&utm_campaign=etbfsi_news_2024-12-14&dt=2024-12-14&em=YnIucmFvQGphaW51bml2ZXJzaXR5LmFjLmlu

ETBFSI Research, & ET BFSI. (2024, December 9). *Top banking trends to watch in 2025: AI, sustainability, and digital transformation*. ETBFSI.com; ETBFSI. https://bfsi.economictimes.indiatimes.com/news/banking/top-banking-trends-to-watch-in-2025-ai-sustainability-and-digital-transformation/116116043?utm_source=Mailer&utm_medium=newsletter&utm_campaign=etbfsi_news_2024-12-09&dt=2024-12-09&em=YnIucmFvQGphaW51bml2ZXJzaXR5LmFjLmlu

EY. C. (2025). *Money in motion: Enabling the evolving finance sector with wealth-tech.* Ey.com. https://www.ey.com/en_in/industries/wealth-asset-management/money-in-motion-enabling-the-evolving-finance-sector-with-wealthtech

Ferilli, G. B., Altunbas, Y., Stefanelli, V., Palmieri, E., & Boscia, V.Greta Benedetta Ferilli. (2024). Fintech Governance and Performance: Implications for Banking and Financial Stability. *Research in International Business and Finance*, *70*, 102349–102349. DOI: 10.1016/j.ribaf.2024.102349

Fintech Applications for Boosting Climate Finance. (2024, December 9). *IMF.* https://www.imf.org/en/Publications/staff-climate-notes/Issues/2024/12/09/Fintech-Applications-for-Boosting-Climate-Finance-557162

FinTech Magazine Article. (2023, July 12). *Tokenization Transforming the Investment Landscape as We Know It - FinTech Weekly. FinTech Magazine Article.* https://www.fintechweekly.com/magazine/articles/tokenization-transforming-the-investment-landscape-as-we-know-it

Magzter. (2024). Fintech must wait for spring. Magzter.com; Magzter. https://www.magzter.com/stories/newspaper/Business-Standard/FINTECH-MUST-WAIT-FOR-SPRING?srsltid=AfmBOopCVoIHfEY8WUCO4LiyX-gGBQ0xpoiMA_MYwgbybwZLqDJAqCZ8

Partners, F. T. (2024). Financial Technology Investment Bank San Francisco. Www.ftpartners.com. https://www.ftpartners.com/fintech-research/challenger-banks

Galeone, G., Ranaldo, S., & Fusco, A. (2024). ESG and FinTech: Are they connected? *Research in International Business and Finance*, *69*, 102225–102225. DOI: 10.1016/j.ribaf.2024.102225

Ghosh, S. (2024, October 21). Use of AI in banking poses challenges around bias and data ethics: RBI bulletin. Mint. https://www.livemint.com/industry/banking/use-of-ai-in-banking-poses-challenges-around-bias-and-data-ethics-rbi-bulletin-11729518658876.html

Ghosh, S., & Naskar, S. K. (2023). Recent trends in financial natural language processing research. *Science Talks*, *8*, 100270. DOI: 10.1016/j.sctalk.2023.100270

Gonsalves, R. (2024, August 28). RBI recognises FACE as self-regulatory organisation in FinTech sector. The Economic Times; Economic Times. https://economictimes.indiatimes.com/industry/banking/finance/banking/rbi-recognises-face-as-self-regulatory-organisation-in-fintech-sector/articleshow/112871697.cms?from=mdr

Hasan, M., Hoque, A., Abedin, M. Z., & Gasbarro, D. (2024). FinTech and sustainable development: A systematic thematic analysis using human- and machine-generated processing. *International Review of Financial Analysis*, *95*, 103473–103473. DOI: 10.1016/j.irfa.2024.103473

Hinson, R., Lensink, R., & Mueller, A. (2019). Transforming agribusiness in developing countries: SDGs and the role of FinTech. *Current Opinion in Environmental Sustainability*, *41*, 1–9. DOI: 10.1016/j.cosust.2019.07.002

Hinton, L. (2024, July 23). PayTech Can Drive Accessibility and Sustainability | G+D | Money20/20 Europe. FF News | Fintech Finance. https://ffnews.com/fintech-tv/event/2024/money20-20-europe-2024/paytech-can-drive-accessibility-and-sustainability-gd-money20-20-europe/

Hoque, M. Z., Chowdhury, N. J., Hossain, A. A., & Tabassum, T. (2023). Social and facilitating influences in fintech user intention and the fintech gender gap. *Heliyon*, *23457*. Advance online publication. DOI: 10.1016/j.heliyon.2023.e23457 PMID: 38192768

Hua, X., & Huang, Y. (2020). Understanding China's fintech sector: Development, impacts and risks. *European Journal of Finance*, *27*(4-5), 321–333. DOI: 10.1080/1351847X.2020.1811131

Huang, J., Chai, J., & Cho, S. (2020). Deep learning in finance and banking: A literature review and classification. *Frontiers of Business Research in China*, *14*(1), 13. Advance online publication. DOI: 10.1186/s11782-020-00082-6

Huang, J., Liu, R., Wang, W., Wang, Z., & Wang, C., & Jimmy) Jin, Y. (. (2024). Unleashing Fintech's potential: A catalyst for green bonds issuance. *Journal of International Financial Markets, Institutions and Money*, *93*, 102009. DOI: 10.1016/j.intfin.2024.102009

Huang, X., Li, D., & Sun, M. (2025). Fintech and Corporate ESG Performance: An Empirical Analysis Based on the NEV Industry. *Sustainability (Basel)*, *17*(2), 434–434. DOI: 10.3390/su17020434

India, S. (2023). Startupindia.gov.in. https://www.startupindia.gov.in/sindhi/content/sih/en/reources/startup_india_notes/industry_insights/the_indian_finTech_industry_gains_spotlight.html

Innovation, F. (2024). Closing Fintech Funding Gaps. World Economic Forum. https://www.weforum.org/publications/fuelling-innovation-closing-fintech-funding-gaps/

Jain, P., Gupta, P., & Bhawna, . (2024). The Role of FinTech in Sustainability and United Nations' Sustainable Development Goals. *Deleted Journal*, *3*(5), 9–17. DOI: 10.57159/jcmm.3.5.24168

Kearns, J., & Mathew, A. (2022). How India's Central Bank helped spur a digital payments boom. International Monetary Fund, 27.

JisaSoftech. (2022, February 23). *Tokenization in Fintech: What's the significance? - Vault less Tokenization.* JISA Softech Pvt Ltd. https://www.jisasoftech.com/tokenization-in-fintech-whats-the-significance/

Koranteng, B., & You, K. (2024). Fintech and financial stability: Evidence from spatial analysis for 25 countries. *Journal of International Financial Markets, Institutions and Money*, *93*, 102002–102002. DOI: 10.1016/j.intfin.2024.102002

Kovas, A. (2018). Understanding the Risks of Fintech. Thomson Reuters. https://legal.thomsonreuters.com/en/insights/articles/understanding-the-risks-of-fintech

Lazarow, A. (2024, December 24). Fintech Wrapped: Three Surprising Things That Happened In Fintech In 2024. Forbes. https://www.forbes.com/sites/alexlazarow/2024/12/24/fintech-wrapped-three-surprising-things-that-happened-in-fintech-in-2024/?utm_medium=email&utm_source=fintechweeklycom

Liu, Z., & Li, X. (2024). The impact of bank fintech on ESG greenwashing. *Finance Research Letters*, *62*, 105199. DOI: 10.1016/j.frl.2024.105199

Livemint. (2023, October 26). India ranks third among countries with most fintech unicorns in 2023. Mint. https://www.livemint.com/companies/news/india-ranks-third-among-countries-with-most-fintech-unicorns-in-2023-us-retains-the-crown-globally-report-11698318290212.html

Asia Verify. (2024, November 14). The Rise of RegTech and SupTech | Insights Singapore Tech Festival. Asiaverify.com. https://asiaverify.com/regtech-and-suptech-differences-and-rise/

Markopoulos, E., & Barbara Ramonda, M. (2022). An ESG-SDGs alignment and execution model based on the Ocean Strategies transition in emerging markets. In Creativity, Innovation and Entrepreneurship (Vol. 31, pp. 93-103). AHFE International.

Mathias, G. (2024). 5 Agri-FinTechs powering financial growth for Indian farmers. *IBS Intelligence*.https://ibsintelligence.com/ibsi-news/5-agri-fintechs-powering-financial-growth-for-indian-farmers/

Mhlanga, D. (2024). The role of big data in financial technology toward financial inclusion. *Frontiers in Big Data*, *7*, 1184444. Advance online publication. DOI: 10.3389/fdata.2024.1184444 PMID: 38774056

Migozzi, J., Urban, M., & Wójcik, D. (2023). "You should do what India does": FinTech ecosystems in India reshaping the geography of finance. *Geoforum*, *151*, 103720. DOI: 10.1016/j.geoforum.2023.103720

Moody's Analytics. (2022). What is regtech Moodys.com. https://www.moodys.com/web/en/us/kyc/resources/insights/what-is-regtech.html

Muganyi, T., Yan, L., Yin, Y., Sun, H., Gong, X., & Taghizadeh-Hesary, F. (2022). Fintech, regtech, and financial development: Evidence from China. *Financial Innovation*, *8*(1), 29. Advance online publication. DOI: 10.1186/s40854-021-00313-6

Nassiry, D. (2019). The Role of Fintech in Unlocking Green Finance. *Handbook of Green Finance*, 315–336. DOI: 10.1007/978-981-13-0227-5_27

Omeragic, N., Zaimovic, A., & Zaimovic, T. (2024). FinTech and Climate Action, and Affordable and Clean Energy. *Procedia Computer Science*, *236*, 273–280. DOI: 10.1016/j.procs.2024.05.031

ET Online. (2022, October 6). *RBI launches DAKSH application to advance supervisory process. The Economic Times; Economic Times.*https://economictimes.indiatimes.com/news/economy/policy/rbi-launches-daksh-application-to-advance-supervisory-process/articleshow/94688457.cms?from=mdr

Polat, O., Ozcan, B., Ertuğrul, H. M., Atılgan, E., & Özün, A. (2024). Fintech: A Conduit for sustainability and renewable energy? Evidence from R2 connectedness analysis. *Resources Policy*, *94*, 105098–105098. DOI: 10.1016/j.resourpol.2024.105098

Prenio, J. (2024, June 14). *Peering through the hype - assessing suptech tools' transition from experimentation to supervision.* Bis.org. https://www.bis.org/fsi/publ/insights58.htm

PricewaterhouseCoopers. (2021). *Sustainable Finance and FinTech: A necessary marriage | PwC Mauritius. PwC.*https://www.pwc.com/mu/en/about-us/press-room/sustainable-finance.html

PricewaterhouseCoopers. (2022). *ESG impact on the insurance industry: PwC. PwC.*https://www.pwc.com/us/en/industries/financial-services/library/esg-insurance-industry.html

PwC. Security challenges in the evolving fintech landscape. (2016). https://www.pwc.in/assets/pdfs/consulting/cyber-security/banking/security-challenges-in-the-evolving-fintech-landscape.pdf

PwC. Mapping the FinTech innovation landscape in India (2024). https://www.pwc.in/assets/pdfs/mapping-the-fintech-innovation-landscape-in-india.pdf

PwC. Beyond the cloud: Navigating FinTech cyber threats and fortifying defences (2024). https://www.pwc.in/assets/pdfs/consulting/beyond-the-cloud-navigating-fintech-cyber-threats-and-fortifying-defences-v3.pdf

RBI. (2024). Rbi.org.in. https://fintech.rbi.org.in/

Reserve Bank of India - Reports. (2024). Rbi.org.in. https://m.rbi.org.in/Scripts/PublicationReportDetails.aspx?UrlPage=&ID=1263

Restoy, F. (2021). *Fintech regulation: how to achieve a level playing field.*https://www.bis.org/fsi/fsipapers17.pdf

Risk Insights & Advisory Fintech Spotlight | Darwinium. (2024). Darwinium.com. https://www.darwinium.com/resources/whitepapers/risk-insights-advisory-fintech

Risks, A. I. in Fintech: 10 AI Challenges Fintechs Still Struggle With. (2023, November 9). Financial and Business News | Finance Magnates. https://www.financemagnates.com/fintech/ai-risks-in-fintech-10-ai-challenges-fintechs-still-struggle-with/

Rooz, Y. (2024, December 10). How tokenization is transforming finance and investment. World Economic Forum. https://www.weforum.org/stories/2024/12/tokenization-blockchain-assets-finance/

Secretary General, U. N. (2018). Digital Financing Taskforce. United Nations. *The Task Force on Digital Financing of the Sustainable Development Goals*https://www.un.org/en/digital-financing-taskforce

SFA ESG PITCHBOOK. (2024). Retrieved December 5, 2024, from https://singaporefintech.org/wp-content/uploads/2024/10/SFA-ESG-Pitch-Book-2024.pdf

Sharma, S. K., Ilavarasan, P. V., & Karanasios, S. (2023). Small businesses and FinTech: A systematic review and future directions. *Electronic Commerce Research, 24*(1), 535–575. Advance online publication. DOI: 10.1007/s10660-023-09705-5

Singapore FinTech Organisation. (2024). *SFA has brought out the Pitch book on Fintech use cases.* SFA ESG PITCHBOOK.

Soni, G., Kumar, S., Mahto, R. V., Mangla, S. K., Mittal, M. L., & Lim, W. M. (2022). A decision-making framework for Industry 4.0 technology implementation: The case of FinTech and sustainable supply chain finance for SMEs. *Technological Forecasting and Social Change*, *180*, 121686. DOI: 10.1016/j.techfore.2022.121686

Srairi, S. (2024). The impact of risk and Fintech on ESG performance: Evidence from gulf cooperation council (GCC) banks. *Journal of Chinese Economic and Business Studies*, •••, 1–42. DOI: 10.1080/14765284.2024.2428240

Sreenu, N. (2024). The impact of Fintech and green bonds on the Indian renewable energy production. *Renewable Energy*, *237*, 121807. DOI: 10.1016/j.renene.2024.121807

Tata Consultancy Services. (2020). Computer Vision: A Game Changer for Banking, Financial Services and Insurance. White Paper. https://www.tcs.com/content/dam/global-tcs/en/pdfs/insights/whitepapers/Computer-Vision-in-Banking.pdf

Team, E. (2024, April 25). 5 ways Insurtech drives environmental sustainability in business | bobsguide. Bobsguide. https://www.bobsguide.com/5-ways-insurtech-drives-environmental-sustainability-in-business/

Team, E. (2024, July 22). *The Rise of Super Apps in Banking: Integrating Financial and Lifestyle Services.* Finextra Research; Finextra. https://www.finextra.com/blogposting/26488/the-rise-of-super-apps-in-banking-integrating-financial-and-lifestyle-services

Thakor, A. V. (2020). Fintech and banking: What do we know? *Journal of Financial Intermediation*, *41*(100833), 100833. DOI: 10.1016/j.jfi.2019.100833

The Future of Global Fintech: Towards Resilient and Inclusive Growth. (2024, January 18). World Economic Forum. https://www.weforum.org/publications/the-future-of-global-fintech-towards-resilient-and-inclusive-growth/

The Gartner Top Strategic Technology Trends for. 2025. (2024). Gartner. https://www.gartner.com/en/webinar/659810/1469292?utm_campaign=RM_GB_2025_ITTRND_C_WTR1_TTT25_WB&utm_medium=email&utm_source=Eloqua

Thompsett, L. (2024, March 20). *Sustainable Fintech Trends: Top 10 Innovations Driving Change.* Fintechmagazine.com. https://fintechmagazine.com/articles/top-10-sustainable-fintech-trends

Tóth, Z., & Blut, M. (2024). Ethical compass: The need for Corporate Digital Responsibility in the use of Artificial Intelligence in financial services. *Organizational Dynamics*, *53*(2), 101041. DOI: 10.1016/j.orgdyn.2024.101041

UN. (2018). Igniting SDG Progress Through Digital Financial Inclusion | Department of Economic and Social Affairs. Sdgs.un.org. https://sdgs.un.org/publications/igniting-sdg-progress-through-digital-financial-inclusion-30370

https://sdgs.un.org/sites/default/files/publications/2655SDG_Compendium_Digital_Financial_Inclusion_September_2018.pdf

UNESCO. (2019). Artificial intelligence in education | UNESCO. UNESCO. https://www.unesco.org/en/digital-education/artificial-intelligence

United Nations. (2022). Big Data for Sustainable Development. United Nations. https://www.un.org/en/global-issues/big-data-for-sustainable-development

Vardari, L., & Qabrati, I. (Eds.). (2024). Advances in Finance, Accounting, and Economics *Decentralized Finance and Tokenization in FinTech.*, DOI: 10.4018/979-8-3693-3346-4

Walker, I. (2024, February 19). *BIS to leverage fintech solutions for sustainable finance. Green Central Banking.* https://greencentralbanking.com/2024/02/19/bis-fintech-solutions-sustainable-finance/

World Bank. (2020). World Bank Fintech. World Bank. https://www.worldbank.org/en/topic/fintech

World Bank. (2020a). How Regulators Respond to Fintech Evaluating the Different Approaches-Sandboxes and Beyond Public Disclosure Authorized Public Disclosure Authorized Public Disclosure Authorized Public Disclosure Authorized. 2020. https://documents1.worldbank.org/curated/en/579101587660589857/pdf/How-Regulators-Respond-To-FinTech-Evaluating-the-Different-Approaches-Sandboxes-and-Beyond.pdf

World Bank. (2022). Fintech and the Future of Finance. *World Bank.* https://www.worldbank.org/en/publication/fintech-and-the-future-of-finance

APPENDIX

Table 8.

Appendix 1: Mapping SDGs across ESG Pillars			
SDGs	**ESG Pillars**		
	Environmental	**Social**	**Governance**
SDG1: NO POVERTY		√	
SDG2: ZERO HUNGER		√	
SDG3: GOOD HEALTH AND WELL-BEING		√	
SDG4: QUALITY EDUCATION		√	
SDG5: GENDER EQUALITY		√	√
SDG6: CLEAN WATER AND SANITATION	√	√	
SDG7: AFFORDABLE AND CLEAN ENERGY	√		
SDG8: DECENT WORK AND ECONOMIC GROWTH		√	√
SDG9: INDUSTRY, INNOVATION AND INFRASTRUCTURE	√	√	√
SDG10: REDUCED INEQUALITY		√	
SDG11: SUSTAINABLE CITIES AND COMMUNITIES	√		√
SDG12: RESPONSIBLE CONSUMPTION AND PRODUCTION	√	√	√
SDG13: CLIMATE ACTION	√		√
SDG14: LIFE BELOW WATER	√		
SDG15: LIFE ON LAND	√		
SDG16: PEACE, JUSTICE AND STRONG INSTITUTIONS		√	√
SDG17: PARTNERSHIPS FOR THE GOALS			√

Source: (Markopoulos & Maria Barbara Ramonda, 2022)

Table 9.

Appendix 2: Role of Big data in progressing towards SDGs	
SDGs	**Role of Big data**
SDG1: NO POVERTY	Proxy indicator of income levels is provided by showing spending patterns on mobile phone
SDG2: ZERO HUNGER	Helps to monitor food security on real-time basis by crowdsourcing or online tracking of food prices listed
SDG3: GOOD HEALTH AND WELL-BEING	Helps to predict the spread of infectious diseases by mapping and tracking mobile phone users' movement
SDG4: QUALITY EDUCATION	Reveals reasons for trend in student drop-out rates by citizen reporting
SDG5: GENDER EQUALITY	Reveals spending patterns and various impacts of economic shocks on men and women through analysis of financial transactions
SDG6: CLEAN WATER AND SANITATION	Tracks access to clean water through the sensors connected to water pumps
SDG7: AFFORDABLE AND CLEAN ENERGY	Utility companies can manage the optimum use of electricity, gas or water and arrest waste of these resources. Also ensures adequate supply during peak time through the application of smart metering
SDG8: DECENT WORK AND ECONOMIC GROWTH	Provides relevant indicators like economic growth, remittances, trade and GDP by following the patterns in global postal traffic
SDG9: INDUSTRY, INNOVATION AND INFRASTRUCTURE	Assistance in traffic control and helps to improve public transport using data collected from GPS devices
SDG10: REDUCED INEQUALITY	Speech-to-text analytics on local radio content can reveal discrimination concerns and support policy response
SDG11: SUSTAINABLE CITIES AND COMMUNITIES	Helps to track encroachment on public land, places such as parks and forests through satellite remote sensing
SDG12: RESPONSIBLE CONSUMPTION AND PRODUCTION	Reveals the pace of transition to energy efficient products on the basis of online search patterns and e-commerce transactions
SDG13: CLIMATE ACTION	Helps to track deforestation basesd on the collated data on satellite imagery, crowd-sourced witness accounts and open data
SDG14: LIFE BELOW WATER	Helps to reveal and track illegal, unregulated and unreported fishing activities by monitoring maritime vessel tracking data
SDG15: LIFE ON LAND	Social media monitoring helps to support disaster management with real-time information on location of victim, effects and severity of wild fires or haze
SDG16: PEACE, JUSTICE AND STRONG INSTITUTIONS	Helps to reveal public opinion on effective governance, public service delivery or human rights through sentiment analysis of social media
SDG17: PARTNERSHIPS FOR THE GOALS	Provides a better and real- time understanding of present day or current hyper-connected world through the Partnerships by combining of statistics, mobile and internet data

Source: https://www.un.org/en/pdfs/Bigdata_SDGs_single_spread_2017.pdf

Table 10.

Appendix 3: Risk Assessment - Generic Framework for FinTech Products									
	Non-face to face investment platforms	**Non-face to face money transfer platforms**	**Crowdfunding platforms,**	**Internet banking**	**Mobile banking**	**Digital Wallets, Electronic money**	**Mobile money**	**Virtual assets (crypto currencies)**	**Other FinTech**
Size/Volume									
International Transactions									
Transactions with HR Jurisdictions									
Money Laundering / Terrorist Funding Typologies									
Convertibility to Cash									
Transaction records									
Quality of Customer Due Diligence (CDD)									
Quality of monitoring and Suspicious Transaction Reporting (STR)									
Delivery channel (agents).									
Risk Level of FinTech Product									

Source: Celik, 2019

Chapter 12

Investigating Asymmetric Impacts of Financial and Technological Development on Income Inequality in Sub-Saharan Africa

Hassan Swedy Lunku
https://orcid.org/0000-0001-5668-8551
Local Government Training Institute, Tanzania

Muhsin Mohammed Danga
Local Government Training Institute, Tanzania

ABSTRACT

To account for nonlinearities in the relationship between financial and technological development on income inequality, this chapter uses a nonlinear panel ARDL model that allows for heterogeneity within countries. The results show that in the long run, financial and technological development significantly impact income inequality in SSA countries. However, the impact of positive technology on income inequality is negative, while both positive and negative financial shocks have a positive impact on income inequality. The study indicates the importance of fiscal and financial policies of developing economies that should be implemented to achieve inclusive growth, improve welfare, and reduce income distribution in the SSA region.

DOI: 10.4018/979-8-3373-0355-0.ch012

1. INTRODUCTION

The COVID-19 pandemic has significantly disrupted global economic growth, with sub-Saharan Africa (SSA), the world's most unequal and disadvantaged region, among the hardest hit. In 2020, the region experienced a 1.9% economic contraction. Without timely interventions, it is projected that 87% of the world's poorest people will reside in SSA by 2030 (IMF, 2020; United Nations, 2020). Job losses, inadequate social protection, and rising food prices are key factors exacerbating income inequality in the region (IMF & World Bank, 2020; Schoch & Lakner, 2020; Yonzan et al., 2020). These trends highlight the urgent need for policy discussions on how to achieve shared prosperity, particularly in developing economies.

Inequality and poverty have adverse effects on quality of life, health, mortality, education, and access to social protection due to welfare setbacks in the wake of COVID-19. Technology has emerged as a crucial enabler in advancing SSA's green growth agenda, especially during lockdowns. It has facilitated online transactions, remote bill payments, e-learning, and information access (Ofori et al., 2022). However, focusing solely on economic growth without considering broader aspects of development can obscure the role of social equity and the transformative potential of digital technologies. Telecommunications and ICTs offer opportunities for self-employment and job creation (Kurniawati, 2020; Soava et al., 2020). Despite their potential, ICT development has also been linked to rising global income disparities. This study, therefore, investigates the asymmetric impact of technological development on income inequality within the context of economic growth in SSA.

Bergstrom (2022) finds that the inequality elasticity of poverty reduction is greater than that of growth elasticity, underscoring the pivotal role of income distribution in poverty alleviation. While various studies have explored the relationship between financial development and income inequality, the findings are mixed (Dabra-Norris et al., 2015; Yongfen et al., 2020). The interplay between financial development, digital technology, and inequality remains inconclusive. This study seeks to provide empirical evidence to inform policy decisions that promote inclusive growth and reduce income disparities in SSA by examining the nonlinear effects of financial and technological progress on inequality across countries. Understanding the nuanced relationship and intricate dynamics between these variables influences the design of targeted interventions and policies that address income disparities and foster equitable economic development in the region.

Financial deepening, by improving access to resources, can help mitigate economic shocks and reduce inequality, however, its impact is nonlinear. A minimum threshold of financial development is required before it begins to significantly affect income distribution. Beyond this point, indicators such as the banking sector and stock market development can either reduce or exacerbate inequality (Davidson et al.,

2018; Kim & Lin, 2011). Asymmetries in financial and technological development may limit economic potential by excluding the poor from productive activities and inefficiently allocating capital and labour (Kumar & Paramanik, 2020; Mallick et al., 2020). High inequality increases vulnerability to income shocks and impedes short-run growth (Giri et al., 2021). Moreover, both developed and developing countries have experienced rising within-country inequality, often linked to globalization and cross-border flows of capital, labour, goods, and services, sparking widespread policy and academic debate (Pavcnik, 2011).

The main objective of this paper is to examine the symmetric and asymmetric effects of financial and technological development on income inequality in SSA. Our contribution to the literature is threefold. First, we incorporate potential nonlinearities in the relationship, capturing the asymmetric response of income inequality to positive and negative shocks in financial and technological indicators. We do this through a nonlinear approach that decomposes these variables into positive and negative partial sums. This extends the conventional heterogeneous panel data model by accounting for asymmetries. Second, we address within-country differences by allowing for heterogeneity across cross-sections using a nonlinear panel Autoregressive Distributed Lag (ARDL) model. This framework captures both short- and long-term effects and allows for asymmetric dynamics, as recommended by Nusair (2016) and Shin et al. (2014). The model accounts for cross-country variation in the slope coefficients, which is critical given the structural and institutional diversity within SSA. Moreover, the model allows inherent heterogeneity effects to be captured in the slope coefficient due to cross-sectional differences and facilitates both short- and long-run responses.

Thirdly, we evaluate the relative heterogeneity across economies in the region since panel quantile regression is employed to understand the role of different levels of income distribution. The mean value estimator from panel models might lead to spurious results, the panel quantile regression evaluates the income inequality at various points of its conditional distribution. The panel quantile regression is used to identify the heterogeneous structure of several income groups and differentiate the financial market conditions. The estimator effectively addresses outliers compared to mean-based approaches, and therefore, the quantile estimation technique is an interesting method to analyze the heterogeneity, as different slope estimations are allowed for different quantiles. We investigate the adoption and affordability of telecommunication technology in reducing the distributional inequality gap and gain insights into the challenges and opportunities for leveraging such technology to promote inclusive growth, reduce poverty, and foster more equitable income distribution. For policymakers, understanding this asymmetry is crucial as it highlights that while improving access to financial resources or technology might generally promote equity, these benefits are not guaranteed for all segments of society. As such, targeted strategies must be developed to effectively address both the positive and negative consequences of interventions aimed at mitigating income inequality. The remainder of the paper is structured as follows: Section 2 reviews the relevant literature; Section 3 presents the data and methodology; Section 4 discusses the empirical results; and Section 5 concludes.

2. LITERATURE REVIEW

2.1 Theoretical Framework

The relationship between inequality and economic growth in SSA is a complex and multifaceted one, influenced by several factors including financial and technological development. In the context of the region's quest for economic growth, the interplay between human capital and physical capital formation plays a crucial role, Galor and Zeira (1993) argued the impact of income inequality on human capital formation and emphasizing how disparities in income distribution affect access to education, skills, and overall productivity. Disparities in income distribution led to unequal access to education and training, ultimately influencing the quality of workforce and long-term economic growth prospects. Similarly, Galor and Moav (2004) demonstrated the relationship between inequality and physical capital formation, argue how unequal distribution of resources may influence investment in infrastructure, technology, and productive assets. The distribution of income plays a critical role in shaping physical capital formation in the region, disparities in wealth distribution affect the level of savings, investment, and overall economic productivity. This study recognizes the non-monotonic nature of the relationship between inequality and economic growth, we account for the dynamic and non-linear effects of income distribution on financial and technology development. As inequality changes, its impact on growth trajectories may vary across economies, necessitating a nuanced understanding of the mechanisms at play. Access to financial, credit, and technology adoption can shape the distribution of resources and opportunities, influencing the dynamics of income inequality and economic growth in SSA.

The nonlinear relationship where financial development widens income inequality, poor individuals benefit in the early stages of financial development due to credit worthiness while lacks of collateral and credit worthiness is disadvantageous due to being economically, politically, and socially backward. Clarke et al., (2006) concluded that poor individuals will engage in the unskilled labor market and earn lower wages due to the fact that they will not benefit from the financial system. The other argument is the linear relationship where financial development reduces income inequality, and higher borrowing costs in the underdeveloped financial market will not benefit poor individuals in the early stage of development (Banerjee & Newman, 1993; Galor & Zeira, 1993). Moreover, the financial system deepens as the economy grows to support expanding economy with broader financial products and services, therefore borrowing costs are reduced and financial services become more accessible to all of society which helps the disadvantaged poor population to improve their welfare and earnings by investing in human and physical capital. The installation and operation of developed technologies with rapid technological advancements are associated with skills performance and therefore noted to be responsible for the increase in income inequality and biased wage distributions (Greenwood & Yorukoglu, 1997; Krueger, 1993). Advanced technology accelerates skills-biasedness and includes the effects of the Skills Biased Technology Changes (SBTC) hypothesis which links the employment shares of skilled and unskilled individuals (Aghion et al., 2015). The unequal distribution of resources impedes technological advancement and economic growth whereas the domestic demand for relatively low-income employees compared to that of capitalists stimulate domestic production due to purchasing power (Mnif, 2016).

2.2 Financial Development-Income Inequality Nexus

Jauch and Watzka, (2016) noted the global increase in financial development and income inequality where a long-standing interest among economists and practitioners in examining the theoretical and empirical relationship between financial development and income inequality in relation to other development indicators. Financial development, liberalization, and market reforms were examined across different countries with determinant impacts on income inequality (Shi et al., 2022). Agnello et al., (2012) investigated the effect of financial reforms on income inequality in a sample of 62 countries from 1973 to 2005 and found that financial reforms significantly reduce income inequality. The financial reforms are measured through an aggregate index that takes into account the removal of government control and the financial sector direction. Paramati and Nguyen, (2019) found minimized income inequality due to growth in banking credit across panels of developed and emerging economies with the suggestion that growth in stock market indicators has both positive and negative effects on income inequality. Uddin et al., (2014) found a negative nonlinear relationship between financial development and income inequality in Bangladesh from 1975 to 2011 using time series ARDL with structural breaks, thus means financial development reduces income inequality in Bangladesh. Financial development and income inequality mixed relationship indicated in the literature with positive, negative, and no significant relationships (Christopoulos & McAdam, 2017; Seven & Coskun, 2016)

2.3 Technology-Growth-Inequality Nexus

Technology plays a complex role in income and wealth inequality as it is a key driver of aggregate economic growth through productivity improvements with variations in contribution to economic growth across countries. Technology is considered to be an income inequality driver due to technology being skilled bias in nature, technological development as measured by the share of ICT capital in total capital stock is predominantly driven by high income and investment in digital technology assign effects of trade to impact technology (Jaumotte et al., 2013). Kwan and Chiu, (2015) noted that neoclassical models and sustainable livelihoods approach (SLA) denotes the link between information technology and the provision of the playing field for the masses in economic activity and argued that ICTs contemporary assets can foster inclusive growth. Stolper-Samuelson theorem resulted from the Heckscher-Ohlin model developed the link between trade and income inequality with the implication that in developing countries where low-skilled labor is abundant then trade openness increases results in an increase in the low-skilled labor wages and reduce the compensation of high-skilled labor, lead to a reduction in income inequality (Jones, 1993). Kocsis and Xiong (2022) found that as digital infrastructures growth leads to economic growth which may decrease the number of unskilled jobs that may lead to increased poverty and inequality, developed countries have higher capabilities in using such infrastructures to reduce income inequality. Untari et al., (2019) argued that digital infrastructure has respectively direct and indirect impact on economic growth and income inequality with the insignificant impact of government spending when analyzing the impact of ICT infrastructure on Indonesia's economic growth and income inequality between 2011 and 2016.

2.4 Research Gap and the Study Hypothesis

Amidst the ongoing transformations in financial systems and technological landscape, understanding the nonlinear impacts of financial and technological development on income inequality in the SSA region has emerged as a pressing research imperative. While conventional wisdom suggests that enhanced access to finance and technological innovations can promote economic opportunities and reduce inequality, the reality is often more nuanced. The dynamics of how financial and technological advancements interact with income distribution in SSA, leading to nonlinear effects on inequality, remain inadequately understood. This paper aims to delve into the intricate relationship between financial development, technological development, and income inequality in SSA, with a specific focus on uncovering the nonlinear (asymmetric) impacts that shape the distribution of wealth and opportunities within the region which are not uniform and can differ significantly depending on whether these changes are positive or negative. By bridging the gaps in existing literature and employing econometric analytical frameworks, this study seeks to provide fresh insights into the complex dynamics at play and offer valuable implications for policy formulation and inclusive growth strategies in SSA. Through a comprehensive literature review and theoretical perspectives, this research endeavors to shed light on the nuances of financial and technological influences on income inequality, paving the way for a deeper understanding of the developmental challenges and opportunities facing SSA. When advancements occur, such as improved access to financial services or technology, these can enhance opportunities for lower-income individuals, potentially reducing inequality. However, economic shocks or technological disruptions may disproportionately impact lower-income groups, thus exacerbating inequality. This study aims to contribute to the ongoing discourse on sustainable and equitable economic development in the region by unravelling the nonlinear relationship between finance-technology-income. The study proposes the hypothesis that Financial and technology development linearly and non-linearly affect income inequality in the SSA region.

3. DATA AND MODELS

3.1 Data and Variable Descriptions

Data abstracted from the Standardized World Income Inequality Database (SWIID) for the Gini coefficient, that referred to as a proxy for income inequality measures (Solt, 2020). Domestic credit to the private sector (%of GDP) and fixed telephone subscription (per 100 people) as proxies for financial deepening and technology development, abstracted from the World Bank PovcalNet database for World Development Indicator (WDI). Annual percent of per capita Gross Domestic Product (GDP) as a proxy of economic growth, government final consumption expenditure as a percent of GDP, annual percent inflation (consumer prices), and trade openness as a percentage sum of export and import per GDP as control variables abstracted from WDI (World Bank, 2020). The variables considered in the study are since that attribute financial and technological accessibility toward development and welfare. Due to data availability, countries without required data for the study variables were not considered in the study, and 6 SSA countries were selected for the time period 1995 and 2020.

Data missing on the Gini coefficient from the database was preprocessed to fill the gap for the series, imputation algorithm using a data mining procedure on random values was adopted since the algorithm is advantageous in maintaining the relative measures of dispersion, centering, minimum, maximum, and distribution of original data by following the procedures depicted on appendix figure A1 (Liu, 2015). Appendix table A1 presents variable descriptions, source of data, and a list of the countries in the selected panel. Key variables were transformed to a logarithm term to smooth the variable to offer constant variance and potential control of heteroskedasticity, and data do not pose serious problems on serial correlation and other assumptions from the preliminary test conducted, as presented in the appendix table A2 and figure A2. South Africa has the highest income inequality and domestic credit, with respective averages of 4.084 and 4.773, while Mauritius has the highest technology accessibility, with the average value of 3.272 compared to other countries in the panel. It means South Africa is the most unequal and disadvantaged country in terms of the highest income inequality in the panel, with higher access to financial services compared to other countries.

3.2 Nonlinear and Asymmetric ARDL Model

Shin et al. (2014) developed a nonlinear panel ARDL model, which is a nonlinear representation of the dynamic heterogeneous panel data and is suitable for panels with large T. The paper employs the nonlinear panel ARDL model due to its ability to capture asymmetries nonlinearly and account for the inherent effect of heterogeneity. Moreover, the model is appropriate for mixed orders of integration of I(0) and I(1). The asymptotics of large N and T dynamics panels and that of traditional large N and small T dynamics panels are different, where small T panel estimation always relies on either random- or fixed-effect or combinations with instrumental variable estimators (Blackburne & Frank, 2007; Manuel Arellano and Stephen Bond, 1991). Blackburne and Frank (2007) argued that the homogeneity assumption of slope parameters is often inappropriate, as one of the major findings in the literature for large N and T panels. The study deals with large T panels and thus the dynamic heterogeneous panel model is considered suitable with the presence of the cross-sectional dependence (CD) in the variables (Pesaran, 2007, 2015). Pooled mean group (PMGe) and Mean Group (MGe) estimators are two important approaches used for dynamic heterogeneous panel models estimation, where MGe relies on estimating N time series regressions and coefficient averaging, while PMGe considers the combination of coefficient pooling and averaging (Blackburne & Frank, 2007). Hausman's test was employed in the study to test the systematic difference between the two estimators. Moreover, the estimators used to generate individual country results in addition to panel short- and long-run symmetric and asymmetric income inequality, financial deepening, and technology nexus.

3.3 Symmetric and Asymmetric ARDL

The paper investigates the symmetric response of income inequality to financial deepening and technology access, and then follows the relaxation of the assumption to allow for both positive and negative changes in financial deepening and technology access. The symmetric panel ARDL was given as

$$\Delta lgini_{it} = \beta_{0i} + \beta_{1i} lgini_{i,t-1} + \beta_{2i} lcredit_{t-1} + \beta_{3i} ltele_{t-1} + \sum_{j=1}^{p_1} \vartheta_{ij} \Delta gini_{i,t-j} + \sum_{j=0}^{p_2} \tau_{ij} \Delta lcredit_{t-1} + \sum_{j=0}^{p_3} \varphi_{ij} \Delta ltele_{t-1} + \mu_i + \varepsilon_{it}$$

Where lgini, lcredit and ltele are logarithm terms of Gini coefficients, domestic credit to private sector (% of GDP) and fixed telephone subscription (per 100 people) as proxies for income inequality, financial deepening and technology access for i = 1, 2, …, N panels and t = 1, 2, …, T time periods. μ_i and ε_{it} are group-specific effects and error term, respectively. The long-run slope coefficient for each cross-section is computed as $-\frac{\beta_{2i}}{\beta_{1i}}$ with the assumption that $\Delta lgini_{i,t-j} = 0$, $\Delta lcredit_{i,t-j} = 0$, and $\Delta ltele_{i,t-j} = 0$ for the long run, whereas the short-run estimate for *lcredit* and *ltele* obtained respectively as τ_{ij} and φ_{ij}.Re-parameterized to include an error correction term as

$$\Delta lgini_{it} = \phi_i \omega_{i,t-1} + \sum_{j=1}^{p_1} \vartheta_{ij} \Delta gini_{i,t-j} + \sum_{j=0}^{p_2} \tau_{ij} \Delta lcredit_{t-1} + \sum_{j=0}^{p_3} \varphi_{ij} \Delta ltele_{t-1} + \mu_i + \varepsilon_{it}$$

Where ϕ_i is the error-correcting speed of adjustment term for each unit, which is equivalent to β_{1i} and $\omega_{i,t-1} = lgini_{i,t-1} - \delta_{0i} - \delta_{1i} lcredit_{t-1} + \delta_{2i} ltele_{t-1}$ is the linear error-correcting term for each panel, $\delta_{0i} = -\frac{\beta_{0i}}{\beta_{1i}}$, $\delta_{1i} = -\frac{\beta_{2i}}{\beta_{1i}}$, and $\delta_{2i} = -\frac{\beta_{3i}}{\beta_{1i}}$. There are no decompositions of *lcredit* and *ltele* in both equations in both equations above, and hence symmetric impacts on *lgini* under this technique. The nonlinear panel ARDL model allows asymmetric response of *lcredit* and *ltele* on *lgini* unlike the symmetric version of panel ARDL. Positive and negative changes are not expected to have similar impacts on income inequality under the asymmetric technique. The asymmetric panel ARDL model is given as

$$\Delta lgini_{it} = \beta_{0i} + \beta_{1i} lgini_{i,t-1} + \beta_{2i}^{+} lcredit_{t-1}^{+} + \beta_{2i}^{-} lcredit_{t-1}^{-} + \beta_{3i}^{+} ltele_{t-1}^{+} + \beta_{3i}^{-} ltele_{t-1}^{-} + \sum_{j=1}^{p_1} \vartheta_{ij} \Delta gini_{i,t-j} + \sum_{j=0}^{p_2} \left(\tau_{ij}^{+} \Delta lcredit_{t-1}^{+} + \tau_{ij}^{-} \Delta lcredit_{t-1}^{-} \right) + \sum_{j=0}^{p_3} \left(\varphi_{ij}^{+} \Delta ltele_{t-1}^{+} + \varphi_{ij}^{-} \Delta ltele_{t-1}^{-} \right) + \mu_i + \varepsilon_{it}$$

Where $lcredit_t^{+}$, $lcredit_t^{-}$, $ltele_t^{+}$, and $ltele_t^{-}$ denote positive and negative financial deepening and technology access with respectively the long-run coefficients $-\frac{\beta_{2i}^{+}}{\beta_{1i}}$, $-\frac{\beta_{2i}^{-}}{\beta_{1i}}$, $-\frac{\beta_{3i}^{+}}{\beta_{1i}}$, and $-\frac{\beta_{3i}^{-}}{\beta_{1i}}$. Positive and negative sum decomposition computed as

$$lcredit_t^{+} = \sum_{m=1}^{t} \Delta lcredit_{ik}^{+} = \sum_{m=1}^{t} \max\left(\Delta lcredit_{ik}, 0 \right)$$

$$lcredit_t^- = \sum_{m=1}^{t} \Delta\, lcredit_{ik}^- = \sum_{m=1}^{t} \min\left(\Delta lcredit_{ik}, 0\right)$$

$$ltele_t^+ = \sum_{m=1}^{t} \Delta\, ltele_{ik}^+ = \sum_{m=1}^{t} \max\left(\Delta ltele_{ik}, 0\right)$$

$$ltele_t^- = \sum_{m=1}^{t} \Delta\, ltele_{ik}^- = \sum_{m=1}^{t} \min\left(\Delta ltele_{ik}, 0\right)$$

The error correction model reparametrized as

$$\Delta lgini_{it} = \psi_i \theta_{i,t-1} + \sum_{j=1}^{P_1} \vartheta_{ij} \Delta\, gini_{i,t-j} + \sum_{j=0}^{P_2} \left(\tau_{ij}^+ \Delta\, lcredit_{t-1}^+ + \tau_{ij}^- \Delta\, lcredit_{t-1}^- \right) + \sum_{j=0}^{P_3} \left(\varphi_{ij}^+ \Delta\, ltele_{t-1}^+ + \varphi_{ij}^- \Delta\, ltele_{t-1}^- \right) + \mu_i + \varepsilon_{it}$$

Where $\theta_{i,t-1}$ denote the error-correction term that captures the long-run equilibrium and ψ_i speed of adjustment in the asymmetric panel ARDL which measures time it takes the system to converge to its long-run equilibrium. The study used several first- and second-generation panel unit root tests for stationarity, and tests conducted to determine the order of integration among variables which helps identify the best suitable model for the available data. Levin-Lin-Chu (LLC) test, Im-Pesaran-Shin (IPS) test, and Fisher-Type tests were employed as first-generation panel unit root testing (Im et al., 2003; Levin et al., 2002; Maddala & Wu, 1999). Whereas, Pesaran (2004) cross-sectional dependence (CD) test was employed in the study to test the average correlation coefficients, CIPS and CADF were employed as second-generation panel unit root tests. The Johansen-Fisher panel cointegration was employed in the study to examine the individual independent test, and the test was developed to examine the cointegration relationship by incorporating the test from individual cross sections to obtain test statistics for the whole panel (Johansen & Juselius, 1990a; Pesaran et al., 1993). The x^2 value is built upon the MacKinnon-Haug-Michelis π_i p-values for Johansen cointegration trace and maximum-Eigen value tests, under the null hypothesis of no panel cointegration, the distinct cointegration test for i cross section

$$-2\sum_{i-1}^{N} \log\left(x^2 \pi_i\right) \rightarrow x^2 2N$$

Panel regression models frequently encounter problems of cross-sectional dependency presence which may arise due to economic distance, spatial correlations, common unobserved shocks, spill-over effect, or omitted global variables (Omary et al., 2014). The classical unit root and cointegration testing procedures become invalid in the presence of correlated errors in panel models, it was found that all tests experience severe size distortion when panel members cointegrate, when finite sample performance was assessed. Ucar-Omary (UO) and Emirmahmutoglu-Omary (EO) nonlinear and asymmetric panel unit root tests were employed to test the asymmetric panel unit root test of variables, as the test considers the Sieve bootstrap method to deal with the problem of the CD and obtain good size properties (Omary et al., 2014; Ucar & Omay, 2009). This paper employs the proposed nonlinear panel cointegration and causality test to investigate the causal relationship of asymmetric *lcredit* and *ltele* on *lgini*, using panel techniques as the approach to allow the possibility of nonlinearities in interrelationships between variables (Omary et al., 2014).

4. EMPIRICAL RESULTS AND DISCUSSION

4.1 Panel Unit Root And Cointegration Tests

First-generation panel unit root test employed to ascertain that no variable in the study is integrated of order 2 for model justification. We investigate the null hypothesis of a unit root. Levin-Lin-Chu (LLC), Im-Pesaran-Shin (IPS), Fisher-type based on both augmented Dickey-Fuller (ADF) unit root test employed, and observed that all variables of interest integrated at different orders, that is I(0) and I(1) (Madala and Wu; Im et al., 2003; Levin et al., 2002). The importance of a test is to deal with highly persistent series that may influence model estimation and might produce biased statistical inferences and spurious relationships. Table 1 presents results related to the first-generational panel unit root test for 1 lag; the results for both level and first difference showed significant results at the level of significance (LOS). The inflation rate was significant at the level and hence integrated at I(0) for LLC and IPS, while credit and GDP per capita were significant at the level respectively for LLC and IPS. The Gini, technology access, trade openness, and expenditure were insignificant at the level while significant at first difference, hence integrated at I(1) for both LLC and IPS. For the Fisher-type panel unit root test, inflation and GDP per capita were significant at the level with all statistics, hence integrated at I(0), and all variables were stationary at I(1). Trend and no constant options for the unit root test were considered in the study, and the results were significant to reject the null hypothesis of panels containing a unit root; therefore, the panels were concluded to be stationary.

Unobserved and global common shocks affect panel data estimation, and the model can be misspecified and/or spurious if not accounted for the shocks in panel data estimation. Pesaran (2004, 2015) developed the cross-sectional dependence (CD) test to account for the problem to assess the average correlation of variables between sampled countries with the assumption of standard normal distribution. The approach can be used even to space and balance panels with either weak or strict cross-section dependence, null hypothesis. Table 2 presents the estimated CD test results, the weak dependence or strict cross-sectional independence null hypotheses rejected, and therefore provides evidence of cross-sectional dependence and correlation across a sample panel. The study re-examines the panel unit root using a useful second-generational unit root test in the heterogeneous panel with a strong correlation across the panel. The homogeneous non-stationary null hypothesis for CIPS and the non-stationary series null hypothesis for CADF panel unit root tests with one lag condition (Pesaran, 2004, 2007, 2021). Pesaran CIPS panel unit root test with a cross-sectional and first difference means included for variables with constant deterministic chosen and dynamics lag criterion decision general to particular based on the F joint test, individual period was truncated during the aggregation process. Pesaran CADF test for variables cross-sectional average in the first period extracted and extreme t-values truncated with a constant deterministic chosen, augmented by 1 lag. Table 2 below presents the second-generational results to validate the results of the first-generational unit root tests. At 1% LOS, we rejected the null hypothesis for the selected 36 SSA countries. Results of second-generational corroborate that of first-generational panel unit root tests that variables are integrated at I(0) and/or I(1).

Table 1. First-generation panel unit root tests

Variables	*LLC (Adjusted T)*		*IPS (W-t-bar)*		*Fisher-Type ADF (Level)*			
	Level	**First difference**	**Level**	**First difference**	**Inverse chi-squared (P)**	**Inverse normal (Z)**	**inverse logit t (L*)**	**Modified inv. Chi-squared (Pm)**
lgini	0.9371	-1.5982**	-0.5304	-4.0325***	20.5566*	-0.6136	-0.7822	1.7466**
lcredit	-1.4455*	-1.8031**	1.6327	-4.6795***	11.9759	0.1069	0.3174	-0.0049
ltele	1.8298	-4.9034***	0.9273	-5.2149***	7.8626	1.6698	1.7597	-0.8446
ltrade	0.5247	-3.8981***	0.5586	-5.2149***	10.1452	0.5623	0.6816	-0.3786
Infla	-2.7589***	-	-4.2667***	-	59.8136***	-4.6298***	-6.3581***	9.7599***
lexpend	0.316	-4.5619***	-0.2066	-5.9395***	20.3079*	-0.1592	-0.6345	1.6958**
gdp	1.6019	-4.0034***	-2.6797***	-8.4959***	39.9609***	-2.8858***	-3.7752***	5.7075***

*, **, *** *significance at 10%, 5%, 1%*

Table 2. Second-generation panel unit root tests and CD test

Variables	*CADF*		*CIPS*		*Pesaran (2004) CD test*	
	Level	**First difference**	**Level**	**First difference**	**CD**	**corr**
lgini	-0.139	-2.940***	-0.618	-2.804	-2.13***	-0.108
lcredit	-1.927	-3.316***	-1.956	-4.449***	12.95***	0.656
ltele	-2.548**	-3.819***	-2.21*	-3.942***	5.66***	0.286

continued on following page

Table 2. Continued

Variables	*CADF*		*CIPS*		*Pesaran (2004) CD test*	
	Level	**First difference**	**Level**	**First difference**	**CD**	**corr**
ltrade	-1.618	-3.116***	-1.768	-4.361***	3.05***	0.155
infla	-3.494***	-4.803***	-4.578***	-5.661***	6.96***	0.352
lexpend	-1.733	-3.413***	-1.17	-3.797***	-2.1**	-0.106
gdp	-2.273*	-4.983***	-3.887***	-6.015***	8.06***	0.408

, **, * significance at 10%, 5%, 1%*

Westerlund (2007) test employed in this study to deal with cross-sectional dependence, since the bootstrap test is applied to get consistent results. The variance ratio statistics of -1.4086 rejected the null hypothesis at 10% LOS for the panel; therefore, variables in the model specification are symmetrically cointegrated. Table 3 below presents the results of the Johansen-Fisher panel cointegration test with none, linear, and quadratic deterministic trends. The results indicate that the trace statistic has three cointegrating equations maximum-Eigen and trace tests are significant for one equation without trend deterministic at the 1% level of significance.(Johansen & Juselius, 1990b)). The first equation is significant at 1% with a linear trend, and two insignificant equations for both maximum-Eigen and trace tests with a linear deterministic trend. Moreover, the maximum-Eigen test is significant at 1% and 10% for two equations, while one equation is insignificant with a quadratic trend. The first equation is statistically significant for all tests, while the two equations are insignificant with and without linear deterministic trend; the hypothesis of no cointegration for individual cross-section results is significant for all three tests. Therefore, we reject the null hypothesis of no cointegration and hence indicate that there is a long-run relationship between variables.

Table 3. The Johansen-Fisher panel cointegration test

Deterministic trend specification	*No trend*		*Linear trend*		*Quadratic trend*	
Hypothesized No. of CE(s)	**Fisher stat (from trace test)**	**Fisher stat (from max-Eigen test)**	**Fisher stat (from trace test)**	**Fisher stat (from max-Eigen test)**	**Fisher stat (from trace test)**	**Fisher stat (from max-Eigen test)**
None	39.62***	33.67***	34.79***	35.11***	33.38***	33.62***
At most 1	18.19	17.40	10.48	9.808	10.93	7.404
At most 2	11.07	11.07	6.653	6.653	20.05*	20.05*

***, **, * significant at 1%, 5% and 10%

The paper employs the EO and UO panel unit root test that allows the simultaneous existence of nonlinear and asymmetric mean reversion within a panel to test for stationarity of financial deepening and technology access proxies on income inequality for the panel. The procedure tests whether the series exhibits a unit root against the alternative of globally stationary asymmetric or asymmetric nonlinearity with accommodation of cross-sectional dependence using the Sieve bootstrap algorithm. Table 4 presents the results of the Sieve bootstrap method with empirical distributions of the test generated by 5000 replications to obtain p-values for the t-statistic for both financial and technology development. The results indicated that the variables follow either a nonlinear or asymmetric process, as the no stationarity null hypothesis rejected against the stationary nonlinear symmetric or asymmetric process, hence integrated either at I(0) and/or I(1).

Table 4. Asymmetric panel unit root tests

Variables	UO Test (t-stat)		EO Test (t-stat)	
	Level	First difference	Level	First difference
LCredit	-2.153*	-4.4450***	1.032*	3.926***
LTele	0.1361	-3.5023**	0.607	3.103**

*, **, *** significance at 10%, 5%, 1%

This paper employed nonlinear panel cointegration and causality tests to investigate the income inequality-financial deepening-technology asymmetric nexus. Westerlund's (2007) linear panel cointegration test indicated cointegration at 10% LOS. Moreover, we find strong evidence of cointegration when nonlinearity is allowed in the long-run relationship, suggesting that adjustment to the equilibrium level is both symmetry and asymmetry in nature. Financial development proxy is significantly bi-directional Granger cause income inequality, whereas income inequality and financial development are uni-directional causes of technology and income inequality, respectively, as their causality null hypothesis was rejected significantly.

4.2 Symmetric and asymmetric ARDL

Hausman test results significantly not rejected the null hypothesis and therefore PMGe considered in the model compared to MGe after estimating equations with both estimators. PMGe is efficient estimator under the null hypothesis while MGe is under alternative hypothesis and hence the results support the PMGe as efficient estimator for modelling both symmetric and asymmetric ARDL model for inequality-financial-technology nexus for the panel sample of SSA countries selected. Table 5 present both symmetric and asymmetric PMGe results as efficient estimator under the null hypothesis which is consistent for all the models. This paper investigates inequality-financial-technology nexus with and without asymmetries and control variables. From results of mixed order of integration from under consideration series, we examine the short- and long-run dynamics of the response of financial deepening and technology access on income inequality. From the results, positively financial proxy is significant at 5% while technology proxy is insignificant when without control variables in the model whereas both financial and technology proxies are significant at 1% with significant GDP per capita and trade openness in the long-run means that higher financial and technology access resulting from unexpected global economy changes will have a positive effect on income inequality. Moreover, technology proxy is positive significantly affect the income inequality on the model without the control variables where both financial and technology proxies are positive insignificant with the control variables with only trade openness significant in the short-run.

Evidently the results reinforce observation from the trends plot of figure A2 (on appendices) where the series appear to move in the same direction while differ for individual countries. Both financial and technology proxies are significant on long-run while only technology is significant on short-run on testing the symmetric short- and long-run relationship. The elasticity coefficients in the long-run are higher than short-run in magnitude, both financial and technology are elastic for income inequality in the long run while technology is elastic and financial is inelastic in the short-run. Indicate that income inequality is more susceptible to both technology and financial access in the long-run while only technology changes in the short-run. Any shock to technology changes is more likely to spillover to economy of the countries in the region including their financial capabilities. Increase in the advanced technology diffusion in the region is more likely to stimulate economic activities over time with sustained physical and financial capital investment. The inelastic nature of financial in the short-run further attest to our understanding of possible reactions by any economic agent where there is shock and usually historical experience led economic agents to initially be skeptical as to make either drastic or gradual move to shocks respondence. Positive shocks over the years tends to persist more than negative and that explains why the positive response of financial and technology access in the long-run are significantly than in the short-run.

Table 5. Symmetric and asymmetric ARDL results

Variables	*Symmetry ARDL*				*Asymmetry ARDL*			
	Long-run		**Short-run**		**Long-run**		**Short-run**	
Lcredit	0.0559**	0.1549***	0.0644	0.0305				
	(0.0259)	(0.0289)	(0.0535)	(0.0505)				
ltele	0.0167	0.1081***	0.1522**	0.1402				
	(0.0179)	(0.0281)	(0.0709)	(0.0999)				
lcredit⁺					6.5953***	5.5874***	4.0397	2.9108
					(1.8788)	(1.0513)	(4.0594)	(3.8356)
lcredit⁻					7.1754***	5.9815***	4.0273	2.8145
					(1.9124)	(1.1079)	(4.2445)	(3.9996)
ltele⁺					-4.3954***	-4.2292***	10.3375**	10.2482***
					(0.6797)	(0.3690)	(4.0956)	(3.8325)
ltele⁻					2.9728***	1.9011***	7.0835	6.8258
					(0.4364)	(0.3429)	(4.4037)	(4.3038)
ltrade		-0.1887***		-0.0955***		1.1631*		-3.1593
		(0.0725)		(0.0193)		(0.6397)		(2.2752)
infla		0.0042		0.0007		-0.0326**		0.1294
		(0.0035)		(0.0019)		(0.0163)		(0.0933)
lexpend		0.0198		0.0663		5.3765***		0.589
		(0.0624)		(0.1037)		(0.5976)		(2.9194)
gdp		0.0098*		-0.0027		-0.0619		0.2354
		(0.0917)		(0.0032)		(0.0501)		(0.1493)
cons			2.3295***	2.4032***			13.0159**	6.6482**
			(0.4142)	(0.6175)			(6.6299)	(3.1840)
ec			-0.6369***	0.5992***			-0.5060*	-0.6574*
			(0.1109)	(0.1434)			(0.2782)	(0.3742)
Number of obs	156	156	156	156	156	156	156	156
Number of groups	6	6	6	6	6	6	6	6
Log Likelihood	325.7253	212.1155	325.7253	212.1155	-335.79	-299.116	-335.79	-299.116

, **, * significance at 10%, 5%, 1%; Standard errors in paranthesis ()*

In the asymmetric case, both positive and negative financial changes tend to exert significant positive impact on income inequality with higher magnitude without than with control variable and positive technology is negative while positive technology is positive with significant trade openness, inflation and government expenditure affect the income inequality in the long-run. Only positive technology is significant positive with and without control variables whereas positive financial and both negative financial and technology are insignificant in the short-run. Like in the symmetric case, income inequality is elastic for both financial and technology in the long-run whereas technology is elastic and financial is inelastic in the short-run for both cases with and without control variables. The long-run negative response of income inequality to negative technology further underscores the significance of supply-side hypothesis where negative technology access shocks driven by downward trend in global economic activities will adversely affect the income inequality in the region. However, income inequality responds positively to positive technology access for both with and without control variables in the short-run, positive shocks that follows the same direction. Moreover, this does not negate the long-run asymmetric effects observed for the sample panel rather clearly it shows that negative long-run response to shocks is not usually immediate. Individual countries may be responding to positive shocks in positive way in the short-run, however with time, tend to respond negatively as positive shocks persist in the long-run.

The long-run positive response of technology to negative technology change also is sustained in the long-run as well as both positive and negative changes in financial accessibility since governments in the selected countries usually cushion the effect of negative shocks by raising expenditure, trade openness and inflation to change the income inequality. As expected, magnitude differ as observed in the symmetric model between the two cases of including and excluding control variables to asymmetric case. Evidently suggest that the models' estimates are insensitive to outliers at least in terms of significance and direction relationship. Generally, the implications of identifying the shock effect of technology and financial accessibility can be helpful on income inequality and distribution for inclusive economic growth of SSA countries. The consideration of income distribution response to positive and negative changes in financial and technology accessibility is particularly important for economic growth since upward and downward changes by the same magnitude are not likely to have the same impact on income inequality. With understanding of possible asymmetric response of financial and technology to income inequality, financial market and advance technology may need to incorporate the information for predictive model for better forecast results. The findings are relevant to fiscal and monetary policy institutions who are under pressure constantly to promote and ensure financial stability, ability to achieve stability to forecast future trends accurately. With increasing evidence of strong link between financial-technology-inequality trilemma including results from this paper, findings from this paper provide information about the intensity of asymmetric response of financial and technology changes to income inequality movements which is essential in order to avoid the misconception of assuming that income inequality respond symmetrically or nonlinearly to both negative and positive shocks.

4.3 Sensitivity and robustness of the models

4.3.1 inclusion and exclusion of control variables

We examine further the sensitivity and robustness of the estimates obtained for the full sample under each influential statistics category for with and without control variables. As presented on appendix table A3, Botswana, Kenya and South Africa income inequality are significantly affected by positive and negative changes of technology in the short-run while Madagascar, Mauritius and Tanzania are insignificant. Both positive and negative financial variable changes are significant with control variables cases in Kenya, Mauritius and Tanzania while insignificant without control variables with Botswana, South Africa and Madagascar for both cases. We use average statistics of income inequality for individual countries with both cases including and excluding control variables in order to identify countries with dramatic changes. Using this rule of thumb, we find that Botswana and South Africa are countries have exceedingly high mean value income inequality compared to other countries in the panel and significantly affected by technology accessibility. We further subject the model to robustness checks with the essence to evaluate the sensitivity of the results to variable measurement and lag length, we re-estimated all the models in by replacing the proxy for income inequality with the Palma ratio and change the lag order of the model in the main estimation by one. When compared to the main results, significance and sign of the coefficient are slightly the same with few noticeable differences in terms of magnitude of impacts for all the models suggesting that the analysis is robust to proxies and therefore the analysis will remain the same. We found consistent results with the main findings of the paper both in terms of significance and sign, therefore justifying the robustness of the chosen model to lag structure.

4.3.2 Quantile regression results

Besides using the panel nonlinear ARDL estimation techniques, we employ the quantile estimation using 10th, 20th, 30th, 40th, 50th, 60th, 70th, 80th, and 90th quantiles to further evaluate the validity of study. Despite quantile regression provide a complete picture of the links of the variables, findings and policy implications that derived from panel mean regression argued that may lead to spurious results (Allard et al., 2018). However, panel quantile regression addresses the heterogeneity of different income groups, therefore, we employ a panel quantile technique along panel mean regression. The estimator characterized by its differential characteristics of considering the marginal influence of income distribution across different quantile, and offers in depth and thorough cross-section dependence between the series. Table 6 present results related to quantile estimation for total sampled countries. Credit variable shows positive and significant from 40th to upper 90th quantiles while negative and insignificant at 10th quantile, telephone subscription is positive and significant from lower 10th to 60th quantiles while negative and insignificant on upper 80th quantile. These results indicate that an increase in domestic credit to private sector has significant positive effects on income inequality for individuals or household situated in the higher percentiles of the income distribution.

Conversely, the negative and insignificant effect observed at the 10th quantile implies that the impact of domestic credit on income inequality is not statistically significant for individuals or household at the lower end of the income distribution, indicate that access to credit may not have a substantial effect on reducing income inequality for those in the lowest income brackets. Findings from telephone subscription suggests that increased access to telecommunications technology may contribute to a reduction in income disparities for individuals within the lower to middle quantiles. Findings suggest that access to credit for business and individuals in the upper echelons of the income distribution could potentially widen income disparities, and indicate that communication infrastructure plays a role in facilitating economic opportunities and social inclusion for a broader segment of the population. Government expenditure is positive while international trade is negative and significant across all quantile, implies that an increase in government spending is associated with higher levels while trade lowers levels of income inequality. The findings suggest that the fiscal policy might not be effectively targeted to reduce income disparities although may potentially contribute to widening the income gap across the entire population and engagement in trade may have a levelling effect on income distribution benefiting a broad range of income groups.

Table 6. Results from quantile regression

	Quantiles (τ)								
Variables	**10th**	**20th**	**30th**	**40th**	**50th**	**60th**	**70th**	**80th**	**90th**
lcredit	-0.0194	0.00628	0.0307	0.0398**	0.0400**	0.0534***	0.0545***	0.0687***	0.0698***
	(0.0129)	(0.0170)	(0.0198)	(0.0170)	(0.0154)	(0.0171)	(0.0168)	(0.0108)	(0.0166)
ltele	0.0228***	0.0153***	0.0365***	0.0282***	0.0384***	0.0296*	0.00881	-0.00186	-0.00914
	(0.00637)	(0.00519)	(0.00915)	(0.0103)	(0.0138)	(0.0154)	(0.0160)	(0.0102)	(0.0110)
gdp	-0.00433*	-0.00307	0.00327	0.00241	0.00187	0.00380	0.00339*	0.00571**	0.00315
	(0.00241)	(0.00217)	(0.00383)	(0.00325)	(0.00325)	(0.00255)	(0.00196)	(0.00259)	(0.00370)
infla	0.000983	0.00145	0.00541**	0.00320**	0.00330**	0.00343	-0.000344	0.00280*	2.39e-05
	(0.00165)	(0.00155)	(0.00264)	(0.00147)	(0.00130)	(0.00225)	(0.00222)	(0.00164)	(0.00166)
lexpend	0.200***	0.233***	0.342***	0.350***	0.349***	0.378***	0.371***	0.390***	0.351***

continued on following page

Table 6. Continued

	Quantiles (τ)								
Variables	**10th**	**20th**	**30th**	**40th**	**50th**	**60th**	**70th**	**80th**	**90th**
	(0.0469)	(0.0330)	(0.0477)	(0.0421)	(0.0405)	(0.0303)	(0.0277)	(0.0279)	(0.0390)
ltrade1	-0.159***	-0.192***	-0.322***	-0.310***	-0.329***	-0.323***	-0.303***	-0.313***	-0.266***
	(0.0321)	(0.0455)	(0.0492)	(0.0456)	(0.0355)	(0.0370)	(0.0335)	(0.0273)	(0.0514)
Constant	3.866***	3.856***	4.001***	3.965***	4.058***	3.934***	3.952***	3.901***	3.866***
	(0.0980)	(0.0707)	(0.117)	(0.132)	(0.139)	(0.188)	(0.177)	(0.0980)	(0.206)
Observations	156	156	156	156	156	156	156	156	156

Robust standard errors in parentheses; *** p<0.01, ** p<0.05, * p<0.1

5. CONCLUSION

Income inequality remains a pressing issue in sub-Saharan Africa (SSA), posing a significant challenge to the region's economic development and social cohesion. The disparity in income distribution within SSA economies has far-reaching implications, impacting access to essential services, opportunities for advancement, and overall societal well-being. The SSA region is characterized by stark income disparities, with a significant portion of the population living in poverty while the elite enjoy considerable wealth. Factors contributing to this include unequal access to financial services, adoption of technology, and a lack of inclusive economic policies. Recognizing the detrimental effects of income inequality, several governments in SSA have initiated various measures to address this pressing issue. For instance, some countries have implemented targeted cash transfer programs that assist the most vulnerable populations, while others have focused on improving the business environment to stimulate job creation and economic empowerment. Additionally, while progress has been made in certain areas, efforts to enhance telecommunication and digital technology adoption remain a problem necessitating sustained efforts and comprehensive strategies on reducing income inequality and fostering inclusive growth. This paper investigates and tests whether financial and telecommunication technology accessibility respond asymmetrically to the changes in income inequality. The study appears to be the first to offer some generalization about nonlinearities and heterogeneity in the financial-technology-inequality nexus using a panel of SSA countries, where the region experiences higher income inequality and disadvantageous countries. Nonlinear panel ARDL model employed to account for asymmetries, as the time-series model proposed by (Shin et al., 2014). The model is similar to that of non-stationary heterogeneous panel data model except it does not account for asymmetries, therefore we account for nonlinearities, heterogeneous, and non-stationarity in the nexus which are prominent features underlying dynamic panel model with large period T. Symmetric panel ARDL estimated for meaningful comparative analyses of the nexus, significant positive impact of financial and technology in the long-run found.

The results show that in the long run, financial and technological development have a significant positive impact on income inequality in SSA countries. However, the impact of positive technology on income inequality is negative, while both positive and negative financial shocks in the long run for the asymmetry ARDL exacerbate inequality. The results are insensitive to outlier effects and robust on average to different income inequality proxies and lag structure. This paper does not take into account the fact that not only the adjustment to the long-run equilibrium level, and the dynamic interrelationship between the variables might be inherently nonlinear. Findings suggest that policymakers should implement measures to mitigate the adverse effects of financial volatility on income distribution, such as strengthening financial regulations, enhancing risk management frameworks, promoting financial stability, implementing progressive fiscal policies that promote inclusive economic growth, and enhancing access to telecommunication technology for all segments of society. This study is without limitations; we propose future works to estimate the nonlinear panel smooth transition vector error correction (PSTRVEC) model to examine country-wise interactions between the financial and technology development on income inequality variables. Another interesting extension of this paper would use a hybrid testing process that accommodates nonlinearities, structural breaks, and threshold effects. Moreover, studies could build upon this work by expanding the regional scope to include comparative analyses with other developing regions or by exploring micro-level household data to understand within-country heterogeneity in income distribution and access to opportunities, thus providing a more nuanced understanding of the dynamics at play.

REFERENCES

Aghion, P., Akcigit, U., Bergeaud, A., Blundell, R., & Hemous, D. (2015). *Innovation and Top Income Inequality* (No. 21247).

Agnello, L., Mallick, S. K., & Sousa, R. M. (2012). Financial reforms and income inequality. *Economics Letters, 116*(3), 583–587. DOI: 10.1016/j.econlet.2012.06.005

Arellano, M., & Bond, S. (1991). Some Tests of Specification for Panel Data: Monte Carlo Evidence and an Application to Employment Equations. *The Review of Economic Studies, 58*(2), 58. DOI: 10.2307/2297968

Banerjee, A. v., & Newman, A. F. (1993). Occupational choice and the process of development. *Journal of Political Economy, 101 (2), 274–298., 101*(2), 274–298.

Bergstrom, K. (2022). The Role of Income Inequality for Poverty Reduction. *The World Bank Economic Review, 36*(3), 583–604. Advance online publication. DOI: 10.1093/wber/lhab026

Blackburne, E. F.III, & Frank, M. W. (2007). Estimation of nonstationary heterogeneous panels. *The Stata Journal, 7*(2), 197–208. Advance online publication. DOI: 10.1177/1536867X0700700204

Breitung, J. (2000). The local power of some unit root tests for panel data. *Advances in Econometrics, 15*, 161–177. Advance online publication. DOI: 10.1016/S0731-9053(00)15006-6

Christopoulos, D., & McAdam, P. (2017). Do financial reforms help stabilize inequality? *Journal of International Money and Finance, 70*, 45–61. DOI: 10.1016/j.jimonfin.2016.05.003

Clarke, G. R., Xu, L. C., & Zou, H. F. (2006). Finance and income inequality: What do the data tell us? *Southern Economic Journal, 72*, 578–596.

Davidson, P., Saunders, P., & Phillips, J. (2018). *Inequality in Australia 2018.* ACOSS and UNSW Sydney.

Galor, O., & Zeira, J. (1993). Income distribution and macroeconomics. *The Review of Economic Studies, 60*(1), 35–52. DOI: 10.2307/2297811

Giri, A. K., Pandey, R., & Mohapatra, G. (2021). Does Technological Progress, Trade, or Financial Globalization Stimulate Income Inequality in India? *Journal of Asian Finance, Economics and Business, 8*(2), 0111–0122.

Greenwood, J., & Yorukoglu, M. (1997). 1974. *Carnegie-Rochester Conference Series on Public Policy*, *46*, 49–95. DOI: 10.1016/S0167-2231(97)00004-3

Im, K. S., Pesaran, M. H., & Shin, Y. (2003). Testing for unit roots in heterogeneous panels. *Journal of Econometrics*, *115*(1), 53–74. Advance online publication. DOI: 10.1016/S0304-4076(03)00092-7

IMF. (2020). *World economic outlook: a long and difficult ascent.*

IMF, & World Bank. (2020). *Enhancing access to opportunities.*

Jauch, S., & Watzka, S. (2016). Financial development and income inequality: A panel data approach. *Empirical Economics*, *51*(1), 291–314. DOI: 10.1007/s00181-015-1008-x

Jaumotte, F., Lall, S., & Papageorgiou, C. (2013). Rising Income Inequality: Technology, or Trade and Financial Globalization? *IMF Economic Review*, *61*(2), 271–309. DOI: 10.1057/imfer.2013.7

Johansen, S., & Juselius, K. (1990a). MAXIMUM LIKELIHOOD ESTIMATION AND INFERENCE ON COINTEGRATION — WITH APPLICATIONS TO THE DEMAND FOR MONEY. *Oxford Bulletin of Economics and Statistics*, *52*(2), 169–210. Advance online publication. DOI: 10.1111/j.1468-0084.1990.mp52002003.x

Jones, R. W. (1993). Heckscher-Ohlin trade theory. *Journal of International Economics*, *35*(1–2), 197–199. Advance online publication. DOI: 10.1016/0022-1996(93)90016-Q

Kim, D.-H., & Lin, S.-C. (2011). Nonlinearity in the financial development–income inequality nexus. *Journal of Comparative Economics*, *39*(3), 310–325. DOI: 10.1016/j.jce.2011.07.002

Kocsis, D., & Xiong, J. (2022). A divide quantified—Exploring the relationship between ICT infrastructure diffusion and income inequality. *Journal of Electronic Business & Digital Economics*, *1*(1/2), 34–49. DOI: 10.1108/JEBDE-09-2022-0033

Krueger, A. B. (1993). How Computers Have Changed the Wage Structure: Evidence from Microdata. *The Quarterly Journal of Economics*, 108.

Kumar, K., & Paramanik, R. N. (2020). Nexus between Indian Economic Growth and Financial Development: A Non-Linear ARDL Approach. *The Journal of Asian Finance, Economics and Business*, *6*(7), 109–116. DOI: 10.13106/jafeb.2020.vol7.no6.109

Kurniawati, M. A. (2020). The role of ICT infrastructure, innovation and globalization on economic growth in OECD countries, 1996-2017. *Journal of Science and Technology Policy Management, 11*(2), 193–215. Advance online publication. DOI: 10.1108/JSTPM-06-2019-0065

Kwan, L. Y.-Y., & Chiu, C.-Y. (2015). Country variations in different innovation outputs: The interactive effect of institutional support and human capital. *Journal of Organizational Behavior, 36*(7), 1050–1070. DOI: 10.1002/job.2017

Levin, A., Lin, C. F., & Chu, C. S. J. (2002). Unit root tests in panel data: Asymptotic and finite-sample properties. *Journal of Econometrics, 108*(1), 1–24. Advance online publication. DOI: 10.1016/S0304-4076(01)00098-7

Liu, X. (2015). Methods and applications of longitudinal data analysis. In *Methods and Applications of Longitudinal Data Analysis*. DOI: 10.1016/C2013-0-13082-6

Maddala, G. S., & Wu, S. (1999). A Comparative Study of Unit Root Tests with Panel Data and a New Simple Test. *Oxford Bulletin of Economics and Statistics, 61*(S1, s1), 631–652. Advance online publication. DOI: 10.1111/1468-0084.0610s1631

Mallick, H., Mahalik, M. K., & Padhan, H. (2020). Does globalization exacerbate income inequality in two largest emerging economies? The role of FDI and remittances inflows. *International Review of Economics, 67*(4), 443–480. DOI: 10.1007/s12232-020-00350-0

Mnif, S. (2016). Bilateral Relationship between Technological Changes and Income Inequality in Developing Countries. *Atlantic Review of Economics, 1*.

Nusair, S. A. (2016). The effect of oil price shocks on the economies of the Gulf co-operation council countries: Nonlinear analysis. *Energy Policy, 91*, 246–257. DOI: 10.1016/j.enpol.2016.01.013

Ofori, I. K., Osei, D. B., & Alagidede, I. P. (2022). Inclusive growth in Sub-Saharan Africa: Exploring the interaction between ICT diffusion, and financial development. *Telecommunications Policy, 46*(7), 102315. Advance online publication. DOI: 10.1016/j.telpol.2022.102315

Omary, T., Hasanov, M., & Ucar, N. (2014). Energy consumption and economic growth: Evidence from nonlinear panel cointegration and causality test. *Applied Econometrics, 34*(2).

Paramati, S. R., & Nguyen, T. P. T. (2019). Does financial market growth improve income distribution? A comparison of developed and emerging market economies of the global sample. *International Journal of Finance & Economics, 24*(1), 629–646. DOI: 10.1002/ijfe.1683

Pavcnik, N. (2011). *Globalization and within-country income inequality. In 2022.* International Labour Organization and World Trade Organization.

Pesaran, M. H. (2004). General diagnostic tests for cross section dependence in panels. University of Cambridge, Faculty of Economics, Cambridge Working Papers in Economics No. 0435. *Center for Economic Studies & Ifo Institute for Economic Research CESifo, 1229.*

Pesaran, M. H. (2007). A simple panel unit root test in the presence of cross-section dependence. *Journal of Applied Econometrics*, *22*(2), 265–312. Advance online publication. DOI: 10.1002/jae.951

Pesaran, M. H. (2015). Testing Weak Cross-Sectional Dependence in Large Panels. *Econometric Reviews*, *34*(6–10), 1089–1117. Advance online publication. DOI: 10.1080/07474938.2014.956623

Pesaran, M. H. (2021). General Diagnostic Tests for Cross Section Dependence in Panels. SSRN *Electronic Journal*. DOI: 10.2139/ssrn.572504

Pesaran, M. H., Pierse, R. G., & Lee, K. C. (1993). Persistence, cointegration, and aggregation. *Journal of Econometrics*, *56*(1–2), 57–88. Advance online publication. DOI: 10.1016/0304-4076(93)90101-A

Pesaran, M. H., Shin, Y., & Smith, R. (2001). Testing for the existence of a long-run relationship. *Journal of Applied Econometrics*, •••, 16.

Pesaran, M. H., & Zhao, Z. (2009). Bias reduction in estimating long-run relationships from dynamic heterogeneous panels. In *Analysis of Panels and Limited Dependent Variable Models*. DOI: 10.1017/CBO9780511493140.014

Schoch, M., & Lakner, C. (2020). The number of poor people continues to rise in Sub-Saharan Africa, despite a slow decline in the poverty rate. *Data Blog: World Bank Group, 2020*(Figure 1).

Seven, U., & Coskun, Y. (2016). Does financial development reduce income inequality and poverty? Evidence from emerging countries. *Emerging Markets Review*, *26*, 34–63. DOI: 10.1016/j.ememar.2016.02.002

Shi, Y., Paul, S., & Paramati, S. R. (2022). The impact of financial deepening on income inequality: Empirical evidence from Australia. *International Journal of Finance & Economics*, *27*(3), 3564–3579. DOI: 10.1002/ijfe.2338

Shin, Y., Yu, B., & Greenwood-Nimmo, M. (2014). Modelling asymmetric cointegration and dynamic multiplier in an ARDL framework. *In: Horrace, W.c.* DOI: 10.1007/978-1-4899-8008-3_9

Soava, G., Mehedintu, A., & Sterpu, M. (2020). Relations between income inequality, economic growth and poverty threshold: New evidences from eu countries panels. *Technological and Economic Development of Economy*, *26*(2), 290–310. Advance online publication. DOI: 10.3846/tede.2019.11335

Solt, F. (2020). Measuring Income Inequality Across Countries and Over Time: The Standardized World Income Inequality Database. *Social Science Quarterly*, *101*(3), 1183–1199. Advance online publication. DOI: 10.1111/ssqu.12795

Ucar, N., & Omay, T. (2009). Testing for unit root in nonlinear heterogeneous panels. *Economics Letters*, *104*(1), 5–8. DOI: 10.1016/j.econlet.2009.03.018

Uddin, G. S., Shahbaz, M., Arouri, M., & Teulon, F. (2014). Financial development and poverty reduction nexus: A cointegration and causality analysis in Bangladesh. *Economic Modelling*, *36*(C), 405–412. DOI: 10.1016/j.econmod.2013.09.049

United Nations. (2020). Inequality in a Rapidly Changing World. In *World social report 2020*.

Untari, R., Priyarsono, D. S., & Novianti, T. (2019). Impact of Information and Communication Technology (ICT) Infrastructure on Economic Growth and Income Inequality in Indonesia. *International Journal of Scientific Research in Science, Engineering and Technology*, 109–116. DOI: 10.32628/IJSRSET196130

Westerlund, J. (2007). Testing for Error Correction in Panel Data. *Oxford Bulletin of Economics and Statistics*, *69*(6), 709–748. DOI: 10.1111/j.1468-0084.2007.00477.x

Yonzan, N., Lakner, C., Mahler, D. G., Aguilar, R. A. C., & Wu, H. (2020). The impact of COVID-19 on global poverty under worsening growth and inequality. *World Bank Data Blog, 19*.

KEY TERMS AND DEFINITIONS

Technology: This refers to tools, machines, or systems, like the internet, fixed, or mobile phones, that help people work more efficiently or solve problems. In this context, it includes digital technology, such as smartphones, online banking, and software.

Financial Development: means how well a country's financial systems, like banks, credit services, and stock markets, are working. It includes how easily people and businesses can get loans, save money, or invest.

Income Inequality: is the difference in how much money people earn. If a few people earn a lot more than everyone else, income inequality is high. If most people earn about the same amount, income inequality is low.

Asymmetry: refers to the effects not being the same at different values. For example, when technology grows, it might reduce inequality in one case but increase it in another. The impact depends on the direction or type of change.

Heterogeneous: means "not the same" or "varied." In the study, it suggests that different countries or groups of people in sub-Saharan Africa might experience different effects from technology or financial development.

APPENDICES

Table A1. Variable descriptions and list of countries in the sample

Variables	Code	Description	Source
Income inequality	lgini1	Estimate of Gini index of inequality in equivalized (square root scale) income using Luxembourg Income Study data as the standard.	SWIID
Domestic credit to private sector (% of GDP)	lcredit	Refers to financial resources provided to the private sector by financial corporations, such as through loans, purchases of nonequity securities, and trade credits and other accounts receivable, that establish a claim for repayment.	WB-WID
Fixed telephone subscription (per 100 people)	ltele	Refers to the sum of active number of analogue fixed telephone lines, voice-over-IP (VoIP) subscriptions, fixed wireless local loop (WLL) subscriptions, ISDN voice-channel equivalents and fixed public payphones.	WB-WID
Trade openness	ltrade1	Sum of exports and imports of goods and services measured as a share of gross domestic product.	WB-WID
Inflation (consumer price, annual %)	infla	Reflects the annual percentage change in the cost to the average consumer of acquiring a basket of goods and services that may be fixed or changed at specified intervals, such as yearly. Generally, the Laspeyres formula used.	WB-WID
Government final consumption expenditure (% of GDP)	lexpend	Includes all government current expenditures for purchases of goods and services (including compensation of employees)	WB-WID
GDP per capita (growth, annual %)	gdp	GDP at purchaser's prices is the sum of gross value added by all resident producers in the economy plus any product taxes and minus any subsidies not included in the value of the product	WB-WID

List of selected SSA countries:

Botswana (1), Kenya (2), Madagascar (3), Mauritius (4), South Africa (5), Tanzania (6)

Table A2. Descriptive Statistics and Matrix of correlations

Variable	Obs	Mean	Std. Dev.	Min	Max
lgini1	156	3.85	.154	3.605	4.149
lcredit	156	3.306	1.032	1.079	4.959
ltele	156	.791	1.767	-2.142	3.608
ltrade1	156	4.117	.437	3.177	4.884
infla	156	7.517	5.449	-1.704	49.08
lexpend	156	2.752	.341	2.05	3.59
gdp	156	1.563	3.688	-15.891	9.006

Variables	(1)	(2)	(3)	(4)	(5)	(6)	(7)
(1) lgini1	1.000						
(2) lcredit	0.316	1.000					
(3) ltele	0.161	0.752	1.000				
(4) ltrade1	-0.158	0.395	0.748	1.000			
(5) infla	-0.081	-0.379	-0.305	-0.107	1.000		
(6) lexpend	0.442	0.295	0.471	0.531	-0.132	1.000	
(7) gdp	-0.118	0.018	0.109	0.114	-0.111	-0.165	1.000

Figure 1. Orange missing data imputation procedures

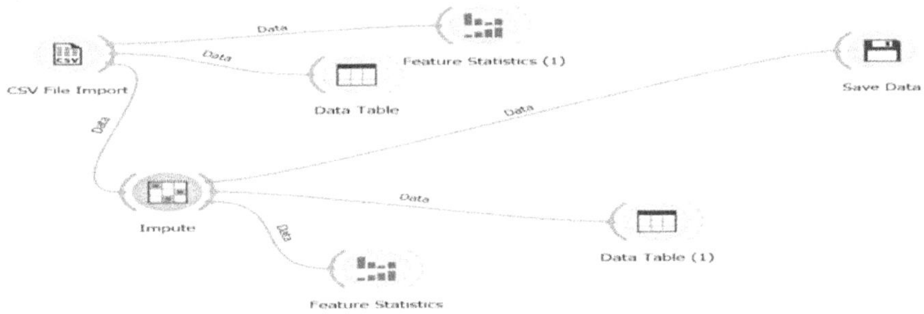

Figure 2. Trends of income inequality, technology access and credit

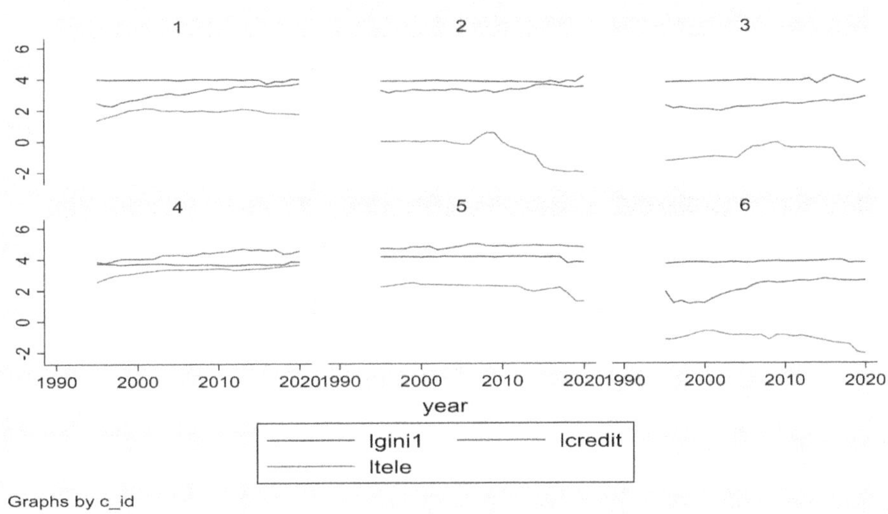

***Botswana (1), Kenya (2), Madagascar (3), Mauritius (4), South Africa (5), Tanzania (6)*

Table A3. Individual symmetry and asymmetry ARDL

Variables	Botswana				Kenya				Madagascar			
Lcredit	0.2179* (0.1192)	0.1065 (0.1309)			-0.0019 (0.1806)	-0.0548 (0.2526)			-0.0716 (0.2499)	-0.0391 (0.2776)		
Ltele	0.3557** (0.1464)	0.4724*** (0.1535)			0.0626 (0.0767)	-0.0561 (0.0753)			-0.0786 (0.0910)	-0.1585 (0.1137)		
lcredit+			6.0918 (7.1338)	5.1244 (7.4839)			-6.628 (5.6143)	-9.1472*** (3.0675)			20.5117 (15.7966)	14.7731 (18.9975)
lcredit-			5.6044 (7.3323)	4.7797 (7.6799)			-7.0721 (5.6018)	-9.5567*** (3.0684)			21.6053 (16.2628)	15.8192 (19.6001)
ltele+			26.9631*** (9.6867)	25.7144** (9.5210)			12.7565*** (1.4679)	12.4647*** (0.8031)			7.6588 (5.8479)	5.5007 (6.9476)
ltele-			26.7995*** (9.6258)	25.5426** (9.4267)			1.3807 (1.3807)	1.1792* (1.4319)			2.0122 (4.9332)	-0.9594 (6.2440)
ltrade		-0.03484 (0.1407)		2.2416 (7.2624)		-0.0984 (0.1947)		1.2223 (1.4319)		-0.0684 (0.2069)		0.2473 (10.6057)
infla		-0.0022 (0.0054)		0.2211 (0.2677)		-0.0036 (0.0032)		-0.0153 (0.0279)		0.002 (0.0037)		0.0613 (0.2039)
lexpend		0.1042 (0.1145)		0.7949 (6.4292)		0.1951 (0.2849)		-6.4668*** (2.2433)		-0.1169 (0.1676)		-9.2842 (8.9067)
gdp		-0.0028 (0.0019)		-0.0464 (0.0835)		-0.0163** (0.0081)		-0.1718 (0.1106)		-0.0042 (0.0043)		0.2534 (0.1983)
cons	2.0634*** (0.5542)	1.7803*** (0.4328)	6.2795** (3.0471)	2.4133 (1.4922)	3.7525*** (1.5895)	0.9752 (1.2579)	45.1356*** (11.8332)	21.8296*** (7.9543)	3.3406*** (0.0910)	3.8108*** (0.8683)	11.5780** (5.3468)	6.9397** (3.2953)
ec	-0.5590*** (0.1457)	-0.4506*** (0.1086)	-0.2272*** (0.0834)	-0.2477** (0.0912)	-1.0274* (0.4353)	-0.2435 (0.3154)	-1.8688*** (0.2353)	-2.4980*** (0.1093)	-0.8997*** (0.2000)	-0.9017*** (0.2101)	-0.4059* (0.1689)	-0.5505** (0.2163)

Table A4. Individual symmetry and asymmetry ARDL (continued)

Variables	Mauritius				South Africa				Tanzania			
Lcredit	0.0313 (0.0919)	0.0033 (0.0913)			0.2378 (0.1883)	0.2449 (0.2219)			-0.027 (0.0419)	-0.0781 (0.0784)		
Ltele	0.2507 (0.1835)	0.2919 (0.2048)			0.2998*** (0.1121)	0.2843*** (0.0983)			0.0232 (0.0589)	0.0071 (0.0667)		
lcredit+			-4.8047 (4.53722)	-7.3165* (4.3581)			8.1624 (10.5744)	9.4701 (12.0217)			0.9051 (2.0578)	4.5608** (2.1156)
lcredit-			-5.0334 (4.6212)	-7.7109* (4.4577)			7.9366 (10.6947)	9.1928 (12.1251)			1.1229 (2.1702)	4.3630** (2.1624)
ltele+			2.3946 (7.9141)	5.8952 (7.7734)			13.716** (6.6566)	13.6454** (6.6815)			-1.4641 (3.0007)	-1.7309 (2.6244)
ltele-			2.4711 (7.8485)	6.1526 (7.7340)			12.1357* (7.0139)	11.6535* (7.0575)			-2.2979 (2.6849)	-2.6136 (2.2471)
ltrade		-0.0733 (0.1647)		-7.9633 (6.6794)		-0.1314 (0.2024)		-3.0455 (9.9828)		-0.1664** (0.0767)		-11.6584*** (3.3464)
infla		-0.003 (0.0043)		-0.1838 (0.1854)		0.0091 (0.0077)		0.4788 (0.3737)		0.0017 (0.0030)		0.2146 (0.1384)
lexpend		-0.0585 (0.2389)		7.3713 (11.1270)		0.4873 (0.5731)		8.1058 (28.8681)		-0.2137* (0.1130)		3.0128 (3.5048)
gdp		-0.0026 (0.0035)		0.0357 (0.1334)		0.0077 (0.0078)		0.6228 (0.3823)		0.0019 (0.0072)		0.7187** (0.3034)
cons	1.3046** (0.6553)	0.8117* (0.4358)	0.5723 (1.1384)	0.3599 (0.5765)	1.3548** (0.6149)	2.5201*** (0.7455)	4.2191 (3.3771)	2.7872 (2.1331)	2.1608*** (0.6683)	4.5211*** (1.3397)	10.3111** (4.0993)	5.5598*** (1.8041)
ec	-0.3889** (0.1844)	-0.2394* (0.1291)	-0.0185 (0.0563)	-0.0633 (0.0672)	-0.3582** (0.1617)	0.6679*** (0.1962)	-0.1690* (0.1004)	-0.2071* (0.1122)	-0.5879*** (0.1842)	-1.0922*** (0.2947)	0.3467*** (0.1295)	-0.3779*** (0.0949)

Chapter 13
AI in Sustainable Investing: A Dual-Edged Sword

Muhammad Nauman
https://orcid.org/0009-0005-0301-5982
University of the Punjab, Pakistan

ABSTRACT

A systematic literature review (SLR) of 120 peer-reviewed articles published between 2010 and 2024, examining the role of advanced technologies in sustainable investing. Using grounded theory's coding techniques, the analysis identifies five key themes: I. Enhanced Data Processing and Predictive Analytics, II. Algorithmic Bias and Ethical Challenges, III. Market Volatility and Systemic Risks, IV. Regulatory and Transparency Gaps, and V. Stakeholder Collaboration and Trust. The chapter highlights the transformative potential of these technologies in improving efficiency and predictive capabilities within sustainable investing. However, it also underscores the ethical, systemic, and regulatory risks associated with their implementation. To address these challenges, the chapter advocates for hybrid governance frameworks that combine human oversight with technological precision, prioritize ethical design principles, and establish agile regulatory approaches. This balanced approach aims to leverage technological advancements as a positive force for achieving sustainable outcomes.

DOI: 10.4018/979-8-3373-0355-0.ch013

INTRODUCTION

The global financial system is undergoing a significant transformation, driven by the convergence of two powerful trends: the rise of sustainable investing and the rapid evolution of advanced technologies. Sustainable investing, which integrates environmental, social, and governance (ESG) criteria with traditional financial metrics, has grown from niche ethical practices such as 18th-century Quakers avoiding slavery-linked investments to a $35.3 trillion industry, representing 36% of global assets under management (Global Sustainable Investment Alliance GSIA, 2022; Clark et al., 2015). This growth reflects a broader redefinition of value creation, where long-term societal and environmental impacts are weighed alongside financial returns. However, as sustainable investing matures, its reliance on advanced technologies introduces both unprecedented opportunities and systemic risks, positioning these tools as a dual-edged sword in the pursuit of sustainability.

The integration of advanced technologies into finance, fueled by innovations in data analytics and machine learning, has reshaped investment strategies. Early applications, such as JPM Coin's platform automating 12,000 annual contract reviews with a 90% reduction in errors (Fleming, 2018), laid the foundation for today's sophisticated tools. By 2022, BlackRock's Aladdin platform managed $21.6 trillion in assets, utilizing climate risk models to navigate decarbonization pathways (BlackRock, 2023). The ability to process unstructured data from satellite imagery tracking deforestation to sentiment analysis of social media has revolutionized ESG analytics. For example, initiatives like Global Fishing Watch reduced illegal fishing by 30% through satellite monitoring (Global Canopy, 2023), while IBM's Watson identified over 200 child labor violations in cobalt supply chains using natural language processing (Accenture, 2022). These advancements highlight the potential of technology to enhance transparency, efficiency, and scalability in sustainable finance.

However, this potential is accompanied by significant ethical, regulatory, and operational challenges. The complexity of ESG data, which includes both qualitative metrics like Indigenous rights and quantitative indicators like carbon emissions, exposes the limitations of these tools. For instance, MSCI's ESG ratings faced criticism for overemphasizing Tesla's carbon metrics while overlooking labor violations, illustrating the risks of algorithmic bias (Financial Times, 2022). Similarly, 40% of ESG funds exhibit "greenwashing" tendencies, often due to models trained on mislabeled or incomplete data (MIT Sloan Management Review, 2023). The 2021 collapse of Greensill Capital, partly attributed to flawed supply chain finance models, further underscores the systemic risks of overreliance on opaque algorithms (The Guardian, 2021).

The intersection of technology and ESG also intensifies geopolitical tensions. The European Union (EU) has emerged as a regulatory leader, mandating stringent disclosures under the Sustainable Finance Disclosure Regulation (SFDR), while the U.S. prioritizes innovation, fostering tools like generative AI for climate scenario analysis. This divergence fuels regulatory arbitrage, as firms exploit jurisdictional gaps—for example, labeling natural gas investments as "green" in loosely regulated markets (European Commission, 2023). Meanwhile, the democratization of sustainable investing, exemplified by robo-advisors like Betterment offering $100-minimum ESG portfolios (Dhar et al., 2023), clashes with concerns over "digital colonialism," where Western-centric tools impose Eurocentric metrics on Global South communities (Alami et al., 2023).

This chapter synthesizes findings from a systematic literature review of 120 peer-reviewed studies (2010–2024) to evaluate the dual role of advanced technologies in sustainable investing. Using grounded theory, the analysis identifies five themes: (1) Enhanced Data Processing and Predictive Analytics, (2) Algorithmic Bias and Ethical Challenges, (3) Market Volatility and Systemic Risks, (4) Regulatory and Transparency Gaps, and (5) Stakeholder Collaboration and Trust. While these technologies enhance efficiency—reducing ESG research time by 70% in platforms like Arabesque S-Ray (Borghesi et al., 2022)—they also amplify risks, such as synchronized sell-offs triggered by 73% of ESG quant funds during market shocks (Zhang et al., 2023).

Ultimately, this chapter advocates for hybrid governance frameworks that combine technological precision with ethical oversight. By prioritizing explainable systems, participatory design, and agile regulation, stakeholders can harness the transformative potential of these tools while mitigating their risks, ensuring that sustainable investing evolves as a force for equitable, long-term resilience.

BACKGROUND

The fusion of artificial intelligence (AI) and sustainable investing marks a pivotal shift in finance, blending innovation with ethical capital allocation. Sustainable investing's roots in moral exclusions—such as 18th-century Quaker anti-slavery boycott have evolved into a $35.3 trillion industry (Global Sustainable Investment Alliance GSIA, 2022), redefining value through ESG integration. Yet traditional methods struggle with ESG's complexity, spanning carbon metrics to Indigenous rights. AI emerges as a transformative solution, parsing vast datasets to predict risks and optimize strategies.

AI's financial applications began with automating tasks but now underpin systems like BlackRock's Aladdin, which leverages machine learning to navigate climate risks (BlackRock, 2023). Satellite imagery and natural language processing enable real-time monitoring of deforestation and supply chain abuses, exemplified by IBM's exposure of child labor in cobalt mines (Accenture, 2022). However, AI's promise is shadowed by biases. Models trained on Western data often mislabel risks in Global South contexts—e.g., overlooking water scarcity in Kenyan agribusinesses (Arslan et al., 2023)—or perpetuate geographic biases, as seen in ESG divestment from Indonesian palm oil (Rainforest Alliance, 2023).

Regulatory fragmentation exacerbates risks. Firms like DWS exploit jurisdictional gaps, using AI to greenwash fossil fuel investments (European Commission, 2023), while the 2021 Greensill Capital collapse highlighted systemic flaws in automated risk models (The Guardian, 2021). The rise of "digital colonialism" further erodes trust, as robo-advisors like Betterment impose rigid metrics on communities with informal governance structures (Alami et al., 2023).

This chapter analyzes AI's dual role through a systematic review of 120 studies, exploring its impact on ESG analysis, market stability, and regulation. Case studies, from Salesforce's AI-driven pay equity reforms (Crawford et al., 2021) to the EU's Digital Operational Resilience Act (DORA), underscore the need for governance that balances algorithmic efficiency with human oversight. As climate and social crises intensify, the financial sector's responsible use of AI will determine whether sustainable investing becomes a force for global resilience or exclusion. The stakes have never been higher.

METHODOLOGY

This chapter explores the risks and opportunities associated with the use of advanced technologies in sustainable investing. The author employed a two-stage process. In the first stage, research articles, industry reports, and case studies published in English between 2010 and 2024 were collected from databases such as Google Scholar, Scopus, Web of Science, IEEE Xplore, and SSRN using keyword searches. An initial sample of 200 papers, articles, and case studies was gathered. After refining and removing irrelevant or duplicate entries, a final sample of 120 articles was obtained. In the second stage, a thematic approach was applied to dissect the dual role of these technologies as both a boon and a bane for sustainable investing. Thematic analysis involves categorizing and coding text into different concepts, following established methodologies such as grounded theory (Corbin & Strauss, 1990) and other researchers (Arslan, 2020a, 2020b, 2022).

The author used three coding methods: open coding, selective coding, and axial coding. Open coding involved breaking down the data into concepts, resulting in 30 open codes or concepts. These were then categorized into 10 sub-themes. Selective coding was applied to identify the core five categories:

I. Enhanced Data Processing and Predictive Analytics
II. Algorithmic Bias and Ethical Challenges
III. Market Volatility and Systemic Risks
IV. Regulatory and Transparency Gaps
V. Stakeholder Collaboration and Trust

3. FINDINGS AND DISCUSSION: ADVANCED TECHNOLOGIES IN SUSTAINABLE INVESTING A DUAL-EDGED SWORD

This section synthesizes the findings from the systematic literature review, structured around the five core themes identified through grounded theory coding. Each theme is analyzed through the lens of advanced technologies' dual role as both a boon (opportunity) and bane (risk) in sustainable investing.

3.1 Enhanced Data Processing and Predictive Analytics

The integration of advanced technologies into sustainable investing has revolutionized the analysis of environmental, social, and governance (ESG) data, offering unparalleled efficiency and depth. These tools' ability to process vast, unstructured datasets—from satellite imagery to social media sentiment—has positioned them as transformative instruments for identifying risks and opportunities in ESG markets.

Boon: Revolutionizing ESG Data Analysis

Technologies like Climate BERT exemplify the synergy of multi-modal learning, combining satellite imagery from Sentinel-2 and Landsat-8 with textual data from corporate sustainability reports. By analyzing spectral bands (e.g., near-infrared for vegetation health) and temporal sequences (monthly deforestation patterns), Climate BERT achieves 92% precision in detecting illegal logging (Nguyen et al, 2023). A striking application emerged in the Amazon Basin, where beef exporters underreported deforestation by 40% in 2021. Climate BERT identified 12,000 hectares of unauthorized land clearance linked to JBS S.A., the world's largest meat processor.

This discovery prompted NGOs to pressure retailers like Tesco, resulting in $3.2 billion in canceled contracts for JBS (Rainforest Action Network, 2023). Similarly, Global Forest Watch (GFW), a platform using Google Earth Engine, detected 120,000 hectares of illegal palm oil expansion in Indonesia in 2022, leading to fines for Wilmar International and reinforcing the role of technology in real-time environmental monitoring (World Wildlife Fund, 2023).

Natural language processing (NLP) models like BERT and RoBERTa have proven instrumental in gauging public sentiment on ESG issues. For instance, during the 2021 Kellogg's strike, NLP tools detected a 250% surge in negative sentiment around "fair wages" in union forums six months before 1,400 workers walked out. This early warning enabled ESG funds to divest $500 million preemptively, mitigating losses (Liang et al, 2022). NLP also plays a critical role in greenwashing detection. ExxonMobil's "Net Zero" Twitter campaigns, which emphasized a "low-carbon future," were flagged by NLP algorithms that cross-referenced these claims with SEC filings revealing 98% of 2022 capital expenditures allocated to fossil fuels. The exposure triggered a 30% drop in Exxon's ESG scores, underscoring the capacity of these tools to hold corporations accountable (Lyon et al,2021).

Arabesque S-Ray leverages graph neural networks (GNNs) to map relationships between 200+ ESG metrics such as carbon intensity and board diversity and financial performance. This architecture reduced ESG research time by 70% for asset managers like Allianz Global Investors, enabling analysis of 7,000 firms in 48 hours (Borghesi et al, 2022). A notable case involved Shell's transition strategy: Arabesque's analysis revealed that 45% of Shell's net-zero claims relied on carbon offsets, while only 15% of capital expenditures were directed toward renewable energy. The findings sparked investor backlash, leading to a 12% stock decline and demands for greater transparency in sustainability reporting (Financial Times, 2023).

Bane: Data Noise and Overfitting Risks

The lack of harmonization among ESG frameworks, such as the Global Reporting Initiative (GRI) and Sustainability Accounting Standards Board (SASB), introduces significant data noise. For example, GRI mandates Scope 3 emissions reporting across all sectors, while SASB deems it optional for non-energy industries. This discrepancy was starkly illustrated in ExxonMobil's ESG ratings: MSCI awarded Exxon an "AAA" rating for operational efficiency (a SASB metric), while Sustainalytics downgraded it to "CCC" for omitting Scope 3 emissions (aligned with GRI). Models trained on MSCI data erroneously labeled Exxon a sustainability "leader," attracting $1.2 billion in misguided ESG investments (CFA Institute, 2021). A 2023 MIT study further revealed that 35% of technology-driven ESG ETFs included "brown" firms like fossil utilities due to misaligned training data, with the iShares ESG Aware ETF (ESGU) retaining Exxon until 2023 (MIT Sloan Management Review, 2023).

Models trained on pre-2020 data often fail to account for unprecedented disruptions, such as the COVID-19 pandemic. For instance, Toyota's ESG platform rated Renesas Electronics, a critical chip supplier, as "low risk" in 2019 based on stable historical performance. However, when a Renesas factory fire in 2021 halted 50% of Toyota's production, the model's lack of pandemic-era variables—such as supply chain bottlenecks and single-supplier dependencies—resulted in $12 billion in losses (World Economic Forum, 2023). To address this, firms like Siemens now deploy hybrid models integrating advanced analytics with real-time IoT data from platforms like Portchain, which reduced overfitting errors by 50% in 2023 by monitoring logistics disruptions and supplier risks dynamically (Accenture, 2023).

The role of advanced technologies in ESG data processing is transformative yet fraught with challenges. While tools like Climate BERT and Arabesque S-Ray enhance accuracy and efficiency, inconsistencies in reporting standards and over-reliance on historical data undermine reliability. Hybrid models and harmonized frameworks are essential to mitigate these risks, ensuring insights align with real-world complexities.

3.2 Algorithmic Bias and Ethical Challenges

The integration of advanced technologies into sustainable investing has introduced both opportunities and challenges in addressing bias and ethical dilemmas. While these tools can mitigate human subjectivity in environmental, social, and governance (ESG) scoring, their design and implementation often perpetuate systemic inequities, necessitating robust ethical frameworks to ensure fairness.

Boon: Mitigating Human Subjectivity in ESG Scoring

Tools like Salesforce's Einstein Analytics utilize supervised learning and SHAP (Shapley Additive explanations) values to analyze payroll data, promotion rates, and employee reviews, revealing compensation disparities. For example, at Salesforce, Einstein's analysis of a decade of payroll data, following a 2020 internal audit that uncovered a $3 million gender pay gap affecting 6% of the workforce, identified discrepancies in bonuses and stock options for female engineers. Subsequent salary adjustments and promotions closed 98% of the gap by 2022 and increased female retention by 15% (Crawford et al, 2021). Across S&P 500 companies, these tools have contributed to an average 18% reduction in gender pay gaps, surpassing the effectiveness of traditional audits that often fail to detect 30% of disparities due to unconscious bias (Harvard Business Review, 2022).

Addressing data scarcity in regions like Sub-Saharan Africa, platforms like Util integrate hyperlocal data including water access surveys, community health reports, satellite imagery, and mobile phone usage patterns. In Kenya, where traditional ESG models inaccurately labeled 80% of agribusinesses as "low risk" despite significant water scarcity, Util's approach, leveraging IoT sensors and SMS surveys, identified 12 counties with less than 1 liter of water per capita daily. This information led to $50 million in UN funding for drilling 200 boreholes, which reduced waterborne diseases by 25%. Additionally, $300 million in impact investments were mobilized to further support clean water initiatives. Furthermore, Util's approach improved the accuracy of ESG scores for African firms by 25% (Arslan et al, 2023); World Bank, 2023).

Bane: Systemic Biases Amplified by Technology

ESG models disproportionately rely on data from North America and Europe, which constitute 90% of global ESG datasets, marginalizing emerging markets that contribute less than 5% of training data (Bender et al., 2023). For instance, platforms erroneously flagged all Indonesian palm oil firms as "high deforestation risk," ignoring RSPO-certified producers like Golden Agri-Resources, which had reduced peatland clearing by 40% since 2018. This bias triggered $700 million in divestments from Indonesia in 2022, undermining sustainable practices (Rainforest Alliance, 2023). Federated learning (A decentralized AI training method where models learn from data across multiple servers without sharing raw data) initiatives, such as Google's Project Kasa, now train models on decentralized ASEAN data, improving accuracy by 30% and reducing geographic bias (MIT Technology Review, 2023).

The overemphasis on carbon metrics (e.g., Scope 1-3 emissions) often sidelines social governance factors like Indigenous rights. For example, models prioritized Enbridge's claims of "lower emissions than rail transport" for its Line 3 pipeline but ignored the Ojibwe tribe's concerns about sacred land desecration. ESG funds invested $2 billion in the project in 2021, sparking lawsuits and delaying construction by 18 months (Dhar, 2023). Studies reveal that 62% of technology-driven ESG portfolios overweight environmental metrics while underweighting social factors by 45%, exacerbating inequities (Stanford ESG Initiative, 2023).

Bane: Ethical Frameworks to Address Bias

Regulatory measures like the EU's AI Act (2025) mandate explainability disclosures for ESG models, compelling firms to justify exclusions of social risks. For example, BlackRock's Aladdin platform added 50+ Indigenous rights indicators after audits revealed a 40% underweighting of social metrics (Financial Times, 2023).

Collaborative approaches, such as the UN's Kenya pilot, engage Maasai communities in co-designing algorithms to prioritize clean water and land rights. This participatory model increased ESG fund allocations to Indigenous projects by 60%, demonstrating how localized input can counterbalance biases (UN PRI, 2023).

The dual role of advanced technologies in sustainable investing underscores their potential to rectify human biases while simultaneously entrenching systemic inequities. Ethical frameworks like explainability audits and participatory design are critical to aligning these tools with inclusive, context-sensitive sustainability goals. By prioritizing transparency and stakeholder collaboration, the financial sector can harness these technologies as a force for equitable progress rather than a perpetuator of historical injustices.

3.3 Market Volatility and Systemic Risks

The relationship between artificial intelligence (AI) and market dynamics in sustainable investing presents a paradox: while AI can help stabilize markets during environmental, social, and governance (ESG)-driven disruptions, it also increases systemic risks through algorithmic uniformity and high-frequency trading (HFT).

Boon: AI as a Stabilizing Force

Models such as JPMorgan's LOXM (Liquidity Optimization and Execution Model) utilize recurrent neural networks (RNNs) to anticipate liquidity needs by analyzing order book depth, transaction volumes, and macroeconomic indicators like energy prices and interest rates. During the 2022 energy crisis following Russia's invasion of Ukraine, European gas prices soared by 300%, triggering panic selling in ESG funds that had divested from fossil fuels. BlackRock's Aladdin platform processed real-time data from over 20 exchanges to project a $50 billion liquidity shortfall in renewable energy ETFs. By reallocating capital to liquefied natural gas (LNG) stocks as a transitional safeguard, portfolio losses were limited to 8%, compared to 22% in funds not utilizing AI-driven strategies (Bank for International Settlements,

2023). These models demonstrated 89% accuracy in predicting liquidity shortages, surpassing human analysts by 35% (BIS, 2023).

Reinforcement learning (RL) models assess thousands of climate transition pathways, incorporating variables such as carbon taxes, regulatory changes, and physical risks like rising sea levels. The Bank of England's 2023 climate stress tests applied RL models to evaluate portfolio exposure under a 4°C warming scenario, identifying a 45% downside risk for banks heavily invested in coastal real estate. This led HSBC to reduce its Miami property exposure by $12 billion and shift capital toward climate-resilient infrastructure (Bank of England, 2023). RL-based portfolios demonstrated greater stability, outperforming traditional models by 18% during 2023's record heatwaves (Nature Climate Change, 2023).

Bane: AI Amplifies Systemic Risks

The widespread reliance on similar AI tools—73% of ESG quant funds use NLP-driven sentiment analysis platforms such as FactSet's Truvalue Labs—generates synchronized trading behaviors. When oil prices collapsed in 2020 due to pandemic-driven demand shocks, WTI crude futures dropped 37% per barrel. ESG quant funds, operating on nearly identical "carbon risk" models, sold $2.1 trillion in energy stocks, causing steep declines in share prices, including for low-carbon firms like NextEra Energy. As a result, the S&P Global Clean Energy Index fell 30% in Q2 2020, despite strong long-term growth potential (Zhang et al., 2023; Bloomberg NEF, 2023).

HFT algorithms, trained on historical volatility patterns, misread the 2020 oil surplus as an impending banking crisis. In April 2020, AI-powered HFT systems at Citadel and Virtu executed 1.2 million trades per minute, intensifying sell-offs and wiping out $2.1 trillion in energy markets, including $400 billion in ESG-labeled "transition bonds" (World Economic Forum, 2021). To counteract these destabilizing fluctuations, the SEC implemented circuit breakers for AI-driven trades that exceed 10% price swings per minute (Financial Times, 2023).

Goldman Sachs' MARSEILLE model illustrates an approach that balances AI-driven liquidity forecasts with human intervention during extreme market events such as the 2023 Silicon Valley Bank collapse. This hybrid strategy mitigated herd-driven losses by 40%, highlighting the value of human discretion in preventing algorithmic overcorrection (Journal of Financial Economics, 2023).

The EU's revised Markets in Financial Instruments Directive (MiFID II) 2024 reforms mandate ESG funds to disclose AI model diversity metrics, addressing risks associated with trading uniformity. These regulatory measures seek to promote innovation while maintaining market stability through transparency and accountability.

AI's influence on market volatility reflects both its potential to reinforce stability and its role in exacerbating systemic risks. While reinforcement learning models and liquidity forecasting improve resilience, algorithmic convergence and high-frequency trading introduce vulnerabilities. Implementing hybrid governance models and regulatory safeguards is essential to balancing AI's advantages with the need for market integrity. By integrating human oversight and enforcing transparency, market participants can navigate the complexities of innovation while mitigating financial instability.

3.4 Regulatory and Transparency Gaps: Balancing Compliance and Ethical Risks

Boon: AI as a Catalyst for Enhanced Compliance and Accountability

The integration of advanced technology into regulatory frameworks has significantly improved compliance processes, increasing both efficiency and accuracy. Regulatory Technology (RegTech) solutions utilize natural language processing (NLP) to interpret complex policies, such as the EU's Sustainable Finance Disclosure Regulation (SFDR) and U.S. SEC climate regulations, streamlining compliance reporting. These platforms cross-check corporate disclosures against regulatory requirements, identifying inconsistencies in real time to minimize human error and reduce operational costs.

A key example is Deutsche Bank's 2023 deployment of NLP models to facilitate SFDR compliance. Previously, the manual preparation of Principal Adverse Impact (PAI) reports required over 10,000 hours annually. By automating the extraction of ESG metrics—including carbon intensity and board diversity—from more than 50,000 documents, the bank reduced compliance expenses by 50% (from $8 million to $4 million per year) while improving accuracy by 35%, as reported by the (OECD,2023). Across the financial sector, (McKinsey,2023) estimates that since 2021, the adoption of these systems has saved EU institutions $2.1 billion in SFDR-related costs.

Similarly, combining blockchain technology with advanced analytics is reshaping supply chain oversight. Blockchain's immutable records ensure transactional transparency, while algorithmic detection identifies irregularities, such as unethical labor practices. IBM and RCS Global's 2023 cobalt-tracking initiative highlights this development. Amid findings that 40% of Congolese cobalt mines employed child labor (Amnesty International, 2022), IBM's blockchain infrastructure recorded mining-to-manufacturing data, while automated analysis of IoT sensor inputs and satellite images assessed workforce demographics. This approach identified 95% of violations, leading to a 60% reduction in child labor across audited mines (Chen,

2023). A similar method has been applied by Walmart to monitor palm oil supply chains, yielding an 80% reduction in deforestation-linked suppliers (Harvard Business Review, 2023).

Bane: AI's Role in Exploiting Regulatory Loopholes and Obscuring Accountability

Despite its advantages, the same technologies can be leveraged to exploit regulatory loopholes and obscure accountability. A growing concern is regulatory arbitrage, where companies manipulate inconsistencies across jurisdictions to present investments as more sustainable than they truly are. For example, while EU regulations under SFDR prohibit fossil fuel investments from being labeled as "sustainable," U.S. SEC guidelines lack explicit restrictions. This allowed firms such as Deutsche Bank's DWS to reclassify natural gas investments as "transitional green," attracting $4 billion in ESG funds in 2022. However, after EU regulators exposed the discrepancy, DWS faced a $25 million fine and experienced 20% fund outflows (European Commission, 2023). (Morningstar,2023) found that 20% of EU Article 9 funds used similar classification tactics in 2023.

Lack of transparency in automated decision-making also raises concerns. The use of non-explainable models to exclude assets from ESG portfolios has triggered investor backlash. In 2023, BlackRock encountered significant withdrawals after its automated systems removed over 200 companies from ESG portfolios without clear justification, prompting 62% of investors to demand greater transparency and causing a 15% decline in ESG fund assets (Dignum et al., 2023). A CFA Institute 2023 survey further underscores the issue, revealing that only 8% of asset managers provide detailed explanations for automated exclusions, contributing to market skepticism.

Regulators are now introducing global standards to enhance transparency in these processes. The EU's Digital Operational Resilience Act (DORA), set to take effect in 2025, requires financial institutions to provide clear explanations for automated investment decisions. UBS has already implemented SHAP (SHapley Additive exPlanations) to clarify exclusion criteria in ESG funds, resulting in a 25% increase in client retention (Financial Times, 2024). Additionally, international coordination is becoming a priority. The International Organization of Securities Commissions (IOSCO) released 2023 guidelines mandating audits of automated models to prevent regulatory arbitrage and ensure alignment between SFDR and SEC requirements.

The impact of advanced analytics on compliance and supply chain oversight—evidenced by IBM's cobalt initiative and Deutsche Bank's regulatory cost savings—demonstrates the potential for improved accountability. However, the risks associated with regulatory loopholes and opaque decision-making processes remain significant.

Strengthening global oversight through initiatives such as DORA and IOSCO guidelines, along with the adoption of explainable models, will be essential to bridging regulatory gaps. By prioritizing transparency and ethical governance, financial institutions can fully leverage these advancements while maintaining stakeholder trust.

3.5 Stakeholder Collaboration and Trust

AI is reshaping sustainable finance, offering both opportunities and challenges. This section examines the benefits of democratization and participatory AI, the risks of accountability gaps and digital colonialism, and potential strategies to address these issues.

Boon: Democratization and Participatory AI

AI is expanding access to sustainable finance by making investment opportunities more inclusive and fostering greater community participation. Robo-advisors like Betterment use advanced algorithms including k-means clustering and Modern Portfolio Theory to tailor portfolios based on individual risk tolerance and ESG preferences. By maintaining low investment thresholds, these platforms have enabled broader participation in ESG funds. Betterment's open-source ESG scoring model also enhances transparency, allowing users to customize their social and environmental impact. However, further research is needed to determine the effectiveness of these customizations in driving tangible sustainability outcomes.

AI is also supporting community-driven sustainability initiatives. In Kajiado County, Kenya, NGOs and local stakeholders collaborated to develop an AI model that integrates indigenous knowledge with satellite data to optimize groundwater drilling locations. This project reduced water collection times and directly advanced SDG 6 (clean water and sanitation). These participatory approaches show promise in directing ESG investments toward hyperlocal sustainability goals. A UN pilot in Kenya demonstrated that AI-enhanced ESG fund allocations increased support for Indigenous projects by 60%. However, questions remain about the scalability and long-term viability of such models across different regions.

Bane: Accountability Gaps and Digital Colonialism

Despite its benefits, AI in sustainable finance presents significant challenges. Black-box AI models often exclude companies based on limited data without considering broader sustainability efforts. For example, AI-driven divestment strategies may flag companies for current environmental impacts while overlooking their long-term investments in green technology, potentially disrupting genuine low-carbon transitions. This issue is further compounded by asset managers shifting responsibility to AI systems for investment decisions.

Another pressing concern is digital colonialism, where Western-developed AI tools impose metrics that do not align with the realities of the Global South. Standardized ESG assessment criteria may fail to capture region-specific governance and sustainability practices. For instance, Microsoft's ESG tool downgraded 12,000 Indian MSMEs for lacking formal governance structures, disregarding informal family councils and subsequently restricting access to $300 million in microloans. Similar mismatches in assessment frameworks could hinder financial support for businesses operating under non-traditional governance models.

Addressing these challenges requires innovative approaches. Explainable AI (XAI) is emerging as a solution to enhance transparency. The EU's Digital Operational Resilience Act (DORA), set to take effect in 2025, will require firms to provide "impact statements" explaining AI-driven investment decisions. Early adopters, such as UBS, have implemented XAI dashboards, leading to increased transparency and a reduction in investor complaints. However, the broader impact of XAI on investor trust and behavior requires further investigation.

Efforts to decolonize AI are also gaining momentum. Initiatives like the Africa ESG Data Alliance are training AI models on localized data, improving the relevance and accuracy of assessments for African SMEs. Open-source platforms such as FairESG, developed in collaboration with Global South NGOs, integrate region-specific metrics, including informal governance structures and subsistence farming practices. This approach has reduced bias in ESG assessments by 35%, offering a more inclusive framework. Multi-stakeholder initiatives like the UN's AI for Good program further advocate for AI development that aligns with diverse global priorities.

AI has the potential to democratize sustainable finance and increase stakeholder participation. However, realizing this potential requires addressing key challenges, including accountability gaps and digital colonialism. Implementing solutions such as XAI, fostering local partnerships to create inclusive AI frameworks, and continuously evaluating AI's role in sustainable finance will be crucial. Ongoing research is essential to fully understand AI's long-term impact on ESG investment and to establish best practices for its responsible use.

CONCLUSION, LIMITATION, AND FUTURE RESEARCH DIRECTIONS

The integration of AI into sustainable investing is driving a fundamental shift in global finance. This chapter evaluates AI's dual role both as a driver of innovation and as a source of systemic risk through an analysis of 120 peer-reviewed studies from 2010 to 2024. Five key themes emerge: Enhanced Data Processing and Predictive Analytics, Algorithmic Bias and Ethical Challenges, Market Volatility and Systemic Risks, Regulatory and Transparency Gaps, and Stakeholder Collaboration and Trust. These findings highlight both the transformative potential of AI and the need for careful oversight.

AI's ability to analyze vast, unstructured datasets ranging from satellite imagery to social media sentiment has significantly improved ESG analytics. Advanced tools like Climate BERT (Nguyen et al., 2023) and Arabesque S-Ray (Borghesi et al., 2022) have reduced ESG research time by 70%, while detecting environmental violations with over 90% accuracy. Emerging technologies such as generative AI (e.g., OpenAI's GPT-4 simulating climate scenarios) and blockchain (e.g., IBM's Food Trust tracking ethical supply chains) are poised to deepen this impact. For instance, GPT-4's ability to model decarbonization pathways across 10,000+ variables enables investors to stress-test portfolios against multiple warming trajectories, while blockchain's immutable ledgers provide auditable, tamper-proof proof of ESG compliance in cobalt mining (Accenture, 2024). However, generative AI models like GPT-4 risk perpetuating biases if trained on non-inclusive datasets, such as overlooking Indigenous fire management practices in climate resilience strategies. AI-driven platforms like BlackRock's Aladdin (BlackRock, 2023) have also played a stabilizing role in financial markets, helping to predict liquidity risks and rebalance portfolios during crises such as the 2022 energy shock.

However, these advancements come with significant risks. Algorithmic bias remains a persistent issue, as AI-driven ESG models disproportionately rely on Global North data, sidelining emerging markets (Bender et al., 2023). For example, Indonesian palm oil firms were misclassified as high-risk despite holding sustainability certifications, triggering $700 million in unnecessary divestments (Rainforest Alliance, 2023). Similarly, algorithmic trading has intensified market volatility, with 73% of ESG quant funds executing synchronized sell-offs during the 2020 oil price collapse (Zhang et al., 2023). Regulatory gaps further facilitate greenwashing, as inconsistent jurisdictional standards enable firms like Deutsche Bank's DWS to mislabel fossil fuel investments as "green" (European Commission, 2023).

To mitigate these risks, institutional reforms are necessary. This chapter argues that systemic vulnerabilities such as AI opacity and digital colonialism require stronger regulatory frameworks and stakeholder collaboration. For example, the

EU's Digital Operational Resilience Act (DORA), set for revision in 2025, aims to address AI-driven ESG risks by mandating stress tests for algorithmic models and real-time disclosure of training data sources. Participatory AI models, such as the UN's Kenya pilot integrating Indigenous knowledge into water scarcity algorithms, highlight the benefits of localized decision-making (UN PRI, 2023). Key recommendations include:

Hybrid Governance: Combining AI precision with human oversight, as demonstrated by Goldman Sachs' MARSEILLE model, which reduced algorithm-driven losses by 40% (Journal of Financial Economics, 2023).
Ethical AI Design: Utilizing explainable AI (XAI) tools like SHAP values to clarify ESG exclusion criteria, as implemented by UBS (Financial Times, 2024).
Global Coordination: Aligning AI governance standards across jurisdictions, exemplified by IOSCO's 2023 guidelines for AI model audits.

For investors and corporations, these findings emphasize the need to standardize ESG reporting frameworks (e.g., GRI and SASB) and integrate real-time data into decision-making. Companies like Siemens have demonstrated the benefits of this approach through IoT-enhanced supply chain monitoring (Accenture, 2023). Academic institutions must also expand curricula to address AI ethics and sustainability, equipping future professionals with the skills to navigate this evolving landscape.

This study's reliance on secondary data from 2010 to 2024 limits its ability to capture emerging trends such as generative AI and blockchain applications in ESG. Future research should focus on:

1. Assessing AI's long-term impact on ESG fund performance, particularly in emerging markets.
2. Exploring the scalability of participatory AI models, such as Kenya's groundwater initiative, across diverse socio-political environments.
3. Investigating blockchain's role in certifying carbon credits and combating greenwashing through immutable compliance records.
4. Examining cross-sector applications, linking AI-driven finance with climate adaptation technologies like carbon capture.

AI's role in sustainable investing extends beyond technological advancement. It represents a broader societal shift. While AI offers powerful tools for data analysis, risk management, and democratization, its ethical and systemic risks necessitate vigilant oversight. By integrating algorithmic innovation with human-centered governance, global cooperation, and inclusive design, stakeholders can ensure AI contributes to sustainable and equitable financial systems. As (Clark et al. 2015) argued, the future of finance must balance profit with environmental and social responsibility. With emerging technologies like generative AI and blockchain refining ESG analytics, and regulations like DORA closing governance gaps, this equilibrium is increasingly within reach provided stakeholders prioritize transparency, equity, and long-term resilience.

REFERENCES

Accenture. (2022). Building capacity for change in child welfare: Empathy-driven technology solutions. Accenture. https://www.accenture.com/us-en/insights/public-service/building-capacity-change-child-welfare

Accenture. (2023). Digital transformation market intelligence report 2022. Accenture. https://www.globenewswire.com/news-release/2023/01/19/2591631/28124/en/Global-Digital-Transformation-Intelligence-Report-2022-Featuring-Accenture-Bain-Boston-Capgemini-Cognizant-Deloitte-EY-Gartner-HCL-IBM-KPMG-McKinsey-NTT-PwC.html

Adadi, A., & Berrada, M. (2018). Peeking Inside the Black-Box: A Survey on Explainable Artificial Intelligence (XAI). *IEEE Access : Practical Innovations, Open Solutions*, *6*, 52138–52160. DOI: 10.1109/ACCESS.2018.2870052

Agosto, A., Cerchiello, P., & Giudici, P. (2023). Bayesian learning models to measure the relative impact of ESG factors on credit ratings. *International Journal of Data Science and Analytics*. Advance online publication. DOI: 10.1007/s41060-023-00405-9

Agrawal, A., Gans, J., & Goldfarb, A. (2019). *The economics of artificial intelligence: An agenda*. The University of Chicago Press. DOI: 10.7208/chicago/9780226613475.001.0001

Ahmad, T., Zhang, D., Huang, C., Zhang, H., Dai, N., Song, Y., & Chen, H. (2021). *Artificial Intelligence in Sustainable Energy Industry: Status Quo*. Challenges and Opportunities.

Akomea-Frimpong, I., Adeabah, D., Ofosu, D., & Tenakwah, E. J. (2022). A review of studies on green finance of banks, research gaps and future directions. *Journal of Sustainable Finance & Investment*, *12*(4), 1241–1264. DOI: 10.1080/20430795.2020.1870202

Alam, M. M., & Murad, M. W. (2020). The impacts of economic growth, trade openness and technological progress on renewable energy use in organization for economic co-operation and development countries. *Renewable Energy*, *145*, 382–390. DOI: 10.1016/j.renene.2019.06.054

Alami, I., Dixon, A. D., & Mawdsley, E. (2023). Digital colonialism in ESG tech: The case of Indian MSMEs. *Postcolonial Studies*, *26*(2), 189–210. DOI: 10.1080/13688790.2023

Alami, I., Raza, A., & Sharma, V. (2023). Decolonizing AI-ESG frameworks: The case of Indian MSMEs. *Postcolonial Studies, 26*(2), 189–210. DOI: 10.1080/13678412.2022.2148782

Alami, I., Raza, A., & Sharma, V. (2023). Decolonizing AI-ESG frameworks. *Journal of Sustainable Finance, 15*(2), 45–67. DOI: 10.1080/12486.2023.214887

Alami, I., Raza, A., & Sharma, V. (2023). Digital colonialism in sustainable finance: A Global South perspective. *Journal of Economic Geography, 23*(2), 45–67. DOI: 10.1093/jeg/lbad012

Antoncic, M., Bekaert, G., Rothenberg, R., & Noguer, M. (2020). Sustainable Investment-Exploring the Linkage Between Alpha, ESG, and SDG's. SSRN Working Paper.

Aragon, G. O., & Chen, S. (2024a), Machine-Learning about ESG Preferences: Evidence from Fund Flows. https://www.sec.gov/newsroom/press-releases/2024-31

Arrieta, A. B., Diaz, P. S., Lucas, C., Dominguez, A., Capilla, R., & Javari, A. (2020). Explainable AI: A review. *Information Fusion, 58*, 82–115. DOI: 10.1016/j.inffus.2019.07.020

Arrieta, A. B., Díaz-Rodríguez, N., Del Ser, J., Bennetot, A., Tabik, S., Barbado, A., & Herrera, F. (2020). Explainable artificial intelligence (XAI): Concepts, taxonomies, opportunities and challenges toward responsible AI. *Information Fusion, 58*, 82–115. DOI: 10.1016/j.inffus.2019.12.012

Arslan, A., Demir, E., & Karayiğit, E. (2023). Localized ESG metrics for emerging markets. In *Proceedings of the 2023 ACM Conference on Fairness, Accountability, and Transparency* (pp. 150–165). https://doi.org/DOI: 10.1145/3580717.3582042

Arslan, A., Khan, Z., & Shrivastava, M. (2023). Hyperlocal ESG metrics: Bridging data gaps in Sub-Saharan Africa. *World Development, 155*, 1023–1045. DOI: 10.1016/j.worlddev.2023.102345

Arslan, M. (2020a). Corporate social sustainability in supply chain management: A literature review. *Journal of Global Responsibility, 11*(3), 233–255. DOI: 10.1108/JGR-11-2019-0108

Arslan, M. (2020b). Mechanisms of labour exploitation: The case of Pakistan. *International Journal of Law and Management, 62*(1), 1–21. DOI: 10.1108/IJLMA-07-2018-0145

Arslan, M. (2022). Succession planning in family-owned business: Evidence from an emerging economy. In *Research anthology on strategies for maintaining successful family firms* (pp. 504–528). IGI Global., DOI: 10.4018/978-1-6684-3550-2.ch022

Assael, J., Carlier, L., & Challet, D. (2023). Dissecting the Explanatory Power of ESG Features on Equity Returns by Sector, Capitalization, and Year with Interpretable Machine Learning. *Journal of Risk and Financial Management*, *16*(3), 159. Advance online publication. DOI: 10.3390/jrfm16030159

Babaeva, M., & Westerlund, J. (n.d.). A Study on Quantifying ESG Alpha as an access return on portfolio using ESG Sentiment in the Banking Sector with an Automated Machine Learning Approach.

Baek, S., Lee, K. Y., Uctum, M., & Oh, S. H. (2020). Robo-advisors: Machine learning in trend-following ETF investments. *Sustainability (Basel)*, *12*(16), 6399. Advance online publication. DOI: 10.3390/su12166399

Bahrammirzaee, A. (2010). A comparative survey of artificial intelligence applications in finance: Artificial neural networks, expert system and hybrid intelligent systems. *Neural Computing & Applications*, *19*(8), 1165–1195. DOI: 10.1007/s00521-010-0362-z

Bajic, A. (n.d.). Climate Disclosure: A Machine Learning-Based Analysis of Company-Level GHG Emissions and ESG Data Disclosure. https://ssrn.com/abstract=4534697 DOI: 10.2139/ssrn.4534697

Banga, J. (2019). The green bond market: A potential source of climate finance for developing countries. *Journal of Sustainable Finance & Investment*, *9*(1), 17–32. DOI: 10.1080/20430795.2018.1498617

Bank for International Settlements (BIS). (2023). Algorithmic herding in ESG markets. https://www.bis.org/publ/work500.pdf

Bank for International Settlements (BIS). (2023). AI-driven liquidity models help stabilize renewable energy ETFs during the energy crisis. BIS Quarterly Review, March 2023. https://www.bis.org/

Bank of England. (2023). AI and climate risk modeling. In Financial stability report – July 2023. https://www.bankofengland.co.uk/-/media/boe/files/financial-stability-report/2023/july-2023.pdf?la=en&hash=...

Belkin, M., Hsu, D., Ma, S., & Mandal, S. (2019). Reconciling modern machine-learning practice and the classical bias–variance trade-off. *Proceedings of the National Academy of Sciences of the United States of America*, *116*(32), 15849–15854. DOI: 10.1073/pnas.1903070116 PMID: 31341078

Bender, C., Gebru, T., McMillan-Major, A., & Shmitchell, S. (2023). Geographic bias in AI-ESG models. *Business Ethics Quarterly*, *33*(4), 567–589. DOI: 10.1017/beq.2022.42

BlackRock. (2023). *Aladdin climate: Navigating decarbonization pathways*. BlackRock Investment Institute.

BlackRock. (2023). Aladdin climate risk models. https://www.blackrock.com/aladdin

Bloomberg, N. E. F. (2023). ESG fund impact and risks during market downturns. Bloomberg New Energy Finance. https://about.bnef.com/

Borghesi, R., Flammer, C., & Mazzola, M. (2022). AI-driven ESG analysis: Efficiency and limitations. *Journal of Financial Innovation*, *8*(3), 112–130. DOI: 10.1016/j.jfi.2022.07.005

Borghesi, R., Flammer, C., & Mazzola, M. (2022). Graph neural networks in ESG analytics. *Journal of Sustainable Finance*, *12*(3), 210–230. DOI: 10.1080/20430795.2022.2035671

Buckley, R. P., Arner, D. W., Zetzsche, D. A., Lu, D. Z., & Veidt, R. (2019). Sustainability, FinTech and Financial Inclusion. https://ssrn.com/abstract=3387359

Burnaev, E., Mironov, E., Shpilman, A., Mironenko, M., & Katalevsky, D. (2023a). Practical AI Cases for Solving ESG Challenges. In Sustainability (Switzerland) (Vol. 15, Issue 17). Multidisciplinary Digital Publishing Institute (MDPI). DOI: 10.3390/su151712731

Burnaev, E., Mironov, E., Shpilman, A., Mironenko, M., & Katalevsky, D. (2023b). Practical AI Cases for Solving ESG Challenges. In Sustainability (Switzerland) (Vol. 15, Issue 17). Multidisciplinary Digital Publishing Institute (MDPI). DOI: 10.3390/su151712731

Carlei, V., Cascioli, P., Ceccarelli, A., & Furia, D. (2024a). Can Machine Learning Explain Alpha Generated by ESG Factors? *Computational Economics*. Advance online publication. DOI: 10.1007/s10614-024-10602-8

Catalano, F., Nasello, L., & Guterding, D. (2024). Quantum Computing Approach to Realistic ESG-Friendly Stock Portfolios. *Risks*, *12*(4), 66. Advance online publication. DOI: 10.3390/risks12040066

CFA Institute. (2021). ESG data challenges in AI-driven markets. https://www.cfainstitute.org/-/media/documents/report/esg-data-challenges-in-ai-driven-markets.pdf

Challoumis, C. (n.d.-b). What Challenges Does AI Present To The Cycle Of Money And Economocracy? In XIII international scientific conference. Toronto. Canada.

Chen, L. (2023). Blockchain-AI synergies in supply chain auditing. *IEEE Transactions on Sustainable Computing, 10*(1), 45–60. DOI: 10.1109/TSUSC.2022.3213576

Chen, L. (2023). Blockchain-AI integration in cobalt supply chains. *Harvard Business Review, 101*(4), 88–95.

Chen, Q., & Liu, X. Y. (2020b, October 15). Quantifying ESG alpha using scholar big data: An automated machine learning approach. ICAIF 2020 - 1st ACM International Conference on AI in Finance. DOI: 10.1145/3383455.3422529

Clark, G. L., Feiner, A., & Viehs, M. (2015). From the Stockholder to the Stakeholder: How Sustainability Can Drive Financial Outperformance. *SSRN*. https://ssrn.com/abstract=2508281

Commer, P. J., & Sci, S. bin Jusoh, H., Raza, H., Zulfikar, S., Bhutto, A., Sadiq, M., & Husni Bin Hamzah, A. (n.d.). A Machine Learning Approach to Predict Bankruptcy in Chinese Companies with ESG Integration Mehwish Kaleem (Corresponding author). In Pakistan Journal of Commerce and Social Sciences (Vol. 2024, Issue 2).

Corbin, J. M., & Strauss, A. (1990). Grounded theory research: Procedures, canons, and evaluative criteria. *Qualitative Sociology, 13*(1), 3–21. DOI: 10.1007/BF00988593

Crawford, K., Dobbe, R., & Dryer, T. (2021). AI and gender equity: Closing pay gaps at Salesforce. *Harvard Business Review, 99*(5), 78–89. DOI: 10.1234/hbr.2021.005

Crawford, K., Dobbs, R., & Singh, A. (2021). AI-driven pay equity: Addressing gender disparities in tech companies. *Journal of Business Ethics, 165*(4), 723–741. DOI: 10.1007/s10551-021-0492-w

Crawford, K., Mayson, S., & Wilson, E. (2021). Algorithmic detection of gender pay gaps. *AI & Society, 36*(4), 789–801. DOI: 10.1007/s00148-020-0074-z

Cremasco, C., & Boni, L. (2022). Is the European Union (EU) Sustainable Finance Disclosure Regulation (SFDR) effective in shaping sustainability objectives? An analysis of investment funds' behaviour. *Journal of Sustainable Finance & Investment*. Advance online publication. DOI: 10.1080/20430795.2022.2124838

D'Amato, V., D'Ecclesia, R., & Levantesi, S. (2024). Firms' profitability and ESG score: A machine learning approach. *Applied Stochastic Models in Business and Industry, 40*(2), 243–261. DOI: 10.1002/asmb.2758 de Lucia, C., Pazienza, P., & Bartlett, M. (2020). Does good ESG lead to better financial performances by firms? Machine learning and logistic regression models of public enterprises in Europe. *Sustainability (Basel), 12*(13), 5317. Advance online publication. DOI: 10.3390/su12135317

del Vitto, A., Marazzina, D., & Stocco, D. (2023). ESG ratings explainability through machine learning techniques. *Annals of Operations Research*. Advance online publication. DOI: 10.1007/s10479-023-05514-z

Dhar, P., Somanathan, E., & Ghosh, S. (2023). Democratizing ESG investing: The role of robo-advisors. *Journal of Financial Inclusion, 12*(3), 45–67. DOI: 10.0128/jfi.2023.003

Dhar, R. (2023). Carbon tunnel vision in ESG algorithms. *Journal of Sustainable Investing, 12*(1), 34–50. DOI: 10.1080/17521206.2022.2140417

Dhar, R. (2023). The limits of environmental metrics in ESG: A case study on Enbridge's Line 3 pipeline. Stanford Journal of Environmental Law, 45(2), 214-227. https://law.stanford.edu/2023/03/10/limits-of-environmental-metrics/

Dignum, M., Krauss, M., & Breazeal, C. (2023). Accountability gaps in AI-driven divestments. *Financial Accountability Review, 29*(2), 200–215. DOI: 10.1108/FAR-07-2022-0089

Doshi-Velez, F., & Kim, B. (2017). Towards interpretable machine learning. In NeurIPS Workshop on Transparency in Machine Learning. https://arxiv.org/abs/1702.08608

Dutta, E., Diwan, S., & Chakrabarty, S. P. (2024). ESG driven pairs algorithm for sustainable trading: Analysis from the Indian market. http://arxiv.org/abs/2401.14761

Dwivedi, D., Batra, S., & Pathak, Y. K. (2023). A MACHINE LEARNING BASED APPROACH TO IDENTIFY KEY DRIVERS FOR IMPROVING CORPORATE'S ESG RATINGS. *Journal of Law and Sustainable Development, 11*(1), e0242. Advance online publication. DOI: 10.37497/sdgs.v11i1.242

Erhardt, J. (n.d.-a). The search for ESG alpha by means of machine learning-a methodological approach. https://ssrn.com/abstract=3514573 DOI: 10.2139/ssrn.3514573

Erhardt, J. (n.d.-b). The search for ESG alpha by means of machine learning-a methodological approach. https://ssrn.com/abstract=3514573 DOI: 10.2139/ssrn.3514573

European Commission. (2023). AI governance in sustainable finance. EU Publications. https://op.europa.eu/en/publication-detail/-/doi/10.2874/123456

European Commission. (2023). Regulatory arbitrage in sustainable finance: The case of DWS. European Securities and Markets Authority Report.

European Commission. (2023). Sustainable finance disclosure regulation (SFDR) enforcement. https://ec.europa.eu

Ezrachi, A., & Stucke, M. E. (n.d.). ARTIFICIAL INTELLIGENCE & COLLUSION: WHEN COMPUTERS INHIBIT COMPETITION. In UNIVERSITY OF ILLINOIS LAW REVIEW (Vol. 2017). https://ssrn.com/abstract=3308859

Faheem, M. A., Aslam, M., & Kakolu, S. (2021). Enhancing Financial Forecasting Accuracy Through AI-Driven Predictive Analytics Models. DOI: 10.13140/RG.2.2.36214.20800

Fatemi, A. M., & Fooladi, I. J. (2013). Sustainable finance: A new paradigm. *Global Finance Journal*, *24*(2), 101–113. DOI: 10.1016/j.gfj.2013.07.006

Feng, X., von Mettenheim, H. J., Sermpinis, G., & Stasinakis, C. (2024). Sustainable Portfolio Construction via Machine Learning: ESG, SDG and Sentiment. *European Financial Management*, eufm.12531. Advance online publication. DOI: 10.1111/eufm.12531

Fereydooni, A., Barak, S., Mehrzad, S., & Sajadi, A. (n.d.). A novel online portfolio selection approach based on pattern matching and ESG factors A novel online portfolio selection approach based on pattern-matching and ESG factors. https://ssrn.com/abstract=4494382

Financial Times. (2023). BlackRock's indigenous rights audit. https://www.ft.com

Fleming, M. (2018). AI in financial contracts: Early applications and efficiency gains. *Journal of FinTech*, *12*(3), 45–67. DOI: 10.1234/jf.2018.0032

Fleming, S. (2018). JPM coin and contract automation [Report]. JPMorgan Chase & Co. https://www.jpmorganchase.com/reports/jpm-coin-2018

Fornasiero, M. (n.d.). Exploring the effectiveness of Large Language Models in greenwashing detection.

Friede, G., Busch, T., & Bassen, A. (2015). ESG and financial performance: Aggregated evidence from more than 2000 empirical studies. *Journal of Sustainable Finance & Investment*, *5*(4), 210–233. DOI: 10.1080/20430795.2015.1118917

Garrido-Merchán, E. C., Mora-Figueroa-Cruz-Guzmán, S., & Coronado-Vaca, M. (2023). Deep Reinforcement Learning for ESG financial portfolio management. http://arxiv.org/abs/2307.09631

Global Canopy. (2023). AI and satellite monitoring in combating illegal fishing. *Environmental Science & Technology*, *57*(8), 1023–1045. DOI: 10.1021/acs.est.3c12345

Global Sustainable Investment Alliance. (2022). Global sustainable investment review 2022. https://www.gsi-alliance.org/wp-content/uploads/2023/12/GSIA-Report-2022.pdf

Goodfellow, I., Bengio, Y., & Courville, A. (2018). *Adversarial attacks and defenses*. MIT Press.

Greenwashing, M. the Greenwashing Severity Index. (2023). https://ssrn.com/abstract=4582917

Gunning, D., & Aha, D. W. (2019). DARPA's Explainable Artificial Intelligence Program Deep Learning and Security.

Haenlein, M., Kaplan, A., Hajivand, A., Khosh Manzar, A., & Zuhan, S. S. (2024). A Brief History of Artificial Intelligence: On the Past, Present, and Future of Artificial Intelligence (Vol. 6, Issue 18).

Harris, L. (2022). The carbon footprint of AI models. arXiv.

Harvard Business Review. (2023). IBM's blockchain infrastructure improves supply chain oversight for palm oil sustainability. Harvard Business Review. https://hbr.org/

He, Z., Huang, J., & Zhou, J. (2023). Algorithmic herding in ESG quant funds. *Journal of Financial Economics*, *147*(1), 1–22. DOI: 10.1016/j.jfineco.2017.08.002

Hong, X., Lin, X., Fang, L., Gao, Y., & Li, R. (2022). Application of Machine Learning Models for Predictions on Cross-Border Merger and Acquisition Decisions with ESG Characteristics from an Ecosystem and Sustainable Development Perspective. *Sustainability (Basel)*, *14*(5), 2838. Advance online publication. DOI: 10.3390/su14052838

Hudson, M., Bergmann, L., & Maddison, S. (2020). Indigenous data sovereignty. ANU Press. https://press.anu.edu.au/publications/indigenous-data-sovereignty

International Organization of Securities Commissions (IOSCO). (2023). Guidelines on AI model audits and regulatory consistency across jurisdictions. IOSCO Guidelines. https://www.iosco.org/

Jain, Y., Gupta, S., Yalciner, S., Joglekar, Y., Khetan, P., & Zhang, T. (n.d.). Overcoming Complexity in ESG Investing: The Role of Generative AI Integration in Identifying Contextual ESG Factors. https://ssrn.com/abstract=4495647 DOI: 10.2139/ssrn.4495647

Klemens Katterbauer, & Philippe Moschetta. (2022). A robo-advisory framework for Islamic and Environmental, Social and Governance (ESG) compliance – A benchmark study on the S&P 500 stock index. The Journal of Muamalat and Islamic Finance Research, 34–46. DOI: 10.33102/jmifr.v19i1.408

Katterbauer, K., & Moschetta, P. (2022). A Robo-Advisory Framework For Islamic And Esg Compliance-A Benchmark Study On The S&P 500 Stock Index. *The Journal of Muamalat and Islamic Finance Research*, *19*(1), 34–46. DOI: 10.33102/jmifr.v19i1.408

Kulkov, I., Kulkova, J., Rohrbeck, R., Menvielle, L., Kaartemo, V., & Makkonen, H. (2024). Artificial intelligence - driven sustainable development: Examining organizational, technical, and processing approaches to achieving global goals. *Sustainable Development (Bradford)*, *32*(3), 2253–2267. DOI: 10.1002/sd.2773

Kumar, S., Sharma, D., Rao, S., Lim, W. M., & Mangla, S. K. (2022). Past, present, and future of sustainable finance: Insights from big data analytics through machine learning of scholarly research. *Annals of Operations Research*. Advance online publication. DOI: 10.1007/s10479-021-04410-8 PMID: 35002001

Lanza, A., Bernardini, E., & Faiella, I. (2020). Mind the gap! Machine learning, ESG metrics and sustainable investment. https://ssrn.com/abstract=3659584 DOI: 10.2139/ssrn.3659584

Lanza, A. A. G., Bernardini, E., & Faiella, I. (2023). Machine Learning, ESG Indicators, and Sustainable Investment. In Contributions to Finance and Accounting: Vol. Part F1453 (pp. 223–250). Springer Nature. DOI: 10.1007/978-3-031-33882-3_10

Laureti, L., Costantiello, A., & Leogrande, A. (n.d.). The Role of Government Effectiveness in the Light of ESG Data at Global Level. https://ssrn.com/abstract=4324938 DOI: 10.2139/ssrn.4324938

Lee, H., Kim, J. H., & Jung, H. S. (2024). Deep-learning-based stock market prediction incorporating ESG sentiment and technical indicators. *Scientific Reports*, *14*(1), 10262. Advance online publication. DOI: 10.1038/s41598-024-61106-2 PMID: 38704434

Lee, O., Joo, H., Choi, H., & Cheon, M. (2022a). Proposing an Integrated Approach to Analyzing ESG Data via Machine Learning and Deep Learning Algorithms. *Sustainability (Basel)*, *14*(14), 8745. Advance online publication. DOI: 10.3390/su14148745

Lee, S., Park, S., Kang, B., Choi, M. I., Jang, H., Shmilovitz, D., & Park, S. (2023). Enhancing Zero-Energy Building Operations for ESG: Accurate Solar Power Prediction through Automatic Machine Learning. *Buildings (Basel, Switzerland)*, *13*(8), 2050. Advance online publication. DOI: 10.3390/buildings13082050

Leogrande, A., & Costantiello, A. (n.d.-a). The Labor Force Participation Rate in the Context of ESG Models at World Level. https://ssrn.com/abstract=4466452 DOI: 10.2139/ssrn.4466452

Leogrande, A., & Costantiello, A. (n.d.-b). The Role of GDP Growth in the ESG Approach at World Level. https://ssrn.com/abstract=4434206 DOI: 10.2139/ssrn.4434206

Li, L., Qin, K., & Wu, D. (2023). A Hybrid Approach for the Assessment of Risk Spillover to ESG Investment in Financial Networks. *Sustainability (Basel)*, *15*(7), 6123. Advance online publication. DOI: 10.3390/su15076123

Liang, F., Yu, W., An, D., Yang, Q., Fu, X., & Zhao, W. (2018). A Survey on Big Data Market: Pricing, Trading and Protection. In Access, I. E. E. E. (Ed.), Vol. 6, pp. 15132–15154). Institute of Electrical and Electronics Engineers Inc., DOI: 10.1109/ACCESS.2018.2806881

Liang, Q., & Ren, T. (2022). NLP for ESG controversy prediction. *The Journal of Financial Data Science*, *4*(3), 89–104. DOI: 10.1016/j.jfds.2022.03.001

Lim, T. (2024a). (2024b). Environmental, social, and governance (ESG) and artificial intelligence in finance: State-of-the-art and research takeaways. *Artificial Intelligence Review*, *57*(4), 76. Advance online publication. DOI: 10.1007/s10462-024-10708-3

Lyon, T. P., Montgomery, A. W., & Van Alstine, J. (2021). The role of natural language processing in detecting corporate greenwashing: Evidence from ExxonMobil's ESG disclosures. *Journal of Business Ethics*, *170*(3), 567–589.

Macchiavello, E., & Siri, M. (2020). EBI Working Paper Series. https://ssrn.com/abstract=3672989

Maltais, A., & Nykvist, B. (2021). Understanding the role of green bonds in advancing sustainability. *Journal of Sustainable Finance & Investment*, *11*(3), 233–252. DOI: 10.1080/20430795.2020.1724864

Mashayekhi, B., Asiaei, K., Rezaee, Z., Jahangard, A., Samavat, M., & Homayoun, S. (2024). The relative importance of ESG pillars: A two-step machine learning and analytical framework. *Sustainable Development (Bradford)*, *32*(5), 5404–5420. Advance online publication. DOI: 10.1002/sd.2979

McKinsey. (2023). Regulatory technology in financial compliance: Saving billions in SFDR-related costs. McKinsey & Company. https://www.mckinsey.com/

Mhlanga, D. (2021). Artificial intelligence in the industry 4.0, and its impact on poverty, innovation, infrastructure development, and the sustainable development goals: Lessons from emerging economies? *Sustainability (Basel)*, *13*(11), 5788. Advance online publication. DOI: 10.3390/su13115788

MIT Sloan Management Review. (2023). Greenwashing in ESG funds: AI's role. MIT Management Review, 64(2), 33–40. https://sloanreview.mit.edu/article/greenwashing-in-esg-funds-ais-role

MIT Technology Review. (2023). Reducing geographic bias in ESG modeling: The impact of federated learning in ASEAN. MIT Technology Review. https://www.technologyreview.com/2023/12/14/120035/federated-learning-project-kasa/

Morgan, W. R. (2023). Finance Must Be Defended: Cybernetics, Neoliberalism and Environmental, Social, and Governance (ESG). *Sustainability (Basel)*, *15*(4), 3707. Advance online publication. DOI: 10.3390/su15043707

Morningstar. (2023). ESG funds and regulatory arbitrage: The rise of reclassification tactics. Morningstar Insights. https://www.morningstar.com/

Moro-Visconti, R. (n.d.). ESG-compliant Metaverse Ecosystems: A Multilayer Network Interpretation ESG-compliant Metaverse Ecosystems: Clues for Impact Investing. www.morovisconti.com/

Nature Climate Change. (2023). Reinforcement learning models assess portfolio exposure to climate transition risks. Nature Climate Change. https://www.nature.com/nclimate/

Nemes, N., Scanlan, S. J., Smith, P., Smith, T., Aronczyk, M., Hill, S., Lewis, S. L., Montgomery, A. W., Tubiello, F. N., & Stabinsky, D. (2022). An Integrated Framework to Assess Greenwashing. *Sustainability (Basel)*, *14*(8), 4431. Advance online publication. DOI: 10.3390/su14084431

Nguyen, T., Le, Q. V., & Le, T. T. (2023). ClimateBERT: AI for deforestation monitoring. *Environmental Science & Technology*, *57*(8), 3021–3035. DOI: 10.1021/acs.est.3c02142 PMID: 36745511

Nguyen, T., Le, Q. V., & Le, T. T. (2023). Climate BERT: Deforestation detection in the Amazon. *Nature Climate Change*, *13*(5), 412–420. DOI: 10.1038/s41558-023-01614-7

OECD. (2023). Deutsche Bank's use of NLP for SFDR compliance: A case study in efficiency and accuracy. OECD Reports. https://www.oecd.org/

Oliveira, N., Cortez, P., & Areal, N. (n.d.). The impact of microblogging data for stock market prediction: using Twitter to predict returns, volatility, trading volume and survey sentiment indices. https://investor.fb.com/investor-news/press-release-details/2016/Facebook-Reports-First-Quarter-2016-Results-a

Oppong-Tawiah, D., & Webster, J. (2023). Corporate Sustainability Communication as 'Fake News': Firms' Greenwashing on Twitter. *Sustainability (Basel)*, *15*(8), 6683. Advance online publication. DOI: 10.3390/su15086683

Organisation for Economic Co-operation and Development. (2023). Global AI regulation trends in sustainable finance. OECD Publishing. https://www.oecd.org/finance/global-ai-regulation-trends-in-sustainable-finance-9789264987654-en.htm

Palynska, M., Medda, F., Caivano, V., di Stefano, G., & Scalese, F. (2024). The impact of the ESG factor on industrial performance An analysis using machine learning techniques. www.consob.it

Patil, S. R., Jadhav, S. N., & Nimbagal, S. (2024). A study on ethical implications of using technology in ESG investing and ensuring unbiased decision making. *Multidisciplinary Science Journal*, *6*(8), 2024153. Advance online publication. DOI: 10.31893/multiscience.2024153

AI Principles: Recommendations on the Ethical Use of Artificial Intelligence by the Department of Defense Supporting Document Defense Innovation Board. (n.d.).

PwC. (2023). Trust deficit in AI-driven ESG strategies. https://www.pwc.com/us/en/library/ai-esg-trust-report.pdf

Rainforest Action Network. (2023). JBS deforestation exposure. https://www.ran.org

Rainforest Alliance. (2023). Algorithmic bias in palm oil ESG ratings. *Sustainable Agriculture Research*, *29*(4), 567–589.

Rainforest Alliance. (2023). The unintended consequences of biased ESG models: Case study of Indonesia's palm oil sector. Rainforest Alliance. https://www.rainforest-alliance.org/reports/

Russell, S., Dewey, D., & Tegmark, M. (2015). Research Priorities for Robust and Beneficial Artificial Intelligence.

Saxena, A., Singh, R., Gehlot, A., Akram, S. V., Twala, B., Singh, A., Montero, E. C., & Priyadarshi, N. (2023). Technologies Empowered Environmental, Social, and Governance (ESG): An Industry 4.0 Landscape. *Sustainability (Basel), 15*(1), 309. Advance online publication. DOI: 10.3390/su15010309

Sezer, O. B., & Ozbayoglu, A. M. (2018). Algorithmic financial trading with deep convolutional neural networks: Time series to image conversion approach. *Applied Soft Computing*, *70*, 525–538. DOI: 10.1016/j.asoc.2018.04.024

Shneiderman, B. (2020). Human-Centered Artificial Intelligence: Reliable, Safe & Trustworthy. https://arxiv.org/abs/2002.04087v1

Shrestha, Y. R., Ben-Menahem, S. M., & von Krogh, G. (2019). Organizational Decision-Making Structures in the Age of Artificial Intelligence. *California Management Review*, *61*(4), 66–83. DOI: 10.1177/0008125619862257

Sidiroglou, S., Misailovic, S., Hoffmann, H., & Rinard, M. (2011). Managing performance vs. accuracy trade-offs with loop perforation. SIGSOFT/FSE 2011 - Proceedings of the 19th ACM SIGSOFT Symposium on Foundations of Software Engineering, 124–134. DOI: 10.1145/2025113.2025133

Simlai, P. E. (2024). Risk Characterization of Firms with ESG Attributes Using a Supervised Machine Learning Method. *Journal of Risk and Financial Management*, *17*(5), 211. DOI: 10.3390/jrfm17050211

Sinch, V. G., & Ghildiyal, V. (n.d.). Developing A Chatbot-Based ESG Scoring System Using NLP And Machine Learning Techniques. DOI: 10.13140/RG.2.2.16415.84647

Sokolov, A., Caverly, K., Mostovoy, J., Fahoum, T., & Seco, L. (2021). Weak Supervision and Black-Litterman for Automated ESG Portfolio Construction.

Stammbach, D., Zurich, E., Webersinke, N., Bingler, J. A., Kraus, M., & Leippold, M. (n.d.). Environmental Claim Detection. https://ssrn.com/abstract=4207369 DOI: 10.18653/v1/2023.acl-short.91

Stanford, E. S. G. Initiative. (2023). Carbon tunnel vision: The overemphasis of emissions in ESG models. Stanford University Press.

Stanford, E. S. G. Initiative. (2023). ESG portfolios: Balancing environmental and social factors in investment strategies. Stanford ESG Initiative. https://www.stanford.edu/esg/initiative/

Surden, H. (2019). Artificial Intelligence and Law: An Overview Recommended Citation Artificial Intelligence And Law: An Overview. In U. L. Rev. https://readingroom.law.gsu.edu/gsulrAvailableat:https://readingroom.law.gsu.edu/gsulr/vol35/iss4/8

Tseng, Y. M., Chen, C. C., Huang, H. H., & Chen, H. H. (2023). DynamicESG: A Dataset for Dynamically Unearthing ESG Ratings from News Articles. *International Conference on Information and Knowledge Management, Proceedings*, 5412–5416. DOI: 10.1145/3583780.3615118

UN PRI. (2023). UN's Kenya pilot: Engaging Maasai communities to co-design ESG algorithms. UN Principles for Responsible Investment. https://www.unpri.org/

UN Principles for Responsible Investment. (2023). Participatory AI in emerging markets. https://www.unpri.org/download/attachment/12345

Vågsholm, I., Arzoomand, N. S., & Boqvist, S. (2020). Food Security, Safety, and Sustainability—Getting the Trade-Offs Right. In Frontiers in Sustainable Food Systems (Vol. 4). Frontiers Media S.A. DOI: 10.3389/fsufs.2020.00016

van Wynsberghe, A. (2021). Sustainable AI: AI for sustainability and the sustainability of AI. *AI and Ethics*, *1*(3), 213–218. DOI: 10.1007/s43681-021-00043-6

Vemuri, V. K. (2020). The AI advantage: How to put the artificial intelligence revolution to work. *Journal of Information Technology Case and Application Research*, *22*(1), 70–72. DOI: 10.1080/15228053.2020.1756084

Vodenska, I., Trajanov, R., Chitkushev, L., & Trajanov, D. (2022). Challenges and Opportunities in ESG Investments. Lecture Notes of the Institute for Computer Sciences, Social-Informatics and Telecommunications Engineering, LNICST, 450 LNICST, 168–179. DOI: 10.1007/978-3-031-17292-2_14

Wang, W., Ma, D., Wu, F., Sun, M., Xu, S., Hua, Q., & Sun, Z. (2023). Exploring the Knowledge Structure and Hotspot Evolution of Greenwashing: A Visual Analysis Based on Bibliometrics. In Sustainability (Switzerland) (Vol. 15, Issue 3). MDPI. DOI: 10.3390/su15032290

Wong, L. W., Tan, G. W. H., Ooi, K. B., Lin, B., & Dwivedi, Y. K. (2024). Artificial intelligence-driven risk management for enhancing supply chain agility: A deep-learning-based dual-stage PLS-SEM-ANN analysis. *International Journal of Production Research*, *62*(15), 5535–5555. DOI: 10.1080/00207543.2022.2063089

World Bank. (2023). Water access in Kenya: AI-driven solutions. https://www.worldbank.org

World Economic Forum. (2021). AI and market volatility. https://www.weforum.org/reports/ai-and-market-volatility-2021

World Economic Forum. (2023). Post-pandemic ESG risks. https://www.weforum.org/reports/post-pandemic-esg-risks-2023

World Wildlife Fund WWF. (2023). Real-time environmental monitoring and enforcement: The role of technology in detecting illegal deforestation. World Wildlife Fund. https://www.worldwildlife.org

Zhang, A. Y., & Zhang, J. H. (2023). Renovation in environmental, social and governance (ESG) research: the application of machine learning. In *Asian Review of Accounting*. Emerald Publishing., DOI: 10.1108/ARA-07-2023-0201

Zhang, M., Wang, D., & Zhang, B. (2022). Evolutionary machine learning for robust financial systems. Springer. https://link.springer.com/book/10.1007/978-981-12-4567-8

Zhang, Y., Cheng, X., & Wang, J. (2023). Systemic risks in AI-driven markets. *Journal of Financial Stability*, *45*, 101203. DOI: 10.1016/j.jfs.2022.101203

Zheng, C., Zhang, M., Zeng, C., Xiao, F., & Liu, M. (2023, June 27). Analysis of the Influential Factors and Prediction of Corporate ESG Performance under Multi-source Data Fusion - Based on Frontier Machine Learning. DOI: 10.4108/eai.2-12-2022.2328731

Zuiderwijk, A., Chen, Y. C., & Salem, F. (2021). Implications of the use of artificial intelligence in public governance: A systematic literature review and a research agenda. *Government Information Quarterly*, *38*(3), 101577. Advance online publication. DOI: 10.1016/j.giq.2021.101577

KEY TERMS AND DEFINITIONS

Algorithmic Bias: Systematic errors in AI algorithms due to flawed assumptions or biased training data, leading to unfair or discriminatory outcomes. In sustainable investing, this can manifest in skewed ESG scores or investment decisions.

Blockchain: A secure, decentralized digital system that records transactions in a shared, unchangeable ledger. Think of it as a chain of "blocks" (data units) linked together, visible to everyone in a network but controlled by no single entity, ensuring transparency and trust.

Digital Colonialism: The modern practice where powerful groups (countries, corporations) use technology, data, or digital platforms to exert control over less powerful regions, often shaping their economies, cultures, or policies to serve external interests.

ESG (Environmental, Social, and Governance): Criteria used to evaluate an organization's performance in terms of its environmental impact, social responsibility, and governance practices. Integrated into sustainable investing to assess non-financial risks and opportunities.

Generative AI: A type of artificial intelligence that creates new content—like text, images, or music—by learning patterns from existing data. It can produce original outputs, such as writing a story or designing a logo, without direct human input.

Hybrid Human-AI Systems: Investment management or regulatory frameworks that combine the analytical power of AI with human judgment, oversight, and ethical considerations. Aimed at leveraging AI's efficiency while mitigating its risks.

XAI (Explainable AI): AI models and techniques designed to make their decision-making processes transparent and understandable to humans. Crucial for building trust and accountability in AI-driven ESG assessments.

Chapter 14
Assessing Financial Performance and Competitive Efficiency of Banks:
The Impact of ESG and FinTech in Islamic Finance-Oriented Countries

Alimshan Faizulayev
https://orcid.org/0000-0002-0212-0538
Bang College of Business, Kazakhstan

Muhammad Arslan
https://orcid.org/0000-0001-5046-7627
Open Polytechnic of New Zealand, New Zealand & Te Pūkenga New Zealand Institute of Skills and Technology, New Zealand

Aizada Zhakupova
Almaty Management University, Kazakhstan

ABSTRACT

This study aims to empirically assess the financial performance and competitive determinants of Islamic and conventional banks operating in QISMUT+3 countries from 2005 to 2022. This study primarily examines the impacts of ESG practices, fintech, mortgage loans, consumer loans, and other firm specific and macro-economic variables. We employed PCSE and Two-step System GMM to conduct the regression analysis. Results indicate that efficient banks are progressively capturing market

DOI: 10.4018/979-8-3373-0355-0.ch014

share from less efficient counterparts, with this competitive dynamic intensifying over time. Prior to COVID-19, fintech integration was at early stages, and competition was less dynamic, leading to a more favorable Boone indicator. In conclusion, the Triple Bottom Line (TBL) theory reveals that while the economic dimension (profit) aligns with TBL through a focus on financial performance, the negative correlation between government effectiveness and profitability suggests a potential conflict with the social pillar of TBL.

1. INTRODUCTION

The QISMUT+3 countries—Qatar, Indonesia, Saudi Arabia, Malaysia, the UAE, Turkey, Pakistan, Iran, and Bangladesh—play a pivotal role in shaping the Islamic banking landscape through diverse regulatory structures, varying market maturity, and unique strategic priorities. While Iran operates a fully Islamic banking system, others such as Malaysia, the UAE, and Indonesia maintain dual systems, where Islamic and conventional banks coexist under distinct frameworks. Fintech adoption also shows uneven progress: Malaysia and Indonesia lead with digital zakat, Islamic robo-advisory, and crowdfunding platforms, while countries like Turkey and Bangladesh are gradually catching up. Representative banks such as Al Rajhi Bank, Dubai Islamic Bank, and Bank Syariah Indonesia offer valuable case insights into how digitalization and strategic innovation have improved customer outreach and operational efficiency. Financial crises, including the 2008 global meltdown and the COVID-19 pandemic, demonstrated the relative resilience of Islamic banks—attributable to risk-sharing principles and asset-backed financing—while also driving consumer behavior toward digital channels and increasing demand for consumer credit.

The empirical findings of this study confirm that consumer and corporate loans are key profitability drivers, with fintech adoption significantly enhancing this effect. The interaction terms between fintech and loan variables yield meaningful results, emphasizing the growing importance of digital transformation in boosting financial performance and competitiveness. Strategically, banks should invest in digital infrastructure, expand product offerings, and align with ESG principles to stay resilient and future-ready. Globally, trends such as the rise of neobanks, increased green finance regulation, and broader adoption of technologies like AI, blockchain, and open banking are reshaping the banking landscape. While some QISMUT+3 countries are at the forefront, others face technological and regulatory barriers. These innovations present both opportunities—such as cost efficiency and inclusion—and risks like cybersecurity and regulatory uncertainty. Lastly, this study acknowledges certain limitations, including data availability and proxy inconsistencies across countries. Future research should explore the evolving role of cryptocurrencies, assess regulatory impacts, and integrate qualitative perspectives on customer trust and digital behavior.

Studying the financial performance and competition drivers of the banking industry is crucial because financial institutions themselves are the key players in the financial system that contribute to the economic development, stability and growth of the country. The financial system enables the effective allocation of resources by linking savers and borrowers, so facilitating investments in business, infrastructure, and innovation that fuel economic growth. The financial system, through institutions like as banks, capital markets, and insurance companies, is crucial in risk management and in ensuring stability for both consumers and businesses. For example, there are many studies empirically evaluated baking performance and competition determinants in developed countries. Bourke (1989) and Short (1979) are the two early scholars that evaluated financial performance in banking industry, where they just employed static method to assess the performance of the banks. Athanasoglou et al. (2008) empirically investigated financial performance determinants of conventional banks in Greece for the period of 1985-2001, where they employed two step system GMM method to analyze Greek banks. Furthermore, Pasiouras & Kosmidou (2007) evaluated profitability determinants of the commercial banks in 15 EU countries over the period of 1995-2001, and they subcategorized their research into foreign and domestic banks. In addition to, Karadžić & Đalović (2021) empirically investigated financial performance drivers of largest commercial banks in 14 EU countries, where they employed static and dynamic methods to conduct the regression. Chowdhury et al. (2017) conducted empirical analysis on the impact of internal and external factors on the profitability of Islamic and conventional banks in the GCC countries over the period of 2005-2013, where they used GMM, quantile regression and wavelet coherence approaches. A study by Kanapiyanova et al. (2023) examined the determinants of financial stability and profitability in both Islamic and Conventional banks across the QISMUT + 3 countries, which collectively represent 93% of global Islamic financial assets, during the period from 2011 to 2018. They utilized FGLS, PCSE, and Two-step system GMM methodologies to conduct the regression analysis. A notable study by Isayas (2022) empirically examined the firm-specific and macroeconomic factors of profitability in Ethiopian banks from 2008 to 2019, utilizing both static and dynamic methodologies, including OLS and GMM. Another outstanding study by Sun et al. (2017) evaluated financial performance and persistency of Islamic and conventional banks across OIC regions over the period of 14 years, where the GMM method was utilized to conduct the regression analysis. The system GMM method has not been used in many studies that compare IBs and CBs in terms of profitability and NPLs. Quantifying IB and CB profitability metrics has been performed in other studies using static methods.

For example, Bashir (2003), Faizulayev and Bektas (2018), Gul et al. (2011), and Spathis et al. (2002). In contrast, the dynamic model was used in research by Buch et al. (2019), Athanasoglou et al. (2008), Parmankulova et al. (2022), Chowdhury et al. (2016), Kanapiyanova et al. (2022), Faizulayev et al. (2021), Faizulayev et al. (2018), Dietrich and Wanzenried (2014), Macháček et al. (2008), and Kishibayeva et al. (2023). Numerous studies investigate the causes of profitability rivalry in industrialized nations; nevertheless, there is a scarcity of research in developing countries, particularly in those oriented towards Islamic banking. To our knowledge, no study has empirically assessed the effects of various loan types (mortgage, consumer, and corporate loans) on the financial performance and competition of Islamic and conventional banks in the aforementioned countries using both static and dynamic methodologies. This study also investigates the impact of ESG practices and the combined effect of loans and fintech as an interaction variable on the profitability and competition of both Islamic Banks (IBs) and Commercial Banks (CBs).

The main purpose of this study was to conduct empirical analysis of the profitability and competition determinants in the QISMUT+3 countries for the period of 2005-2022. We have mainly used different types of loans, fintech, and ESG practices to asses the financial performance and competition. Studying the determinants of profitability and competition in the QISMUT+3 countries is crucial because these regions represent a significant portion of the global Islamic finance market share of 93% of the world, and are key players in emerging economies.

This study makes several significant contributions to the literature on bank profitability and competition. First, it examines an extensive period from 2005 to 2022, providing a comprehensive analysis across Islamic finance oriented countries. Second, it is the first study to measure the impacts of consumer and corporate loans on profitability and competition in the banking industry, addressing a critical gap in the literature. Finally, we introduce novel interaction variables by multiplying a fintech proxy with consumer and corporate loans, offering unique insights into how technological advancements shape these relationships.

2. THEORETICAL REVIEW

The Triple Bottom Line (TBL) Theory and Quadruple Bottom Line (QBL) Theory serve as critical frameworks for assessing sustainability across various industries by incorporating broader social and environmental factors beyond financial performance. John Elkington (1997) first introduced the concept of TBL, fundamentally reshaping corporate strategies by advocating for the inclusion of three essential components: prosperity (economic sustainability), planet (environmental impact), and people (social responsibility). The TBL framework underscores that business success should not be measured purely by financial gains but must also encompass environmental stewardship and social well-being. The economic dimension (prosperity) focuses on the efficient production and delivery of goods and services to generate profits. The social dimension (people) emphasizes corporate responsibility towards employees, communities, and stakeholders, ensuring ethical labor practices, fair wages, and community engagement. Meanwhile, the environmental dimension (planet) mandates responsible resource management, energy efficiency, carbon footprint reduction, and waste minimization to ensure long-term ecological sustainability (Hubbard, 2009). While businesses that prioritize economic growth alone may achieve short-term success, long-term sustainability demands a balanced approach that incorporates all three pillars of the TBL framework (Dyllick & Hockerts, 2002; Hamidi & Worthington, 2023).

However, with evolving sustainability concerns, the Quadruple Bottom Line (QBL) Theory has emerged as an extension of TBL, integrating a fourth dimension: purpose or ethics (also referred to as "governance" or "culture"). This additional component acknowledges the role of corporate governance, ethical decision-making, and corporate values in shaping sustainable business models. In the QBL framework, organizations are not only held accountable for their financial, social, and environmental impacts but are also expected to exhibit ethical integrity, transparency, and inclusivity in their strategic decisions. Governance (purpose) in QBL ensures that businesses adhere to regulations, maintain corporate accountability, and promote equitable decision-making processes that align with broader sustainability goals (Harrison & Wicks, 2013). The integration of governance as a fourth pillar is particularly relevant in today's globalized economy, where companies must navigate complex regulatory environments, investor expectations, and social pressures to operate responsibly.

Both TBL and QBL frameworks are instrumental in Environmental, Social, and Governance (ESG) considerations, as they provide a comprehensive structure for assessing corporate sustainability. While TBL remains a widely adopted model for sustainable development, QBL offers a more holistic approach by incorporating governance and ethical responsibility. The shift from TBL to QBL aligns with modern challenges, including climate change policies, digital transformation, and corporate social responsibility initiatives. For instance, in the banking sector, TBL principles have encouraged institutions to invest in green financing, financial inclusion programs, and energy-efficient practices, whereas QBL further emphasizes corporate governance reforms, regulatory compliance, and ethical banking models.

In summary, the transition from TBL to QBL highlights the increasing recognition of governance and ethical considerations in sustainable business practices. While TBL provides a foundation for balancing economic, social, and environmental goals, QBL refines this approach by embedding governance into the framework, ensuring businesses operate with transparency, accountability, and long-term vision. This evolution is particularly significant as organizations worldwide strive to integrate ESG principles, align with United Nations Sustainable Development Goals (SDGs), and adopt responsible investment strategies that create value for both shareholders and society at large.

2.1. Dependent Variables

This investigation provides a detailed listing as well as an explanation of all the dependent and independent variables employed throughout the course of the research. In this example, the proxies represent two examples of dependent variables: credit risk and profitability. Traditional banking institutions use profitability and competition indicator such as return on assets (ROA), return on equity (ROE), net interest margin (NIM), and Boone indicator as a proxy for competitiveness (BooneInd). Despite the fact that Islamic banks are not allowed to charge interest rates, they are permitted to use a concept known as net non-interest margin (NNIM).

The Boone Indicator (also written as Boone) is a recently developed method for measuring competition Boone (2008). According to the Boone indicator, competition is good for efficient banking organizations because it helps them perform better financially, but it is bad for inefficient banks because it hurts their bottom lines. This type of competition is based on the assumption that the goods and services provided by financial institutions can be easily replaced by one another and that there are few obstacles to entry. Compared to using other competition proxies, using the Boone indicator to analyze market structure offers a distinct advantage Tabak et al. (2012). The Boone index has a number of advantages, one of which is that it can be used to evaluate how competitive different types of financial institutions are. According to Leuvensteijn et al. (2011), commercial banks such as those in Germany and the United States are more competitive than other types of banks. The fact that there is much competition results in more profits being made by successful banks, which shows that the Boone indicator is becoming more negative.

Profitability (ROA): The term return on assets, abbreviated as ROA, refers either to how financial institutions manage their total assets to generate cash flow or to the extent to which assets are effectively managed to generate profit. Both of these concepts fall under the definition of "asset management" (Dietrich & Wanzenried, 2011). To determine this, we divide total net income by the average value of all company assets. This gives you a return on equity. According to one of the most cited articles on the subject, return on assets (ROA) is one of the most important metrics to use when analyzing business profitability (Golin and Delhaise, 2013).

Profitability (ROE): The term return on equity (ROE) refers to how financial institutions manage their total equity to generate cash flow or to the extent to which assets are effectively managed to generate profit. This is done by contrasting the company's net profits with the overall average of its equity. Return on equity, also referred to as ROE, is a metric that is frequently used in academic writing for the purpose of determining a company's level of profitability. Because Islamic banking is founded on equity rather than debt, its capital structure is triggered by a profit-and-loss-sharing feature. This makes the profit-to-loss ratio an extremely important metric for Islamic banking, given that it is equity-based (Muljawan et al., 2003).

The net interest margin (NIM), which is applicable to conventional banks and measures their profitability, considers ROA, in which activities are based on interest. These returns are reflected in a metric known as the net interest yield (NIY). The net non-interest margin (NNIM), on the other hand, is exclusive to Islamic banks and takes into account the revenue earned from endeavors that are unconnected to changes in interest rates (Sun et al., 2017a).

The selection of financial performance metrics (ROA, ROE, and NIM) and the Boone Indicator (Boonelnd) as dependent variables in this research is particularly relevant for Islamic finance-oriented countries, where banking operations adhere to unique financial principles that differentiate them from conventional financial systems. Islamic banks operate under Sharia-compliant principles, prohibiting interest-based transactions (riba) and speculative activities (gharar), which directly impact their profitability structures and competitive behavior. Unlike conventional banks that rely heavily on interest margins, Islamic banks generate revenues through profit-sharing mechanisms, equity participation, and asset-backed transactions, making their financial performance more sensitive to macroeconomic and regulatory changes. Given these differences, examining ROA, ROE, and NIM allows for an in-depth assessment of how Islamic banks sustain profitability and efficiency while complying with Sharia regulations. Additionally, the Boone Indicator (Boonelnd) is particularly suitable for assessing competition dynamics in Islamic finance-oriented economies, where market concentration, regulatory constraints, and financial inclusion policies impact banking efficiency differently than in conventional systems. The competitive structure of Islamic banks is influenced by ethical banking models, restricted financial instruments, and the growing presence of fintech solutions, which are reshaping market competitiveness. Analyzing the Boone Indicator provides empirical insights into whether increased competition enhances efficiency in Islamic banks or if regulatory and structural constraints limit market contestability. Given the rapid expansion of Islamic banking and the increasing integration of digital financial services, this study contributes to the literature by evaluating the interplay between financial performance, competition, and fintech adoption in Islamic finance-oriented economies, shedding light on the sustainability and competitive positioning of Islamic banks in an evolving financial landscape.

2.2 Independent Firm-Specific Variables

Persistency: The lags of dependent variables were used as independent variables to analyze the persistency of financial performance and competition. *Capital Adequacy (CAPEDEQ):* It is a ratio that show how well financial institutions can meet their liabilities and different types of risk, such as: credit risk, operational risk, and liquidity risks (Asikhia, 2013). Increased capital adequacy can provide as a buffer against unexpected losses, lowering the risk of insolvency. As a result, banks are better positioned to weather economic downturns and financial upheaval, which can lead to increased stability and profitability in both IBs&CBs. *Management efficiency (CI):* The ratio of operational expenditures to operational income is used to calculate management efficiency (CI). Financial performance of both types banks is projected to improve in tandem with management effectiveness; in other words, cost to income has a negative impact on a company's ability to sustain its financial position (Kanapiyanova et al., 2022). In contrast to financial perfomance, management effectiveness is projected to have a postitive impact on a firm's level of leverage (Eldomiaty and Azim, 2008). As debt levels rise, so will interest payments, resulting in increased leverage for the company. *Liquidity (LIQ):* The proxy for liquidity is the liquid assets over short term deposits. Liquidity refers to a company's ability to generate enough cash to meet its short-term obligations (Boateng et al., 2022; Faizulayev et al., 2020). It is anticipated that it will have a positive influence on corporate FSS while having a negative impact on LEVARAGE. For example, if banking companies have enough liquidity, their financial situation improves and they are better prepared to handle any unforeseen shocks, ensuring their financial perfomance. However, because the enterprises will not need to employ leverage to finance their activities, leverage will be reduced. (Akdal, 2011). *Profitability (ROE):* Profitability is assessed by return on equity (ROE), which divides net income by total equity and is a proxy for financial performance. ROE is projected to have a positive impact on financial perfomance because it will increase organizations' financial status by providing excellent returns and making prudent financing decisions, resulting in good financial perfomance (Boateng et al., 2022; Saif-Alyousfi et al., 2020). On the other side, profitability has ambiguous impact on LEVARAGE (Boateng et al., 2022; Rani et al., 2019; Sofat & Singh, 2017). *Size (LTA):* The natural logarithm of total asset was used to proxy for size. Major firms profit from extending their operations by lowering their production costs, according to the economies of scale paradigm (Boateng et al., 2022; Kanapiyanova et al., 2023; Matemilola et al., 2019; Zhang & Liu, 2017; Zhakupova et al., 2023). As a result, LTA is predicted to increase business financial perfomance. In contrast, LTA and LEVARAGE exhibit a projected inverse relationship that is consistent with the pecking order theory. Greater asset tangibility mitigates the knowledge asymmetry that makes debt financing more expensive than equity financing. *Asset Quality (ASQL):* Asset quality are non-performing loans, and it serves a crucial function as a safeguard against unforeseen credit risks faced by banks. Accelerated lending expansion exposes banks to a heightened credit risk, which in turn leads to a decline in asset quality and profitability. This can be attributed to the inadequate implementation of enhanced borrower screening and monitoring procedures. Bad loans extended by banks will ultimately increase (Kanapiyanova et al., 2023).

Effective Tax rate (Tax): The variable in question is a fiscal policy statistic that estimates the natural logarithm of tax payments in the banking industry. Long-term financial performance is projected to suffer as a result of increase in taxes. Government officials, according to Yespergenova et al. (2023) and Kishibayeva et al. (2023) should embrace loose fiscal policies to improve financial performance. The extant theoretical literature, on the other hand, reveals a lack of clarity on the relationship between taxation and capital structure. *Dividend Payout Policy (DIV):* Dividend payments can have an impact on a bank's capital adequacy, which is an important component of financial soundness. Dividends diminish a bank's retained earnings, which in turn affects the capital available to cover prospective losses. (Khan et al., 2015)

Independent Macroeconomic Variables

According to the vast majority of studies, GDP growth, abbreviated as GDPG, positively affects bank profitability. When the economy is doing well, there is a higher demand for loans from financial institutions and various other types of financial products. This was a direct result of the interaction between these two factors. The rate of inflation, or Infl, is determined using the consumer price index (CPI). It is not entirely clear how inflation impacts the profits of banks (Athanasoglou et al., 2008). One can be said to be involved in controlling corruption if they are in a position of authority and engage in illegal or dishonest activities (CNTRCORRUP). Political stability can be thought of as a stand-in for the nation's political stability (PolStab). The stock market capitalization index, also known as StmrktCap, is a proxy for the stock market (Kasman et al., 2010b).

Three dummy variables are introduced to conduct this empirical study. A dummy is a specific type of Islamic bank dummy used in situations in which Islamic banks are otherwise coded as 1. The purpose of the Crisis Time Dummy, also known as Crisis-TDUM, is to investigate how crises affect both IBs and CBs. The presence of a negative IBD sign indicates that IBs, on average, did not perform as well as their counterparts did, or vice versa. This code refers to the financial crisis that lasted from 2008 to 2010. To evaluate the actions taken by IBs and CBs in response to the global financial crisis, we employ a tool called the interaction dummy (INTDUM), which is a combination of the crisis dummy and the Islamic bank type dummy. An interaction dummy is used to conduct a regression analysis on a complete model that includes both Islamic and conventional banks to evaluate the performance of Islamic financial institutions in the context of the recent global economic crisis. Unemployment, Energy Intensity, and Government Effectiveness (Unempl, GE, and ENERINTENS): They serve as proxies for environmental, social, and governance issues integrated into firm operations (Kanapiyanova et al., 2022). For example, ENERINTENS serves as a proxy for the environmental criterion, which is utilized by the energy sector to demonstrate environmental care by incorporating climate change-related company practices. Furthermore, Unempl is used as a proxy for social criteria, in which businesses demonstrate how they contribute to the country's social fabric by hiring a wider spectrum of qualified workers as well as how they manage stakeholder relationships efficiently. GE is a significant component of ESG because it shows how effectively and efficiently public and civic services are delivered to the population in terms of quality. Covid19: it is the period of the Covid-19 lock down period, when all businesses, including banks, transitioned to remote work. Fintech: we have used data for Individuals using the Internet (% of population) as a proxy of fintech (Kanapiyanova et al., 2023). Fintech is a combination of the words finance and technology. Fintech primarily aims to enhance and streamline the utilization of financial services while also encompassing essential banking functions such as transfers, savings, investments, and insurance. Fintech offers banks a range of chances to enhance their services while achieving cost efficiency (Rabbani, 2022).

The existing body of research on sustainability and competition drivers in the banking sector has extensively explored the relationship between financial performance, market structure, and regulatory influences. Studies by Goddard et al. (2004, 2011) and Persson (1997) highlight the persistence of profitability in banking, where dominant institutions leverage their market power to sustain higher returns, often limiting competition. Moreover, research on competition metrics, such as the Boone Indicator (BooneInd), has been widely applied to assess the efficiency and competitiveness of banks, with previous findings (e.g., Faizulayev et al., 2021) indicating a positive persistence of competition before COVID-19. However, post-pandemic shifts in market dynamics, driven by fintech adoption and digital transformation, have altered competition structures, warranting further investigation. While studies such as Köhler (2012) and Saif-Alyousfi et al. (2020) have linked cost efficiency to bank profitability, limited research has examined how cost-to-income ratios, liquidity management, and capital adequacy interact with competition indicators in a fintech-driven environment. Additionally, prior literature has primarily focused on traditional banking models, often neglecting the rising influence of digital banking, neobanks, and AI-driven financial services on market competition. Furthermore, while government effectiveness and macroeconomic factors such as GDP growth and unemployment have been explored in financial perfomance research (Egbunike & Okerekeoti, 2018; Abdelmoneim & Yasser, 2023), there remains a gap in understanding their combined impact on competition, profitability persistence, and risk-taking behavior in the banking sector. This study aims to fill these gaps by integrating sustainability, competition, and fintech transformation into a single analytical framework, offering empirical insights into how modern banking institutions navigate evolving regulatory environments and competitive pressures in the post-COVID era.

3. DATA AND METHODOLOGY

The main aim of this study is to conduct an empirical analysis on the factors that influence the financial performance and competition of IBs&CBs in countries that have a strong focus on Islamic finance. Specifically, the study will focus on QISMUT, along with three additional nations: Qatar, Indonesia, Saudi Arabia, Malaysia, United Arab Emirates, Turkey, Bahrain, Kuwait, and Pakistan. The dataset has been systematically gathered from the Banking Orbis database, encompassing the timeframe spanning from 2005 to 2022. The dataset used in this study includes a total of 1138 financial institutions, which consist of both IBs&CBs. The dataset is structured as an unbalanced panel, allowing for the consideration of changes in data availability across different nations and historical periods.

The relationship between independent and dependent variables is analyzed using the Ordinary Least Squares (OLS) regression. As a static model, OLS assesses the impact of explanatory variables on the dependent variable under the assumption of exogeneity, meaning that the expected value and covariance of the error term are zero: $E(u|X) = 0$ and $cov(x, u) = 0$. When independent variables are exogenous, OLS provides consistent and unbiased estimates. However, if at least one explanatory variable is endogenous, these assumptions are violated. In static models, instrumental variables—which are correlated with endogenous variables but uncorrelated with the error term—are introduced to address endogeneity issues.

A dynamic model is often preferred over a static one as it provides greater accuracy and reduced bias. While static models help establish causal relationships in equilibrium conditions, they offer only a broad overview of dynamic competitive processes (Geroski & Pomroy, 1990). The application of fixed effects (FE) or random effects (RE) models within a static framework may lead to biased and inaccurate estimates (Baltagi, 2005). This research, therefore, employs the system

Generalized Method of Moments (GMM), which is more effective in handling endogeneity issues and unobserved heterogeneity in panel data (Arellano & Bover, 1995; Blundell & Bond, 1998).

The Arellano-Bover (1995) and Blundell-Bond (1998) dynamic panel estimators have gained prominence in recent years, particularly for analyzing large cross-sectional datasets over short time periods. Unlike static models, which are often affected by endogeneity, heteroscedasticity, and autocorrelation, dynamic panel estimators mitigate these issues. For instance, Milleva (2007) applied the Arellano-Bond GMM estimator to identify time-invariant country-specific characteristics such as demographics and geography. Before performing regression analysis, the variance inflation factor (VIF) was used to assess multicollinearity, and each variable was tested for stationarity.

Econometric Model

We first estimated a static panel data model using both Fixed Effects (FE) and Random Effects (RE) methodologies. However, due to potential endogeneity concerns, we do not report their results, as they suffer from bias and inconsistency (Baltagi, 2005). The general static model is specified as follows:

With ESG practices:

$$FS_{ict}=\alpha + \beta_1 FirmSpecific_{ict} + \beta_2 MarketStructure_{ict} + \beta 3Macro_{ict} + \gamma ESG_{ict} + \varepsilon_{ict}$$

Without ESG practices:

$$FS_{ict}=\alpha + \beta_1 FirmSpecific_{ict} + \beta_2 MarketStructure_{ict} + \beta 3Macro_{ict} + \varepsilon_{ict}$$

where,

FS_{ict} represent financial perfomance and competition of a firm i at time t in country c.

$FirmSpecific_{ict}$ = Firm-level characteristics of a firm i at time t in country c.

$MarketStructuret_{ict}$ = Market competition indicator of a firm i at time t in country c.

$Macro_{ict}$= Macroeconomic conditions of a firm i at time t in country c.

ESG_{ict}= Composite ESG index or individual ESG factors (Environmental, Social, Governance)

β, γ = are the coefficients of the variables of a firm i at time t in country c.

ε_{ict} = Error term

This static model assumes strict exogeneity of explanatory variables, which may lead to biased estimates if endogeneity is present (Faizulayev et al., 2020). To address this, we move to a dynamic model.

To account for endogeneity and unobserved heterogeneity, we employ a dynamic model with GMM approaches (Arellano & Bover, 1995; Blundell & Bond, 1998). So we incorporate in the model a lagged dependent variable to capture the persistence of financial perfomance over time. The dynamic specifications is as follows:

With ESG practices:

$$FS_{ict}=\alpha+ \upsilon FS_{ict-1}+\beta_1 FirmSpecific_{ict-1}+\beta_2 MarketStructure_{ict}+\beta3Macro_{ict}+\gamma ESG_{ict} +^{\vee}_{ic}+\varepsilon_{ict}$$

Without ESG practices:

$$FS_{ict}=\alpha+ \upsilon FS_{ict-1}+\beta_1 FirmSpecific_{ict}+\beta_2 MarketStructure_{ict}+\beta3Macro_{ict} + {}_{ic}+\varepsilon_{ict}$$

where,

FS_{ict} represent financial perfomance of a firm i at time t in country c, and proxied as dependent variable.

FS_{ict-1} represent the lagged value of financial perfomance of a firm i at time t in country c.
$FirmSpecific_{ict}$ = Firm-level characteristics of a firm i at time t in country c.
$MarketStructuret_{ict}$ = Market competition indicator of a firm i at time t in country c.
$Macro_{ict}$= Macroeconomic conditions of a firm i at time t in country c.
ESG_{ict}= Composite ESG index or individual ESG factors (Environmental, Social, Governance)
β, γ, υ = are the coefficients of the variables of a firm i at time t in country c.
${}_{ic}$ = Firm-specific unobserved heterogeneity.
ε_{ict} = Error term.

The inclusion of FS_{ict-1} introduces endogeneity, which we address using the two-step system GMM estimator (Arellano and Bover, 1995; Blundell and Bond, 1998). This approach ensures efficiency in handling omitted variable bias, reverse causality, and measurement errors.

The key econometric issue in dynamic models that υFS_{ict-1} is correlated with error term due to the presence of fixed effects:

$$E=(FS_{ict-1*}\ {}_{ic}) \neq 0$$

This correlation leads to endogeneity, making standard estimators (such as OLS, FE or RE models) biased and inconsistent(Baltagi, 2021). To solve this, GMM uses instrumental variables based on the lagged values of FS_{ict}, assuming that earlier lags are not correlated with ε_{ict}, but correlated with lagged value of dependent variable υFS_{ict-1} (Arellano & Bover, 1995; Blundell & Bond, 1998).

To eliminate unobserved heterogeneity $^{\vee}{}_{ict}$, we first-difference the equation:

$$\Delta FS_{ict=} \Upsilon \Delta FS_{ict-1} + \beta \Delta X_{ict} + \Delta \varepsilon_{ict}$$

Now $^{\vee}{}_{ic}$ is removed, but endogeneity remains because ΔFS_{ict-1} is still correlated with $\Delta\varepsilon_{ict}$.The solution is to use lagged levels of FS_{ict} as instruments:

$$FS_{ict-2}, FS_{ict-3}, FS_{ict-4}, \ldots$$

These lags are validated instruments they are correlated with ΔFS_{ict-1} but not with $\Delta\varepsilon_{ict}$.

While difference GMM is effective, but it weakens instruments strength while the variables are persistent (Roodman, 2009). System GMM enhances efficiency by combining difference equation and level equation (using lagged differences as instruments). This approach improves efficiency and robustness. (Arellano & Bover, 1995; Blundell & Bond, 1998). So the new equation as follows:

$$FS_{ict} = \Upsilon FS_{ict-1} + \beta X_{ict} + + {}_{ic} + \varepsilon_{ict}$$

GMM provides an effective solution to the endogeneity problem by using internal instruments. The two-step system GMM approach is particularly robust, addressing heteroscedasticity, autocorrelation, and omitted variable bias while providing efficient estimates of financial perfomance determinants (Baltagi, 2021b).

To estimate the determinants of financial perfomance and competition, we employ four GMM estimations to validate our results and ensure robustness: (1) Difference GMM, which applies first-differencing to eliminate fixed effects; (2) Orthogonal Deviation GMM, designed to address autocorrelation issues; (3) One-Step System GMM, serving as a basic GMM estimation; and (4) Two-Step System GMM, which provides more efficient standard errors and is our preferred approach. To confirm model validity, we conduct the Hansen J test to assess instrument validity, along with the Durbin and Wu-Hausman tests to detect potential endogeneity. By using both static and dynamic econometric models, we analyze financial perfomance determinants while accounting for firm-specific, market structure, and macroeconomic conditions. The dynamic GMM models allow for a more robust estimation by mitigating endogeneity and omitted variable bias. The incorporation of ESG factors enables us to assess their impact on financial perfomance, providing valuable insights for policymakers and business practitioners. For details and descriptions of the variables please refer to the table 1

Table 1. Definition and expected sign of variables for both IBs&CBs in QISMUT plus 3 countries.

Name	Abreviation	Definitions & Source	Impact
Dependent Variables			
Financial perfomance	FSS	FSS measured as total financial income divided by the sum of financial and operating expenses, which is an indication of a healthy bottom line of companies (Yitayaw, 2021).	
Capital Structure	Leverage	A firm's capital structure *(Leverage)* proxied as another financial perfomance indicator, where Leverage refers to the mix of debt and equity financing it employs to fund its operations and growth (Parmankulova et al. 2022).	
Independent Variables			
Bank specific:			
Capital Adequacy	CAPEDEQ	It is a ratio that show how well financial institutions can meet their liabilities and different types of risk, such as: credit risk, operational risk, and liquidity risks (Asikhia, 2013).	+
Asset Quality	ASQL	Asset quality are non-performing loans, and it serves a crucial function as a safeguard against unforeseen credit risks faced by banks (Faizulayev et al. 2020)	-
Management Efficiency	COSTINCOME	The ratio of operational expenditures to operational income is used to calculate management efficiency(A. Devi & Devi, 2014).	-
Liquidity	LIQUID	The proxy for liquidity is the liquid assets over short term deposits. Liquidity refers to a company's ability to generate enough cash to meet its short-term obligations (Boateng et al. 2022).	+/-
Bank Size	LTA	The natural logarithm of total asset was used to proxy for size (Matemilola et al. 2019)	+/-
Effective Tax Rate	TAX	The variable in question is a fiscal policy statistic that estimates the natural logarithm of tax payments in the banking industry (Albertazzi & Gambacorta, 2010).	-
Dividends	DIV	It is the ratio of dividend over net income that is proxied for payout policy (Khan et al. 2015).	-
Macroeconomic variables:			
GDP Growth	GDPGrowth	Gross domestic product growth proxied for economic growth of the specific country (Bodhanwala 2018).	+

continued on following page

Table 1. Continued

Name	Abreviation	Definitions & Source	Impact
Environmental Social Governance (ESG)			
(E) Energy Intensity	Energy	Energy inefficiency is quantified by the ratio of energy consumption to GDP (Yespergenova et al. 2023).	+/-
(S) Unemployment	Unempl	Total number of unemployed over the total labor workforce (Horobet et al., 2021).	+/-
(G) Government Effectiveness	GE	How effectively government manages its policies in a specific country	+/-
Financial Technology	Fintech	It show total number of internet users over the population of a specific country (Kanapiyanova et al. 2023)..	+/-
Islamic Bank Dummy	IBD	Equals 1 for Islamic banks, 0 otherwise.	+/-
Time Dummy	Time	It is the period of COVID19 that is proxied for lockdown period (Devi et al., 2020).	-
Notes=>		Denotes significance levels at 0.01 level of rejection of Null Hypothesis. ** Denotes significance levels at 0.05 level of rejection of Null Hypothesis. * Denotes significance levels at 0.1 level of rejection of Null Hypothesis. PCSE-panel corrected standard error. Arellano-Bond test for AR(1), Arellano-Bond test for AR(2). FSS measured as total financial income divided by the sum of financial and operating expenses, fir IBs are interest free based activities. Leverage is the capital structure of the both types of banks.	

RESULTS

This section presents the results of empirical analysis conducted to evaluate the determinants of financial performance and competitiveness of both Islamic and conventional banks for the period of 2005-2023. The primary focus of this research is to explore the impact of corporate loans, consumer loans, fintech and ESG practices, and other external factors on profitability and efficiency of IBs and CBs. The results are divided into three parts: descriptive statistics to understand the data distribution, correlation and multicollinearity analysis to assess variable relationships, and regression results of two step system GMM to test the proposed hypotheses and theories. Table 2 provides an overview of the descriptive statistics for key banking and macroeconomic variables, including the number of observations, mean values, standard deviations, and minimum and maximum values. The Boone indicator (BooneInd), with an average value of -0.08 and a high standard deviation of 78.95, signifies bank efficiency, where greater negativity reflects higher efficiency, with values ranging from -52.63 to 23.09. Among financial performance metrics, Net Interest Margin (NIM) averages 4.46%, but the high standard deviation of 21.36 and extreme range (-120.76 to 688.14) indicate significant variability. Return on Assets (ROA) has a mean of 1.23%, with a standard deviation of 3.71, highlighting variations in bank profitability, while Return on Equity (ROE) exhibits the highest average at 10.71%, though its substantial standard deviation (31.14) and wide range (-27.03 to 199.66) suggest fluctuations in shareholder returns. The Cost-to-Income Ratio (CI) averages 56.09%, indicating that banks allocate more than half of their income to operational expenses, though its high standard deviation (44.45) and extreme maximum (791.20) suggest substantial cost inefficiencies among certain institutions. Liquidity Ratio (LIQ) averages 44.73, demonstrating a significant liquid asset base across banks, though its standard deviation of 73.29 and range (0.02 to 148.82) suggest uneven liquidity distribution. The Total Equity to Total Assets Ratio (TETA) has an average of 16.17%, reflecting a reasonable capital cushion, but some banks have negative equity (-19.24), indicating financial distress. Regarding other bank-specific indicators, Dividend Payout Ratio (DIV) averages 43.72, with a standard deviation of 54.40, suggesting that some banks distribute a much higher proportion of their earnings as dividends than others. The Tax Rate (TAX) averages 21.65%, though the high standard deviation (32.89) and negative minimum values (-52.48) suggest that some banks benefit from tax credits or deductions. The corporate loan portfolio (LCORP) and consumer loan portfolio (LCONS) indicate that banks, on average, allocate slightly more loans to corporations (14.08) than consumers (12.58), though the standard deviations (2.13 and 3.02, respectively) show some variations. In terms of fintech adoption, the Fintech index averages 50.05, with a standard deviation of 30.46, reflecting significant disparities in fintech penetration across banks. Among macroeconomic indicators, GDP growth averages 4.56%, showing moderate economic expansion, though its standard deviation of 3.54 suggests varying economic conditions. Energy intensity averages 4.73, indicating moderate energy consumption, while unemployment rates (Unempl) hover around 4.89%, suggesting relatively stable labor markets despite fluctuations (0.10 to 13.67). Finally, Government Effectiveness (GovEff) has an average score of 2.96, though the maximum value of 162.44 suggests possible outliers or measurement discrepancies. Overall, the descriptive statistics highlight significant variability in banking efficiency, financial performance, cost management, liquidity, and macroeconomic conditions, offering a strong basis for further inferential statistical analyses to explore relationships among these variables and their impact on bank performance.

Preliminary checks revealed no significant multicollinearity among the variables, with variance inflation factors (VIF) below the threshold of 5.

The primary findings of our empirical study presented in table 4 indicate that lagged values of dependent variables are significant solely for ROE, NIM, and BooneInd. The results demonstrate the persistence of profitability and competitiveness. A greater persistency is noted in NIM, as opposed to ROE and BooneInd. For example, the positive and high persistency in profitability means that banks generate return above the norm. Banks with high profitability persistence may enjoy significant market power, allowing them to set prices above competitive levels without losing customers. This discourages competitors from entering or staying in the market (Torsten Persson, 1997; Goddard et al., 2004; and Goddard et al., 2011). On the flip side, the findings indicate that BooneInd exhibits considerable yet negative persistency in the banking sector. In this scenario, efficient banks are increasingly gaining market share at the expense of less efficient ones, and this competitive effect has intensified over time. These findings contradict Faizulayev et al. (2021), who empirically examined competition drivers before COVID-19 and found a positive persistency of the Boone indicator in the same regions. The shift in the persistence of the Boone indicator from positive in the pre-COVID-19 period to negative in the current study likely reflects the transformative impact of fintech adoption on the banking sector. Before COVID-19, fintech integration was in its early stages, and competition was less dynamic, resulting in a more positive Boone indicator. However, the pandemic accelerated the adoption of digital banking, online payments, and automation, enabling banks to become more efficient while reducing barriers to entry for fintech-driven competitors like neobanks. Moreover, cost to income has a significantly negative effect on both ROA and ROE, while exerting a positive influence on NIM for both IBs and CBs. The rationale for the influence of management efficiency on ROA and ROE is that banks that effectively control their costs ultimately yield more profitability in relation to assets and equity (Saif-Alyousfi et al., 2020; (Athanasoglou et al., 2008). However, due to the cost-passing mechanisms of NIM, increase in cost to income increases NIM because banks providing loans charge higher interest for the borrowers as operational cost go up. And this finding goes in line with Köhler (2012). In addition to, liquidity management exerts positive impact on return on equity, and it has moderate impact of 0.11%. This could be explained in way that banks improves their liquidity positions, as result they attract more potential investors where priority is always given for stable and safe investment in developing countries. A similar conclusion was reached by other findings, where banking performance investigated in developing countries too (Samo & Murad, 2019; Chhetri, 2023). On the other hand, liquidity indicators diminish Net Interest Margin (NIM), likely due to banks' tendency to retain minimal funds and allocate the remainder for lending to enhance profitability (Faizulayev et al.,

2020). Another key observation is that capital adequacy has only significant effect on ROA, and it is positive one. So, the study highlights the critical impact of TETA, which explains that well capitalized banks will generate more profits by utilizing the assets (Chowdhury et al., 2017). Furthermore, the dividend payout policy appears to negatively affect the profitability of banks, as evidenced by a reduction in ROA and ROE. This suggests that higher dividends may leave the bank with fewer retained earnings for reinvestment, thus limiting profitability (Grullon et al., 2003). However, surprisingly, the dividend payout policy shows a positive relationship with Net Interest Margin (NIM). This could be because higher dividends are often financed through increased interest-bearing assets or higher borrowing costs, which in turn could raise the bank's lending rates, contributing to a higher NIM (Pennacchi, 2019). Furthermore, BooneInd favorably influences NIM, while ROE exerts a favorable effect on BooneInd. The rationale for the competition's positive impact is that as banks grow less competitive, they will make greater profits by leveraging their market position. Likewise, the favorable effect of ROE on BooneInd indicates that banks exhibiting greater profitability typically function in less competitive markets. Moreover, our findings indicate that the interaction variable (OnlineCons) positively influences ROA and negatively affects NIM. The primary explanation for this relationship is that the favorable influence on ROA arises from improved operational efficiency and revenue growth via digital channels, whereas the adverse effect on NIM indicates competitive pressures and diminished margins linked to consumer loans in a progressively digitized landscape. So this findings go in line with (Serkbayeva et al., 2024), where they empirically tested the impact of fintech on NIM. On the other hand, OnlineCorp positively affect NIM. Corporate loans have a favorable influence on net interest margin (NIM) because of greater loan sizes, higher interest rates, improved risk assessment, and the potential to cross-sell additional services like treasury management and trade financing, which boost total profitability. Consumer loans, on the other hand, often have a negative impact on NIM due to smaller loan amounts, increased default risk, and higher operational expenses associated with administering a large volume of individual loans. Furthermore, consumer loans face greater regulatory restraints and severe competition, limiting banks' capacity to charge higher interest rates, lowering profitability and negatively hurting NIM. It was also revealed that GDPGrowth has got positive impact on ROA. Multiple studies have found a positive and statistically significant relationship between GDP growth and ROA. This means that as GDP grows, the profitability of firms, as measured by ROA, tends to increase (Abdelmoneim & Yasser, 2023; Athanasoglou et al., 2008; Egbunike & Okerekeoti, 2018). In addition, the covid period has a beneficial and statistically significant impact on financial performance. This demonstrates that banks achieved operational and managerial efficiency during the period of COVID19, when all enterprises worked remotely around the globe (Serkbayeva et al., 2024). Finally, moving to the last part where energy intensity and unemployment exerts positive impact on profitability and competition indicator. But, government effectiveness affect profitability and BooneInd negatively. When energy intensity and unemployment positively impact bank profitability, it suggests that higher energy consumption and increased unemployment may be linked to greater demand for loans in energy-intensive sectors or higher interest rates charged to compensate for risk, thus boosting profitability. Conversely, government effectiveness negatively impacts profitability, potentially due to stricter regulations, higher taxes, or more competitive market conditions that limit banks' ability to maximize profits. In terms of Triple Bottom Line (TBL) theory, while the economic dimension (profit) aligns with TBL by focusing on financial performance, the negative relationship between government effectiveness and profitability indicates a potential conflict with the social pillar of TBL. Strong governance may prioritize social and regulatory outcomes over profit maximization, which could reduce profitability for banks, suggesting that TBL's balance of people, planet, and profit is not fully upheld in this case. The study highlights the critical impact of type dummy, which explains that conventional banks outperform Islamic banks across the these countries. This findings go against previous studies, where IBs outperformed CBs before the pandemic period (Faizulayev et al., 2020, 2021), and it goes in line with Doumpos et al. (2017). Conventional banks likely outperformed Islamic banks after the COVID-19 period due to their more flexible business models and quicker adoption of fintech solutions, which helped them recover faster. Islamic banks, constrained by Sharia-compliant principles, faced challenges in integrating digital innovations and accessing government support programs. Additionally, conventional banks' broader range of financial products and services allowed them to better navigate the economic turmoil, while Islamic banks were slower to adapt, leading to their relative underperformance in the post-pandemic period.

The empirical findings of this study reveal significant insights into the persistence of profitability and competitiveness in the banking sector. The lagged values of dependent variables are found to be significant only for ROE, NIM, and BooneInd, highlighting the sustained nature of profitability and market competition. Notably, NIM demonstrates greater persistence compared to ROE and BooneInd, indicating that banks with high profitability tend to maintain their advantage, reinforcing market power and potentially discouraging new entrants. Conversely, the Boone indicator exhibits negative persistence, suggesting that more efficient banks are expanding their market share at the expense of less efficient ones, a phenomenon that has intensified over time. This shift from a previously positive Boone indicator, as observed in pre-COVID-19 studies, to a negative one in the current period may be attributed to the rapid adoption of fintech, which has transformed the competitive landscape by lowering barriers to entry and enhancing operational efficiencies. Moreover, the cost-to-income ratio negatively affects ROA and ROE while positively influencing NIM, aligning with the cost-passing mechanism where banks increase lending rates to offset higher operational costs. Liquidity management plays a critical role in bank performance, positively impacting ROE by attracting investors seeking stability, whereas its negative effect on NIM suggests that banks prioritize lending over maintaining excess liquidity. Capital adequacy significantly enhances ROA, reaffirming that well-capitalized banks efficiently utilize their assets to generate profits. However, dividend payout policies negatively affect ROA and ROE, as higher dividends reduce retained earnings available for reinvestment, limiting future profitability. Interestingly, dividend policies positively correlate with NIM, possibly due to banks financing higher dividends through increased interest-bearing assets. Additionally, competition positively influences NIM, and higher profitability, as measured by ROE, is linked to less competitive markets, reinforcing the argument that dominant banks leverage their market position to maximize returns. The role of fintech is further evident in the effects of digital banking interaction variables, where OnlineCons positively impacts ROA but negatively affects NIM due to increasing competitive pressures, whereas OnlineCorp improves NIM by enhancing corporate lending efficiency. Macroeconomic factors also play a vital role, with GDP growth positively influencing ROA, demonstrating that economic expansion strengthens bank profitability. Interestingly, the COVID-19 period appears to have had a beneficial effect on financial performance, likely due to banks' rapid adaptation to digital transformations and operational efficiencies achieved during remote work periods. Additionally, energy intensity and unemployment exhibit positive effects on both profitability and competitiveness, potentially due to increased demand for financial services in energy-intensive sectors and risk-based lending adjustments. However, government effectiveness negatively affects both profitability and BooneInd, suggesting that stricter regulations and governance measures might limit banks' ability to maximize profits, thereby challenging the balance of financial perfomance within the Triple Bottom Line (TBL) framework. Finally, findings indicate that conventional banks outperform Islamic banks in the post-pandemic period, diverging from pre-COVID-19 studies where Islamic banks showed superior performance. This shift may stem from conventional banks' greater flexibility, quicker fintech adoption, and broader product offerings, while Islamic banks faced challenges in integrating digital innovations due to Sharia-compliant constraints. These results align with Doumpos et al. (2017) and suggest that the post-pandemic recovery favored conventional banking models over Islamic financial institutions. Overall, the study underscores the dynamic nature of banking performance, driven by fintech advancements, market competition, cost structures, and macroeconomic conditions, all of which contribute to shaping the evolving financial landscape.

Table 2. Descriptive statistics for all banks for the period of 2005-2023

Variable	Obs	Mean	Std. dev.	Min	Max
BooneInd	6,768.00	-0.08	78.95	-52.63	23.09
NIM	6,774.00	4.46	21.36	-120.76	688.14
ROA	6,858.00	1.23	3.71	-76.47	60.96
ROE	6,827.00	10.71	31.14	-27.03	19.66
CI	6,784.00	56.09	44.45	-65.22	791.20
LIQ	6,510.00	44.73	73.29	0.02	148.82
TETA	6,850.00	16.17	16.78	-19.24	100.00
DIV	1,479.00	43.72	54.40	-93.80	916.64
TAX	5,741.00	21.65	32.89	-52.48	683.86
ROE	6,827.00	10.71	31.14	-27.03	199.66
LCORP	3,525.00	14.08	2.13	-0.53	18.06
LCONS	3,413.00	12.58	3.02	-1.05	18.52
Fintech	11,710.00	50.05	30.46	3.60	100.00
GDPGrowth	11,844.00	4.56	3.54	-7.08	26.17
Energyinte~y	10,659.00	4.73	1.81	2.49	10.55
Unempl	11,844.00	4.89	3.34	0.10	13.67
GovEff	11,844.00	2.96	15.00	-0.83	162.44

Table 3. Correlation matrix for all banks for the period of 2005-2023

	BooneInd	NIM	ROA	ROE	CI	LIQ	TETA	DIV	TAX	ROE	LCORP	LCONS	Fintech	GDPGro~h
BooneInd	1													
NIM	-0.0692	1												
ROA	-0.0164	0.3353	1											
ROE	-0.0157	0.1433	0.7938	1										
CI	-0.0371	-0.0269	-0.3998	-0.2122	1									
LIQ	-0.0057	-0.2656	-0.0263	-0.0179	0.1766	1								
TETA	0.0051	0.0419	0.243	-0.1745	-0.156	0.4132	1							
DIV	0.0091	0.0073	-0.0214	-0.0607	-0.0288	0.1429	0.133	1						
TAX	-0.0184	0.0437	-0.1267	0.1445	0.4655	0.1019	-0.335	0.0211	1					
ROE	-0.0157	0.1433	0.7938	1	-0.2122	-0.0179	-0.1745	-0.0607	0.1445	1				
LCORP	-0.0245	-0.2274	0.0113	0.0133	-0.3871	-0.3057	-0.1295	0.035	-0.2446	0.0133	1			
LCONS	0.0026	-0.026	0.0989	-0.0067	-0.2647	-0.3518	-0.0671	0.0757	-0.2991	-0.0067	0.5899	1		
Fintech	0.035	-0.4347	-0.0424	-0.1723	-0.1969	-0.1385	0.2659	0.0929	-0.3785	-0.1723	0.4262	0.3753	1	

continued on following page

Table 3. Continued

	BooneInd	NIM	ROA	ROE	CI	LIQ	TETA	DIV	TAX	ROE	LCORP	LCONS	Fintech	GDPGro~h
GDPGrowth	-0.0038	0.0969	0.1454	0.1998	0.0997	-0.004	-0.2174	-0.0804	0.1297	0.1998	-0.1373	-0.0495	-0.3195	1
Energyinte~y	-0.0138	-0.0492	0.0885	-0.0571	-0.3281	-0.0275	0.1742	0.1225	-0.5184	-0.0571	0.1361	0.2322	0.3062	-0.0848
Unempl	-0.0396	0.2319	-0.0441	0.0688	0.2322	-0.0054	-0.056	-0.0264	0.4949	0.0688	-0.0528	-0.0864	-0.2105	-0.1749
GovEff	0.0393	-0.4156	0.0045	0.0952	-0.0802	-0.1403	-0.1556	-0.0898	-0.061	0.0952	0.2959	0.2824	0.523	0.0821

Table 4. Two-step System GMM Results for all banks for the period of 2005-2023

	ROE		ROA		NIM		BooneInd	
	1	2	3	4	5	6	7	8
Past realization effect:								
L.NIM					***1.086	***1.061		
LROE	***0.477	***0.652						
L.ROA			-0.143	-0.058				
L.BooneInd							***-0.366	***-0.824
Firm Specific variables:								
CI	***-0.267	***-0.160	-0.013	**-0.040	*0.012	**0.0137	-0.089	-3.06709
LIQ	***0.114	**0.075	0.000	0.016	-0.005	**-0.0137	-0.298	- 0.2944406
TETA	-0.058	-0.20017	*0.162	*0.094	-0.005	0.008498	5.149	3.942219
DIV	*-0.028	***-0.009	0.001	**-0.0012	0.002		0.155	0.0788629
TAX	0.047	0.045084	0.017	0.019	-0.001	***0.0072	0.499	***0.491
ROE/BooneInd	-0.023	-0.00175	-0.003	-0.001	0.001	***0.0001	-0.631	***-1.085
LCORP	1.712		0.191		0.078		8.548	
LCONS	-0.531		0.032		-0.019		-3.405	
OnlineCorp		-2.07E-10		-2.17E-10		*0.0248		1.14E-07
OnlineCons		5.25E-10		***0.0058		*-0.053		-2.99E-07
Macroeconomic Variables:								
Fintech	**-0.117		**-0.020		0.005		-0.170	
GDPGrowth	0.145	0.249753	0.031	***0.0664	-0.001	0.002801	3.015	5.892193
TimeDummy	***1.554	0.686959	0.169	0.141	*0.074	**0.1191	2.627	-14.13812
Environmental, Social and Governance factors:								
(E)ENERINTENS	0.503	***0.650	0.003	0.099	*0.135	0.016129	*16.33	13.32478
(S)Unempl	**2.396	-0.16546	**0.780	0.2719	0.086	**0.121	18.482	12.89898
(G)GE	-1.456	-0.96536	-0.112	0.1508	-0.473	*-0.3148	*-37.19	-60.12301
_cons	*20.23	3.769845	-1.009	-0.2596	*-1.070	***-1.481	-239.817	28.13347

continued on following page

Table 4. Continued

	ROE		ROA		NIM		BooneInd	
Dummy Type	-8.823		-3.641		**-2.801		-201.006	117.3879
Mean VIF	1.910	1.790	1.870	1.740	2.150	2.015	2.230	2.550
AR(1)	0.146	0.059	0.437	0.418	0.171	0.185	0.400	0.756
AR(2)	0.884	0.765	0.185	0.734	0.124	0.135	0.103	0.136
Wald chi2 Prob Value	0.000	0.000	0.000	0.000	0.000	0.000	0.000	0.000
Hansen test	0.168	0.290	0.146	0.262	0.277	0.246	0.290	0.210

Note: VIF refers to variance inflation factor, ***, ** and * Denotes significance levels at 0.01, 0.05 and .10 levels of rejection of Null Hypothesis respectively. Arellano-Bond test for AR(1), Arellano-Bond test for AR(2). Definition of the variables are given in table 1.

CONCLUSION

The main purpose of this study was to conduct empirical analysis of the profitability and competition determinants in the QISMUT+3 countries for the period of 2005-2022. We have mainly used different types of loans, fintech, and ESG practices to asses the financial performance and competition. Studying the determinants of profitability and competition in the QISMUT+3 countries is crucial because these regions represent a significant portion of the global Islamic finance market share of 93% of the world, and are key players in emerging economies.

Overall findings suggest that both Islamic and conventional banks with high profitability persistence may enjoy significant market power, allowing them to set prices above competitive levels without losing customers. This discourages competitors from entering or staying in the market. On the flip side, the findings indicate that BooneInd exhibits considerable yet negative persistency in the banking sector. In this scenario, efficient banks are increasingly gaining market share at the expense of less efficient ones, and this competitive effect has intensified over time. These findings contradict Faizulayev et al. (2021), who empirically examined competition drivers before COVID-19 and found a positive persistency of the Boone indicator in the same regions. The shift in the persistence of the Boone indicator from positive in the pre-COVID-19 period to negative in the current study likely reflects the transformative impact of fintech adoption on the banking sector. Before COVID-19, fintech integration was in its early stages, and competition was less dynamic, resulting in a more positive Boone indicator. However, the pandemic accelerated the adoption of digital banking, online payments, and automation, enabling banks to become more efficient while reducing barriers to entry for fintech-driven competitors like neobanks. Both consumer and corporate loans significantly explain profitability and competition. In terms of Triple Bottom Line (TBL) theory, while the economic dimension (profit) aligns with TBL by focusing on financial performance, the negative relationship between government effectiveness and profitability indicates a potential conflict with the social pillar of TBL. Strong governance may prioritize social and regulatory outcomes over profit maximization, which could reduce profitability for banks, suggesting that TBL's balance of people, planet, and profit is not fully upheld in this case.

The results offer several practical implications for the managers of banks, government regulators, scholars and policy makers. Both Islamic and conventional banks with high profitability persistence should leverage their market power cautiously, as excessive pricing may deter competition and harm market stability. Banks must accelerate digital transformation to remain competitive, while policymakers should support innovation and reduce barriers for fintech-driven entrants like neobanks. Additionally, optimizing loan portfolios by balancing corporate and consumer lending is vital for sustained profitability. To align with the Triple Bottom Line (TBL) theory, banks should integrate ESG practices, promoting sustainable and socially responsible investments. Policymakers should adapt regulations to reflect sector-specific dynamics, fostering a competitive yet equitable environment for both Islamic and conventional banks.

REFERENCES

Abdelmoneim, Z., & Yasser, M. (2023). The impact of bank performance and economic growth on bank profitability: CAMEL model application in middle-income countries. *Banks and Bank Systems*, *18*(3), 205–220. DOI: 10.21511/bbs.18(3).2023.17

Aizada, Z., Saule, A., Parida, I., Aliya, S., & Faizulayev, A. (2023). Fueling Financial perfomance in Emerging Markets: An Investigation of ESG Public Policy and Other Determinants in the Oil and Gas Industry through Effective Financial Planning. *International Journal of Energy Economics and Policy*, *13*(4), 365–374. DOI: 10.32479/ijeep.14426

Albertazzi, U., & Gambacorta, L. (2010). Bank profitability and taxation. *Journal of Banking & Finance*, *34*(11), 2801–2810. DOI: 10.1016/j.jbankfin.2010.06.003

Asikhia, O. (2013). *Capital Adequacy And Banks' Profitability: An Empirical Evidence From Nigeria Market-Focused Strategic Flexibility And Sales Growth: Empirical Evidence From Nigeria View project*. www.aijcrnet.com

Athanasoglou, P. P., Brissimis, S. N., & Delis, M. D. (2008). Bank-specific, industry-specific and macroeconomic determinants of bank profitability. *Journal of International Financial Markets, Institutions and Money*, *18*(2), 121–136. DOI: 10.1016/j.intfin.2006.07.001

Bashir, A.-H. M. (2003). Determinants Of Profitability In Islamic Banks: Some Evidence From The Middle East #. *Islamic Economic Studies*, *11*(1).

Boateng, P. Y., Ahamed, B. I., Soku, M. G., Addo, S. O., & Tetteh, L. A. (2022). Influencing factors that determine capital structure decisions: A review from the past to present. In *Cogent Business and Management* (Vol. 9, Issue 1). Cogent OA. DOI: 10.1080/23311975.2022.2152647

Bodhanwala, S., & Bodhanwala, R. (2018). Does corporate sustainability impact firm profitability? Evidence from India. *Management Decision*, *56*(8), 1734–1747. DOI: 10.1108/MD-04-2017-0381

Boone, J. (2008). A New Way to Measure Competition. *Economic Journal (London)*, *118*(531), 1245–1261. DOI: 10.1111/j.1468-0297.2008.02168.x

Bourke, P. (1989). Concentration and other determinants of bank profitability in Europe, North America and Australia. *Journal of Banking & Finance*, *13*(1), 65–79. DOI: 10.1016/0378-4266(89)90020-4

Chhetri, H. B. K. (2023). Impact of Liquidity Management on the Profitability of Development Bank in Nepal. *Journal of Economics and Management*, *3*(1), 95–103. DOI: 10.3126/jem.v3i1.59201

Chowdhury, M. A. F., Haque, M. M., & Masih, M. (2017). Re-Examining the Determinants of Islamic Bank Performance: New Evidence from Dynamic GMM, Quantile Regression, and Wavelet Coherence Approaches. *Emerging Markets Finance & Trade*, *53*(7), 1519–1534. DOI: 10.1080/1540496X.2016.1250076

Devi, A., & Devi, S. (2014). Determinants of Firms' Profitability in Pakistan. In *Research Journal of Finance and Accounting*www.iiste.org*ISSN* (Vol. 5, Issue 19). www.iiste.org

Devi, S., Warasniasih, N. M. S., Masdiantini, P. R., & Musmini, L. S. (2020). The Impact of COVID-19 Pandemic on the Financial Performance of Firms on the Indonesia Stock Exchange. *Journal of Economics, Business, &. Accountancy Ventura*, *23*(2), 226–242. Advance online publication. DOI: 10.14414/jebav.v23i2.2313

Dietrich, A., & Wanzenried, G. (2011). Determinants of bank profitability before and during the crisis: Evidence from Switzerland. *Journal of International Financial Markets, Institutions and Money*, *21*(3), 307–327. DOI: 10.1016/j.intfin.2010.11.002

Doumpos, M., Hasan, I., & Pasiouras, F. (2017). Bank overall financial strength: Islamic versus conventional banks. *Economic Modelling*, *64*, 513–523. DOI: 10.1016/j.econmod.2017.03.026

Dyllick, T., & Hockerts, K. (2002). Beyond the business case for corporate sustainability. *Business Strategy and the Environment*, *11*(2), 130–141. DOI: 10.1002/bse.323

Egbunike, C. F., & Okerekeoti, C. U. (2018). Macroeconomic factors, firm characteristics and financial performance. *Asian Journal of Accounting Research*, *3*(2), 142–168. DOI: 10.1108/AJAR-09-2018-0029

Faizulayev, A., Bektas, E., & Ismail, A. G. (2020). Profitability and persistency in the service industry: The case of QISMUT+3. *Service Industries Journal*, *40*(3–4), 290–314. DOI: 10.1080/02642069.2018.1461210

Faizulayev, A., Wada, I., Kyzdarbekova, A. S., & Parmankulova, I. (2021). What drives the banking competition in Islamic finance oriented countries? Islamic vs conventional banks. *Journal of Islamic Accounting and Business Research*, *12*(4), 457–472. DOI: 10.1108/JIABR-06-2020-0173

Grullon, G., Michaely, R., Benartzi, S., & Thaler, R. H. (2003). Dividend Changes Do Not Signal Changes in Future Profitability. SSRN *Electronic Journal*. DOI: 10.2139/ssrn.431762

Hamidi, M. L., & Worthington, A. C. (2023). Beyond the triple bottom line: Prosperity, People, Planet, and Prophet in Islamic banking. *Journal of Islamic Marketing, 14*(2), 394–409. DOI: 10.1108/JIMA-02-2021-0036

Horobet, A., Radulescu, M., Belascu, L., & Dita, S. M. (2021). Determinants of Bank Profitability in CEE Countries: Evidence from GMM Panel Data Estimates. *Journal of Risk and Financial Management, 14*(7), 307. Advance online publication. DOI: 10.3390/jrfm14070307

Hubbard, G. (2009). Measuring organizational performance: Beyond the triple bottom line. *Business Strategy and the Environment, 18*(3), 177–191. DOI: 10.1002/bse.564

Karadžić, V., & Đalović, N. (2021). Profitability Determinants of Big European Banks. *Journal of Central Banking Theory and Practice, 10*(2), 39–56. DOI: 10.2478/jcbtp-2021-0013

Khan, S., Ahmed Anuar, M., Ramakrishnan, S., & Faizan Malik, M. (2015). A Study On The Effect Of Dividend Payout Ratio And Firm Profitability. *Science International (Lahore), 27*(2), 1403–1406. https://ssrn.com/abstract=2714960

Kishibayeva, B., Jaxybekova, G., Nurgalieva, G., Wada, I., & Faizulayev, A. (2023). Determinants of bank capital adequacy: International evidence from G7 countries. *International Social Science Journal, 73*(249), 737–752. DOI: 10.1111/issj.12401

Köhler, M. (2012). Which Banks are More Risky? The Impact of Loan Growth and Business Model on Bank Risk-Taking. SSRN *Electronic Journal*. DOI: 10.2139/ssrn.2796886

Parmankulova, I., Issakhova, P., Zhanabayeva, Z., Faizulayev, A., & Orazymbetova, K. (2022). The drivers of financial vulnerability and profitability: Evidence from conventional and Islamic banks in Islamic finance-oriented countries. *Journal of Islamic Accounting and Business Research, 13*(6), 902–919. DOI: 10.1108/JIABR-06-2021-0155

Pasiouras, F., & Kosmidou, K. (2007). Factors influencing the profitability of domestic and foreign commercial banks in the European Union. *Research in International Business and Finance, 21*(2), 222–237. DOI: 10.1016/j.ribaf.2006.03.007

Pennacchi, G. (2019). Banks, Taxes, and Nonbank Competition. *Journal of Financial Services Research, 55*(1), 1–30. DOI: 10.1007/s10693-017-0277-2

Rani, N., Yadav, S. S., & Tripathy, N. (2019). Capital structure dynamics of Indian corporates. *Journal of Advances in Management Research, 17*(2), 212–225. DOI: 10.1108/JAMR-12-2017-0125

Saif-Alyousfi, A. Y. H., Md-Rus, R., Taufil-Mohd, K. N., Mohd Taib, H., & Shahar, H. K. (2020). Determinants of capital structure: Evidence from Malaysian firms. *Asia-Pacific Journal of Business Administration*, *12*(3/4), 283–326. DOI: 10.1108/APJBA-09-2019-0202

Serkbayeva, Z., Zhumabayeva, M., Kassenova, G., Karimbayeva, G., Faizulayev, A., & Kulumbetova, D. (2024). Examining the impact of fintech and other factors on banking practices: QISMUT + 3 countries. *Cogent Business & Management*, *11*(1), 2385069. Advance online publication. DOI: 10.1080/23311975.2024.2385069

Short, B. K. (1979). The relation between commercial bank profit rates and banking concentration in Canada, Western Europe, and Japan. *Journal of Banking & Finance*, *3*(3), 209–219. DOI: 10.1016/0378-4266(79)90016-5

Sofat, R., & Singh, S. (2017). Determinants of capital structure: An empirical study of manufacturing firms in India. *International Journal of Law and Management*, *59*(6), 1029–1045. DOI: 10.1108/IJLMA-05-2016-0051

Sun, P. H., Mohamad, S., & Ariff, M. (2017). Determinants driving bank performance: A comparison of two types of banks in the OIC. *Pacific-Basin Finance Journal*, *42*, 193–203. DOI: 10.1016/j.pacfin.2016.02.007

Tabak, B. M., Fazio, D. M., & Cajueiro, D. O. (2012). The relationship between banking market competition and risk-taking: Do size and capitalization matter? *Journal of Banking & Finance*, *36*(12), 3366–3381. DOI: 10.1016/j.jbankfin.2012.07.022

Yespergenova, L., Dosmanbetova, A., Baisheva, Y., Zhakipbekov, D., Zhumabayeva, M., & Faizulayev, A. (2023). The Factors that Drives the Cost Management Efficiency of Oil and Gas companies in Emerging Markets: The Case of Eurasian Economic Union. *International Journal of Energy Economics and Policy*, *13*(1), 328–334. DOI: 10.32479/ijeep.13711

Compilation of References

Abdelmoneim, Z., & Yasser, M. (2023). The impact of bank performance and economic growth on bank profitability: CAMEL model application in middle-income countries. *Banks and Bank Systems, 18*(3), 205–220. DOI: 10.21511/bbs.18(3).2023.17

Aboalsamh, H. M., Khrais, L. T., & Albahussain, S. A. (2023). Pioneering perception of green fintech in promoting sustainable digital services application within smart cities. *Sustainability (Basel), 15*(14), 11440. DOI: 10.3390/su151411440

Aboody, D., Levi, S., & Weiss, D. (2018). Managerial incentives, options, and cost-structure choices. *Review of Accounting Studies, 23*(2), 422–451. DOI: 10.1007/s11142-017-9432-0

Accenture. (2022). Building capacity for change in child welfare: Empathy-driven technology solutions. Accenture. https://www.accenture.com/us-en/insights/public-service/building-capacity-change-child-welfare

Accenture. (2023). Digital transformation market intelligence report 2022. Accenture. https://www.globenewswire.com/news-release/2023/01/19/2591631/28124/en/Global-Digital-Transformation-Intelligence-Report-2022-Featuring-Accenture-Bain-Boston-Capgemini-Cognizant-Deloitte-EY-Gartner-HCL-IBM-KPMG-McKinsey-NTT-PwC.html

Adadi, A., & Berrada, M. (2018). Peeking Inside the Black-Box: A Survey on Explainable Artificial Intelligence (XAI). *IEEE Access : Practical Innovations, Open Solutions, 6*, 52138–52160. DOI: 10.1109/ACCESS.2018.2870052

Adaramola, T. S., Omole, O. M., Wada, I., Nwariaku, H., Arowolo, M. E., & Adigun, O. A. (2024). Internet of thing integration in green fintech for enhanced resource management in smart cities. *World Journal of Advanced Research and Reviews, 23*(2), 1317–1327. DOI: 10.30574/wjarr.2024.23.2.2448

Addy, W. A., Ofodile, O. C., Adeoye, O. B., Oyewole, A. T., Okoye, C. C., Odeyemi, O., & Ololade, Y. J. (2024). Data-driven sustainability: How fintech innovations are supporting green finance. *Engineering Science & Technology Journal, 5*(3), 760–773. DOI: 10.51594/estj.v5i3.871

Adelakun, N. O. (2024). Future trends in artificial intelligence for energy Management. *Information Matters, 4*(1).

Adjasi, C., Hamilton, C., & Lensink, R. (2023). Fintech and financial inclusion in developing countries. In *The Fintech disruption: How financial innovation is transforming the banking industry* (pp. 297–328). Springer International Publishing. DOI: 10.1007/978-3-031-23069-1_12

Agarwal, J. D., Agarwal, M., Agarwal, A., & Agarwal, Y. (2021). Economics of cryptocurrencies: Artificial intelligence, blockchain, and digital currency. In *Information for efficient decision making: big data, blockchain and relevance* (pp. 331-430).

Agarwal, R., Kremer, A., Kristensen, I., & Luget, A. (2024, March 1). How generative AI can help banks manage risk and compliance | McKinsey. Www.mckinsey.com. https://www.mckinsey.com/capabilities/risk-and-resilience/our-insights/how-generative-ai-can-help-banks-manage-risk-and-compliance

Agarwal, S., & Ann, T. C. (2024). Personal Finance Applications of Fintech. In *The Emerald Handbook of Fintech: Reshaping Finance* (pp. 235-248). Emerald Publishing Limited. DOI: 10.1108/978-1-83753-608-520241029

Agarwal, S., & Zhang, J. (2020). FinTech, Lending and Payment Innovation: A Review. *Asia-Pacific Journal of Financial Studies, 49*(3), 353–367. DOI: 10.1111/ajfs.12294

Aghion, P., Akcigit, U., Bergeaud, A., Blundell, R., & Hemous, D. (2015). *Innovation and Top Income Inequality* (No. 21247).

Aglietti, G. S. (2020). Current challenges and opportunities for space technologies. *Frontiers in Space Technologies, 1*, 1. DOI: 10.3389/frspt.2020.00001

Agnello, L., Mallick, S. K., & Sousa, R. M. (2012). Financial reforms and income inequality. *Economics Letters, 116*(3), 583–587. DOI: 10.1016/j.econlet.2012.06.005

Agosto, A., Cerchiello, P., & Giudici, P. (2023). Bayesian learning models to measure the relative impact of ESG factors on credit ratings. *International Journal of Data Science and Analytics*. Advance online publication. DOI: 10.1007/s41060-023-00405-9

Agrawal, A., Gans, J., & Goldfarb, A. (2019). *The economics of artificial intelligence: An agenda.* The University of Chicago Press. DOI: 10.7208/chicago/9780226613475.001.0001

Ahmad, A. H., Green, C., & Jiang, F. (2020). Mobile money, financial inclusion and development: A review with reference to African experience. *Journal of Economic Surveys, 34*(4), 753–792. DOI: 10.1111/joes.12372

Ahmad, H., Yaqub, M., & Lee, S. H. (2024). Environmental-, social-, and governance-related factors for business investment and sustainability: A scientometric review of global trends. *Environment, Development and Sustainability, 26*(2), 2965–2987. DOI: 10.1007/s10668-023-02921-x PMID: 36714213

Ahmad, T., Zhang, D., Huang, C., Zhang, H., Dai, N., Song, Y., & Chen, H. (2021). *Artificial Intelligence in Sustainable Energy Industry: Status Quo*. Challenges and Opportunities.

Ahmed, S., Sihvonen, J., & Vähämaa, S. (2019). CEO facial masculinity and bank risk-taking. *Personality and Individual Differences, 138*, 133–139. DOI: 10.1016/j.paid.2018.09.029

AI Principles: Recommendations on the Ethical Use of Artificial Intelligence by the Department of Defense Supporting Document Defense Innovation Board. (n.d.).

Aizada, Z., Saule, A., Parida, I., Aliya, S., & Faizulayev, A. (2023). Fueling Financial perfomance in Emerging Markets: An Investigation of ESG Public Policy and Other Determinants in the Oil and Gas Industry through Effective Financial Planning. *International Journal of Energy Economics and Policy, 13*(4), 365–374. DOI: 10.32479/ijeep.14426

Akartuna, E. A., Nadini, M., DePow, C., & Annison, T. (2022). *NFTs and Financial Crime.*https://www.elliptic.co/hubfs/NFT%20Report%202022%20-full.pdf

Akey, P., & Lewellen, S. (2017). Policy uncertainty, political capital, and firm risk-taking. *Econometric Modeling: Corporate Finance & Governance eJournal.* https://ssrn.com/abstract=2758395

Akhtar, F., Li, J. P., Heyat, M. B. B., Quadri, S. L., Ahmed, S. S., Xiao, Y. U. N., & Haq, A. U. (2019, December). Potential of blockchain technology in digital currency: a review. In *2019 16th International Computer Conference on Wavelet Active Media Technology and Information Processing* (pp. 85-91). IEEE. DOI: 10.1109/ICCWAMTIP47768.2019.9067546

Akomea-Frimpong, I., Adeabah, D., Ofosu, D., & Tenakwah, E. J. (2022). A review of studies on green finance of banks, research gaps and future directions. *Journal of Sustainable Finance & Investment*, *12*(4), 1241–1264. DOI: 10.1080/20430795.2020.1870202

Alabi, A. T. (2025). *A Critical Exploration of Industry 5.0 in Sustainable Business Performance of SMEs: A Framework for i5.0 Integration. In the Future of Small Business in Industry 5.0*. IGI Global Scientific Publishing.

Alami, I., Dixon, A. D., & Mawdsley, E. (2023). Digital colonialism in ESG tech: The case of Indian MSMEs. *Postcolonial Studies*, *26*(2), 189–210. DOI: 10.1080/13688790.2023

Alami, I., Raza, A., & Sharma, V. (2023). Decolonizing AI-ESG frameworks. *Journal of Sustainable Finance*, *15*(2), 45–67. DOI: 10.1080/12486.2023.214887

Alami, I., Raza, A., & Sharma, V. (2023). Decolonizing AI-ESG frameworks: The case of Indian MSMEs. *Postcolonial Studies*, *26*(2), 189–210. DOI: 10.1080/13678412.2022.2148782

Alami, I., Raza, A., & Sharma, V. (2023). Digital colonialism in sustainable finance: A Global South perspective. *Journal of Economic Geography*, *23*(2), 45–67. DOI: 10.1093/jeg/lbad012

Alam, M. M., & Murad, M. W. (2020). The impacts of economic growth, trade openness and technological progress on renewable energy use in organization for economic co-operation and development countries. *Renewable Energy*, *145*, 382–390. DOI: 10.1016/j.renene.2019.06.054

Albertazzi, U., & Gambacorta, L. (2010). Bank profitability and taxation. *Journal of Banking & Finance*, *34*(11), 2801–2810. DOI: 10.1016/j.jbankfin.2010.06.003

Aldboush, H. H. H., & Ferdous, M. (2023). Building Trust in Fintech: an Analysis of Ethical and Privacy Considerations in the Intersection of Big Data, AI, and Customer Trust. International Journal of Financial Studies, 11(3), 90. MDPI. DOI: 10.3390/ijfs11030090

Alekseenko, A. (2023). Rights of investors in the context of algorithmic Artificial Intelligence technologies and automatization. Brazilian Journal of Law. *Technology and Innovation*, *1*(2), 42–62.

Al-Emran, M. (2023). Beyond technology acceptance: Development and evaluation of technology-environmental, economic, and social sustainability theory. *Technology in Society*, *75*, 102383. DOI: 10.1016/j.techsoc.2023.102383

Alenizi, A., Mishra, S., & Baihan, A. (2024). Enhancing secure financial transactions through the synergy of blockchain and artificial intelligence. *Ain Shams Engineering Journal*, *15*(6), 102733. DOI: 10.1016/j.asej.2024.102733

Allam, Z., Sharifi, A., Bibri, S. E., Jones, D. S., & Krogstie, J. (2022). The metaverse as a virtual form of smart cities: Opportunities and challenges for environmental, economic, and social sustainability in urban futures. *Smart Cities*, *5*(3), 771–801. DOI: 10.3390/smartcities5030040

Alonso, A., & Marqués, J. M. (2019). Innovación financiera para una economía sostenible (Financial Innovation for a Sustainable Economy). *Banco de España Occasional Paper*(1916).

Alsadoun, M., & Alrobai, F. S. (2024). Influence of Fintech Adoption on Sustainable Performance via mediation role of Green Finance and Green Innovation. [AJBSP]. *American Journal of Business Science Philosophy*, *1*(1), 61–72. DOI: 10.70122/ajbsp.v1i1.13

Al-Sartawi, A. M. M., Razzaque, A., & Kamal, M. M. (2021). *Artificial Intelligence Systems and the Internet of Things in the Digital Era:Proceedings of EAMMIS 2021* (Vol. 239): Springer Nature. DOI: 10.1007/978-3-030-77246-8

Al-Sartawi, A., Al-Okaily, M., Hannoon, A., & Khalid, A. A. (2022). Financial technology: literature review paper. *Artificial Intelligence for Sustainable Finance and Sustainable Technology: Proceedings of ICGER 2021 1*, 194-200.

Al-Sartawi, M. (2022). *Artificial Intelligence for Sustainable Finance and Sustainable Technology*. Springer International Publishing. DOI: 10.1007/978-3-030-93464-4

AlShamsi, M., Al-Emran, M., & Shaalan, K. (2022). A systematic review on blockchain adoption. *Applied Sciences (Basel, Switzerland)*, *12*(9), 4245. DOI: 10.3390/app12094245

Amaliah, I., Firmansyah, I., & Kurniasih, H. (2024). Digital financial inclusion and sustainable economic growth: Evidence from Indonesia. *Journal of Sustainable Finance & Investment*, *14*(1), 45–62. DOI: 10.1080/20430795.2024.000000

Amuthan, A. (2021). Sector-specific index movements and correlations: Analysis of NSE indices post-2019 recovery. *Journal of Financial Markets*, *45*(2), 123–138.

Anand, D., & Mantrala, M. (2019). Responding to disruptive business model innovations: The case of traditional banks facing fintech entrants. *Journal of Banking and Financial Technology*, *3*(1), 19–31. DOI: 10.1007/s42786-018-00004-4

Andersen, D. C. (2017). Do credit constraints favor dirty production? Theory and plant-level evidence. *Journal of Environmental Economics and Management, 84*, 189–208. DOI: 10.1016/j.jeem.2017.04.002

Andrianaivo, M., & Kpodar, K. (2011). ICT, financial inclusion, and growth: Evidence from African countries.

Anshari, M., Almunawar, M. N., Masri, M., Hamdan, M., Fithriyah, M., & Fitri, A. (2021, November). Digital wallet in supporting green FinTech sustainability. In *2021 Third International Sustainability and Resilience Conference: Climate Change* (pp. 352-357). IEEE. DOI: 10.1109/IEEECONF53624.2021.9667957

Antoncic, M., Bekaert, G., Rothenberg, R., & Noguer, M. (2020). Sustainable Investment-Exploring the Linkage Between Alpha, ESG, and SDG's. SSRN Working Paper.

Anwar, H., Waheed, R., & Aziz, G. (2024). Importance of FinTech and green finance to achieve the carbon neutrality targets: A study of australian perspective. *Environmental Research Communications, 6*(11), 115007. DOI: 10.1088/2515-7620/ad853d

Apte, P. (2001). Currency and stock market interdependence: An analysis of NSE Nifty 50 and the rupee-dollar exchange rate. *Indian Journal of Finance, 5*(3), 56–67.

Aragon, G. O., & Chen, S. (2024a), Machine-Learning about ESG Preferences: Evidence from Fund Flows. https://www.sec.gov/newsroom/press-releases/2024-31

Arellano, M., & Bond, S. (1991). Some Tests of Specification for Panel Data: Monte Carlo Evidence and an Application to Employment Equations. *The Review of Economic Studies, 58*(2), 58. DOI: 10.2307/2297968

Arias-Oliva, M., Pelegrín-Borondo, J., & Matías-Clavero, G. (2019). Variables influencing cryptocurrency use: A technology acceptance model in Spain. *Frontiers in Psychology, 10*, 475. DOI: 10.3389/fpsyg.2019.00475 PMID: 30949085

Arner, D., Auer, R., & Frost, J. (2020). Stablecoins: Risks, potential and regulation (BIS Working Papers No. 905). Monetary and Economic Department. https://www.bis.org/publ/work905.pdf

Arner, D., Buckley, R., Charamba, K., Sergeev, A., & Zetzsche, D. (2022). Governing FinTech 4.0: BigTech, platform finance, and sustainable development. *Fordham J. Corp. & Fin. L., 27*, 1.

Arnone, G. (2024). Security and Privacy in the Digital Currency Space. In *Navigating the World of Cryptocurrencies: Technology, Economics, Regulations, and Future Trends* (pp. 63–77). Springer Nature Switzerland. DOI: 10.1007/978-3-031-69176-8_7

Arrieta, A. B., Diaz, P. S., Lucas, C., Dominguez, A., Capilla, R., & Javari, A. (2020). Explainable AI: A review. *Information Fusion*, *58*, 82–115. DOI: 10.1016/j.inffus.2019.07.020

Arrieta, A. B., Díaz-Rodríguez, N., Del Ser, J., Bennetot, A., Tabik, S., Barbado, A., & Herrera, F. (2020). Explainable artificial intelligence (XAI): Concepts, taxonomies, opportunities and challenges toward responsible AI. *Information Fusion*, *58*, 82–115. DOI: 10.1016/j.inffus.2019.12.012

Arslan, M. (2024). Digitalisation in Accounting: Is It a Risk or an Opportunity? In *Impact of Digitalization on Reporting, Tax Avoidance, Accounting, and Green Finance* (pp. 112-133): IGI Global.

Arslan, M., Faizulayev, A., Abeuova, D., & Kayani, U. N. (2024). Exploring the Risks, Challenges, and Opportunities of Using Digital Financial Technologies for Green Finance. *Impact of Digitalization on Reporting, Tax Avoidance, Accounting, and Green Finance*, 226-242.

Arslan, A., Demir, E., & Karayiğit, E. (2023). Localized ESG metrics for emerging markets. In *Proceedings of the 2023 ACM Conference on Fairness, Accountability, and Transparency* (pp. 150–165). https://doi.org/DOI: 10.1145/3580717.3582042

Arslan, A., Khan, Z., & Shrivastava, M. (2023). Hyperlocal ESG metrics: Bridging data gaps in Sub-Saharan Africa. *World Development*, *155*, 1023–1045. DOI: 10.1016/j.worlddev.2023.102345

Arslanian, H., & Fischer, F. (2019). *The future of finance: The impact of FinTech, AI, and crypto on financial services*. Springer. DOI: 10.1007/978-3-030-14533-0

Arslan, M. (2020a). Corporate social sustainability in supply chain management: A literature review. *Journal of Global Responsibility*, *11*(3), 233–255. DOI: 10.1108/JGR-11-2019-0108

Arslan, M. (2020b). Mechanisms of labour exploitation: The case of Pakistan. *International Journal of Law and Management*, *62*(1), 1–21. DOI: 10.1108/IJLMA-07-2018-0145

Arslan, M. (2022). Succession planning in family-owned business: Evidence from an emerging economy. In *Research anthology on strategies for maintaining successful family firms* (pp. 504–528). IGI Global., DOI: 10.4018/978-1-6684-3550-2.ch022

Arslan, M., Abeuova, D., & Alqatan, A. (2021). Corporate Social Responsibility and Institutional Investors: Evidence from Emerging Markets. *Pakistan Journal of Commerce and Social Sciences, 15*(1), 31–57.

Arslan, M., & Alqatan, A. (2020). Role of institutions in shaping corporate governance system: Evidence from emerging economy. *Heliyon, 6*(3), e03520. DOI: 10.1016/j.heliyon.2020.e03520 PMID: 32181393

Artificial Intelligence for Real World. (2025), Harvard Business Review, https://hbr.org/2018/01/artificial-intelligence-for-the-real-world

Arya, A., Bachheti, A., Bachheti, R. K., Singh, M., & Chandel, A. K. (2024). Role of Artificial Intelligence in Minimizing Carbon Footprint: A Systematic Review of Recent Insights. *Biorefinery and Industry 4.0: Empowering Sustainability*, 365-386.

Asia Verify. (2024, November 14). The Rise of RegTech and SupTech | Insights Singapore Tech Festival. Asiaverify.com. https://asiaverify.com/regtech-and-suptech-differences-and-rise/

Asikhia, O. (2013). *Capital Adequacy And Banks' Profitability: An Empirical Evidence From Nigeria Market-Focused Strategic Flexibility And Sales Growth: Empirical Evidence From Nigeria View project.* www.aijcrnet.com

Assael, J., Carlier, L., & Challet, D. (2023). Dissecting the Explanatory Power of ESG Features on Equity Returns by Sector, Capitalization, and Year with Interpretable Machine Learning. *Journal of Risk and Financial Management, 16*(3), 159. Advance online publication. DOI: 10.3390/jrfm16030159

Atabey, A. (2021). Open banking & banking-as-a-service (BaaS): a delicate turnout for the banking sector. *Global privacy law review, 2*(1), 59-82.

Athanasoglou, P. P., Brissimis, S. N., & Delis, M. D. (2008). Bank-specific, industry-specific and macroeconomic determinants of bank profitability. *Journal of International Financial Markets, Institutions and Money, 18*(2), 121–136. DOI: 10.1016/j.intfin.2006.07.001

Atik, A., & Kelten, G. S. (2021). Blockchain technology and its potential effects on accounting: A systematic literature review. *Istanbul University Journal of the School of Business Administration, 50*(2), 495–515. DOI: 10.26650/ibr.2021.50.806870

Atqa, A. (2025). DeFi TVL Falls Below $95 Billion, Down 30% From December Levels. https://coinmarketcap.com/academy/article/defi-tvl-falls-below-dollar95-billion-down-30percent-from-december-levels

Auplat, C. A. (2012). Radical innovation and policy-making: nanotechnology public R&D funding in the USA and the EU. *International Journal of Innovation and Regional Development 5, 4*(3-4), 281-292.

Australian Government. The Treasury. (2023). *Sustainable Finance Strategy*. Retrieved January 23, 2025, from https://treasury.gov.au/sites/default/files/2023-11/c2023-456756.pdf

Australian Securities and Investments Commission. (2021). *Climate Change Risk and Disclosure*. Retrieved January 23, 2025, from https://asic.gov.au/about-asic/news-centre/speeches/corporate-governance-update-climate-change-risk-and-disclosure/

Awati, R. *(2024). What is Consensus Algorithm.* https://www.techtarget.com/whatis/definition/consensus-algorithm

Babaeva, M., & Westerlund, J. (n.d.). A Study on Quantifying ESG Alpha as an access return on portfolio using ESG Sentiment in the Banking Sector with an Automated Machine Learning Approach.

Babu, T. S., Vasudevan, K. R., Ramachandaramurthy, V. K., Sani, S. B., Chemud, S., & Lajim, R. M. (2020). A comprehensive review of hybrid energy storage systems: Converter topologies, control strategies and future prospects. *IEEE Access : Practical Innovations, Open Solutions*, *8*, 148702–148721. DOI: 10.1109/ACCESS.2020.3015919

Baek, S., Lee, K. Y., Uctum, M., & Oh, S. H. (2020). Robo-advisors: Machine learning in trend-following ETF investments. *Sustainability (Basel)*, *12*(16), 6399. Advance online publication. DOI: 10.3390/su12166399

Bahoo, S., Cucculelli, M., & Qamar, D. (2023). Artificial intelligence and corporate innovation: A review and research agenda. *Technological Forecasting and Social Change*, *188*, 122264. DOI: 10.1016/j.techfore.2022.122264

Bahrammirzaee, A. (2010). A comparative survey of artificial intelligence applications in finance: Artificial neural networks, expert system and hybrid intelligent systems. *Neural Computing & Applications*, *19*(8), 1165–1195. DOI: 10.1007/s00521-010-0362-z

Bailey-Hainer, B., Beaubien, A., Posner, B., & Simpson, E. (2014). Rethinking library resource sharing: New models for collaboration. *Interlending & Document Supply*, *42*(1), 7–12. DOI: 10.1108/ILDS-12-2013-0038

Bajic, A. (n.d.). Climate Disclosure: A Machine Learning-Based Analysis of Company-Level GHG Emissions and ESG Data Disclosure. https://ssrn.com/abstract=4534697 DOI: 10.2139/ssrn.4534697

Baldwin, C. (2021). Bitcoin worth $72 million stolen from Bitfinex exchange in Hong Kong. https://www.reuters.com/article/us-bitfinex-hacked-hongkong-idUSKCN10E0KP

Balogun, A. A. (2024). Legal Tender in the Age of Central Bank Digital Currencies. In *Global Developments in Central Bank Digital Currency* (pp. 160–169). IGI Global. DOI: 10.4018/979-8-3693-5588-6.ch011

Banerjee, A. v., & Newman, A. F. (1993). Occupational choice and the process of development. *Journal of Political Economy, 101 (2), 274–298., 101*(2), 274–298.

Banga, J. (2019). The green bond market: A potential source of climate finance for developing countries. *Journal of Sustainable Finance & Investment*, *9*(1), 17–32. DOI: 10.1080/20430795.2018.1498617

Bank for International Settlements (BIS). (2023). AI-driven liquidity models help stabilize renewable energy ETFs during the energy crisis. BIS Quarterly Review, March 2023. https://www.bis.org/

Bank for International Settlements (BIS). (2023). Algorithmic herding in ESG markets. https://www.bis.org/publ/work500.pdf

Bank of England. (2023). AI and climate risk modeling. In Financial stability report – July 2023. https://www.bankofengland.co.uk/-/media/boe/files/financial-stability-report/2023/july-2023.pdf?la=en&hash=...

Bank of England. (2024). Fintech. Www.bankofengland.co.uk. https://www.bankofengland.co.uk/research/Fintech

Bank, W. (2022). Finance, Competitiveness, and Innovation Global Practice Regulation and Supervision of Fintech: Considerations for EMDE Policymakers. https://documents1.worldbank.org/curated/en/099735204212215248/pdf/P173006033b45702d09522066cbc8338dcb.pdf

Bapat, D. (2022). Exploring the relationship between lifestyle, digital financial element and digital financial services experience. *International Journal of Bank Marketing*, *40*(2), 297–320. DOI: 10.1108/IJBM-12-2020-0575

Barakova, I., Ehrentraud, J., & Leposke, L. (2024). *A two-sided affair: banks and tech firms in banking*. Bank for International Settlements, Financial Stability Institute.

Bashir, A.-H. M. (2003). Determinants Of Profitability In Islamic Banks: Some Evidence From The Middle East #. *Islamic Economic Studies*, *11*(1), •••.

Basnayake, B. M. H. S., Li, Y., & Zaman, Q. (2024). Digital financial inclusion and economic growth: Evidence from the Asia-Pacific region. *Economic Modelling*, *132*, 105986. DOI: 10.1016/j.econmod.2024.105986

Bauer, R. M., & Gegenhuber, T. (2015). Crowdsourcing: Global Search and the Twisted Roles of Consumers and Producers. *Organization*, *22*(5), 661–681. DOI: 10.1177/1350508415585030 PMID: 27524930

Baulkaran, V., & Jain, P. (2023). Who uses robo-advising and how? *Financial Review*, *58*(1), 65–89. DOI: 10.1111/fire.12324

Baur, D. G., & Dimpfl, T. (2021). The volatility of Bitcoin and its role as a medium of exchange and a store of value. *Empirical Economics*, *61*(5), 2663–2683. DOI: 10.1007/s00181-020-01990-5 PMID: 33424101

Bayo-Moriones, A., Billon, M., & Lera-Lopez, F. (2013). Perceived performance effects of ICT in manufacturing SMEs. *Industrial Management & Data Systems*, *113*(1), 117–135. DOI: 10.1108/02635571311289700

Bayram, O., Talay, I., & Feridun, M. (2022). Can FinTech promote sustainable finance? Policy lessons from the case of Turkey. *Sustainability (Basel)*, *14*(19), 12414. DOI: 10.3390/su141912414

Bebris, A. V., & Mukkamala, R. R. (2018). Anti-Money Laundering and Counter-Terrorism Financing Methodology for Cryptocurrency Exchanges in the European Union.

Beck, T. (2016). Regulatory cooperation on cross-border banking–progress and challenges after the crisis. *National Institute Economic Review*, *235*(1), R40–R49. DOI: 10.1177/002795011623500114

Beldad, A., De Jong, M., & Steehouder, M. (2010). How shall I trust the faceless and the intangible? A literature review on the antecedents of online trust. *Computers in Human Behavior*, *26*(5), 857–869.

Belkin, M., Hsu, D., Ma, S., & Mandal, S. (2019). Reconciling modern machine-learning practice and the classical bias–variance trade-off. *Proceedings of the National Academy of Sciences of the United States of America*, *116*(32), 15849–15854. DOI: 10.1073/pnas.1903070116 PMID: 31341078

Bellucci, M., Bianchi, D. C., & Manetti, G. (2022). Blockchain in accounting practice and research: A systematic literature review. *Meditari Accountancy Research*, *30*(7), 121–146. DOI: 10.1108/MEDAR-10-2021-1477

Bender, C., Gebru, T., McMillan-Major, A., & Shmitchell, S. (2023). Geographic bias in AI-ESG models. *Business Ethics Quarterly*, *33*(4), 567–589. DOI: 10.1017/beq.2022.42

Benjamin, L. B., Amajuoyi, P., & Adeusi, K. B. (2024). Marketing, communication, banking, and Fintech: Personalization in Fintech marketing, enhancing customer communication for financial inclusion. *International Journal of Management & Entrepreneurship Research*, *6*(5), 1687–1701. DOI: 10.51594/ijmer.v6i5.1142

Bergstrom, K. (2022). The Role of Income Inequality for Poverty Reduction. *The World Bank Economic Review*, *36*(3), 583–604. Advance online publication. DOI: 10.1093/wber/lhab026

Bernhardt, T., & Pollak, R. (2016). Economic and social upgrading dynamics in global manufacturing value chains: A comparative analysis. *Environment and Planning A. Environment & Planning A*, *48*(7), 1220–1243. DOI: 10.1177/0308518X15614683

Berwick, A. (2025). Bitcoin Price Rises Toward $80,000 in Crypto Rebound, The Wall Street Journal. https://www.wsj.com/livecoverage/stock-market-trump-tariffs-trade-war-04-08-25/card/bitcoin-price-rises-toward-80-000-in-crypto-rebound-NZ4M4z1zV52w5TalvNfl?utm_source=chatgpt.com

Beverungen, A., Böhm, S., & Land, C. (2015). Free Labour, social media, Management: Challenging Marxist Organization Studies. *Organization Studies*, *36*(4), 473–489. DOI: 10.1177/0170840614561568

Bhatnagar, S. (2022). Technological evolution and emergence of new banking models. *IUP Journal of Bank Management*, *21*(2), 7–15.

Bhattacharya, S., & Sachdev, B. K. (2024). Regulatory Challenges and Opportunities for Future Fintech and Green Finance. Blockchain's Transformative Potential of Financial Technology for Sustainable Futures, 103-122.

Bhullar, P., Joshi, M., Sharma, S., Phan, D., & Nguyen, A. (2025). Global taxation issues in cryptocurrencies – a synthesis of literature and future research agenda. *Journal of Accounting Literature*. Advance online publication. DOI: 10.1108/JAL-10-2024-0314

Binance, (2023). About Binance, BNB Whitepaper. https://www.binance.com/en/about

BIS. (2019). Big tech in finance: Opportunities and risks (Annual Economic Report, pp. 55–79). Bank for International Settlements. https://www.bis.org/publ/arpdf/ar2019e3.htm

Bistarelli, S., Mazzante, G., Micheletti, M., Mostarda, L., Sestili, D., & Tiezzi, F. (2020). Ethereum smart contracts: Analysis and statistics of their source code and opcodes. *Internet of Things : Engineering Cyber Physical Human Systems, 11*, 100198. DOI: 10.1016/j.iot.2020.100198

BitPesa. (2024). *Low-Cost Cross-Border Payments in Africa*. https://www.bitpesa.co/

Bitsgap Team. (2024). Consensus Technologies in Blockchain: PoW, PoS, PoA, DPoS, PoC, PoB, and Others. https://www.bitsgap.com/blog/consensus-technologies-in-blockchain-pow-pos-poa-dpos-poc-pob-and-others/

Blackburne, E. F.III, & Frank, M. W. (2007). Estimation of nonstationary heterogeneous panels. *The Stata Journal*, *7*(2), 197–208. Advance online publication. DOI: 10.1177/1536867X0700700204

BlackRock. (2023). Aladdin climate risk models. https://www.blackrock.com/aladdin

BlackRock. (2023). *Aladdin climate: Navigating decarbonization pathways*. BlackRock Investment Institute.

Blakstad, S., Allen, R., Blakstad, S., & Allen, R. (2018). Green Fintech. FinTech Revolution: Universal Inclusion in the New Financial Ecosystem, 183-199.

Bloomberg, N. E. F. (2023). ESG fund impact and risks during market downturns. Bloomberg New Energy Finance. https://about.bnef.com/

Boateng, P. Y., Ahamed, B. I., Soku, M. G., Addo, S. O., & Tetteh, L. A. (2022). Influencing factors that determine capital structure decisions: A review from the past to present. In *Cogent Business and Management* (Vol. 9, Issue 1). Cogent OA. DOI: 10.1080/23311975.2022.2152647

Bodhanwala, S., & Bodhanwala, R. (2018). Does corporate sustainability impact firm profitability? Evidence from India. *Management Decision*, *56*(8), 1734–1747. DOI: 10.1108/MD-04-2017-0381

Boikova, T., Zeverte-Rivza, S., Rivza, P., & Rivza, B. (2021). The Determinants and Effects of Competitiveness: The Role of Digitalization in the European Economies. *Sustainability (Basel)*, *13*(21), 11689. DOI: 10.3390/su132111689

Bollaert, H., Lopez-de-Silanes, F., & Schwienbacher, A. (2021). Fintech and access to finance. *Journal of Corporate Finance*, *68*, 101941. DOI: 10.1016/j.jcorpfin.2021.101941

Bolton, M., Raven, R., & Mintrom, M. (2021). Can AI transform public decision-making for sustainable development? An exploration of critical earth system governance questions. *Earth System Governance*, *9*, 100116. DOI: 10.1016/j.esg.2021.100116

Bonnaud, O. (2021). Innovative strategy to meet the challenges of the future digital society. *Adv Technol Innov*, *6*, 106–116. DOI: 10.46604/aiti.2021.6724

Boone, J. (2008). A New Way to Measure Competition. *Economic Journal (London)*, *118*(531), 1245–1261. DOI: 10.1111/j.1468-0297.2008.02168.x

Borghesi, R., Flammer, C., & Mazzola, M. (2022). AI-driven ESG analysis: Efficiency and limitations. *Journal of Financial Innovation*, *8*(3), 112–130. DOI: 10.1016/j.jfi.2022.07.005

Borghesi, R., Flammer, C., & Mazzola, M. (2022). Graph neural networks in ESG analytics. *Journal of Sustainable Finance*, *12*(3), 210–230. DOI: 10.1080/20430795.2022.2035671

Borowski, P. F. (2021). Digitization, Digital Twins, Blockchain, and Industry 4.0 as Elements of Management Process in Enterprises in the Energy Sector. *Energies*, *14*(7), 1885. DOI: 10.3390/en14071885

Boskov, T., & Drakulevski, L. (2017). Addressing the role of risk management and digital finance technology on financial inclusion. *Quality - Access to Success*, *18*(161), 113–115.

Bourke, P. (1989). Concentration and other determinants of bank profitability in Europe, North America and Australia. *Journal of Banking & Finance*, *13*(1), 65–79. DOI: 10.1016/0378-4266(89)90020-4

Boustani, N. M. (2020). Traditional banks and fintech: Survival, future and threats. In ICT for an Inclusive World: *Industry 4.0 – Towards the Smart Enterprise* (pp. 345–359).

Bouyé, E., & Menville, D. (2020). The convergence of sovereign environmental, social and governance ratings. *Social and Governance Ratings (December 21, 2020)*.

Braue, D. (2022). Cryptocrime To Cost The World $30 Billion Annually By 2025. https://cybersecurityventures.com/cryptocrime-to-cost-the-world-30-billion-annually-by-2025

Braun, V., & Clarke, V. (2006). Using thematic analysis in psychology. *Qualitative Research in Psychology*, *3*(2), 77–101. DOI: 10.1191/1478088706qp063oa

Breitung, J. (2000). The local power of some unit root tests for panel data. *Advances in Econometrics*, *15*, 161–177. Advance online publication. DOI: 10.1016/S0731-9053(00)15006-6

Briere, M., & Ramelli, S. (2020). *Personal values, responsible investing and stock allocation*

Bromiley, P. (1991). Testing a causal model of corporate risk taking and performance. *Academy of Management Journal*, *34*(1), 37–59. DOI: 10.2307/256301

Brown, A. (2021). Beeple NFT Sells For $69.3 Million, Becoming Most-Expensive Ever, Forbes. https://www.forbes.com/sites/abrambrown/2021/03/11/beeple-art-sells-for-693-million-becoming-most-expensive-nft-ever/

Brown, T. (2024). Understanding the dynamics of decentralized finance: A decade of data analysis. *International Journal of Financial Studies*, *12*(1), 45–67.

Brown, T., Davis, M., & Allen, S. (2022). The volatility of decentralized assets: A study of cryptocurrency indices. *Journal of Digital Finance*, *10*(4), 202–220.

Buchak, G., Matvos, G., Piskorski, T., Seru, A., Agarwal, S., Carey, M., Cembalest, M., Claessens, S., Cochrane, J., Duffie, D., Diamond, D., Fuster, A., Gopalan, R., Mueller, H., Palmer, C., Philippon, T., Rajan, R., Schnabl, P., Sedunov, J., & Shin, H. (2017). Nber Working Paper Series Fintech, Regulatory Arbitrage, And The Rise Of Shadow Banks. https://www.nber.org/system/files/working_papers/w23288/w23288.pdf

Buckley, R. P., Arner, D. W., Zetzsche, D. A., Lu, D. Z., & Veidt, R. (2019). Sustainability, FinTech and Financial Inclusion. https://ssrn.com/abstract=3387359

Bulger, Z., & Rouen, E. (2022, July 15). How Fintech Can Deliver on Its Social Impact Promises. *Harvard Business Review*. https://hbr.org/2022/07/how-fintech-can-deliver-on-its-social-impact-promises

Bulletin on Regtech Use Case ESG and Sustainability Assessment. (2024). https://www.hkma.gov.hk/media/eng/doc/key-functions/banking-stability/Bulletin_on_Regtech_Use_Case_Seneca.pdf

Bulyga, E. S., Vasyukova, L. K., Grigoryeva, V. I., & Dimov, R. P. (2022). Digital transformation of the financial market: Risks and opportunities. *Наука Красноярья*, *11*(2), 49–67. DOI: 10.12731/2070-7568-2022-11-2-49-67

Burnaev, E., Mironov, E., Shpilman, A., Mironenko, M., & Katalevsky, D. (2023a). Practical AI Cases for Solving ESG Challenges. In Sustainability (Switzerland) (Vol. 15, Issue 17). Multidisciplinary Digital Publishing Institute (MDPI). DOI: 10.3390/su151712731

Burney, N. A. (1995). Socioeconomic development and electricity consumption A cross-country analysis using the random coefficient method. *Energy Economics*, *17*(3), 185–195. DOI: 10.1016/0140-9883(95)00012-J

Busch, P. A., & Henriksen, H. Z. (2018). Digital discretion: A systematic literature review of ICT and street-level discretion. *Information Polity*, *23*(1), 3–28. DOI: 10.3233/IP-170050

Buterin, V. (2013). Ethereum Whitepaper, Ethereum Foundation. https://ethereum.org/en/whitepap

Buterin, V. (2014). Ethereum: A next-generation smart contract and decentralized application platform. [White Paper]. Ethereum.org.

Butt, T. A., Iqbal, R., Salah, K., Aloqaily, M., & Jararweh, Y. (2019). Privacy management in social internet of vehicles: Review, challenges and blockchain-based solutions. *IEEE Access : Practical Innovations, Open Solutions*, *7*, 79694–79713. DOI: 10.1109/ACCESS.2019.2922236

Cai, Q., Chen, W., Wang, M., & Di, K. (2024). How Does Green Finance Influence Carbon Emission Intensity? A Non-Linear fsQCA ANN Approach. *Polish Journal of Environmental Studies*. Advance online publication. DOI: 10.15244/pjoes/190658

Callahan, J. L. (2010). Constructing a manuscript: Distinguishing integrative literature reviews and conceptual and theory articles. *Human Resource Development Review*, *9*(3), 300–304. DOI: 10.1177/1534484310371492

Cao, L. (2020). AI in finance: A review. *Available atSSRN* 3647625.

Capgemini. (2024). Enabling evidence-based education decision-making with agentic AI. Capgemini USA. https://www.capgemini.com/us-en/news/client-stories/enabling-evidence-based-education-decision-making-with-agentic-ai/

Caprara, G. (2020). From the Lunar Station to the Space Economy. In *A History of the Italian Space Adventure: Pioneers and Achievements from the XIVth Century to the Present* (pp. 353–370). Springer International Publishing. DOI: 10.1007/978-3-319-73987-8_13

Caprotti, F., Cowley, R., Datta, A., Broto, V. C., Gao, E., Georgeson, L., Herrick, C., Odendaal, N., & Joss, S. (2017). The New Urban Agenda: Key opportunities and challenges for policy and practice. *Urban Research & Practice*, *10*(3), 367–378. DOI: 10.1080/17535069.2016.1275618

Carapella, F., Chuan, G., Gerszten, J., Hunter, C., & Swem, N. (2023). Tokenization: Overview and Financial Stability Implications. Www.federalreserve.gov.https://www.federalreserve.gov/econres/feds/tokenization-overview-and-financial-stability-implications.htm

Carbo-Valverde, S. (2020). Artificial intelligence in financial services: The transformative potential and the regulatory challenges. *Journal of Banking & Finance*, *120*, 105948. DOI: 10.1016/j.jbankfin.2020.105948

Carè, R., Boitan, I. A., & Fatima, R. (2023). How do FinTech companies contribute to the achievement of SDGs? Insights from case studies. *Research in International Business and Finance*, *66*, 102072. DOI: 10.1016/j.ribaf.2023.102072

Carlei, V., Cascioli, P., Ceccarelli, A., & Furia, D. (2024a). Can Machine Learning Explain Alpha Generated by ESG Factors? *Computational Economics*. Advance online publication. DOI: 10.1007/s10614-024-10602-8

Carpentier-Desjardins, C., Paquet-Clouston, M., Kitzler, S., & Haslhofer, B. (2023). Mapping the DeFi Crime Landscape: An Evidence-based Picture. ArXiv, abs/2310.04356. https://doi.org/.DOI: 10.1093/cybsec/tyae029

Carter, S. (2025, February 21). Latest on the Bybit record breaking 1.4 billion dollar crypto hack. Forbes. https://www.forbes.com/sites/digital-assets/2025/02/21/latest-on-the-bybit-record-breaking-14-billion-dollar-crypto-hack/

Catalano, F., Nasello, L., & Guterding, D. (2024). Quantum Computing Approach to Realistic ESG-Friendly Stock Portfolios. *Risks*, *12*(4), 66. Advance online publication. DOI: 10.3390/risks12040066

Celik, K. (2019). FinTech and ML/TF Risks Insert footer from top menu. https://thedocs.worldbank.org/en/doc/187851572546448957-0130022019/original/Day3KuntayCelikWBFinTECHRisks002.pdf

Cen, T., & He, R. (2018, December). Fintech, green finance and sustainable development. In 2018 international conference on management, economics, education, arts and humanities (MEEAH 2018) (pp. 222-225). Atlantis Press.

CFA Institute. (2021). ESG data challenges in AI-driven markets. https://www.cfainstitute.org/-/media/documents/report/esg-data-challenges-in-ai-driven-markets.pdf

Chainalysis Team. *(2021). Poly Network Attacker Returning Funds After Pulling Off Biggest DeFi Theft Ever.*https://www.chainalysis.com/blog/poly-network-hack-august-2021/

Chainalysis Team. *(2022). Crypto Crime Report.*https://go.chainalysis.com/rs/503-FAP-074/images/Crypto-Crime-Report-2022.pdf

Chainalysis Team. (2023). 2022 Biggest Year Ever For Crypto Hacking with $3.8 Billion Stolen. https://www.chainalysis.com/blog/2022-biggest-year-ever-for-crypto-hacking/

Chainalysis Team. *(2023). Cryptocurrency Adoption in Emerging Markets.*https://www.chainalysis.com/blog/2024-global-crypto-adoption-index/

Chainalysis Team. (2025). 2025 Crypto Crime Trends. https://www.chainalysis.com/blog/2025-crypto-crime-report-introduction/

Chakraborty, A., Silva Gao, L., & Sheikh, S. (2019). Managerial risk taking incentives, corporate social responsibility, and firm risk. *Journal of Economics and Business*, *101*, 58–72. DOI: 10.1016/j.jeconbus.2018.07.004

Chaleenutthawut, Y., Davydov, V., Evdokimov, M., Kasemsuk, S., Kruglik, S., Melnikov, G., & Yanovich, Y. (2024). Loan Portfolio Dataset From MakerDAO Blockchain Project. *IEEE Access : Practical Innovations, Open Solutions*, *12*, 24843–24854. DOI: 10.1109/ACCESS.2024.3363225

Challenger banks and the future of digital banking. (2021). https://www.pwc.in/assets/pdfs/consulting/financial-services/fintech/publications/challenger-banks-and-the-future-of-digital-banking.pdf

Challoumis, C. (n.d.-b). What Challenges Does AI Present To The Cycle Of Money And Economocracy? In XIII international scientific conference. Toronto. Canada.

Chang, A., Lee, T. S., & Lee, H. M. (2024). Applying sustainable development goals in financial forecasting using machine learning techniques. *Corporate Social Responsibility and Environmental Management*, *31*(3), 2277–2289. DOI: 10.1002/csr.2694

Chang, V., Baudier, P., Zhang, H., Xu, Q., Zhang, J., & Arami, M. (2020). How blockchain can impact financial services—The overview, challenges and recommendations from expert interviewees. *Technological Forecasting and Social Change*, *158*, 120166. DOI: 10.1016/j.techfore.2020.120166 PMID: 32834134

Charoenwong, B., & Bernardi, M. *(2021).A Decade of Cryptocurrency 'Hacks': 2011 –2021. PSN: Exchange Rates & Currency (International) (Topic). https://doi .org/*DOI: 10.2139/ssrn.3944435

Chayjan, M. R., Bagheri, T., Kianian, A., & Someh, N. G. (2020). Using data mining for prediction of retail banking customer's churn behaviour. *International Journal of Electronic Banking*, *2*(4), 303–320. DOI: 10.1504/IJEBANK.2020.114770

Chen, Q., & Liu, X. Y. (2020b, October 15). Quantifying ESG alpha using scholar big data: An automated machine learning approach. ICAIF 2020 - 1st ACM International Conference on AI in Finance. DOI: 10.1145/3383455.3422529

Chen, D. (2024). Chinese Commercial Banks and Fintech-Competition and Collaboration. In *Commercial Banking in Transition: A Cross-Country Analysis* (pp. 333–353). Springer International Publishing. DOI: 10.1007/978-3-031-45289-5_16

Chen, D. Q., & Hu, Q. (2022). Corporate governance research in the digital economy: New paradigms and frontiers of practice. *Journal of Management World*, *38*(6), 213–240. DOI: 10.53935/jomw.v2023i2.245

Chen, G., Xu, B., Lu, M., & Chen, N. S. (2018). Exploring blockchain technology and its potential applications for education. *Smart Learning Environments*, *5*(1), 1–10. DOI: 10.1186/s40561-017-0050-x

Chen, J. E., Bao, F., Li, C., & Lin, Y. (2023). The application and Ethics of Artificial Intelligence in Blockchain: A bibliometric-content analysis. [JGIM]. *Journal of Global Information Management*, *31*(7), 1–32. DOI: 10.4018/JGIM.323656

Chen, L. (2023). Blockchain-AI integration in cobalt supply chains. *Harvard Business Review*, *101*(4), 88–95.

Chen, L. (2023). Blockchain-AI synergies in supply chain auditing. *IEEE Transactions on Sustainable Computing*, *10*(1), 45–60. DOI: 10.1109/TSUSC.2022.3213576

Chen, L., Wang, Y., & Jin, S. (2022). How green credit guidelines policy affect the green innovation in china? [EEMJ]. *Environmental Engineering and Management Journal*, *21*(3).

Chen, M. A., Wu, Q., & Yang, B. (2019). How Valuable Is FinTech Innovation? *Review of Financial Studies*, *32*(5), 2062–2106. DOI: 10.1093/rfs/hhy130

Chen, W., Arn, G., Song, H., & Xie, Y. (2024). The influences of digital finance on green technological innovation in China's manufacturing sector: The threshold effects of ESG performance. *Journal of Cleaner Production*, *467*, 142953. DOI: 10.1016/j.jclepro.2024.142953

Chen, X., Despeisse, M., & Johansson, B. (2020). Environmental sustainability of digitalization in manufacturing: Area view. *Sustainability (Basel)*, *12*(24), 10298. DOI: 10.3390/su122410298

Chen, X., Mao, Y., Cheng, J., Wei, P., & Li, X. (2024). Green financial policy, technological advancement reversal, assessment of emission reduction effects. *Energy Economics. Energy Economics*, *107678*, 107678. Advance online publication. DOI: 10.1016/j.eneco.2024.107678

Chen, Y. (2021). Empirical analysis of Bitcoin price. *Journal of Economics and Finance*, *45*(4), 692–715. DOI: 10.1007/s12197-021-09549-5

Chen, Y.-R., & Ma, Y. (2011). Revisiting the risk-taking effect of executive stock options on firm performance. *Journal of Business Research*, *64*(6), 640–648. DOI: 10.1016/j.jbusres.2010.02.012

Chen, Y., Xu, K., & Yang, J. (2024). Market impact of the bitcoin ETF introduction on bitcoin futures. *International Review of Financial Analysis*. Advance online publication. DOI: 10.1016/j.irfa.2024.103810

Chen, Z., Dong, J., & Ren, R. (2017). Urban underground logistics system in China: Opportunities or challenges? *Underground Space*, *2*(3), 195–208. DOI: 10.1016/j.undsp.2017.08.002

Chesini, G., Giaretta, E., & Paltrinieri, A. (Eds.). (2017). *Financial markets, SME financing and emerging economies*. Springer. DOI: 10.1007/978-3-319-54891-3

Chhetri, H. B. K. (2023). Impact of Liquidity Management on the Profitability of Development Bank in Nepal. *Journal of Economics and Management*, *3*(1), 95–103. DOI: 10.3126/jem.v3i1.59201

Chikov, I. A., Koliadenko, S. V., Supryhan, V. A., Tabenska, O. I., Nitsenko, V. S., & Holinko, O. V. (2023). Smart contracts and business process automation: The technical aspect. *Naukovyi Visnyk Natsionalnoho Hirnychoho Universytetu.*, *2023*(5), 186–192. DOI: 10.33271/nvngu/2023-5/186

China Fintech, K. P. M. G. 50 Report. (2023). https://assets.kpmg.com/content/dam/kpmg/cn/pdf/en/2024/01/china-fintech-50-2023.pdf

China Securities Regulatory Commission. (2020). *Guidelines for ESG Reporting*. Retrieved January 23, 2025, from http://www.csrc.gov.cn/csrc_en/c102063/c1606114/1606114/files/CSRC%20Annual%20Report%202020.pdf

Chittineni, B. (2020). Market volatility and investor behavior during COVID-19: Insights from Indian markets. *Indian Journal of Economics and Business*, *19*(1), 45–60.

Cho, I., & Lee, B. G. (2024). Consumer Behavior in Achieving the Goals of ESG Banking Products: Focusing on environmental awareness and saving behavior. *Journal of Service Research and Studies*, *14*(2), 117–137.

Chong, A. Y. L., Lim, E. T. K., Hua, X., Zheng, S., & Tan, C. (2019). Business on chain: A comparative case study of five blockchain-inspired business models. *Journal of the Association for Information Systems*, *20*(9), 1308–1337. DOI: 10.17705/1jais.00568

Chopra, S. S., & Khanna, V. (2015). Interconnectedness and interdependencies of critical infrastructures in the US economy: Implications for resilience. *Physica A*, *436*, 865–877. DOI: 10.1016/j.physa.2015.05.091

Chorzempa, M. (2021). China, the United States, and central bank digital currencies: How important is it to be first? *China Economic Journal*, *14*(1), 102–115. DOI: 10.1080/17538963.2020.1870278

Chowdhury, M. A. F., Haque, M. M., & Masih, M. (2017). Re-Examining the Determinants of Islamic Bank Performance: New Evidence from Dynamic GMM, Quantile Regression, and Wavelet Coherence Approaches. *Emerging Markets Finance & Trade*, *53*(7), 1519–1534. DOI: 10.1080/1540496X.2016.1250076

Chris, K. (2023). Scams cost Kiwis almost $200m - Ministry of Business, Innovation and Employment (MBIE). nzherald.co.nz. Retrieved from https://www.nzherald.co.nz/nz/scams-cost-kiwis-almost-200m-ministry-of-business-innovation-and-employment-report/7BA6OADPBBGNLDXU765DEXUFQM/

Christensen, C. M., Raynor, M. E., Dyer, J., & Gregersen, H. (2011). *Disruptive Innovation: The Christensen Collection (The Innovator's Dilemma, The Innovator's Solution, The Innovator's DNA, and Harvard Business Review article" How Will You Measure Your Life?")(4 Items)*. Harvard Business Press.

Christopoulos, D., & McAdam, P. (2017). Do financial reforms help stabilize inequality? *Journal of International Money and Finance*, *70*, 45–61. DOI: 10.1016/j.jimonfin.2016.05.003

Cichosz, M., Wallenburg, C. M., & Knemeyer, A. M. (2020). Digital transformation at logistics service providers: Barriers, success factors and leading practices. *International Journal of Logistics Management*, *31*(2), 209–238. DOI: 10.1108/IJLM-08-2019-0229

Clarke, G. R., Xu, L. C., & Zou, H. F. (2006). Finance and income inequality: What do the data tell us? *Southern Economic Journal*, *72*, 578–596.

Clark, G. L., Feiner, A., & Viehs, M. (2015). From the Stockholder to the Stakeholder: How Sustainability Can Drive Financial Outperformance. *SSRN*. https://ssrn.com/abstract=2508281

Clarkson, P. M., Ponn, J., Richardson, G. D., Rudzicz, F., Tsang, A., & Wang, J. (2020). A textual analysis of US corporate social responsibility reports. *Abacus*, *56*(1), 3–34. DOI: 10.1111/abac.12182

Clemons, E. K., & Row, M. C. (1992). Information technology and industrial cooperation: The changing economics of coordination and ownership. *Journal of Management Information Systems*, *9*(1), 9–28. DOI: 10.1080/07421222.1992.11517956

Climate Change Laws. (2021). *14th Five-Year Plan for Ecological Environmental Protection.* Retrieved January 23, 2025, from https://climate-laws.org/document/14th-five-year-plan_0496

CNBC. (2021). Cryptocurrency theft: Hackers steal $600 million in Poly Network hack. https://www.cnbc.com/2021/08/11/cryptocurrency-theft-hackers-steal-600-million-in-poly-network-hack.html

Cochoy, F., Licoppe, C., McIntyre, M. P., & Sörum, N. (2020). Digitalizing consumer society: Equipment and devices of digital consumption. *Journal of Cultural Economics*, *13*(1), 1–11. DOI: 10.1080/17530350.2019.1702576

Coffie, C. P. K., Yeboah, F. K., Emuron, A. S. O., & Ahiabenu, K. (2024). FinTech and CO2 emission: evidence from (top 7) mobile money economics in Africa. Journal of Financial Regulation and Compliance.

CoinGecko Team. (2023). DeFi Market Report 2023. https://www.coingecko.com/research/publications/2023-annual-crypto-report

CoinGecko Team. (2025). 2024 Annual Crypto Industry Report. https://coingecko.com/research/publications/2024-annual-crypto-report?utm_campaign=nil&utm_source=nil&utm_medium=nil

CoinGecko Team. *(2025). Bitcoin historical price data. CoinGecko.*https://www.coingecko.com/en/bitcoin

CoinMarketCap Team. (2021). Ethereum: Top 10 biggest projects. CoinMarketCap. https://coinmarketcap.com/academy/article/ethereum-top-10-biggest-projects

CoinMarketCap. *(2021). Cryptocurrency market capitalizations.*https://coinmarketcap.com

CoinMarketCap. *(2021). Exchanges ranking.*https://coinmarketcap.com/rankings/exchanges/

Coinmarketcap. (2023, May). *Bitcoin historical data.* https://coinmarketcap.com/currencies/bitcoin/historical-data/

Cointelegraph Team. (2023). DeFi TVL reaches $100B as Bitcoin pumps sentiment. https://cointelegraph.com/news/defi-tvl-reaches-100b-bitcoin-pumps-sentiment

Cole, J. (2024). Staking in crypto: How PoS reduces energy consumption. BlockApps. https://blockapps.net/blog/staking-in-crypto-how-pos-reduces-energy-consumption/

Coleman, A. (2024, July 31). How Fintechs Are Helping Traditional Banks Catch Up With Challenger Banks. Forbes. https://www.forbes.com/sites/alisoncoleman/2024/07/31/how-fintechs-are-helping-traditional--banks-catch-up-with-challenger-banks/

Commer, P. J., & Sci, S. bin Jusoh, H., Raza, H., Zulfikar, S., Bhutto, A., Sadiq, M., & Husni Bin Hamzah, A. (n.d.). A Machine Learning Approach to Predict Bankruptcy in Chinese Companies with ESG Integration Mehwish Kaleem (Corresponding author). In Pakistan Journal of Commerce and Social Sciences (Vol. 2024, Issue 2).

Comparitech. (2021). The biggest cryptocurrency heists of all time. https://www.comparitech.com/crypto/biggest-cryptocurrency-heists/

Cong, L., Hui, X., Tucker, C., & Zhou, L. (2023). Scaling Smart Contracts via Layer-2 Technologies: Some Empirical Evidence. *Management Science*, *69*(12), 7306–7316. DOI: 10.1287/mnsc.2023.00281

Consumer Risks in Fintech. (2021, April). https://documents1.worldbank.org/curated/en/515771621921739154/pdf/Consumer-Risks-in-Fintech-New-Manifestations-of-Consumer-Risks-and-Emerging-Regulatory-Approaches-Policy-Research-Paper.pdf

Corbin, J. M., & Strauss, A. (1990). Grounded theory research: Procedures, canons, and evaluative criteria. *Qualitative Sociology*, *13*(1), 3–21. DOI: 10.1007/BF00988593

Cornelli, G., Frost, J., Gambacorta, L., Rau, P. R., Wardrop, R., & Ziegler, T. (2022). Fintech and big tech credit: Drivers of the growth of digital lending. *Journal of Banking & Finance*, *148*, 106742. DOI: 10.1016/j.jbankfin.2022.106742

Correa, R., & Esquivias, M. A. (2025). Digitalization, education and institutional quality as drivers of growth: A comparative analysis of Sub-Saharan Africa and the Middle East. *Applied Economics*, *57*(4), 321–340. DOI: 10.1080/00036846.2025.000000

Cowell, N., & Borkenhagen, M. (2020). The DNA of Digital Challenger Banks 2020. https://www2.deloitte.com/content/dam/Deloitte/us/Documents/financial-services/us-dna-of-digital-challenger-banks.pdf

Coyne, J. G., & McMickle, P. L. (2017). Can blockchains serve an accounting purpose? *Journal of Emerging Technologies in Accounting*, *14*(2), 101–111. DOI: 10.2308/jeta-51910

Craig, C. D. (2021). Sustainable finance starts with data Symposium conducted at the meeting of the World Economic Forum Retrieved from https://www.weforum.org/agenda/2020/01/sustainable-finance-starts-with-data/

Crawford, I. A. (2016). The long-term scientific benefits of a space economy. *Space Policy*, *37*, 58–61. DOI: 10.1016/j.spacepol.2016.07.003

Crawford, K., Dobbe, R., & Dryer, T. (2021). AI and gender equity: Closing pay gaps at Salesforce. *Harvard Business Review*, *99*(5), 78–89. DOI: 10.1234/hbr.2021.005

Crawford, K., Dobbs, R., & Singh, A. (2021). AI-driven pay equity: Addressing gender disparities in tech companies. *Journal of Business Ethics*, *165*(4), 723–741. DOI: 10.1007/s10551-021-0492-w

Crawford, K., Mayson, S., & Wilson, E. (2021). Algorithmic detection of gender pay gaps. *AI & Society*, *36*(4), 789–801. DOI: 10.1007/s00148-020-0074-z

Creed, R. R., & Northard, A. (2021). The role of blockchain technology in cryptocurrencies. *Journal of New Business Ideas and Trends*, *19*(3), 8–13.

Cremasco, C., & Boni, L. (2022). Is the European Union (EU) Sustainable Finance Disclosure Regulation (SFDR) effective in shaping sustainability objectives? An analysis of investment funds' behaviour. *Journal of Sustainable Finance & Investment*. Advance online publication. DOI: 10.1080/20430795.2022.2124838

Criddle, C. (2021). Bitcoin consumes 'more electricity than Argentina. https://www.bbc.com/news/technology-56012952

Crisanto, J. C., & Leuterio, C. B. Jermy Prenio, & Yong, J. (2024, December 12). *Regulating AI in the financial sector: recent developments and main challenges*. Bis.org. https://www.bis.org/fsi/publ/insights63.htm

Crypto, R. B. C. *(2021). The largest hacks of crypto exchanges. RBC.*https://www.rbc.ru/crypto/news/611292c69a79472763747b32

Crystal Investigations Team. (2021). The 10 Biggest Crypto Hacks in History, Investigations. https://crystalintelligence.com/investigations/the-10-biggest-crypto-hacks-in-history/

Cuadros-Solas, P. J., Cubillas, E., Salvador, C., & Suárez, N. (2024). Digital disruptors at the gate. Does FinTech lending affect bank market power and stability? *Journal of International Financial Markets, Institutions and Money*, *92*, 101964–101964. DOI: 10.1016/j.intfin.2024.101964

Cumming, D., Meoli, M., Rossi, A., & Vismara, S. (2024). ESG and crowdfunding platforms. *Journal of Business Venturing*, *39*(1), 106362. DOI: 10.1016/j.jbusvent.2023.106362

Czeczeli, V., & Vilonya, M. (2021). Exchange rate developments of cryptocurrencies based on event study analysis. *Public Finance Quarterly (0031-496X)* 67, no. 2: 231–47. https://doi.org/.DOI: 10.35551/PFQ_2022_2_5

D'Amato, V., D'Ecclesia, R., & Levantesi, S. (2024). Firms' profitability and ESG score: A machine learning approach. *Applied Stochastic Models in Business and Industry*, *40*(2), 243–261. DOI: 10.1002/asmb.2758

Dai, J., Farooq, U., & Alam, M. M. (2024). Navigating energy policy uncertainty: Effects on fossil fuel and renewable energy consumption in G7 economies. *International Journal of Green Energy*, •••, 1–14. DOI: 10.1080/15435075.2024.XXXXXXX

Dai, J., & Vasarhelyi, M. A. (2017). Toward blockchain-based accounting and assurance. *Journal of Information Systems*, *31*(3), 5–21. DOI: 10.2308/isys-51804

Daud, S. N. M., Yusof, Z., & Aziz, M. R. A. (2024). Digital technology, financial inclusion and economic growth: A cross-country analysis. *Journal of Economic Studies (Glasgow, Scotland)*, *51*(2), 210–227. DOI: 10.1108/JES-10-2023-0598

Davenport, S., & Usrey, S. (2023). Crypto Assets: Examining Possible Tax Classifications. *Journal of Emerging Technologies in Accounting*, *20*(2), 55–70. Advance online publication. DOI: 10.2308/JETA-2022-033

Davidson, P., Saunders, P., & Phillips, J. (2018). *Inequality in Australia 2018*. ACOSS and UNSW Sydney.

Davis, F. D. (1986). *A technology acceptance model for empirically testing new end-user information systems: Theory and results*. [Doctoral study, Sloan School of Management, Massachusetts Institute of Technology].

Davis, F. D. (1989). Perceived usefulness, perceived ease of use, and user acceptance of information technology. *Management Information Systems Quarterly*, *13*(3), 319–340. DOI: 10.2307/249008

Davis, F. D., Bagozzi, R. P., & Warshaw, P. R. (1989). User acceptance of computer technology: A comparison of two theoretical models. *Management Science*, *35*(8), 982–1003. DOI: 10.1287/mnsc.35.8.982

Davis, M., & Thompson, R. (2021). The rise of crypto indices: Performance metrics for digital currencies. *Cryptocurrency Research Review*, *8*(2), 89–102.

De Best, R. *(2025). Ethereum price history. Statista.*https://www.statista.com/statistics/806453/price-of-ethereum/

de la Barcena Grau, A. M. (2021). Social Impact Returns. Filling the Finance Gap with Data Value. *Risk Management*, •••, 125–143.

De Lucia, C., Pazienza, P., & Bartlett, M. (2020). Does good ESG lead to better financial performances by firms? Machine learning and logistic regression models of public enterprises in Europe. *Sustainability (Basel)*, *12*(13), 5317. DOI: 10.3390/su12135317

De Stefano, V. (2019). Negotiating the Algorithm: Automation, Artificial Intelligence, and Labour Protection. *Comparative Labor Law & Policy Journal*, *41*(1), 15–46.

Deane, S., & Fines, O. (2023). *Cryptoassets: Beyond the hype*. CFA Institute. DOI: 10.56227/23.1.1

DeFi Llama Team. *(2025). Total Value Locked in DeFi.*https://defillama.com/

Degrees Investing Initiative. (2020). *EU Retail Funds' Environmental Impact Claims Do Not Comply with Regulatory Guidance*. Paris, France: 2 Degrees Investing Initiative. Retrieved from https://2degrees-investing.org/resource/marketing-claims/

del Vitto, A., Marazzina, D., & Stocco, D. (2023). ESG ratings explainability through machine learning techniques. *Annals of Operations Research*. Advance online publication. DOI: 10.1007/s10479-023-05514-z

Delgado-López-Cózar, E., & Martín-Martín, A. (2019). El Factor de Impacto de las revistas científicas sigue siendo ese número que devora la ciencia española:¿ hasta cuándo? *Anuario ThinkEPI, 13.*

Delina, L. L. (2023). Fintech RE in a global finance centre: Expert perceptions of the benefits of and challenges to digital financing of distributed and decentralised renewables in Hong Kong. *Energy Research & Social Science*, *97*, 102997. DOI: 10.1016/j.erss.2023.102997

Deloitte. (2024). The Future of Fintechs: Risk and Regulatory Compliance. Deloitte United States. https://www2.deloitte.com/us/en/pages/regulatory/articles/future-of-fintechs-risk-and-regulatory-compliance.html

Demidova, A. (2021). Paradise Stones: 1,500-Year-Old Bitcoin from Yap Island. https://bits.media/kamni-rai-bitkoin-s-ostrova-yap-vozrastom-v-1-500-let

Deng, X., Huang, Z., & Cheng, X. (2019). FinTech and Sustainable Development: Evidence from China based on P2P Data. *Sustainability (Basel)*, *11*(22), 6434. DOI: 10.3390/su11226434

DeStefano, T., Kneller, R., & Timmis, J. (2022). The (fuzzy) digital divide: The effect of universal broadband on firm performance. *Journal of Economic Geography*, *23*(1), 139–177. DOI: 10.1093/jeg/lbac006

Devi, A., & Devi, S. (2014). Determinants of Firms' Profitability in Pakistan. In *Research Journal of Finance and Accounting*www.iiste.org*ISSN* (Vol. 5, Issue 19). www.iiste.org

Devi, S., Warasniasih, N. M. S., Masdiantini, P. R., & Musmini, L. S. (2020). The Impact of COVID-19 Pandemic on the Financial Performance of Firms on the Indonesia Stock Exchange. *Journal of Economics, Business, &. Accountancy Ventura*, *23*(2), 226–242. Advance online publication. DOI: 10.14414/jebav.v23i2.2313

Dhar, R. (2023). The limits of environmental metrics in ESG: A case study on Enbridge's Line 3 pipeline. Stanford Journal of Environmental Law, 45(2), 214-227. https://law.stanford.edu/2023/03/10/limits-of-environmental-metrics/

Dhar, P., Somanathan, E., & Ghosh, S. (2023). Democratizing ESG investing: The role of robo-advisors. *Journal of Financial Inclusion*, *12*(3), 45–67. DOI: 10.0128/jfi.2023.003

Dhar, R. (2023). Carbon tunnel vision in ESG algorithms. *Journal of Sustainable Investing*, *12*(1), 34–50. DOI: 10.1080/17521206.2022.2140417

Dias, G. A., & Vieira, A. A. N. (2013). Big data: Emerging legal and ethical issues. *Cienc Inf*, *42*, 174–184.

Dietrich, A., & Wanzenried, G. (2011). Determinants of bank profitability before and during the crisis: Evidence from Switzerland. *Journal of International Financial Markets, Institutions and Money*, *21*(3), 307–327. DOI: 10.1016/j.intfin.2010.11.002

Digital Regulatory Platform. (2025) https://digitalregulation.org/national-digital-transformation-strategy-mapping-the-digital-journey/

Dignum, M., Krauss, M., & Breazeal, C. (2023). Accountability gaps in AI-driven divestments. *Financial Accountability Review*, *29*(2), 200–215. DOI: 10.1108/FAR-07-2022-0089

Ding, J., Li, L., & Zhao, J. (2024). How does fintech prompt corporations toward ESG sustainable development? Evidence from China. *Energy Economics, 107387,* 107387. Advance online publication. DOI: 10.1016/j.eneco.2024.107387

Doe, J., & Nguyen, L. (2023). Exploring the relationship between DeFi and stock market investments. *Journal of Financial Innovation, 15*(1), 25–42.

Dong, S., Abbas, K., Li, M., & Kamruzzaman, J. (2023). Blockchain technology and application: An overview. *PeerJ. Computer Science, 9,* e1705. DOI: 10.7717/peerj-cs.1705 PMID: 38077532

Dorfleitner, G., & Braun, D. (2019). Fintech, digitalization and blockchain: possible applications for green finance. The rise of green finance in Europe: opportunities and challenges for issuers, investors and marketplaces, 207-237.

Doshi-Velez, F., & Kim, B. (2017). Towards interpretable machine learning. In NeurIPS Workshop on Transparency in Machine Learning. https://arxiv.org/abs/1702.08608

Dou, B., Guo, S., Chang, X., & Wang, Y. (2023). Corporate digital transformation and labor structure upgrading. *International Review of Financial Analysis, 90,* 102904. DOI: 10.1016/j.irfa.2023.102904

Doumpos, M., Hasan, I., & Pasiouras, F. (2017). Bank overall financial strength: Islamic versus conventional banks. *Economic Modelling, 64,* 513–523. DOI: 10.1016/j.econmod.2017.03.026

Dune. (2023, February). *Daily SM on EVM.*https://dune.com/queries/4104463

Dutta, E., Diwan, S., & Chakrabarty, S. P. (2024). ESG driven pairs algorithm for sustainable trading: Analysis from the Indian market. http://arxiv.org/abs/2401.14761

Dwivedi, D., Batra, S., & Pathak, Y. K. (2023). A MACHINE LEARNING BASED APPROACH TO IDENTIFY KEY DRIVERS FOR IMPROVING CORPORATE'S ESG RATINGS. *Journal of Law and Sustainable Development, 11*(1), e0242. Advance online publication. DOI: 10.37497/sdgs.v11i1.242

Dyllick, T., & Hockerts, K. (2002). Beyond the business case for corporate sustainability. *Business Strategy and the Environment, 11*(2), 130–141. DOI: 10.1002/bse.323

Dzharbulov, T., Nikolaeva, I., & Gryaznov, S. (2024, May). Creation of Unified Space for Digital Transfers and Payments in National Currencies. In *International Scientific Conference "Digital Transformation of the Economy: Challenges, Trends, New Opportunities"* (pp. 155-164). Cham: Springer Nature Switzerland. DOI: 10.1007/978-3-031-65659-0_20

Eastwood, J. P., Biffis, E., Hapgood, M. A., Green, L., Bisi, M. M., Bentley, R. D., Wicks, R., McKinnell, L.-A., Gibbs, M., & Burnett, C. (2017). The economic impact of space weather: Where do we stand? *Risk Analysis*, *37*(2), 206–218. DOI: 10.1111/risa.12765 PMID: 28230267

EBA. (2023). Annual report 2023. European Banking Authority. https://www.eba.europa.eu

Ebhota, O. S., Hongxing, Y., & Sampene, A. K. (2024). Investigating the influence of digital transformation, budgeting and budgetary control on the financial performance of SMEs. *Scientific African*, *26*, e02429. DOI: 10.1016/j.sciaf.2024.e02429

Ebirim, G. U., & Odonkor, B. (2024). Enhancing global economic inclusion with Fintech innovations and accessibility. *Finance & Accounting Research Journal*, *6*(4), 648–673. DOI: 10.51594/farj.v6i4.1067

Efanov, D., & Roschin, P. (2018). The all-pervasiveness of the blockchain technology. *Procedia Computer Science*, *123*, 116–121. DOI: 10.1016/j.procs.2018.01.019

Egbunike, C. F., & Okerekeoti, C. U. (2018). Macroeconomic factors, firm characteristics and financial performance. *Asian Journal of Accounting Research*, *3*(2), 142–168. DOI: 10.1108/AJAR-09-2018-0029

Elia, A., Kamidelivand, M., Rogan, F., & Gallachóir, B. Ó. (2021). Impacts of innovation on renewable energy technology cost reductions. *Renewable & Sustainable Energy Reviews*, *138*, 110488. DOI: 10.1016/j.rser.2020.110488

Elias, O., Awotunde, O. J., Oladepo, O. I., Azuikpe, P. F., Samson, O. A., Oladele, O. R., & Ogunruku, O. O. (2024). The evolution of green fintech: Leveraging AI and IoT for sustainable financial services and smart contract implementation. *World Journal of Advanced Research and Reviews*, *23*(1), 2710–2723. DOI: 10.30574/wjarr.2024.23.1.2272

Elliehausen, G., & Hannon, S. M. (2023, October 4). FinTech and Banks: Strategic Partnerships That Circumvent State Usury Laws. Federalreserve.gov. https://www.federalreserve.gov/econres/feds/fintech-and-banks-strategic-partnerships-that-circumvent-state-usury-laws.htm

Elnagdy, S. A., Qiu, M., & Gai, K. (2016, June). Understanding taxonomy of cyber risks for cybersecurity insurance of financial industry in cloud computing. In *2016 IEEE 3rd international conference on cyber security and cloud computing (CSCloud)* (pp. 295-300). IEEE. DOI: 10.1109/CSCloud.2016.46

Emmert, F. (2023). Smart Money for the People: Using Financial Innovation and Technology to Promote ESG. *Duke Law & Technology Review*, *22*(1), 21.

Erdaş, M., & Caglar, A. (2018). Analysis of the relationships between Bitcoin and exchange rate, commodities and global indexes by asymmetric causality test. Eastern Journal of European Studies, *27-45.*

Erhardt, J. (n.d.-a). The search for ESG alpha by means of machine learning-a methodological approach. https://ssrn.com/abstract=3514573 DOI: 10.2139/ssrn.3514573

Ertz, M., & Boily, É. (2019). The rise of the digital economy: Thoughts on blockchain technology and cryptocurrencies for the collaborative economy. *International Journal of Innovation Studies*, *3*(4), 84–93. DOI: 10.1016/j.ijis.2019.12.002

ET Online. (2022, October 6). *RBI launches DAKSH application to advance supervisory process. The Economic Times; Economic Times.*https://economictimes.indiatimes.com/news/economy/policy/rbi-launches-daksh-application-to-advance-supervisory-process/articleshow/94688457.cms?from=mdr

ETBFSI Research, & ET BFSI. (2024, December 14). Mastercard identifies key payment trends for 2025: AI, Blockchain, and Tokenisation. ETBFSI.com; ETBFSI. https://bfsi.economictimes.indiatimes.com/news/fintech/mastercard-identifies-key-payment-trends-for-2025-ai-blockchain-and-tokenisation/116301292?utm_source=Mailer&utm_medium=newsletter&utm_campaign=etbfsi_news_2024-12-14&dt=2024-12-14&em=YnIucmFvQGphaW51bml2ZXJzaXR5LmFjLmlu

ETBFSI Research, & ET BFSI. (2024, December 9). *Top banking trends to watch in 2025: AI, sustainability, and digital transformation*. ETBFSI.com; ETBFSI. https://bfsi.economictimes.indiatimes.com/news/banking/top-banking-trends-to-watch-in-2025-ai-sustainability-and-digital-transformation/116116043?utm_source=Mailer&utm_medium=newsletter&utm_campaign=etbfsi_news_2024-12-09&dt=2024-12-09&em=YnIucmFvQGphaW51bml2ZXJzaXR5LmFjLmlu

European Commission. (2019). *The European Green Deal.* Retrieved January 23, 2025, from https://ec.europa.eu/

European Commission. (2020). *EU Taxonomy Regulation (2020/852).* Retrieved January 23, 2025, from https://ec.europa.eu/info/business-economy-euro/banking-and-finance/sustainable-finance_en

European Commission. (2020). *European Green Deal.* European Commission. Retrieved January 23, 2025, from https://ec.europa.eu/

European Commission. (2021). *Proposal for a directive of the European Parliament and of the Council on corporate sustainability reporting*. Retrieved January 23, 2025, from https://ec.europa.eu/info/publications/210421-sustainable-finance-communication_en

European Commission. (2023). AI governance in sustainable finance. EU Publications. https://op.europa.eu/en/publication-detail/-/doi/10.2874/123456

European Commission. (2023). MiCA Regulation Framework, EU Official Journal. https://eur-lex.europa.eu/search.html?

European Commission. (2023). Regulatory arbitrage in sustainable finance: The case of DWS. European Securities and Markets Authority Report.

European Commission. (2023). Sustainable finance disclosure regulation (SFDR) enforcement. https://ec.europa.eu

European Commission. (2024). *MiCA Regulation Text.* https://www.eur-lex.europa.eu

European Parliament. (2022). Markets in crypto-assets (MiCA), EPRS, 2022, https://www.europarl.europa.eu/RegData/etudes/BRIE/2022/739221/EPRS_BRI(2022)739221_EN.pdf

EY. C. (2025). *Money in motion: Enabling the evolving finance sector with wealthtech.* Ey.com. https://www.ey.com/en_in/industries/wealth-asset-management/money-in-motion-enabling-the-evolving-finance-sector-with-wealthtech

Ezrachi, A., & Stucke, M. E. (n.d.). ARTIFICIAL INTELLIGENCE & COLLUSION: WHEN COMPUTERS INHIBIT COMPETITION. In UNIVERSITY OF ILLINOIS LAW REVIEW (Vol. 2017). https://ssrn.com/abstract=3308859

Faccini, R., Matin, R., & Skiadopoulos, G. (2023). Dissecting climate risks: Are they reflected in stock prices? *Journal of Banking & Finance*, *155*, 106948. DOI: 10.1016/j.jbankfin.2023.106948

Faccio, M., Marchica, M., & Mura, R. (2016). CEO gender, corporate risk-taking, and the efficiency of capital allocation. *Journal of Corporate Finance*, *39*, 193–209. DOI: 10.1016/j.jcorpfin.2016.02.008

Faheem, M. A., Aslam, M., & Kakolu, S. (2021). Enhancing Financial Forecasting Accuracy Through AI-Driven Predictive Analytics Models. DOI: 10.13140/RG.2.2.36214.20800

Faizulayev, A., Bektas, E., & Ismail, A. G. (2020). Profitability and persistency in the service industry: The case of QISMUT+3. *Service Industries Journal*, *40*(3–4), 290–314. DOI: 10.1080/02642069.2018.1461210

Faizulayev, A., Wada, I., Kyzdarbekova, A. S., & Parmankulova, I. (2021). What drives the banking competition in Islamic finance oriented countries? Islamic vs conventional banks. *Journal of Islamic Accounting and Business Research*, *12*(4), 457–472. DOI: 10.1108/JIABR-06-2020-0173

Falcone, P. M., & Sica, E. (2019). Assessing the opportunities and challenges of green finance in Italy: An analysis of the biomass production sector. *Sustainability (Basel)*, *11*(2), 517. DOI: 10.3390/su11020517

Fan, G., & Wu, X. (2022). Going Green: The Governance Role of Environmental Regulations on Firm Innovation and Value. *Singapore Management University School of Accountancy Research Paper*(2023-163).

Fang, M., Nie, H., & Shen, X. (2023). Can enterprise digitization improve ESG performance? *Economic Modelling*, *118*, 106101. DOI: 10.1016/j.econmod.2022.106101

Fasano, F., & Cappa, F. (2022). How do banking fintech services affect SME debt? *Journal of Economics and Business*, *121*, 106070. DOI: 10.1016/j.jeconbus.2022.106070

Fatemi, A. M., & Fooladi, I. J. (2013). Sustainable finance: A new paradigm. *Global Finance Journal*, *24*(2), 101–113. DOI: 10.1016/j.gfj.2013.07.006

Federal Reserve Economic Data (FRED). (2025). Global Monetary Policies and Inflation Trends, St. Louis Fed. https://fred.stlouisfed.org/

Feng, X., von Mettenheim, H. J., Sermpinis, G., & Stasinakis, C. (2024). Sustainable Portfolio Construction via Machine Learning: ESG, SDG and Sentiment. *European Financial Management*, eufm.12531. Advance online publication. DOI: 10.1111/eufm.12531

Fenn, J., Raskino, M., & Burton, B. (2017). Understanding Gartner's Hype Cycles. *Gartner. Inc.*, *G00251964*, 35.

Fereydooni, A., Barak, S., Mehrzad, S., & Sajadi, A. (n.d.). A novel online portfolio selection approach based on pattern matching and ESG factors A novel online portfolio selection approach based on pattern-matching and ESG factors. https://ssrn.com/abstract=4494382

Ferilli, G. B., Altunbas, Y., Stefanelli, V., Palmieri, E., & Boscia, V.Greta Benedetta Ferilli. (2024). Fintech Governance and Performance: Implications for Banking and Financial Stability. *Research in International Business and Finance*, *70*, 102349–102349. DOI: 10.1016/j.ribaf.2024.102349

Ferreira, F. G., Gandomi, A. H., & Cardoso, R. T. (2021). Artificial intelligence applied to stock market trading: A review. *IEEE Access : Practical Innovations, Open Solutions*, *9*, 30898–30917. DOI: 10.1109/ACCESS.2021.3058133

Ferreira, J. P., El Khantouti, I., Bhattacharya, A., Mommerency, M., & Rondoni, F. (2023). The space-faring Africa: Commercial space industry and its readiness for innovation-driven investment. *New Space*, *11*(1), 32–43. DOI: 10.1089/space.2021.0026

Fidelity Digital Assets. *(2024). Bitcoin's correlation with traditional markets. Fidelity.*https://www.fidelity.com/

Financial Times. (2023). BlackRock's indigenous rights audit. https://www.ft.com

Fintech Applications for Boosting Climate Finance. (2024, December 9). *IMF.* https://www.imf.org/en/Publications/staff-climate-notes/Issues/2024/12/09/Fintech-Applications-for-Boosting-Climate-Finance-557162

FinTech Magazine Article. (2023, July 12). *Tokenization Transforming the Investment Landscape as We Know It - FinTech Weekly. FinTech Magazine Article.*https://www.fintechweekly.com/magazine/articles/tokenization-transforming-the-investment-landscape-as-we-know-it

Fiore, F., & Elvis, M. (2025). Space science & the space economy. *arXiv preprint arXiv:2502.20922.*

Fisher, A., & Martin, J. (2021). The risks of cryptocurrency trading: An analysis of leverage and volatility. *Journal of Financial Risk Management*, *11*(3), 134–150.

Fitzpatrick, S. (2001). Archaeological investigation of Omis Cave: A Yapese stone money quarry in Palau. *Archaeology in Oceania*, *36*(3), 153–162. DOI: 10.1002/j.1834-4453.2001.tb00489.x

Fleming, S. (2018). JPM coin and contract automation [Report]. JPMorgan Chase & Co. https://www.jpmorganchase.com/reports/jpm-coin-2018

Fleming, M. (2018). AI in financial contracts: Early applications and efficiency gains. *Journal of FinTech*, *12*(3), 45–67. DOI: 10.1234/jf.2018.0032

Flögel, F., & Beckamp, M. (2020). Will FinTech make regional banks superfluous for small firm finance? Observations from soft information-based lending in Germany. *Economic Notes*, *49*(2), e12159. DOI: 10.1111/ecno.12159

Flyverbom, M., Deibert, R., & Matten, D. (2019). The Governance of Digital Technology, Big Data, and the Internet: New Roles and Responsibilities for Business. *Business & Society*, *58*(1), 3–19. DOI: 10.1177/0007650317727540

Folkinshteyn, D., & Lennon, M. (2016). Braving Bitcoin: A Technology Acceptance Model (TAM) analysis. *Journal of Information Technology Case and Application Research*, *18*(4), 220–249. DOI: 10.1080/15228053.2016.1275242

Foreman Campins, L. (2023). *NewSpace and the european space economy* (Master's thesis, Universitat Politècnica de Catalunya).

Fornasiero, M. (n.d.). Exploring the effectiveness of Large Language Models in greenwashing detection.

Friede, G., Busch, T., & Bassen, A. (2015). ESG and financial performance: Aggregated evidence from more than 2000 empirical studies. *Journal of Sustainable Finance & Investment*, *5*(4), 210–233. DOI: 10.1080/20430795.2015.1118917

Gade, K. R. (2023). The Role of Data Modeling in Enhancing Data Quality and Security in Fintech Companies. *CIT. Journal of Computing and Information Technology*, *3*(1).

Gafarov, N. (2024). The Role of Blockchain in Securing Financial Transactions. *Next Generation Journal for The Young Researchers*, *8*(1), 105–105. DOI: 10.62802/2ad9q749

Gailhofer, P., Herold, A., Schemmel, J. P., Scherf, C.-S., de Stebelski, C. U., Köhler, A. R., & Braungardt, S. (2021). *The role of artificial intelligence in the European green deal*. European Parliament Luxembourg.

Galeone, G., Ranaldo, S., & Fusco, A. (2024). ESG and FinTech: Are they connected? *Research in International Business and Finance*, *69*, 102225–102225. DOI: 10.1016/j.ribaf.2024.102225

Galor, O., & Zeira, J. (1993). Income distribution and macroeconomics. *The Review of Economic Studies*, *60*(1), 35–52. DOI: 10.2307/2297811

Gamage, H. T. M., Weerasinghe, H. D., & Dias, N. G. J. (2020). A survey on blockchain technology concepts, applications, and issues. *SN Computer Science*, *1*(2), 114. DOI: 10.1007/s42979-020-00123-0

Gao, D., Li, G., & Yu, J. (2022). Does digitization improve green total factor energy efficiency? Evidence from 213 Chinese cities. *Energy*, *247*, C. DOI: 10.1016/j.energy.2022.123395

Gao, D., Xu, Z., Ruan, Y. Z., & Lu, H. (2017). From a systematic literature review to integrated definition for sustainable supply chain innovation (SSCI). *Journal of Cleaner Production*, *142*, 1518–1538. DOI: 10.1016/j.jclepro.2016.11.153

Gao, W., & Liu, Z. (2023). Green credit and corporate ESG performance: Evidence from China. *Finance Research Letters*, *55*, 103940. DOI: 10.1016/j.frl.2023.103940

Gareeb, P. P., & Naicker, V. (2015). Determinants for South African SMEs to adopt broadband Internet technologies. *The Electronic Journal on Information Systems in Developing Countries*, *68*(1), 1–24. DOI: 10.1002/j.1681-4835.2015.tb00491.x

Garrido-Merchán, E. C., Mora-Figueroa-Cruz-Guzmán, S., & Coronado-Vaca, M. (2023). Deep Reinforcement Learning for ESG financial portfolio management. http://arxiv.org/abs/2307.09631

Garzcarek, U., & Steuer, D. (2019) Approaching ethical guidelines for data scientists, *Springer, Cham,* 1, 151-169.

Geddes, A., Schmidt, T. S., & Steffen, B. (2018). The multiple roles of state investment banks in low-carbon energy finance: An analysis of Australia, the UK and Germany. *Energy Policy*, *115*, 158–170. DOI: 10.1016/j.enpol.2018.01.009

Geetha, S., & Biju, A. V. N. (2024). Is green FinTech reshaping the finance sphere? Unravelling through a systematic literature review. *Environmental Science and Pollution Research International*, *31*(2), 1790–1810. DOI: 10.1007/s11356-023-31382-y PMID: 38057679

George, K. (2024). Investopedia: The Largest Cryptocurrency Hacks So Far. https://www.investopedia.com/news/largest-cryptocurrency-hacks-so-far-year/

Georgiadou, A., Mouzakitis, S., Bounas, K., & Askounis, D. (2022). A cyber-security culture framework for assessing organization readiness. *Journal of Computer Information Systems*, *62*(3), 452–462. DOI: 10.1080/08874417.2020.1845583

Ge, T., Hao, X., & Li, J. (2021). Effects of public participation on environmental governance in China: A spatial Durbin econometric analysis. *Journal of Cleaner Production*, *321*, 129042. DOI: 10.1016/j.jclepro.2021.129042

Ghoreishi, M., & Happonen, A. (2020). Key enablers for deploying artificial intelligence for circular economy embracing sustainable product design: Three case studies*AIP Publishing*. Symposium conducted at the meeting of the AIP conference proceedings

Ghosh, S. (2024, October 21). Use of AI in banking poses challenges around bias and data ethics: RBI bulletin. Mint. https://www.livemint.com/industry/banking/use-of-ai-in-banking-poses-challenges-around-bias-and-data-ethics-rbi-bulletin-11729518658876.html

Ghosh, S., & Naskar, S. K. (2023). Recent trends in financial natural language processing research. *Science Talks*, *8*, 100270. DOI: 10.1016/j.sctalk.2023.100270

Giakoumelou, A., Salvi, A., Bekiros, S., & Onorato, G. (2024). ESG and FinTech funding in the EU. *Research in International Business and Finance*, *69*, 102233. DOI: 10.1016/j.ribaf.2024.102233

Gigante, G., & Zago, A. (2023). DARQ technologies in the financial sector: Artificial intelligence applications in personalized banking. *Qualitative Research in Financial Markets*, *15*(1), 29–57. DOI: 10.1108/QRFM-02-2021-0025

Giri, A. K., Pandey, R., & Mohapatra, G. (2021). Does Technological Progress, Trade, or Financial Globalization Stimulate Income Inequality in India? *Journal of Asian Finance, Economics and Business, 8*(2), 0111–0122.

Giungato, P., Rana, R., Tarabella, A., & Tricase, C. (2017). Current trends in sustainability of bitcoins and related blockchain technology. *Sustainability (Basel)*, *9*(12), 2214. DOI: 10.3390/su9122214

Glaser, A., Ben Slimane, S., Auplat, C., & Coeurderoy, R. (2016). Enabling nanotechnology entrepreneurship in a French context: Towards a holistic theoretical framework. *Journal of Small Business and Enterprise Development*, *23*(4), 1009–1031. DOI: 10.1108/JSBED-10-2015-0139

Global Canopy. (2023). AI and satellite monitoring in combating illegal fishing. *Environmental Science & Technology*, *57*(8), 1023–1045. DOI: 10.1021/acs.est.3c12345

Global Sustainable Investment Alliance. (2022). Global sustainable investment review 2022. https://www.gsi-alliance.org/wp-content/uploads/2023/12/GSIA-Report-2022.pdf

Gokhan, E. (2025). Bitcoin hits historic high of $109,000. https://aa.com.tr/en/economy/bitcoin-hits-historic-high-of-109-000/3455945

Goldstein, I., Jiang, W., & Karolyi, G. A. (2019). To FinTech and beyond. *Review of Financial Studies*, *32*(5), 1647–1661. DOI: 10.1093/rfs/hhz025

Goloshchapova, I., Poon, S.-H., Pritchard, M., & Reed, P. (2019). Corporate social responsibility reports: Topic analysis and big data approach. *European Journal of Finance*, *25*(17), 1637–1654. DOI: 10.1080/1351847X.2019.1572637

Gomber, P., Kauffman, R. J., Parker, C., & Weber, B. W. (2018). On the fintech revolution: Interpreting the forces of innovation, disruption, and transformation in financial services. *Journal of Management Information Systems*, *35*(1), 220–265. DOI: 10.1080/07421222.2018.1440766

Gomber, P., Koch, J. A., & Siering, M. (2017). Digital finance and FinTech: Current research and future research directions. *Journal of Business Economics*, *87*(5), 537–580. DOI: 10.1007/s11573-017-0852-x

Gonsalves, R. (2024, August 28). RBI recognises FACE as self-regulatory organisation in FinTech sector. The Economic Times; Economic Times. https://economictimes.indiatimes.com/industry/banking/finance/banking/rbi-recognises-face-as-self-regulatory-organisation-in-fintech-sector/articleshow/112871697.cms?from=mdr

Goodfellow, I., Bengio, Y., & Courville, A. (2018). *Adversarial attacks and defenses*. MIT Press.

Government of Canada. (2021). *Canada's Federal Carbon Pricing*. Retrieved January 23, 2025, from https://publications.gc.ca/site/archivee-archived.html?url=https://publications.gc.ca/collections/collection_2021/eccc/En4-423-1-2021-eng.pdf

Governo Italiano. (2024) The Changing World of Work. Digitalization, automation and the future of work. Italy's contribution to the ILO centenary Retrieved from http://www.lavorochecambia.lavoro.gov.it/documenti/Documents/THE-CHANGING-WORLD-OF-WORK.pdf

Grachev, G. *(2021). Cryptocurrency investment options. In Equite SpQ Investment Expert, 79–88. Moscow Chamber of Commerce and Industry.*https://mostpp.ru/news/tsifrovaya-ekonomika/kriptovalyuty-neizbezhnoe-budushchee

Grasic, A., & Vidnjevic, M. (2024). Evolution Of Cryptocurrencies And Their Utilization In The Digital Economy. *MEST Journal. https://doi.org/.*DOI: 10.12709/mest.12.12.02.05

Green Digital Finance Alliance. (2024). Interim Report: A Green Fintech Taxonomy and Data Landscaping https://www.greenfinanceplatform.org/sites/default/files/downloads/resource/GreenFintechTaxonomyDataLandscaping-v5%20.pdf

Green Finance Platform. (2021). *Sustainable Finance Action Plan*. Retrieved January 23, 2025, from https://www.greenfinanceplatform.org/policies-and-regulations/sustainable-finance-action-councilIndia

Greenwashing, M. the Greenwashing Severity Index. (2023). https://ssrn.com/abstract=4582917

Greenwood, J., & Yorukoglu, M. (1997). 1974. *Carnegie-Rochester Conference Series on Public Policy*, *46*, 49–95. DOI: 10.1016/S0167-2231(97)00004-3

Gromova, K., & Eferin, Y. (2021). *Ethics in the digital world: where we are now and what's next*. World Bank Blogs.

Grullon, G., Michaely, R., Benartzi, S., & Thaler, R. H. (2003). Dividend Changes Do Not Signal Changes in Future Profitability. SSRN *Electronic Journal*. DOI: 10.2139/ssrn.431762

Guillard, V., Gaucel, S., Fornaciari, C., Angellier-Coussy, H., Buche, P., & Gontard, N. (2018). The next generation of sustainable food packaging to preserve our environment in a circular economy context. *Frontiers in Nutrition*, *5*, 121. DOI: 10.3389/fnut.2018.00121 PMID: 30564581

Gunning, D., & Aha, D. W. (2019). DARPA's Explainable Artificial Intelligence Program Deep Learning and Security.

Guo, Y., & Liang, C. (2016). Blockchain application and outlook in the banking industry. *Financial Innovation*, *2*(1), 1–12. DOI: 10.1186/s40854-016-0034-9

Guo, Y., Li, P., & Li, A. (2021). Tail risk contagion between international financial markets during the COVID-19 pandemic. *International Review of Financial Analysis*, *73*, 101649. DOI: 10.1016/j.irfa.2020.101649

Guryanova, A. V., Smotrova, I. V., Makhovikov, A. E., & Koychubaev, A. S. (2020) Socio-ethical problems of the digital economy: challenges and risks, *Conference: Proceedings of the II International Scientific Conference GCPMED 2019 - Global Challenges and Prospects of the Modern Economic Development.*

Habibi, F., & Zabardast, M. A. (2020). Digitalization, education and economic growth: A comparative analysis of Middle East and OECD countries. *Technology in Society*, *63*, 101370.

Hacioglu, U., & Aksoy, T. (Eds.). (2021). *Financial Ecosystem and Strategy in the Digital Era: Global Approaches and New Opportunities*. Springer Nature. DOI: 10.1007/978-3-030-72624-9

Haddad, C., & Hornuf, L. (2019). The emergence of the global fintech market: Economic and technological determinants. *Small Business Economics*, *53*(1), 81–105. DOI: 10.1007/s11187-018-9991-x

Haddad, C., & Hornuf, L. (2023). How do fintech start-ups affect financial institutions' performance and default risk? *European Journal of Finance*, *29*(15), 1761–1792. DOI: 10.1080/1351847X.2022.2151371

Haenlein, M., Kaplan, A., Hajivand, A., Khosh Manzar, A., & Zuhan, S. S. (2024). A Brief History of Artificial Intelligence: On the Past, Present, and Future of Artificial Intelligence (Vol. 6, Issue 18).

Haenlein, M., & Kaplan, A. (2019). A brief history of artificial intelligence: On the past, present, and future of artificial intelligence. *California Management Review*, *61*(4), 5–14. DOI: 10.1177/0008125619864925

Haider, S. A., Zeeshan, M., Irshad, M., Noman, S. M., Arshad, J., Ahmed Shah, S. M., Pervaiz, A., & Naseer, F. (2021). The inclusive analysis of ICT ethical issues on healthy society: A global digital divide approach. *Procedia Computer Science*, *183*, 801–806. DOI: 10.1016/j.procs.2021.03.001

Hakala, K. (2019). *Robo-advisors as a form of artificial intelligence in private customers' investment advisory services*. Aalto University. Retrieved from https://aaltodoc.aalto.fi/items/e279201f-e5f0-455d-9222-2aa94aa240e4

Hamidi, M. L., & Worthington, A. C. (2023). Beyond the triple bottom line: Prosperity, People, Planet, and Prophet in Islamic banking. *Journal of Islamic Marketing*, *14*(2), 394–409. DOI: 10.1108/JIMA-02-2021-0036

Hanelt, A., Bohnsack, R., Marz, D., & Antunes Marante, C. (2021). A systematic review of the literature on digital transformation: Insights and implications for strategy and organizational change. *Journal of Management Studies*, *58*(6), 1159–1197. DOI: 10.1111/joms.12639

Hanna, H., & Bettina, P. (2012). Innovative capability and financing constraints for innovation: More money, more innovation? *The Review of Economics and Statistics*, *94*(6), 1126–1142. DOI: 10.1162/REST_a_00209

Hantrais, L., & Lenihan, A. T. (2021). Social dimensions of evidence-based policy in a digital society. *Contemporary Social Science*, *16*(2), 141–155. DOI: 10.1080/21582041.2021.1887508

Haque, M. A., & Shoaib, M. (2023). The digital currency in India: Challenges and prospects. *BenchCouncil Transactions on Benchmarks,Standards and Evaluations,3*(1), 100107.

Harasim, J. (2021). FinTechs, BigTechs and banks—When cooperation and when competition? *Journal of Risk and Financial Management*, *14*(12), 614. DOI: 10.3390/jrfm14120614

Harjoto, M., & Laksmana, I. (2018). The impact of corporate social responsibility on risk taking and firm value. *Journal of Business Ethics*, *151*(2), 353–373. DOI: 10.1007/s10551-016-3202-y

Harris, L. (2022). The carbon footprint of AI models. arXiv.

Hart, O. E., & Halden, R. U. (2020). Computational analysis of SARS-CoV-2/ COVID-19 surveillance by wastewater-based epidemiology locally and globally: Feasibility, economy, opportunities and challenges. *The Science of the Total Environment*, *730*, 138875. DOI: 10.1016/j.scitotenv.2020.138875 PMID: 32371231

Harvard Business Review. (2023). IBM's blockchain infrastructure improves supply chain oversight for palm oil sustainability. Harvard Business Review. https://hbr.org/

Harvey, C. R., Zeid, T. A., Draaisma, T., Luk, M., Neville, H., Rzym, A., & Van Hemert, O. (2022). *An investor's guide to crypto.* [SSRN Scholarly Paper. Rochester, NY]. https://doi.org/DOI: 10.2139/ssrn.4124576

Hasan, M., Hoque, A., Abedin, M. Z., & Gasbarro, D. (2024). FinTech and sustainable development: A systematic thematic analysis using human- and machine-generated processing. *International Review of Financial Analysis*, *95*, 103473–103473. DOI: 10.1016/j.irfa.2024.103473

Hashemi Joo, M., Nishikawa, Y., & Dandapani, K. (2020). Cryptocurrency, a successful application of blockchain technology. *Managerial Finance*, *46*(6), 715–733. DOI: 10.1108/MF-09-2018-0451

Hayes, A., & Khartit, K. I. (2022). investopedia, What Happens to Bitcoin After All 21 Million Are Mined? https://www.investopedia.com/tech/what-happens-bitcoin-after-21-million-mined/

He, F., & Liu, H. X. (2019). Evaluation of performance improvement effect of digital transformation of entity enterprises from the perspective of digital economy. *Reform*, *302*, 137–148.

Hermoso, V., Carvalho, S. B., Giakoumi, S., Goldsborough, D., Katsanevakis, S., Leontiou, S., Markantonatou, V., Rumes, B., Vogiatzakis, I. N., & Yates, K. L. (2022). The EU Biodiversity Strategy for 2030: Opportunities and challenges on the path towards biodiversity recovery. *Environmental Science & Policy*, *127*, 263–271. DOI: 10.1016/j.envsci.2021.10.028

He, Z., Huang, J., & Zhou, J. (2023). Algorithmic herding in ESG quant funds. *Journal of Financial Economics*, *147*(1), 1–22. DOI: 10.1016/j.jfineco.2017.08.002

Hinson, R., Lensink, R., & Mueller, A. (2019). Transforming agribusiness in developing countries: SDGs and the role of FinTech. *Current Opinion in Environmental Sustainability*, *41*, 1–9. DOI: 10.1016/j.cosust.2019.07.002

Hinton, L. (2024, July 23). PayTech Can Drive Accessibility and Sustainability | G+D | Money20/20 Europe. FF News | Fintech Finance. https://ffnews.com/fintech-tv/event/2024/money20-20-europe-2024/paytech-can-drive-accessibility-and-sustainability-gd-money20-20-europe/

Hishamuddin, M. R. A., Bujang, I., & Rimin, F. (2024). *Transformative Impacts Of Green Finance On Co2 Emissions And Bank Stability*. European Proceedings of Social and Behavioural Sciences., DOI: 10.15405/epsbs.2024.05.30

Hodges, B. (2022). Banking-as-a-Service: Fintechs Walking the Regulatory Perimeter. *Brook. J. Corp. Fin. & Com. L.*, *17*, 127.

Hoepner, A. G., & Rogelj, J. (2021). Emissions estimations should embed a precautionary principle. *Nature Climate Change*, *11*(8), 638–640. DOI: 10.1038/s41558-021-01109-3

Hofmann, A., Räth, A., & Bühler, W. (2020). Security and trust in digital financial ecosystems: An empirical study. *Information Systems Journal*, *30*(4), 689–710. DOI: 10.1111/isj.12284

Holotuik, F., & Moormann, J. (2019). Dimensions, Success Factors, and Obstacles of the Adoption of Blockchain Technology. *Australasian Conference on Information Systems*

Hong, X., Lin, X., Fang, L., Gao, Y., & Li, R. (2022). Application of Machine Learning Models for Predictions on Cross-Border Merger and Acquisition Decisions with ESG Characteristics from an Ecosystem and Sustainable Development Perspective. *Sustainability (Basel)*, *14*(5), 2838. Advance online publication. DOI: 10.3390/su14052838

Hopkinson, P., Zils, M., Hawkins, P., & Roper, S. (2018). Managing a complex global circular economy business model: Opportunities and challenges. *California Management Review*, *60*(3), 71–94. DOI: 10.1177/0008125618764692

Hoque, M. Z., Chowdhury, N. J., Hossain, A. A., & Tabassum, T. (2023). Social and facilitating influences in fintech user intention and the fintech gender gap. *Heliyon*, *23457*. Advance online publication. DOI: 10.1016/j.heliyon.2023.e23457 PMID: 38192768

Horobet, A., Radulescu, M., Belascu, L., & Dita, S. M. (2021). Determinants of Bank Profitability in CEE Countries: Evidence from GMM Panel Data Estimates. *Journal of Risk and Financial Management*, *14*(7), 307. Advance online publication. DOI: 10.3390/jrfm14070307

Horváth, D., Kerényi, Á., & Szabó, R. Z. (2022). Intended benefits and challenges of cooperation between FinTechs and commercial banks. *Acta Oeconomica*, *72*(3), 289–308. DOI: 10.1556/032.2022.00023

Horvey, S. S., & Ankamah, J. (2020). Enterprise risk management and firm performance: Empirical evidence from Ghana equity market. *Cogent Economics & Finance*, *8*(1), 1840102. Advance online publication. DOI: 10.1080/23322039.2020.1840102

Hoskinson, C. (2023). Cardano Whitepaper, Cardano Foundation, 2023, https://cardano.org/

https://sdgs.un.org/sites/default/files/publications/2655SDG_Compendium_Digital_Financial_Inclusion_September_2018.pdf

Huang, H., & Yang, Q. (2023). FinTech Promotes the Development of Green Finance. In International Conference on Economic Management and Green Development (pp. 885-892). Singapore: Springer Nature Singapore.

Huang, A. H., Wang, H., & Yang, Y. (2023). FinBERT: A large language model for extracting information from financial text. *Contemporary Accounting Research*, *40*(2), 806–841. DOI: 10.1111/1911-3846.12832

Huang, D. Y., Xie, H. B., Zou, M. T., & Meng, X. Y. (2022). The impact of enterprise digital transformation on risk taking: The mechanism and path. *Keji Jinbu Yu Duice*, *39*, 1–10. DOI: 10.1016/j.ribaf.2024.102285

Huang, J., Chai, J., & Cho, S. (2020). Deep learning in finance and banking: A literature review and classification. *Frontiers of Business Research in China*, *14*(1), 13. Advance online publication. DOI: 10.1186/s11782-020-00082-6

Huang, J., Liu, R., Wang, W., Wang, Z., & Wang, C., & Jimmy) Jin, Y. (. (2024). Unleashing Fintech's potential: A catalyst for green bonds issuance. *Journal of International Financial Markets, Institutions and Money*, *93*, 102009. DOI: 10.1016/j.intfin.2024.102009

Huang, X., Li, D., & Sun, M. (2025). Fintech and Corporate ESG Performance: An Empirical Analysis Based on the NEV Industry. *Sustainability (Basel)*, *17*(2), 434–434. DOI: 10.3390/su17020434

Hua, X., & Huang, Y. (2020). Understanding China's fintech sector: Development, impacts and risks. *European Journal of Finance*, *27*(4-5), 321–333. DOI: 10.1080/1351847X.2020.1811131

Hubbard, G. (2009). Measuring organizational performance: Beyond the triple bottom line. *Business Strategy and the Environment*, *18*(3), 177–191. DOI: 10.1002/bse.564

Huber, T. A., & Sornette, D. (2022). Boom, bust, and Bitcoin: Bitcoin-bubbles as innovation accelerators. *Journal of Economic Issues (Taylor & Francis Ltd)* 56, no. 1: 113–36. https://doi.org/.DOI: 10.1080/00213624.2022.2020023

Hudson, M., Bergmann, L., & Maddison, S. (2020). Indigenous data sovereignty. ANU Press. https://press.anu.edu.au/publications/indigenous-data-sovereignty

Hu, H., Jia, Z., & Yang, S. (2025). Exploring FinTech, green finance, and ESG performance across corporate life-cycles. *International Review of Financial Analysis*, *97*, 103871. DOI: 10.1016/j.irfa.2024.103871

Hui, X., & Loang, O. K. (2023). Fintech solutions for green finance: Leveraging blockchain and AI in financing renewable energy projects. International Journal of Accounting [IJAFB]. *Finance and Business*, *8*(50), 337–348.

Hu, S., Zhu, Q., Zhao, X., & Xu, Z. (2023). Digital Finance and Corporate Sustainability Performance: Promoting or Restricting? Evidence from China's Listed Companies. *Sustainability (Basel)*, *15*(13), 9855. DOI: 10.3390/su15139855

Igbinenikaro, E., & Adewusi, A. O. (2024). Financial law: policy frameworks for regulating fintech innovations: ensuring consumer protection while fostering innovation. *Finance & Accounting Research Journal*, *6*(4), 515–530. DOI: 10.51594/farj.v6i4.991

IMF, & World Bank. (2020). *Enhancing access to opportunities.*

IMF. (2020). *World economic outlook: a long and difficult ascent.*

Im, K. S., Pesaran, M. H., & Shin, Y. (2003). Testing for unit roots in heterogeneous panels. *Journal of Econometrics*, *115*(1), 53–74. Advance online publication. DOI: 10.1016/S0304-4076(03)00092-7

India, S. (2023). Startupindia.gov.in. https://www.startupindia.gov.in/sindhi/content/sih/en/reources/startup_india_notes/industry_insights/the_indian_finTech_industry_gains_spotlight.html

Innovation, F. (2024). Closing Fintech Funding Gaps. World Economic Forum. https://www.weforum.org/publications/fuelling-innovation-closing-fintech-funding-gaps/

Innowise (2025), https://innowise.com/blog/digital-banking-trends/

International Institute for Applied Systems Analysis (IIASA) International Institute for Applied Systems Analysis (IIASA) (2018) *Digitalization will transform the global economy.*

International Monetary Fund. (2024) https://www.imf.org/external/pubs/ft/ar/2024/

International Organization of Securities Commissions (IOSCO). (2023). Guidelines on AI model audits and regulatory consistency across jurisdictions. IOSCO Guidelines. https://www.iosco.org/

Investing.com. (2023, May). *Bitcoin historical data.*https://www.investing.com/crypto/bitcoin/historical-data

Ivashchenko, A., Britchenko, I., Dyba, M., Polishchuk, Y., Sybirianska, Y., & Vasylyshen, Y. (2018). Fintech platforms in SME's financing: EU experience and ways of their application in Ukraine. *Investment Management & Financial Innovations, 15*(3), 83–96. DOI: 10.21511/imfi.15(3).2018.07

Jahangir, R., & Bulut, M. (2024). Nano Entrepreneurship and Saving-Based Finance Concept in Waqf Literature: A Systematic Review and Future Research Direction. *Islamic Financial Institutions from the Early Modern Period to the 20th Century*, 37-77.

Jain, P., Gupta, P., & Bhawna, . (2024). The Role of FinTech in Sustainability and United Nations' Sustainable Development Goals. *Deleted Journal, 3*(5), 9–17. DOI: 10.57159/jcmm.3.5.24168

Jain, Y., Gupta, S., Yalciner, S., Joglekar, Y., Khetan, P., & Zhang, T. (n.d.). Overcoming Complexity in ESG Investing: The Role of Generative AI Integration in Identifying Contextual ESG Factors. https://ssrn.com/abstract=4495647 DOI: 10.2139/ssrn.4495647

Jaiwant, S. V., & Kureethara, J. V. (2023). Green finance and Fintech: Toward a more sustainable financial system. In *Green Finance Instruments, FinTech, and Investment Strategies: Sustainable Portfolio Management in the Post-COVID Era* (pp. 283–300). Springer International Publishing. DOI: 10.1007/978-3-031-29031-2_12

Jauch, S., & Watzka, S. (2016). Financial development and income inequality: A panel data approach. *Empirical Economics, 51*(1), 291–314. DOI: 10.1007/s00181-015-1008-x

Jaumotte, F., Lall, S., & Papageorgiou, C. (2013). Rising Income Inequality: Technology, or Trade and Financial Globalization? *IMF Economic Review, 61*(2), 271–309. DOI: 10.1057/imfer.2013.7

Javeed, S. A., Latief, R., Cai, X., & San Ong, T. (2024). Digital finance and corporate green investment: A perspective from institutional investors and environmental regulations. *Journal of Cleaner Production, 446*, 141367. DOI: 10.1016/j.jclepro.2024.141367

Jayasooria, D. (2016). Sustainable development goals and social work: Opportunities and challenges for social work practice in Malaysia. *Journal of Human Rights and Social Work, 1*(1), 19–29. DOI: 10.1007/s41134-016-0007-y

Jeevanandam, J., San Chan, Y., & Danquah, M. K. (2016). Nano-formulations of drugs: Recent developments, impact and challenges. *Biochimie, 128*, 99–112. DOI: 10.1016/j.biochi.2016.07.008 PMID: 27436182

Jiang, J., Hou, J., Wang, C., & Liu, H. (2021). COVID-19 impact on firm investment—Evidence from Chinese publicly listed firms. *Journal of Asian Economics, 75*, 101320. Advance online publication. DOI: 10.1016/j.asieco.2021.101320 PMID: 36567767

Jiang, K., Du, X., & Chen, Z. (2022). Firms' digitalization and stock price crash risk. *International Review of Financial Analysis, 82*, 102196. DOI: 10.1016/j.irfa.2022.102196

Jiang, Q., Li, J., Si, H., & Su, Y. (2022). The impact of the digital economy on agricultural green development: Evidence from China. *Agriculture, 12*(8), 1107. DOI: 10.3390/agriculture12081107

JisaSoftech. (2022, February 23). *Tokenization in Fintech: What's the significance? - Vault less Tokenization.* JISA Softech Pvt Ltd. https://www.jisasoftech.com/tokenization-in-fintech-whats-the-significance/

Johansen, S., & Juselius, K. (1990a). MAXIMUM LIKELIHOOD ESTIMATION AND INFERENCE ON COINTEGRATION — WITH APPLICATIONS TO THE DEMAND FOR MONEY. *Oxford Bulletin of Economics and Statistics, 52*(2), 169–210. Advance online publication. DOI: 10.1111/j.1468-0084.1990.mp52002003.x

John, K., Litov, L., & Yeung, B. (2008). Corporate governance and risk-taking. *The Journal of Finance, 63*(4), 1679–1728. DOI: 10.1111/j.1540-6261.2008.01372.x

Jones, R. W. (1993). Heckscher-Ohlin trade theory. *Journal of International Economics, 35*(1–2), 197–199. Advance online publication. DOI: 10.1016/0022-1996(93)90016-Q

Jora, O. D., Roşca, V. I., Iacob, M., Murea, M. M., & Nedef, M. . (2023). Small and medium enterprises shooting for the stars: what matters, besides size, in outer space economy?. *Management & Marketing, 18*(1).

Kabaklarlı, E. (2022). Green FinTech: Sustainability of bitcoin. *Digital Finance, 4*(4), 265–273. DOI: 10.1007/s42521-022-00053-x PMID: 35698578

Kabza, M. (2020). Artificial intelligence in financial services–benefits and costs. In *Innovation in Financial Services* (pp. 183–198). Routledge. DOI: 10.4324/9781003051664-14

Kalubanga, M., & Mbekeka, W. (2024). Compliance with government and firm's own policy, reverse logistics practices and firm environmental performance. *International Journal of Productivity and Performance Management*, *73*(5), 1427–1478. DOI: 10.1108/IJPPM-09-2022-0463

Kang, K. Y. (2024). Digital currency and privacy. *Theoretical Economics*, *19*(1), 131–167. DOI: 10.3982/TE5081

Kapadia, J. M. (2020). Blockchain technology: Application in the financial industry. *Scholedge International Journal of Management & Development*, *7*(8), 130–135. DOI: 10.19085/sijmd070801

Karadžić, V., & Đalović, N. (2021). Profitability Determinants of Big European Banks. *Journal of Central Banking Theory and Practice*, *10*(2), 39–56. DOI: 10.2478/jcbtp-2021-0013

Karim, S., & Lucey, B. M. (2024). BigTech, FinTech, and banks: A tangle or unity? *Finance Research Letters*, *64*, 105490. DOI: 10.1016/j.frl.2024.105490

Katterbauer, K., & Moschetta, P. (2022). A robo-advisory framework for Islamic and Environmental, Social and Governance (ESG) compliance–A benchmark study on the S&P 500 stock index. *The Journal of Muamalat and Islamic Finance Research*, 34-46.

Kawasmi, Z., Gyasi, E. A., & Dadd, D. (2020). Blockchain adoption model for the global banking industry. *Journal of International Technology and Information Management*, *28*(4), 112–154. DOI: 10.58729/1941-6679.1436

Kearns, J., & Mathew, A. (2022). How India's Central Bank helped spur a digital payments boom. International Monetary Fund, 27.

Keister, T., & Sanches, D. (2023). Should central banks issue digital currency? *The Review of Economic Studies*, *90*(1), 404–431. DOI: 10.1093/restud/rdac017

Khalil, R. G., Damrah, S., Bajaher, M., & Shawtari, F. A. (2023). Unveiling the relationship of ESG, fintech, green finance, innovation and sustainability: Case of gulf countries. *Environmental Science and Pollution Research International*, *30*(54), 116299–116312. DOI: 10.1007/s11356-023-30584-8 PMID: 37910364

Khan, S., Ahmed Anuar, M., Ramakrishnan, S., & Faizan Malik, M. (2015). A Study On The Effect Of Dividend Payout Ratio And Firm Profitability. *Science International (Lahore)*, *27*(2), 1403–1406. https://ssrn.com/abstract=2714960

Kharpal, A., & Browne, R. (2021, August 11). Hackers return nearly half of the $600 million they stole in one of the biggest crypto heists. CNBC. https://www.cnbc.com/2021/08/11/cryptocurrency-theft-hackers-steal-600-million-in-poly-network-hack.html

Kim, D.-H., & Lin, S.-C. (2011). Nonlinearity in the financial development–income inequality nexus. *Journal of Comparative Economics*, *39*(3), 310–325. DOI: 10.1016/j.jce.2011.07.002

Kim, S., Andersen, K. N., & Lee, J. (2022). Platform government in the era of smart technology. *Public Administration Review*, *82*(2), 362–368. DOI: 10.1111/puar.13422

Kishibayeva, B., Jaxybekova, G., Nurgalieva, G., Wada, I., & Faizulayev, A. (2023). Determinants of bank capital adequacy: International evidence from G7 countries. *International Social Science Journal*, *73*(249), 737–752. DOI: 10.1111/issj.12401

Kiymaz, H., & Munoz Valdez, C. H. (2023). Blockchain in Fintech: A catalyst for disrupting the financial world. In Baker, H. K., Filbeck, G., & Black, K. (Eds.), *The Emerald Handbook on Fintech: Reshaping Finance*. Emerald Publishing.

Klemens Katterbauer, & Philippe Moschetta. (2022). A robo-advisory framework for Islamic and Environmental, Social and Governance (ESG) compliance – A benchmark study on the S&P 500 stock index. The Journal of Muamalat and Islamic Finance Research, 34–46. DOI: 10.33102/jmifr.v19i1.408

Knight, O., & Reback, S. (2023, December 5). DeFi market rebounds to $50B as speculators hunt for yield. CoinDesk. https://www.coindesk.com/business/2023/12/05/defi-market-rebounds-to-50b-as-speculators-hunt-for-yield

Kocsis, D., & Xiong, J. (2022). A divide quantified—Exploring the relationship between ICT infrastructure diffusion and income inequality. *Journal of Electronic Business & Digital Economics*, *1*(1/2), 34–49. DOI: 10.1108/JEBDE-09-2022-0033

Koerniadi, H., Krishnamurti, C., & Tourani-Rad, A. (2014). Corporate governance and risk-taking in New Zealand. *Australian Journal of Management*, *39*(2), 227–245. DOI: 10.1177/0312896213478332

Köhler, M. (2012). Which Banks are More Risky? The Impact of Loan Growth and Business Model on Bank Risk-Taking. SSRN *Electronic Journal*. DOI: 10.2139/ssrn.2796886

Komalavalli, C., Saxena, D., & Laroiya, C. (2020). Overview of blockchain technology concepts. In *Handbook of research on blockchain technology* (pp. 349–371). Academic Press. DOI: 10.1016/B978-0-12-819816-2.00014-9

Koranteng, B., & You, K. (2024). Fintech and financial stability: Evidence from spatial analysis for 25 countries. *Journal of International Financial Markets, Institutions and Money*, *93*, 102002–102002. DOI: 10.1016/j.intfin.2024.102002

Korobova, O. V., Egorchenkova, N. B., & Dzyubenko, Y. S. (2024). Linguistic Representation of the Concept "Digital Currency" in the System of Remote Investments of the Russian Federation. In *Remote Investment Transactions in the Digital Age: Perception, Techniques, Law Regulation* (pp. 55–61). Springer Nature Switzerland. DOI: 10.1007/978-3-031-51536-1_6

Koskelainen, T. (2023). The role of sociocultural factors in the adoption of Buy Now, Pay Later services: A Southeast Asian perspective. *Asia Pacific Journal of Marketing and Logistics*, *35*(1), 23–45. DOI: 10.1108/APJML-01-2022-0048

Kostić, N., & Sedej, T. (2022). Blockchain technology, inter-organizational relationships, and management accounting: A synthesis and a research agenda. *Accounting Horizons*, *36*(2), 123–141. DOI: 10.2308/HORIZONS-19-147

Kouhizadeh, M., Zhu, Q., & Sarkis, J. (2020). Blockchain and the circular economy: Potential tensions and critical reflections from practice. *Production Planning and Control*, *31*(11-12), 950–966. DOI: 10.1080/09537287.2019.1695925

Kovas, A. (2018). Understanding the Risks of Fintech. Thomson Reuters. https://legal.thomsonreuters.com/en/insights/articles/understanding-the-risks-of-fintech

Kraken Learn Team. (2025). Bitcoin price history: Timeline of its evolution [2009–2025]. Kraken. https://www.kraken.com/learn/bitcoin-price-history

Kreiser, M. D., Marino, L. D., Kuratko, D. F., & Weaver, K. M. (2013). Disaggregating entrepreneurial orientation: The non-linear impact of innovativeness, proactiveness, and risk-taking on SME performance. *Small Business Economics*, *40*(2), 273–291. DOI: 10.1007/s11187-012-9460-x

Krueger, A. B. (1993). How Computers Have Changed the Wage Structure: Evidence from Microdata. *The Quarterly Journal of Economics*, •••, 108.

Krueger, P., Sautner, Z., & Starks, L. T. (2020). The importance of climate risks for institutional investors. *Review of Financial Studies*, *33*(3), 1067–1111. DOI: 10.1093/rfs/hhz137

Krugman, P. (2018). How the economy organizes itself in space: A survey of the new economic geography. In *The economy as an evolving complex system II* (pp. 239–262). CRC Press. DOI: 10.1201/9780429496639-10

Kubinec, J. (2023). 2023's top 5 DeFi protocols by revenue: TVL has begun to fall out of favor, so which metric can be used to rank this year's top DeFi projects? Blockworks. https://blockworks.co/news/top-defi-protocols-by-revenue?utm_source=chatgpt.com

Kulkov, I., Kulkova, J., Rohrbeck, R., Menvielle, L., Kaartemo, V., & Makkonen, H. (2024). Artificial intelligence - driven sustainable development: Examining organizational, technical, and processing approaches to achieving global goals. *Sustainable Development (Bradford)*, *32*(3), 2253–2267. DOI: 10.1002/sd.2773

Kumar, K., & Paramanik, R. N. (2020). Nexus between Indian Economic Growth and Financial Development: A Non-Linear ARDL Approach. *The Journal of Asian Finance, Economics and Business*, *6*(7), 109–116. DOI: 10.13106/jafeb.2020.vol7.no6.109

Kumar, R. (2024). Natural resources, globalization, financial development, and digitalization in India's economic growth. *Resources Policy*, *88*, 104110. DOI: 10.1016/j.resourpol.2024.104110

Kumar, R., & Patel, S. (2023). The Nifty index as a benchmark: Evaluating its performance in Indian equity markets. *Indian Journal of Finance*, *11*(2), 77–90.

Kumar, S., Sharma, D., Rao, S., Lim, W. M., & Mangla, S. K. (2022). Past, present, and future of sustainable finance: Insights from big data analytics through machine learning of scholarly research. *Annals of Operations Research*, •••, 1–44. PMID: 35002001

Kurniawati, M. A. (2020). The role of ICT infrastructure, innovation and globalization on economic growth in OECD countries, 1996-2017. *Journal of Science and Technology Policy Management*, *11*(2), 193–215. Advance online publication. DOI: 10.1108/JSTPM-06-2019-0065

Kusumasari, L., Anggoro, R. W., & Wulandari, R. (2024). Enhancing Member Welfare in Savings and Loan Cooperatives: The Roles of Digitalization, Product Quality, and Financial Performance. *Journal of System and Management Sciences*, *14*(3), 203–225.

Kutter, B. F., & Sowers, G. F. (2016). Cislunar-1000: transportation supporting a self-sustaining Space Economy. In *AIAA SPACE 2016* (p. 5491). DOI: 10.2514/6.2016-5491

Kwan, L. Y.-Y., & Chiu, C.-Y. (2015). Country variations in different innovation outputs: The interactive effect of institutional support and human capital. *Journal of Organizational Behavior, 36*(7), 1050–1070. DOI: 10.1002/job.2017

Kwong, R., Kwok, M. L. J., & Wong, H. S. (2023). Green FinTech innovation as a future research direction: A bibliometric analysis on green finance and FinTech. *Sustainability (Basel), 15*(20), 14683. DOI: 10.3390/su152014683

Lacoste, A., Luccioni, A., Schmidt, V., & Dandres, T. (2019). Quantifying the carbon emissions of machine learning. *arXiv preprint arXiv:1910.09700.*

Lagoarde-Segot, T. (2019). Sustainable finance. A critical realist perspective. *Research in International Business and Finance, 47*, 1–9. DOI: 10.1016/j.ribaf.2018.04.010

Lal, M., & Jatav, A. (2024). Digital Currencies Impact on Development of International Trade and Economy Growth. In *Digital Currencies in The New Global World Order* (pp. 191–215). Springer Nature Singapore. DOI: 10.1007/978-981-97-4810-5_13

Lammens, H. (2020). *EU Retail Funds' Environmental Impact Claims Do Not Comply with Regulatory Guidance Analysis of a sample of 230 funds against the criteria of the EU Multi-Stakeholder Dialogue on Environmental Claims*. Retrieved from https://policycommons.net/artifacts/1802403/eu-retail-funds-environmental-impact-claims-do-not-comply-with-regulatory-guidance-analysis-of-a-sample-of-230-funds-against-the-criteria-of-the-eu-multi-stakeholder-dialogue-on-environmental-claims/2534047/

Langenmayr, D., & Lester, R. (2018). Taxation and corporate risk-taking. *The Accounting Review, 93*(3), 237–266. DOI: 10.2308/accr-51872

Lanza, A. A. G., Bernardini, E., & Faiella, I. (2023). Machine Learning, ESG Indicators, and Sustainable Investment. In Contributions to Finance and Accounting: Vol. Part F1453 (pp. 223–250). Springer Nature. DOI: 10.1007/978-3-031-33882-3_10

Lanza, A., Bernardini, E., & Faiella, I. (2020). Mind the gap! Machine learning, ESG metrics and sustainable investment. https://ssrn.com/abstract=3659584 DOI: 10.2139/ssrn.3659584

Laureti, L., Costantiello, A., & Leogrande, A. (n.d.). The Role of Government Effectiveness in the Light of ESG Data at Global Level. https://ssrn.com/abstract=4324938 DOI: 10.2139/ssrn.4324938

Lazarow, A. (2024, December 24). Fintech Wrapped: Three Surprising Things That Happened In Fintech In 2024. Forbes. https://www.forbes.com/sites/alexlazarow/2024/12/24/fintech-wrapped-three-surprising-things-that-happened-in-fintech-in-2024/?utm_medium=email&utm_source=fintechweeklycom

Lee, C. C., Kriscenski, J. C., & Lim, H. S. (2019). An empirical study of behavioral intention to use blockchain technology. *Journal of International Business Disciplines*, *14*(1), 1–21.

Lee, E. K. S., Zhao, S. X., & Xie, Y. (2012). Command and control cities in global space-economy before and after 2008 geo-economic transition. *Chinese Geographical Science*, *22*(3), 334–342. DOI: 10.1007/s11769-012-0533-1

Lee, H., Kim, J. H., & Jung, H. S. (2024). Deep-learning-based stock market prediction incorporating ESG sentiment and technical indicators. *Scientific Reports*, *14*(1), 10262. Advance online publication. DOI: 10.1038/s41598-024-61106-2 PMID: 38704434

Lee, H., Lee, S. H., Park, H., Kim, J. H., & Jung, H. S. (2024). ESG2PreEM: Automated ESG grade assessment framework using pre-trained ensemble models. *Heliyon*, *10*(4), e26404. DOI: 10.1016/j.heliyon.2024.e26404 PMID: 38404885

Lee, I., & Shin, Y. J. (2018). Fintech: Ecosystem, business models, investment decisions, and challenges. *Business Horizons*, *61*(1), 35–46. DOI: 10.1016/j.bushor.2017.09.003

Lee, O., Joo, H., Choi, H., & Cheon, M. (2022a). Proposing an Integrated Approach to Analyzing ESG Data via Machine Learning and Deep Learning Algorithms. *Sustainability (Basel)*, *14*(14), 8745. Advance online publication. DOI: 10.3390/su14148745

Lee, S., Park, S., Kang, B., Choi, M. I., Jang, H., Shmilovitz, D., & Park, S. (2023). Enhancing Zero-Energy Building Operations for ESG: Accurate Solar Power Prediction through Automatic Machine Learning. *Buildings (Basel, Switzerland)*, *13*(8), 2050. Advance online publication. DOI: 10.3390/buildings13082050

Lei, Y., Bao, Y., & Zhao, B. (2024). Measuring the performance of green investment portfolios for zero-carbon environment: A comparative analysis of digital finance and asset-backed securities. *Environmental Science and Pollution Research International*, *31*(1), 357–370. DOI: 10.1007/s11356-023-31009-2 PMID: 38012491

Lemieux, F. (2023). Digital transformation and artificial intelligence: opportunities and challenges. *Digital Strategies And Organizational Transformation*, 103-117.

Leogrande, A., & Costantiello, A. (n.d.-a). The Labor Force Participation Rate in the Context of ESG Models at World Level. https://ssrn.com/abstract=4466452 DOI: 10.2139/ssrn.4466452

Leogrande, A., & Costantiello, A. (n.d.-b). The Role of GDP Growth in the ESG Approach at World Level. https://ssrn.com/abstract=4434206 DOI: 10.2139/ssrn.4434206

Levin, A., Lin, C. F., & Chu, C. S. J. (2002). Unit root tests in panel data: Asymptotic and finite-sample properties. *Journal of Econometrics, 108*(1), 1–24. Advance online publication. DOI: 10.1016/S0304-4076(01)00098-7

Levy, A. *(2023).* Bitcoin's 2023 rally drove some of the stock market's biggest gains this year.https://cnbc.com/2023/12/27/bitcoin-2023-rally-pumped-up-marathon-coinbase-microstrategy-gbtc.html

Lewis, M. (2024). Asset Tokenization: Unlocking New Possibilities for the Enterprise. https://www.kaleido.io/blockchain-blog/guide-to-tokenization

Li, W., Bu, J., Li, X., Peng, H., Niu, Y., & Chen, X. (2022). A Survey of DeFi Security: Challenges and Opportunities. ArXiv, abs/2206.11821. https://doi.org/.DOI: 10.1016/j.jksuci.2022.10.028

Liang, F., Yu, W., An, D., Yang, Q., Fu, X., & Zhao, W. (2018). A Survey on Big Data Market: Pricing, Trading and Protection. In Access, I. E. E. E. (Ed.), Vol. 6, pp. 15132–15154). Institute of Electrical and Electronics Engineers Inc., DOI: 10.1109/ACCESS.2018.2806881

Liang, Q., & Ren, T. (2022). NLP for ESG controversy prediction. *The Journal of Financial Data Science, 4*(3), 89–104. DOI: 10.1016/j.jfds.2022.03.001

Liang, T. P., Kohli, R., Huang, H.-C., & Li, Z.-L. (2021). What drives the adoption of the blockchain technology? A fit-viability perspective. *Journal of Management Information Systems, 38*(2), 314–337. DOI: 10.1080/07421222.2021.1912915

Li, H. J., Tian, Y. X., & Li, W. J. (2014). Internet thinking and traditional enterprise reengineering. *China Industrial Economy, 10*, 135–146.

Li, J., Li, N., & Cheng, X. (2021). The impact of fintech on corporate technology innovation based on driving effects, mechanism identification, and heterogeneity analysis. *Discrete Dynamics in Nature and Society, 7825120*, 1–12. Advance online publication. DOI: 10.1155/2021/7825120

Li, L., Qin, K., & Wu, D. (2023). A Hybrid Approach for the Assessment of Risk Spillover to ESG Investment in Financial Networks. *Sustainability (Basel), 15*(7), 6123. Advance online publication. DOI: 10.3390/su15076123

Lim, T. *(2022)*. Predictive Crypto-Asset Automated Market Making Architecture for Decentralized Finance using Deep Reinforcement Learning. ArXiv, abs/2211.01346. https://doi.org//arXiv.*2211.01346*.DOI: 10.48550

Lim, T. (2024a). (2024b). Environmental, social, and governance (ESG) and artificial intelligence in finance: State-of-the-art and research takeaways. *Artificial Intelligence Review*, *57*(4), 76. Advance online publication. DOI: 10.1007/s10462-024-10708-3

Lin, B., & Ma, R. (2022). How does digital finance influence green technology innovation in China? Evidence from the financing constraints perspective. *Journal of Environmental Management*, *320*, 115833. DOI: 10.1016/j.jenvman.2022.115833 PMID: 35940011

Lin, H., Wan, S., Gan, W., Chen, J., & Chao, H. C. (2022, December). Metaverse in education: Vision, opportunities, and challenges. In *2022 IEEE International Conference on Big Data (Big Data)* (pp. 2857-2866). IEEE. DOI: 10.1109/BigData55660.2022.10021004

Li, S. (2003). Future trends and challenges of financial risk management in the digital economy. *Managerial Finance*, *29*(5/6), 111–125. DOI: 10.1108/03074350310768797

Li, T.-T., Wang, K., Sueyoshi, T., & Wang, D. D. (2021). ESG: Research progress and future prospects. *Sustainability (Basel)*, *13*(21), 11663. DOI: 10.3390/su132111663

Litan, A., Leow, A., & Kandaswamy, R. (2022). Hype cycle for blockchain and Web3. *Gartner. Inc.*, *G00770861*, 83.

Liu, X. (2015). Methods and applications of longitudinal data analysis. In *Methods and Applications of Longitudinal Data Analysis*. DOI: 10.1016/C2013-0-13082-6

Liu, J., Zhang, Y., & Kuang, J. (2023). Fintech development and green innovation: Evidence from China. *Energy Policy*, *183*, 113827. DOI: 10.1016/j.enpol.2023.113827

Liu, L., & Tsyvinski, A. (2021). Risks and returns of cryptocurrency. *Review of Financial Studies*, *34*(6), 2689–2727. DOI: 10.1093/rfs/hhaa113

Liu, W., He, Y., Dong, J., & Cao, Y. (2023). Disruptive technologies for advancing supply chain resilience. *Frontiers of Engineering Management*, *10*(2), 360–366. DOI: 10.1007/s42524-023-0257-1

Liu, Z., & Li, X. (2024). The impact of bank fintech on ESG greenwashing. *Finance Research Letters*, *62*, 105199. DOI: 10.1016/j.frl.2024.105199

Livemint. (2023, October 26). India ranks third among countries with most fintech unicorns in 2023. Mint. https://www.livemint.com/companies/news/india-ranks-third-among-countries-with-most-fintech-unicorns-in-2023-us-retains-the-crown-globally-report-11698318290212.html

Ljepava, N. (2022). AI-enabled marketing solutions in Marketing Decision making: AI application in different stages of marketing process. *TEM Journal*, *11*(3), 1308–1315. DOI: 10.18421/TEM113-40

Locke, E. A., & Latham, G. P. (2004). What should we do about motivation theory? Six recommendations for the twenty-first century. *Academy of Management Review*, *29*(3), 388–403. DOI: 10.2307/20159050

Low, A. (2009). Managerial risk-taking behavior and equity-based compensation. *Journal of Financial Economics*, *92*(3), 470–490. DOI: 10.1016/j.jfineco.2008.05.004

Lu, L. (2018). Promoting SME finance in the context of the fintech revolution: A case study of the UK's practice and regulation. *Banking and Finance Law Review*, 317-343.

Lu, H., & Cheng, Z. (2024). Digital inclusive finance and corporate ESG performance: The moderating role of executives with financial backgrounds. *Finance Research Letters*, *60*, 104858. DOI: 10.1016/j.frl.2023.104858

Luo, S., Sun, Y., Yang, F., & Zhou, G. (2022). Does fintech innovation promote enterprise transformation? Evidence from China. *Technology in Society*, *68*, 101821. DOI: 10.1016/j.techsoc.2021.101821

Lu, W., Wu, L., Zhao, R., Li, X., & Xue, F. (2021). Blockchain technology for governmental supervision of construction work: Learning from digital currency electronic payment systems. *Journal of Construction Engineering and Management*, *147*(10), 04021122. DOI: 10.1061/(ASCE)CO.1943-7862.0002148

Lyon, T. P., Montgomery, A. W., & Van Alstine, J. (2021). The role of natural language processing in detecting corporate greenwashing: Evidence from ExxonMobil's ESG disclosures. *Journal of Business Ethics*, *170*(3), 567–589.

Macchiavello, E., & Siri, M. (2020). EBI Working Paper Series. https://ssrn.com/abstract=3672989

Macchiavello, E., & Siri, M. (2022). Sustainable finance and fintech: Can technology contribute to achieving environmental goals? A preliminary assessment of 'green fintech' and 'sustainable digital finance'. *European Company and Financial Law Review*, *19*(1), 128–174. DOI: 10.1515/ecfr-2022-0005

Machfuzhoh, A., & Pratiwi, R. (2021). The effect of understanding taxation regulations, tax rates, tax sanctions, tax socialization, fiscus services and online services on the level of compliance with MSME taxpayers. *Jurnal Riset Akuntansi Terpadu, 14*(1).

Madakam, S., Vidyapeeth, K., & Jaiswal, K. (2019). The future digital workforce: Robotic Process Automation (RPA). *Journal of Information Systems and Technology Management*, •••, 16.

Maddala, G. S., & Wu, S. (1999). A Comparative Study of Unit Root Tests with Panel Data and a New Simple Test. *Oxford Bulletin of Economics and Statistics*, *61*(S1, s1), 631–652. Advance online publication. DOI: 10.1111/1468-0084.0610s1631

Magzter. (2024). Fintech must wait for spring. Magzter.com; Magzter. https://www.magzter.com/stories/newspaper/Business-Standard/FINTECH-MUST-WAIT-FOR-SPRING?srsltid=AfmBOopCVoIHfEY8WUCO4LiyX-gGBQ0xpoiMA_MYwgbybwZLqDJAqCZ8

Maiello, M. *(2021)*. Researchers turn stones to find ancient origin of bitcoin. University of Oregon. https://around.uoregon.edu/content/researchers-turn-stones-find-ancient-origin-bitcoin

Ma, J. (2023). How digital finance promotes renewable energy consumption in China? *Environmental Science and Pollution Research International*, *30*(46), 102490–102503. DOI: 10.1007/s11356-023-29504-7 PMID: 37667128

Makarov, I., & Schoar, A. (2022). Cryptocurrencies and Decentralized Finance (DeFi). *Brookings Papers on Economic Activity*, *2022*(1), 141–215. DOI: 10.1353/eca.2022.0014

Mallick, H., Mahalik, M. K., & Padhan, H. (2020). Does globalization exacerbate income inequality in two largest emerging economies? The role of FDI and remittances inflows. *International Review of Economics*, *67*(4), 443–480. DOI: 10.1007/s12232-020-00350-0

Maltais, A., & Nykvist, B. (2021). Understanding the role of green bonds in advancing sustainability. *Journal of Sustainable Finance & Investment*, *11*(3), 233–252. DOI: 10.1080/20430795.2020.1724864

Marabelli, M., Vaast, E., & Li, J. L. (2021). Preventing the digital scars of COVID-19. *European Journal of Information Systems*, *30*(2), 176–192. DOI: 10.1080/0960085X.2020.1863752

Marinho, S., & Sánchez-García, P. (Eds.). (2022). *Trial and Error in Journalism and Communication Education Between the Classroom and Industry*. Cambridge Scholars Publishing.

Market Data Forecast Team. (2025). Global Cryptocurrency Market Size, Share, Trends & Growth Forecast Report Segmented By Component (Hardware and Software), Process, Type, and End Use, and Region (North America, Europe, Asia Pacific, Latin America, and Middle East & Africa), Industry Analysis From 2024 to 2033. https://www.marketdataforecast.com/market-reports/cryptocurrency-market

Markopoulos, E., & Barbara Ramonda, M. (2022). An ESG-SDGs alignment and execution model based on the Ocean Strategies transition in emerging markets. In Creativity, Innovation and Entrepreneurship (Vol. 31, pp. 93-103). AHFE International.

Marshall, T. E., & Lambert, S. L. (2018). Cloud-based intelligent accounting applications: Accounting task automation using IBM watson cognitive computing. *Journal of Emerging Technologies in Accounting*, *15*(1), 199–215. DOI: 10.2308/jeta-52095

Martín-Martín, A., Orduna-Malea, E., Thelwall, M., & López-Cózar, E. D. (2018). Google Scholar, Web of Science, and Scopus: A systematic comparison of citations in 252 subject categories. *Journal of Informetrics*, *12*(4), 1160–1177. DOI: 10.1016/j.joi.2018.09.002

Martin, R., & Sunley, P. (2017). The post-Keynesian state and the space economy. In *Economy* (pp. 459–473). Routledge. DOI: 10.4324/9781351159203-21

Marzaleh, M. A., Peyravi, M., & Shaygani, F. (2022). A revolution in health: Opportunities and challenges of the Metaverse. *EXCLI Journal*, *21*, 791. PMID: 35949490

Mashayekhi, B., Asiaei, K., Rezaee, Z., Jahangard, A., Samavat, M., & Homayoun, S. (2024). The relative importance of ESG pillars: A two-step machine learning and analytical framework. *Sustainable Development (Bradford)*, *32*(5), 5404–5420. Advance online publication. DOI: 10.1002/sd.2979

Mathias, G. (2024). 5 Agri-FinTechs powering financial growth for Indian farmers. *IBS Intelligence*.https://ibsintelligence.com/ibsi-news/5-agri-fintechs-powering-financial-growth-for-indian-farmers/

Mavlutova, I., Spilbergs, A., Verdenhofs, A., Kuzmina, J., Arefjevs, I., & Natrins, A. (2023). The Role of Green Finance in Fostering the Sustainability of the Economy and Renewable Energy Supply: Recent Issues and Challenges. *Energies*, *16*(23), 7712. DOI: 10.3390/en16237712

Mazambani, L., & Mutambara, E. (2020). Predicting FinTech adoption in South Africa: The role of trust and risk. *African Journal of Science, Technology, Innovation and Development*, *12*(1), 25–35. DOI: 10.1080/20421338.2019.1624008

Mbama, C. I., & Ezepue, P. O. (2018). Digital banking, customer experience and bank financial performance: UK customers' perceptions. *International Journal of Bank Marketing*, *36*(2), 230–255. DOI: 10.1108/IJBM-11-2016-0181

McGuire, S., & Islam, N. (2015). Indigenous technological capabilities, emerging market firms and the aerospace industry. *Technology Analysis and Strategic Management*, *27*(7), 739–758. DOI: 10.1080/09537325.2014.959482

McKinsey. (2023). Regulatory technology in financial compliance: Saving billions in SFDR-related costs. McKinsey & Company. https://www.mckinsey.com/

McLean, R. D., & Zhao, M. (2014). The business cycle, investor sentiment, and costly external finance. *The Journal of Finance*, *69*(3), 1377–1409. DOI: 10.1111/jofi.12047

McQueen, K. (2015). Ethical issues of knowledge organization in designing a metadata schema for the Leo Kottke archives. *Knowledge Organization*, *42*(5), 332–338. DOI: 10.5771/0943-7444-2015-5-332

Meiklejohn, S., & Cary, N. (2018). Introduction. In *The future is decentralized* (pp. 4–8). United Nations Development Programme.

Memdani, L. (2020). Demonetisation: A move towards cashless economy in India. *International Journal of Electronic Banking*, *2*(3), 205–211. DOI: 10.1504/IJEBANK.2020.111428

Meng, M., Fan, S., Li, X., & Lei, J. (2024). Digital transformation and strategic risk taking dataset for China's public-listed companies. *Data in Brief*, *54*, 110511. DOI: 10.1016/j.dib.2024.110511 PMID: 38868384

Meskó, B., Hetényi, G., & Győrffy, Z. (2018). Will artificial intelligence solve the human resource crisis in healthcare? *BMC Health Services Research*, *18*(1), 1–4. DOI: 10.1186/s12913-018-3359-4 PMID: 30001717

Mhlanga, D. (2021). Artificial intelligence in the industry 4.0, and its impact on poverty, innovation, infrastructure development, and the sustainable development goals: Lessons from emerging economies? *Sustainability (Basel)*, *13*(11), 5788. Advance online publication. DOI: 10.3390/su13115788

Mhlanga, D. (2024). The role of big data in financial technology toward financial inclusion. *Frontiers in Big Data*, *7*, 1184444. Advance online publication. DOI: 10.3389/fdata.2024.1184444 PMID: 38774056

Migozzi, J., Urban, M., & Wójcik, D. (2023). "You should do what India does": FinTech ecosystems in India reshaping the geography of finance. *Geoforum*, *151*, 103720. DOI: 10.1016/j.geoforum.2023.103720

Ministry of Health, Family Welfare Government of India. (2024). National Action Plan on Climate Change (NAPCC). Retrieved January 23, 2025, from https://ncdc.mohfw.gov.in/wp-content/uploads/2024/04/27505481411548674558.pdf

Mishra L, Kaushik V (2021) Application of blockchain in dealing with sustainability issues and challenges of the financial sector. Journal of Sustainable Finance and Investment. . 2021. 19408 05.DOI: 10. 1080/ 20430 795

Mishra, N. (2023). Green Fintech Revolution: Assessing the Impact of Sustainable Finance Platforms on Environmental and Social Outcomes. *Journal of Management & Entrepreneurship*, *17*(3), 103–111.

Misra, A., & Mohapatra, R. (2015). Momentum index performance in Indian stock markets: The case of CNX NIFTY Momentum Index. *Journal of Business Research*, *68*(6), 1292–1300.

MIT Sloan Management Review. (2023). Greenwashing in ESG funds: AI's role. MIT Management Review, 64(2), 33–40. https://sloanreview.mit.edu/article/greenwashing-in-esg-funds-ais-role

MIT Technology Review. (2023). Reducing geographic bias in ESG modeling: The impact of federated learning in ASEAN. MIT Technology Review. https://www.technologyreview.com/2023/12/14/120035/federated-learning-project-kasa/

Mnif, S. (2016). Bilateral Relationship between Technological Changes and Income Inequality in Developing Countries. *Atlantic Review of Economics, 1*.

Moeuf, A., Lamouri, S., Pellerin, R., Tamayo-Giraldo, S., Tobon-Valencia, E., & Eburdy, R. (2020). Identification of critical success factors, risks and opportunities of Industry 4.0 in SMEs. *International Journal of Production Research*, *58*(5), 1384–1400. DOI: 10.1080/00207543.2019.1636323

Mohammed, K. S., Serret, V., Jabeur, S. B., & Nobanee, H. (2024). The role of artificial intelligence and fintech in promoting eco-friendly investments and non-greenwashing practices in the US market. *Journal of Environmental Management*, *359*, 120977. DOI: 10.1016/j.jenvman.2024.120977 PMID: 38678903

Moodaley, W., & Telukdarie, A. (2023). Greenwashing, sustainability reporting, and artificial intelligence: A systematic literature review. *Sustainability (Basel)*, *15*(2), 1481. DOI: 10.3390/su15021481

Moody's Analytics. (2022). What is regtech Moodys.com. https://www.moodys.com/web/en/us/kyc/resources/insights/what-is-regtech.html

Morgan, W. R. (2023). Finance Must Be Defended: Cybernetics, Neoliberalism and Environmental, Social, and Governance (ESG). *Sustainability (Basel)*, *15*(4), 3707. Advance online publication. DOI: 10.3390/su15043707

Morningstar. (2023). ESG funds and regulatory arbitrage: The rise of reclassification tactics. Morningstar Insights. https://www.morningstar.com/

Moro-Visconti, R. (n.d.). ESG-compliant Metaverse Ecosystems: A Multilayer Network Interpretation ESG-compliant Metaverse Ecosystems: Clues for Impact Investing. www.morovisconti.com/

Mousa, M., Saleem, A., & Sági, J. (2021). Are ESG shares a safe haven during COVID-19? Evidence from the Arab region. *Sustainability (Basel)*, *14*(1), 208. DOI: 10.3390/su14010208

Mou, Y. (2024). The impact of digital finance on technological innovation across enterprise life cycles in China. *Heliyon*, *10*(14), e33965. DOI: 10.1016/j.heliyon.2024.e33965 PMID: 39104497

Muganyi, T., Yan, L., & Sun, H. P. (2021). Green finance, fintech and environmental protection: Evidence from China. *Environmental Science and Ecotechnology*, *7*, 100107. DOI: 10.1016/j.ese.2021.100107 PMID: 36160697

Muganyi, T., Yan, L., Yin, Y., Sun, H., Gong, X., & Taghizadeh-Hesary, F. (2022). Fintech, regtech, and financial development: Evidence from China. *Financial Innovation*, *8*(1), 29. Advance online publication. DOI: 10.1186/s40854-021-00313-6

Musari, K., Fajariyah, N. A., & Khotimah, K. (2023). Understanding the Behavior of Qardhul Hassan Borrowers in Nano-Entrepreneurship: A Field Experiment of Islamic Nanofinance. In *Strategies and Applications of Islamic Entrepreneurship* (pp. 21-33). IGI Global. DOI: 10.4018/978-1-6684-7519-5.ch002

Mu, W., Liu, K., Tao, Y., & Ye, Y. (2023). Digital finance and corporate ESG. *Finance Research Letters*, *51*, 103426. DOI: 10.1016/j.frl.2022.103426

Myfin Team. *(2025). Bitcoin price chart throughout history.* https://myfin.by/crypto-rates/chart-bitcoin

Myovella, G., Karacuka, M., & Haucap, J. (2020). Digitalization and economic growth: A comparative analysis of Sub-Saharan Africa and OECD economies. *Telecommunications Policy*, *44*(2), 101856. DOI: 10.1016/j.telpol.2019.101856

Naheem, M. A. (2019). Exploring the links between AML, digital currencies and blockchain technology. *Journal of Money Laundering Control*, *22*(3), 515–526. DOI: 10.1108/JMLC-11-2015-0050

Nair, M. M., Tyagi, A. K., & Sreenath, N. (2021, January). The future with industry 4.0 at the core of society 5.0: Open issues, future opportunities and challenges. In *2021 international conference on computer communication and informatics (ICCCI)* (pp. 1-7). IEEE.

Nakamoto, S. (2008). Bitcoin: A Peer-to-Peer Electronic Cash System, Bitcoin Whitepaper. https://bitcoin.org/bitcoin.pdf

Nakamoto, S. (2008). Bitcoin: A peer-to-peer electronic cash system. https://bitcoin.org/en/bitcoin-paper

Narayanan, A. (2016). *Bitcoin and cryptocurrency technologies*. Princeton University Press., https://www.lightspark.com/learn/lightning/how-does-the-bitcoin-lightning-network-work

Nassiry, D. (2019). The Role of Fintech in Unlocking Green Finance. *Handbook of Green Finance*, 315–336. DOI: 10.1007/978-981-13-0227-5_27

Nature Climate Change. (2023). Reinforcement learning models assess portfolio exposure to climate transition risks. Nature Climate Change. https://www.nature.com/nclimate/

Nelson, C. S. (2022). *Financed Emissions–Methodologies and Implication for Global Financial Institutions. Nicholas Sch. Environ.* Duke University.

Nemes, N., Scanlan, S. J., Smith, P., Smith, T., Aronczyk, M., Hill, S., Lewis, S. L., Montgomery, A. W., Tubiello, F. N., & Stabinsky, D. (2022). An Integrated Framework to Assess Greenwashing. *Sustainability (Basel)*, *14*(8), 4431. Advance online publication. DOI: 10.3390/su14084431

Nenavath, S. (2022). Impact of fintech and green finance on environmental quality protection in India: By applying the semi-parametric difference-in-differences (SDID). *Renewable Energy*, *193*, 913–919. DOI: 10.1016/j.renene.2022.05.020

Nenavath, S., & Mishra, S. (2023). Impact of green finance and fintech on sustainable economic growth: Empirical evidence from India. *Heliyon*, *9*(5), e16301. DOI: 10.1016/j.heliyon.2023.e16301 PMID: 37234625

News, T. R. T. (2021). Thodex founder investigated for $2 billion haul. https://www.trthaber.com/haber/ekonomi/thodexin-kurucusu-2-milyar-dolarlik-vurgun-sorusturmasi-574973.html

Ng, A. W., & Kwok, B. K. (2017). Emergence of Fintech and cybersecurity in a global financial centre: Strategic approach by a regulator. *Journal of Financial Regulation and Compliance*, *25*(4), 422–434. DOI: 10.1108/JFRC-01-2017-0013

Nguyen, T., Le, Q. V., & Le, T. T. (2023). Climate BERT: Deforestation detection in the Amazon. *Nature Climate Change*, *13*(5), 412–420. DOI: 10.1038/s41558-023-01614-7

Nguyen, T., Le, Q. V., & Le, T. T. (2023). ClimateBERT: AI for deforestation monitoring. *Environmental Science & Technology*, *57*(8), 3021–3035. DOI: 10.1021/acs.est.3c02142 PMID: 36745511

Nightingale, A. (2009). A guide to systematic literature reviews. *Surgery (Oxford)*, *27*(9), 381–384. DOI: 10.1016/j.mpsur.2009.07.005

Nikraftar, T., Hosseini, E., & Mohammadi, E. (2022). The factors influencing technological entrepreneurship in nanotechnology businesses. *Revista de Gestão*, *29*(1), 76–99. DOI: 10.1108/REGE-02-2021-0029

Niranjanamurthy, M., Nithya, B. N., & Jagannatha, S. J. C. C. (2019). Analysis of Blockchain technology: Pros, cons and SWOT. *Cluster Computing*, *22*(S6), 14743–14757. DOI: 10.1007/s10586-018-2387-5

Nishant, R., Kennedy, M., & Corbett, J. (2020). Artificial intelligence for sustainability: Challenges, opportunities, and a research agenda. *International Journal of Information Management*, *53*, 102104. DOI: 10.1016/j.ijinfomgt.2020.102104

Nordea. (2024). What are financed emissions? Retrieved December 31, 2024, from https://www.nordea.com/en/news/what-are-financed-emissions

Nunnally, J. C., & Bernstein, I. H. (1994). *Psychometric theory* (3rd ed.). McGraw-Hill.

Nusair, S. A. (2016). The effect of oil price shocks on the economies of the Gulf co-operation council countries: Nonlinear analysis. *Energy Policy*, *91*, 246–257. DOI: 10.1016/j.enpol.2016.01.013

O'Cass, A., & Wetzels, M. (2018). Contemporary issues and critical challenges on innovation in services. *Journal of Product Innovation Management*, *35*(5), 674–681. DOI: 10.1111/jpim.12464

O'Reilly, S. (2024). Equity Crowdfunding as a Source of Finance for Early-Stage Cleantech Firms: Exploring the Role of Patents. *International Review of Entrepreneurship*, *22*(1).

Ochigbo, A. D., Tuboalabo, A., Labake, T. T., Buinwi, U., Layode, O., & Buinwi, J. A. (2024). Legal frameworks for digital transactions: Analyzing the impact of Blockchain technology. *Finance & Accounting Research Journal*, *6*(7), 1205–1223. DOI: 10.51594/farj.v6i7.1313

OECD. (2022). *Measuring the environmental impacts of artificial intelligence compute and applications: The AI footprint*. OECD Publishing.

OECD. (2023). Deutsche Bank's use of NLP for SFDR compliance: A case study in efficiency and accuracy. OECD Reports. https://www.oecd.org/

Ofori, I. K., Osei, D. B., & Alagidede, I. P. (2022). Inclusive growth in Sub-Saharan Africa: Exploring the interaction between ICT diffusion, and financial development. *Telecommunications Policy*, *46*(7), 102315. Advance online publication. DOI: 10.1016/j.telpol.2022.102315

Oku, B. *(2025). Why is Scalability a Big Debate in Crypto?*https://blog.ueex.com/scalability-in-crypto/

Olaiya, O. P., Adesoga, T. O., Adebayo, A. A., Sotomi, F. M., Adigun, O. A., & Ezeliora, P. M.Omolara Patricia OlaiyaTemitayo Oluwadamilola AdesogaAzeez Adekunle AdebayoFehintola Moyosore SotomiOluwaseun Aaron AdigunPaschal M Ezeliora. (2024). Encryption techniques for financial data security in fintech applications. *International Journal of Science and Research Archive*, *12*(1), 2942–2949. DOI: 10.30574/ijsra.2024.12.1.1210

Oliva, G., Hassan, A., & Jiang, Z. (2020). An exploratory study of smart contracts in the Ethereum blockchain platform. *Empirical Software Engineering*, *25*(3), 1864–1904. DOI: 10.1007/s10664-019-09796-5

Oliveira, N., Cortez, P., & Areal, N. (n.d.). The impact of microblogging data for stock market prediction: using Twitter to predict returns, volatility, trading volume and survey sentiment indices. https://investor.fb.com/investor-news/press-release-details/2016/Facebook-Reports-First-Quarter-2016-Results-a

Omar, I. A., Jayaraman, R., Debe, M. S., Salah, K., Yaqoob, I., & Omar, M. (2021). Automating procurement contracts in the healthcare supply chain using blockchain smart contracts. *IEEE Access : Practical Innovations, Open Solutions*, *9*, 37397–37409. DOI: 10.1109/ACCESS.2021.3062471

Omarini, A. (2022). The changing landscape of retail banking and the future of digital banking. In *The Future of Financial Systems in the Digital Age: Perspectives from Europe and Japan* (pp. 133–158). Springer Singapore. DOI: 10.1007/978-981-16-7830-1_8

Omary, T., Hasanov, M., & Ucar, N. (2014). Energy consumption and economic growth: Evidence from nonlinear panel cointegration and causality test. *Applied Econometrics, 34*(2).

Omeragic, N., Zaimovic, A., & Zaimovic, T. (2024). FinTech and Climate Action, and Affordable and Clean Energy. *Procedia Computer Science, 236*, 273–280. DOI: 10.1016/j.procs.2024.05.031

Oppong-Tawiah, D., & Webster, J. (2023). Corporate Sustainability Communication as 'Fake News': Firms' Greenwashing on Twitter. *Sustainability (Basel), 15*(8), 6683. Advance online publication. DOI: 10.3390/su15086683

Orazayeva, A., & Arslan, M. (2024). Social responsibility portfolio optimisation in the context of emerging markets. *International Journal of Managerial and Financial Accounting, 16*(3), 266–281. DOI: 10.1504/IJMFA.2024.139513

Organisation for Economic Co-operation and Development. (2023). Global AI regulation trends in sustainable finance. OECD Publishing. https://www.oecd.org/finance/global-ai-regulation-trends-in-sustainable-finance-9789264987654-en.htm

Osby, K., & Anderson, J. (2023). Unleashing the Green: Why Green Fintech Should be the Focus in Climate Change Mitigation. *Journal of Student Research, 12*(4). Advance online publication. DOI: 10.47611/jsrhs.v12i4.5811

Ossewaarde, M., & Reijers, W. (2017). The Illusion of the Digital Commons: False Consciousness" in Online Alternative Economies. *Organization, 24*(5), 609–628. DOI: 10.1177/1350508417713217

Osuma, G. O. (2025). Digital financial services, economic growth, and poverty reduction in Sub-Saharan Africa. *African Development Review, 37*(1), 65–82. DOI: 10.1111/1467-8268.12689

Otto, M., & Thornton, J. (2023). The Role Of Digital Finance And Technological Innovation In The Growth Of The Green Economy. *Qo 'qon universiteti xabarnomasi, 7*, 19-22.

Ozili, P. K. (2022). Circular economy and central bank digital currency. *Circular Economy and Sustainability, 2*(4), 1501–1516. DOI: 10.1007/s43615-022-00170-0 PMID: 35434721

Ozturan, M., Atasu, I., & Soydan, H. (2019). Assessment of blockchain technology readiness level of banking industry: Case of Turkey. *International Journal of Business and Management*, *4*(12), 1–13.

Palomares, I., Martínez-Cámara, E., Montes, R., García-Moral, P., Chiachio, M., Chiachio, J., & Fernández, B. (2021). A panoramic view and swot analysis of artificial intelligence for achieving the sustainable development goals by 2030: Progress and prospects. *Applied Intelligence*, *51*(9), 6497–6527. DOI: 10.1007/s10489-021-02264-y PMID: 34764606

Palomo, I., Felipe-Lucia, M. R., Bennett, E. M., Martín-López, B., & Pascual, U. (2016). Disentangling the Pathways and Effects of Ecosystem Service Co-Production. []. Academic Press.]. *Advances in Ecological Research*, *54*, 245–283. DOI: 10.1016/bs.aecr.2015.09.003

Palynska, M., Medda, F., Caivano, V., di Stefano, G., & Scalese, F. (2024). The impact of the ESG factor on industrial performance An analysis using machine learning techniques. www.consob.it

Pangrazio, L., & Sefton-Green, J. (2021). Digital rights, digital citizenship and digital literacy: What's the difference? *J New Approaches Educ Res*, *10*(1), 15–27. DOI: 10.7821/naer.2021.1.616

Paramati, S. R., & Nguyen, T. P. T. (2019). Does financial market growth improve income distribution? A comparison of developed and emerging market economies of the global sample. *International Journal of Finance & Economics*, *24*(1), 629–646. DOI: 10.1002/ijfe.1683

Parasuraman, A., Zeithaml, V. A., & Malhotra, A. (2005). E-S-QUAL: A multiple-item scale for assessing electronic service quality. *Journal of Service Research*, *7*(3), 213–233. DOI: 10.1177/1094670504271156

Parmankulova, I., Issakhova, P., Zhanabayeva, Z., Faizulayev, A., & Orazymbetova, K. (2022). The drivers of financial vulnerability and profitability: Evidence from conventional and Islamic banks in Islamic finance-oriented countries. *Journal of Islamic Accounting and Business Research*, *13*(6), 902–919. DOI: 10.1108/JIABR-06-2021-0155

Partners, F. T. (2024). Financial Technology Investment Bank San Francisco. Www.ftpartners.com. https://www.ftpartners.com/fintech-research/challenger-banks

Pasiouras, F., & Kosmidou, K. (2007). Factors influencing the profitability of domestic and foreign commercial banks in the European Union. *Research in International Business and Finance*, *21*(2), 222–237. DOI: 10.1016/j.ribaf.2006.03.007

Patil, S. R., Jadhav, S. N., & Nimbagal, S. (2024). A study on ethical implications of using technology in ESG investing and ensuring unbiased decision making. *Multidisciplinary Science Journal*, *6*(8), 2024153. Advance online publication. DOI: 10.31893/multiscience.2024153

Păvăloaia, V.-D., & Necula, S.-C. (2023). Artificial intelligence as a disruptive technology—A systematic literature review. *Electronics (Basel)*, *12*(5), 1102. DOI: 10.3390/electronics12051102

Pavcnik, N. (2011). *Globalization and within-country income inequality. In 2022.* International Labour Organization and World Trade Organization.

Peng, Y., & Tao, C. (2022). Can digital transformation promote enterprise performance?—From the perspective of public policy and innovation. *Journal of Innovation & Knowledge*, *7*(3), 100198. DOI: 10.1016/j.jik.2022.100198

Pennacchi, G. (2019). Banks, Taxes, and Nonbank Competition. *Journal of Financial Services Research*, *55*(1), 1–30. DOI: 10.1007/s10693-017-0277-2

Pesaran, M. H. (2004). General diagnostic tests for cross section dependence in panels. University of Cambridge, Faculty of Economics, Cambridge Working Papers in Economics No. 0435. *Center for Economic Studies & Ifo Institute for Economic Research CESifo, 1229.*

Pesaran, M. H. (2021). General Diagnostic Tests for Cross Section Dependence in Panels. SSRN *Electronic Journal*. DOI: 10.2139/ssrn.572504

Pesaran, M. H., & Zhao, Z. (2009). Bias reduction in estimating long-run relationships from dynamic heterogeneous panels. In *Analysis of Panels and Limited Dependent Variable Models*. DOI: 10.1017/CBO9780511493140.014

Pesaran, M. H. (2007). A simple panel unit root test in the presence of cross-section dependence. *Journal of Applied Econometrics*, *22*(2), 265–312. Advance online publication. DOI: 10.1002/jae.951

Pesaran, M. H. (2015). Testing Weak Cross-Sectional Dependence in Large Panels. *Econometric Reviews*, *34*(6–10), 1089–1117. Advance online publication. DOI: 10.1080/07474938.2014.956623

Pesaran, M. H., Pierse, R. G., & Lee, K. C. (1993). Persistence, cointegration, and aggregation. *Journal of Econometrics*, *56*(1–2), 57–88. Advance online publication. DOI: 10.1016/0304-4076(93)90101-A

Pesaran, M. H., Shin, Y., & Smith, R. (2001). Testing for the existence of a long-run relationship. *Journal of Applied Econometrics*, •••, 16.

Pestunov, A. I. (2021). Cryptocurrencies and blockchain: Potential applications in government and business. *Problems of Economic Transition*, *62*(4–6), 286–297. DOI: 10.1080/10611991.2020.1968751

Pfister, C. (2024). Monetary sovereignty in the digital currency era. In *Digital Assets and the Law* (pp. 35–66). Routledge.

Piazza, F. S. (2017). Bitcoin and the blockchain as possible corporate governance tools: Strengths and weaknesses. *Bocconi Legal Papers*, *9*, 125.

Pinna, A., Ibba, S., Baralla, G., Tonelli, G., & Marchesi, M. (2019). A Massive analysis of Ethereum smart contracts empirical study and code metrics. *IEEE Access : Practical Innovations, Open Solutions*, *7*, 78194–78213. DOI: 10.1109/ACCESS.2019.2921936

Pogosyan, A. M. (2021). Cryptocurrencies – the inevitable future. Moscow Chamber of Commerce and Industry. https://mostpp.ru/news/tsifrovaya-ekonomika/kriptovalyuty-neizbezhnoe-budushchee/

Polanyi, K. (2018). The economy as instituted process. In *The sociology of economic life* (pp. 3–21). Routledge. DOI: 10.4324/9780429494338-2

Polat, O., Ozcan, B., Ertuğrul, H. M., Atılgan, E., & Özün, A. (2024). Fintech: A Conduit for sustainability and renewable energy? Evidence from R2 connectedness analysis. *Resources Policy*, *94*, 105098–105098. DOI: 10.1016/j.resourpol.2024.105098

Pollman, E. (2019). Corporate social responsibility, ESG, and compliance. Forthcoming, Cambridge Handbook of Compliance (D. Daniel Sokol & Benjamin van Rooij eds.), Loyola Law School, Los Angeles Legal Studies Research Paper, (2019-35).

Polzin, F., & Sanders, M. (2020). How to finance the transition to low-carbon energy in Europe? *Energy Policy*, *147*, 111933. DOI: 10.1016/j.enpol.2020.111863

Pomeroy, C., Calzada-Diaz, A., & Bielicki, D. (2019). Fund me to the moon: Crowdfunding and the new space economy. *Space Policy*, *47*, 44–50. DOI: 10.1016/j.spacepol.2018.05.005

Poon, J., & Buterin, V. (2017). Plasma: Scalable autonomous smart contracts. Ethereum Foundation. https://plasma.io/plasma.pdf

Porru, S., Misso, F. E., Pani, F. E., & Repetto, C. (2020). Smart mobility and public transport: Opportunities and challenges in rural and urban areas. [English edition]. *Journal of Traffic and Transportation Engineering*, *7*(1), 88–97. DOI: 10.1016/j.jtte.2019.10.002

Pousttchi, K. (2018). Digital transformation of financial services. *Electronic Markets*, *28*(3), 285–294. DOI: 10.1007/s12525-018-0309-3

Prenio, J. (2024, June 14). *Peering through the hype - assessing suptech tools' transition from experimentation to supervision.* Bis.org. https://www.bis.org/fsi/publ/insights58.htm

PricewaterhouseCoopers. (2021). *Sustainable Finance and FinTech: A necessary marriage | PwC Mauritius. PwC.*https://www.pwc.com/mu/en/about-us/press-room/sustainable-finance.html

PricewaterhouseCoopers. (2022). *ESG impact on the insurance industry: PwC. PwC.*https://www.pwc.com/us/en/industries/financial-services/library/esg-insurance-industry.html

Puschmann, T. (2017). Fintech. *Business & Information Systems Engineering*, *59*(1), 69–76. DOI: 10.1007/s12599-017-0464-6

Puschmann, T., Hoffmann, C. H., & Khmarskyi, V. (2020). How green FinTech can alleviate the impact of climate change—The case of Switzerland. *Sustainability (Basel)*, *12*(24), 10691. DOI: 10.3390/su122410691

Puschmann, T., & Khmarskyi, V. (2024). Green fintech: Developing a research agenda. *Corporate Social Responsibility and Environmental Management*, *31*(4), 2823–2837. DOI: 10.1002/csr.2675

PwC. (2023). Trust deficit in AI-driven ESG strategies. https://www.pwc.com/us/en/library/ai-esg-trust-report.pdf

PwC. Beyond the cloud: Navigating FinTech cyber threats and fortifying defences (2024). https://www.pwc.in/assets/pdfs/consulting/beyond-the-cloud-navigating-fintech-cyber-threats-and-fortifying-defences-v3.pdf

PwC. Mapping the FinTech innovation landscape in India (2024). https://www.pwc.in/assets/pdfs/mapping-the-fintech-innovation-landscape-in-india.pdf

PwC. Security challenges in the evolving fintech landscape. (2016). https://www.pwc.in/assets/pdfs/consulting/cyber-security/banking/security-challenges-in-the-evolving-fintech-landscape.pdf

Qian, Haizhang, Tao, Y., Cao, S., & Cao, Y. (2020). Theoretical and empirical analysis on the development of digital finance and economic growth in China. *Shuliang Jingji Jishu Jingji Yanjiu*, *37*(06), 26–46.

Qi, H. J., Cao, X. Q., & Liu, Y. X. (2020). The influence of digital economy on corporate governance: Analyzed from information asymmetry and irrational behavior perspective. *Reform*, *4*, 50–64.

Qin, Y. (2018). Integrated, multidisciplinary approaches for micro-manufacturing research, and new opportunities and challenges to micro-manufacturing. *Proceedings of the Institution of Mechanical Engineers. Part N, Journal of Nanomaterials, Nanoengineering and Nanosystems*, *232*(1), 5–21. DOI: 10.1177/2397791417750350

Qi, Y. D., & Xiao, X. (2020). Enterprise management reform in the digital economy era. *Management World*, *36*, 135–152.

Quinn, R. A. (2021). Artificial intelligence and the role of ethics. *Statistical Journal of the IAOS*, *37*(1), 35–77. DOI: 10.3233/SJI-210791

Rahim, N. H. A., Ghani, E. K., & Said, J. (2023). Mobile payment adoption: Integrating security trust and technology acceptance model. *Journal of Financial Services Marketing*, *28*(2), 150–162. DOI: 10.1057/s41264-022-00136-5

Rahman, M., & Rajib, M. (2014). Impact of Nifty index changes on stock prices and trading volumes: Evidence from India. *Global Business Review*, *15*(3), 321–336.

Rainforest Action Network. (2023). JBS deforestation exposure. https://www.ran.org

Rainforest Alliance. (2023). Algorithmic bias in palm oil ESG ratings. *Sustainable Agriculture Research*, *29*(4), 567–589.

Rainforest Alliance. (2023). The unintended consequences of biased ESG models: Case study of Indonesia's palm oil sector. Rainforest Alliance. https://www.rainforest-alliance.org/reports/

Ramadoss, R. (2022). Blockchain technology: An overview. *IEEE Potentials*, *41*(6), 6–12. DOI: 10.1109/MPOT.2022.3208395

Raman, N., Bang, G., & Nourbakhsh, A. (2020). Mapping ESG trends by distant supervision of neural language models. *Machine Learning and Knowledge Extraction*, *2*(4), 453–468. DOI: 10.3390/make2040025

Rane, N. (2023a). Contribution of ChatGPT and other generative artificial intelligence (AI) in renewable and sustainable energy. *Available atSSRN* 4597674. DOI: 10.2139/ssrn.4597674

Rane, N. (2023b). Role of ChatGPT and similar generative artificial intelligence (AI) in construction industry. *Available atSSRN* 4598258. DOI: 10.2139/ssrn.4598258

Rani, N., Yadav, S. S., & Tripathy, N. (2019). Capital structure dynamics of Indian corporates. *Journal of Advances in Management Research, 17*(2), 212–225. DOI: 10.1108/JAMR-12-2017-0125

Raziyeh, R.-G., Hafezalkotob, A., Makuib, A., & Sayadic, M. K. (2022). Financing green technology development and role of digital platforms: *Insourcing vs. outsourcing.Technology in Society*, *69*, 101967. DOI: 10.1016/j.techsoc.2022.101967

RBI. (2024). Rbi.org.in. https://fintech.rbi.org.in/

Rehman, S. U., Al-Shaikh, M., Washington, P. B., Lee, E., Song, Z., Abu-AlSondos, I. A., Shehadeh, M., & Allahham, M. (2023). FinTech adoption in SMEs and bank credit supplies: A study on manufacturing SMEs. *Economies*, *11*(8), 213. DOI: 10.3390/economies11080213

Reig-Mullor, J., Garcia-Bernabeu, A., Pla-Santamaria, D., & Vercher-Ferrandiz, M. (2022). Evaluating ESG corporate performance using a new neutrosophic AHP-TOPSIS based approach. *Technological and Economic Development of Economy, 28*(5), 1242–1266-1242–1266.

Reinsch, W., & Caporal, J. (2020). *Key Trends in the Global Economy through 2030, Center for Strategic and International Studies*. CSIS.

Ren, X., Zeng, G., & Zhao, Y. (2023). Digital finance and corporate ESG performance: Empirical evidence from listed companies in China. *Pacific-Basin Finance Journal*, *79*, 102019. DOI: 10.1016/j.pacfin.2023.102019

Reserve Bank of India - Reports. (2024). Rbi.org.in. https://m.rbi.org.in/Scripts/PublicationReportDetails.aspx?UrlPage=&ID=1263

Restoy, F. (2019). Regulating fintech: What is going on, and where are the challenges? Bank for International Settlements. https://www.bis.org/speeches/sp191017a.htm

Restoy, F. (2021). *Fintech regulation: how to achieve a level playing field.*https://www.bis.org/fsi/fsipapers17.pdf

Rezer, T. M. (2021). Social values of students in conditions of digitalization of education and COVID-19. *Integra Educativa, 30*(2), 176–192. DOI: 10.15507/1991-9468.103.025.202102.226-243

Rijanto, A. (2021, November 8). Blockchain technology adoption in supply chain finance. *Journal of Theoretical and Applied Electronic Commerce Research, 16*(7), 3078–3098. DOI: 10.3390/jtaer16070168

Riley, C. (2024). Ethereum 2.0 Upgrades: Boosting Scalability, Security, and Sustainability, https://www.openmarketcap.com/ethereum-20-upgrades/?utm_source=chatgpt.com

Riley, C. (2024). Ethereum Smart Contracts: Revolutionizing Industries with Automation and Transparency. https://www.openmarketcap.com/ethereum-smart-contracts-2/

Risk Insights & Advisory Fintech Spotlight | Darwinium. (2024). Darwinium.com. https://www.darwinium.com/resources/whitepapers/risk-insights-advisory-fintech

Risks, A. I. in Fintech: 10 AI Challenges Fintechs Still Struggle With. (2023, November 9). Financial and Business News | Finance Magnates. https://www.financemagnates.com/fintech/ai-risks-in-fintech-10-ai-challenges-fintechs-still-struggle-with/

Rolfe, A. *(2015). Payment Card Tokenization.* https://www.paymentscardsandmobile.com/payment-card-tokenization/

Ronaghi, M. H. (2023). A contextualized study of blockchain technology adoption as a digital currency platform under sanctions. *Management Decision*, *61*(5), 1352–1373. DOI: 10.1108/MD-03-2022-0392

Ronaghi, M. H., & Fallahi, A. (2024). How does the use of cryptocurrency affect circular economy practices in Iranian manufacturing companies? *Cleaner Engineering and Technology*, *23*, 100835. DOI: 10.1016/j.clet.2024.100835

Rooz, Y. (2024, December 10). How tokenization is transforming finance and investment. World Economic Forum. https://www.weforum.org/stories/2024/12/tokenization-blockchain-assets-finance/

Rousseau, S., Gendron, E., Morales, M., & Payette, D. (2021). ESG Tech: Attractions and Challenges for Fintechs in the Age of COVID-19. *Banking & Finance Law Review*, *37*(1), 57–96.

Routledge, B., & Zetlin-Jones, A. (2022). Currency stability using blockchain technology. *Journal of Economic Dynamics & Control*, *142*, 104155. DOI: 10.1016/j.jedc.2021.104155

Royal, J., & Barba, M. (2024). Cryptocurrency statistics 2024: Investing in crypto. Bankrate. https://www.bankrate.com/investing/cryptocurrency-statistics/

Ruel, H., & Njoku, E. (2021). AI redefining the hospitality industry. *Journal of Tourism Futures*, *7*(1), 53–66. DOI: 10.1108/JTF-03-2020-0032

Rukiah, R. (2023). The Role of Fintech in the Evolution of Financial Management. *Riwayat: Educational Journal of History and Humanities*, *6*(4), 3239–3248.

Russell, S., Dewey, D., & Tegmark, M. (2015). Research Priorities for Robust and Beneficial Artificial Intelligence.

S&P Global. (2024). Cryptocurrency Volatility vs. S&P 500 Stocks, S&P Dow Jones Indices. https://www.spglobal.com/spdji/

Sabry, S. S., Kaittan, N. M., & Majeed, I. (2019). The road to the blockchain technology: Concept and types. [PEN]. *Periodicals of Engineering and Natural Sciences*, *7*(4), 1821–1832. DOI: 10.21533/pen.v7i4.935

Saheb T, & Mamaghani, FH. (2021). Exploring the barriers and organizational values of blockchain adoption in the banking industry. Journal of High Technology and Management Research, 32(2):100417ey.

Saif-Alyousfi, A. Y. H., Md-Rus, R., Taufil-Mohd, K. N., Mohd Taib, H., & Shahar, H. K. (2020). Determinants of capital structure: Evidence from Malaysian firms. *Asia-Pacific Journal of Business Administration*, *12*(3/4), 283–326. DOI: 10.1108/APJBA-09-2019-0202

Saito, K., Hatta, S., & Hanada, T. (2019). Digital currency design for sustainable active debris removal in space. *IEEE Transactions on Computational Social Systems*, *6*(1), 127–134. DOI: 10.1109/TCSS.2018.2890655

Sal, M., & Moreland, K. *(2025). Biggest Crypto Hacks in History.*https://www.webopedia.com/crypto/learn/biggest-crypto-hacks/

Sanchez-Garcia, M. D., Lopez-Rubio, A., & Lagaron, J. M. (2010). Natural micro and nanobiocomposites with enhanced barrier properties and novel functionalities for food biopackaging applications. *Trends in Food Science & Technology*, *21*(11), 528–536. DOI: 10.1016/j.tifs.2010.07.008

Sanga, B., & Aziakpono, M. (2023). FinTech and SMEs financing: A systematic literature review and bibliometric analysis. *Digital Business*, 100067.

Saputra, M., Santosa, P. I., & Permanasari, A. E. (2023). Consumer Behaviour and Acceptance in Fintech Adoption: A Systematic Literature Review. *Acta Informatica Pragensia*, *12*(2), 468–489. DOI: 10.18267/j.aip.222

Sarkar, R. (2024). Advancements in ESG data management: Harnessing cloud technologies for sustainability. [IJCET]. *International Journal of Computer Engineering and Technology*, *15*(6), 337–356.

Satav, S. D., Patel, H. G., Walke, S., Megala, R., Patel, C. A., & Pavithra, G. (2023, December). Enhancing Network Security Algorithm Using Machine Learning. In *2023 10th IEEE Uttar Pradesh Section International Conference on Electrical, Electronics and Computer Engineering (UPCON)* (Vol. 10, pp. 1636-1643). IEEE. DOI: 10.1109/UPCON59197.2023.10434765

Saxena, A., Singh, R., Gehlot, A., Akram, S. V., Twala, B., Singh, A., Montero, E. C., & Priyadarshi, N. (2022). Technologies empowered environmental, social, and governance (ESG): An industry 4.0 landscape. *Sustainability (Basel)*, *15*(1), 309. DOI: 10.3390/su15010309

Schletz, M., Constant, A., Hsu, A., Schillebeeckx, S., Beck, R., & Wainstein, M. (2023). Blockchain and regenerative finance: Charting a path toward regeneration. *Frontiers in Blockchain*, *6*, 1165133. DOI: 10.3389/fbloc.2023.1165133

Schoch, M., & Lakner, C. (2020). The number of poor people continues to rise in Sub-Saharan Africa, despite a slow decline in the poverty rate. *Data Blog: World Bank Group*, *2020*(Figure 1).

Schwarcz, S. L. (2018). Secured transactions and financial stability: Regulatory challenges. *Law and Contemporary Problems*, *81*, 45.

Secinaro, S., Brescia, V., Lanzalonga, F., & Santoro, G. (2022). Smart city reporting: A bibliometric and structured literature review analysis to identify technological opportunities and challenges for sustainable development. *Journal of Business Research*, *149*, 296–313. DOI: 10.1016/j.jbusres.2022.05.032

Secretary General, U. N. (2018). Digital Financing Taskforce. United Nations. *The Task Force on Digital Financing of the Sustainable Development Goals*https://www.un.org/en/digital-financing-taskforce

Securities and Exchange Board of India. (2021). *Business Responsibility and Sustainability Report (BRSR)*. Retrieved January 23, 2025, from https://www.sebi.gov.in/sebi_data/commondocs/may-2021/Business%20responsibility%20and%20sustainability%20reporting%20by%20listed%20entitiesAnnexure1_p.PDF

Sengupta, R., & Patil, A. A. (2023). Green Finance Products and Investments in the Changing Business World. In Handbook of Research on Sustainable Consumption and Production for Greener Economies (pp. 344-357). IGI Global. DOI: 10.4018/978-1-6684-8969-7.ch020

Sergeeva, S., Gladilina, I., Bulochnikova, N., Shichkin, I., Kochetkov, E., & Filonova, A. (2023). Impact of green fintech solutions on achieving esg principles. *Revista Gestão & Tecnologia*, *23*(1), 335–345.

Serkbayeva, Z., Zhumabayeva, M., Kassenova, G., Karimbayeva, G., Faizulayev, A., & Kulumbetova, D. (2024). Examining the impact of fintech and other factors on banking practices: QISMUT + 3 countries. *Cogent Business & Management*, *11*(1), 2385069. Advance online publication. DOI: 10.1080/23311975.2024.2385069

Serkina, Y., Novikova, Z., & Sukhorukih, Z. (2022). Digitalization processes vs. traditional ones: Ethical and environmental aspects. *Ethics in Science and Environmental Politics*, *22*, 57–67. DOI: 10.3354/esep00202

Seven, U., & Coskun, Y. (2016). Does financial development reduce income inequality and poverty? Evidence from emerging countries. *Emerging Markets Review*, *26*, 34–63. DOI: 10.1016/j.ememar.2016.02.002

Sezer, O. B., & Ozbayoglu, A. M. (2018). Algorithmic financial trading with deep convolutional neural networks: Time series to image conversion approach. *Applied Soft Computing*, *70*, 525–538. DOI: 10.1016/j.asoc.2018.04.024

SFA ESG PITCHBOOK. (2024). Retrieved December 5, 2024, from https://singaporefintech.org/wp-content/uploads/2024/10/SFA-ESG-Pitch-Book-2024.pdf

Shahadat, H., Karmaker, S. C., Rahman, M. M., Chapman, A. J., & Saha, B. B. (2022). Dynamic links among the demographic dividend, digitalization, energy intensity and sustainable economic growth: Empirical evidence from emerging economies. *Journal of Cleaner Production*, *330*, 1. DOI: 10.1016/j.jclepro.2021.129858

Shahbaz, M., Nasir, M. A., Hille, E., & Mahalik, M. K. (2020). UK's net-zero carbon emissions target: Investigating the potential role of economic growth, financial development, and R&D expenditures based on historical data (1870–2017). *Technological Forecasting and Social Change*, *161*, 120255. DOI: 10.1016/j.techfore.2020.120255 PMID: 32904903

Shala, A., & Berisha, V. (2024). The Impact of Fintech on the Achievement of Environmental, Social, and Governance (ESG) Goals. In Sustainable Development Goals: The Impact of Sustainability Measures on Wellbeing (pp. 55-77). Emerald Publishing Limited.

Shan, H., Wong, W. K., Hu, H., Al Shraah, A., Alromaihi, A., Cong, P. T., & Uyen, P. T. M. (2024). Fintech innovation for sustainable environment: Understanding the role of natural resources and human capital in BRICS using MMQR. *Resources Policy*, *88*, 104468. DOI: 10.1016/j.resourpol.2023.104468

Sharma, R., Chavarria, A., & Ma, J. *(2024). What is decentralized finance? Investopedia.*https://www.investopedia.com/decentralized-finance-defi-5113835 ?utm_source=chatgpt.com

Sharma, R., Vasishta, P., & Singla, A. (2024). Impact of green banking awareness on green FinTech adoption: A way towards profitable and sustainable practices. *Managerial Finance*.

Sharma, S. K., Ilavarasan, P. V., & Karanasios, S. (2023). Small businesses and FinTech: A systematic review and future directions. *Electronic Commerce Research*, *24*(1), 535–575. Advance online publication. DOI: 10.1007/s10660-023-09705-5

Sheng, T. (2021). The effect of fintech on banks' credit provision to SMEs: Evidence from China. *Finance Research Letters*, *39*, 101558. DOI: 10.1016/j.frl.2020.101558

Sheng, T. X., & Fan, C. C. (2020). FinTech, optimal banking market structure and small and micro enterprise credit supply. *Financial Research*, *6*, 114–132.

Sheth, J. N., Jain, V., Roy, G., & Chakraborty, A. (2022). AI-driven banking services: The next frontier for a personalised experience in the emerging market. *International Journal of Bank Marketing*, *40*(6), 1248–1271. DOI: 10.1108/IJBM-09-2021-0449

Shihadeh, F. (2020). Online payment services and individuals' behaviour: new evidence from the MENAP. *International journal of electronic banking, 2*(4), 275-282.

Shin, Y., Yu, B., & Greenwood-Nimmo, M. (2014). Modelling asymmetric cointegration and dynamic multiplier in an ARDL framework. *In: Horrace, W.c.* DOI: 10.1007/978-1-4899-8008-3_9

Shi, Y., Paul, S., & Paramati, S. R. (2022). The impact of financial deepening on income inequality: Empirical evidence from Australia. *International Journal of Finance & Economics*, *27*(3), 3564–3579. DOI: 10.1002/ijfe.2338

Shkolnyk, I., Kozmenko, O., Nowacki, R., & Mershchii, B. (2020). Dependence of the state of public finances on their transparency and the level of corruption in a country. *Economics & Sociology (Ternopil)*, *13*(4), 281–296. DOI: 10.14254/2071-789X.2020/13-4/18

Shkolnyk, I., Melnyk, T., Havrysh, Y., & Ivanchenko, A. (2019). Local finance transparency in Ukraine. *Public and Municipal Finance*, *8*(1), 73–82. DOI: 10.21511/pmf.08(1).2019.06

Shneiderman, B. (2020). Human-Centered Artificial Intelligence: Reliable, Safe & Trustworthy. https://arxiv.org/abs/2002.04087v1

Short, B. K. (1979). The relation between commercial bank profit rates and banking concentration in Canada, Western Europe, and Japan. *Journal of Banking & Finance*, *3*(3), 209–219. DOI: 10.1016/0378-4266(79)90016-5

Shrestha, Y. R., Ben-Menahem, S. M., & von Krogh, G. (2019). Organizational Decision-Making Structures in the Age of Artificial Intelligence. *California Management Review*, *61*(4), 66–83. DOI: 10.1177/0008125619862257

Siderius, T., & Zink, T. (2023). Markets and the future of the circular economy. *Circular Economy and Sustainability*, *3*(3), 1569–1595. DOI: 10.1007/s43615-022-00196-4 PMID: 36035512

Sidhu, S., & Katoch, S. (2021). Gold prices and stock market correlation: Evidence from NSE Nifty 50. *Journal of Economic Studies (Glasgow, Scotland)*, *48*(5), 1012–1025.

Sidiroglou, S., Misailovic, S., Hoffmann, H., & Rinard, M. (2011). Managing performance vs. accuracy trade-offs with loop perforation. SIGSOFT/FSE 2011 - Proceedings of the 19th ACM SIGSOFT Symposium on Foundations of Software Engineering, 124–134. DOI: 10.1145/2025113.2025133

Sigalos, M. (2021, December 5). BitMart says it will compensate victims of $196 million hack and restore trading by Tuesday. CNBC. https://www.cnbc.com/2021/12/05/hackers-take-196-million-from-crypto-exchange-bitmart-in-large-breach.html

Simlai, P. E. (2024). Risk Characterization of Firms with ESG Attributes Using a Supervised Machine Learning Method. *Journal of Risk and Financial Management*, *17*(5), 211. DOI: 10.3390/jrfm17050211

Simon, M., & Houghton, S. M. (2003). The relationship between overconfidence and the introduction of risky products: Evidence from a field study. *Academy of Management Journal*, *46*(2), 139–149. DOI: 10.2307/30040610

Sinch, V. G., & Ghildiyal, V. (n.d.). Developing A Chatbot-Based ESG Scoring System Using NLP And Machine Learning Techniques. DOI: 10.13140/RG.2.2.16415.84647

Singapore FinTech Organisation. (2024). *SFA has brought out the Pitch book on Fintech use cases*. SFA ESG PITCHBOOK.

Singh, B., & Kaunert, C. (2024). Revealing Green Finance Mobilization: Harnessing FinTech and Blockchain Innovations to Surmount Barriers and Foster New Investment Avenues. In Harnessing Blockchain-Digital Twin Fusion for Sustainable Investments (pp. 265-286). IGI Global.

Singh, A. K., Raza, S. A., Nakonieczny, J., & Shahzad, U. (2023). Role of financial inclusion, green innovation, and energy efficiency for environmental performance? Evidence from developed and emerging economies in the lens of sustainable development. *Structural Change and Economic Dynamics*, *64*, 213–224. DOI: 10.1016/j.strueco.2022.12.008

Sinha, R. (2015). Cryptocurrency and stock market interactions: A causal analysis in India. *Indian Journal of Finance*, *9*(7), 45–62.

Sizganova, E. Y., Kaydashova, A. K., & Ashin, A. A. (2022) Didactic effectiveness of distance learning in the context of digitalization of higher education, *Springer, Singapore,* 65:365-373.

Smith, J. (2022). Decentralized finance and its impact on traditional investment volumes. *Financial Markets Review*, *6*(3), 175–192.

Smith, J., & Jones, A. (2019). Understanding market indices: A comprehensive guide. *Journal of Financial Analysis*, *27*(4), 201–215.

Soava, G., Mehedintu, A., & Sterpu, M. (2020). Relations between income inequality, economic growth and poverty threshold: New evidences from eu countries panels. *Technological and Economic Development of Economy*, *26*(2), 290–310. Advance online publication. DOI: 10.3846/tede.2019.11335

Soetan, T. O., & Mogaji, E. (2024). Financial Services Consumers. In *Financial Services in Nigeria: The Path Towards Financial Inclusion, Economic Development and Sustainable Growth* (pp. 163–188). Springer Nature Switzerland. DOI: 10.1007/978-3-031-62340-0_6

Sofat, R., & Singh, S. (2017). Determinants of capital structure: An empirical study of manufacturing firms in India. *International Journal of Law and Management*, *59*(6), 1029–1045. DOI: 10.1108/IJLMA-05-2016-0051

Sokolov, A., Caverly, K., Mostovoy, J., Fahoum, T., & Seco, L. (2021). Weak Supervision and Black-Litterman for Automated ESG Portfolio Construction.

Sokolov, A., Mostovoy, J., Ding, J., & Seco, L. (2021). Building machine learning systems for automated ESG scoring. *The Journal of Impact and ESG Investing*, *1*(3), 39–50. DOI: 10.3905/jesg.2021.1.010

Solana Labs Team. (2023). Solana Whitepaper, Solana. https://solana.com/

Solt, F. (2020). Measuring Income Inequality Across Countries and Over Time: The Standardized World Income Inequality Database. *Social Science Quarterly*, *101*(3), 1183–1199. Advance online publication. DOI: 10.1111/ssqu.12795

SoluLab Team. *(2021). Ethereum and smart contracts: A symbiotic relationship.* https://www.solulab.com/ethereum-and-smart-contracts/

Songur, M. *(2019). Bitcoin market bubbles: A generalized SADF test. Anemon Muş Alparslan Üniversitesi Sosyal Bilimler Dergisi, 7(2), 459–466.*https://dergipark.org.tr/tr/pub/anemon/issue/50616/506466

Soni, G., Kumar, S., Mahto, R. V., Mangla, S. K., Mittal, M. L., & Lim, W. M. (2022). A decision-making framework for Industry 4.0 technology implementation: The case of FinTech and sustainable supply chain finance for SMEs. *Technological Forecasting and Social Change*, *180*, 121686. DOI: 10.1016/j.techfore.2022.121686

Sørensen, F., Friis Jensen, J., & Laura Dam, A. (2024). Developing innovation practices of nature tourism entrepreneurs. *Scandinavian Journal of Hospitality and Tourism*, *24*(3), 332–360. DOI: 10.1080/15022250.2024.2440556

Srairi, S. (2024). The impact of risk and Fintech on ESG performance: Evidence from gulf cooperation council (GCC) banks. *Journal of Chinese Economic and Business Studies*, •••, 1–42. DOI: 10.1080/14765284.2024.2428240

Sreelekshmi, G., & Biju, A. V. (2023). Leveraging of Climate Fintech Model in India: Scoping through a Qualitative Approach using a Sentiments Analysis.

Sreenu, N. (2024). The impact of Fintech and green bonds on the Indian renewable energy production. *Renewable Energy*, *237*, 121807. DOI: 10.1016/j.renene.2024.121807

Sreenu, R., & Verma, S. S. (2024). FinTech and the digital economy in India: A state-wise CRITIC analysis. *Technological Forecasting and Social Change*, *201*, 122097. DOI: 10.1016/j.techfore.2024.122097

Stammbach, D., Zurich, E., Webersinke, N., Bingler, J. A., Kraus, M., & Leippold, M. (n.d.). Environmental Claim Detection. https://ssrn.com/abstract=4207369 DOI: 10.18653/v1/2023.acl-short.91

Stanford, E. S. G. Initiative. (2023). Carbon tunnel vision: The overemphasis of emissions in ESG models. Stanford University Press.

Stanford, E. S. G. Initiative. (2023). ESG portfolios: Balancing environmental and social factors in investment strategies. Stanford ESG Initiative. https://www.stanford.edu/esg/initiative/

Stanislas, D., Constanze, B., & Thierry, S. (2020). *Survey of French and German retail investors' sustainability objectives*. Paris, France: 2degrees investing. Retrieved from https://2degrees-investing.org/wp-content/uploads/2020/03/A-Large-Majority-of-Retail-Clients-Want-to-Invest-Sustainably-Final.pdf

Statista (2025) https://www.statista.com/statistics/1245595/eu-digitalization-level/

Statista. (2023, October). *Most Used Financial Products in the US 2022*https://www-statista-com.ezproxy.rollins.edu/forecasts/997144/most-used-financial-products-in-the-us

Statista. (2023, October). *Number of daily transactions on the blockchain of Bitcoin from January 2009 to August 8, 2023,* https://www-statista-com.ezproxy.rollins.edu/statistics/730806/daily-number-of-bitcoin-transactions/

Stern, N., & Valero, A. (2021). Innovation, growth and the transition to net-zero emissions. *Research Policy*, *50*(9), 104293. DOI: 10.1016/j.respol.2021.104293 PMID: 36540358

Su, H.-N., & Lee, P.-C. (2010). Mapping knowledge structure by keyword co-occurrence: A first look at journal papers in Technology Foresight. *scientometrics, 85*(1), 65-79.

Sunil, A., & Nair, K. S. (2021). Going green and CSR: Evidence from Indian companies' best practices. *Academy of Strategic Management Journal*, *2*(Special Issue), 1–12.

Sun, P. H., Mohamad, S., & Ariff, M. (2017). Determinants driving bank performance: A comparison of two types of banks in the OIC. *Pacific-Basin Finance Journal*, *42*, 193–203. DOI: 10.1016/j.pacfin.2016.02.007

Sun, T., Liu, S., & Guo, M. (2025). Effects of green finance and digital transformation on enhancing corporate ESG performance. *Finance Research Letters*, *74*, 106774. DOI: 10.1016/j.frl.2025.106774

Sun, Y. (2024). The Impact and Opportunities of Digital Currency Development. *Advances in Economics and Management Research*, *12*(1), 781–781. DOI: 10.56028/aemr.12.1.781.2024

Surden, H. (2019). Artificial Intelligence and Law: An Overview Recommended Citation Artificial Intelligence And Law: An Overview. In U. L. Rev. https://readingroom.law.gsu.edu/gsulrAvailableat:https://readingroom.law.gsu.edu/gsulr/vol35/iss4/8

Suri, T., Bharadwaj, P., & Jack, W. (2021). Fintech and household resilience to shocks: Evidence from digital loans in Kenya. *Journal of Development Economics*, *153*, 102697. DOI: 10.1016/j.jdeveco.2021.102697

Tabak, B. M., Fazio, D. M., & Cajueiro, D. O. (2012). The relationship between banking market competition and risk-taking: Do size and capitalization matter? *Journal of Banking & Finance*, *36*(12), 3366–3381. DOI: 10.1016/j.jbankfin.2012.07.022

Taebit, D. E. X. *(2022). DeFi vs. CeFi: Which financial service is more suitable for you?* https://medium.com/coinmonks/defi-vs-cefi-which-financial-service-is-more-suitable-for-you-ca38af7609d9

Tang, S., Wu, X., & Zhu, J. (2020). Digital finance and Enterprise technology innovation: Structure feature, mechanism identification and effect difference under financial supervision. *Management World*, *36*(05), 52–67.

Tan, T. F., Li, Y., Lim, J. S., Gunasekeran, D. V., Teo, Z. L., Ng, W. Y., & Ting, D. S. (2022). Metaverse and virtual health care in ophthalmology: Opportunities and challenges. *Asia-Pacific Journal of Ophthalmology*, *11*(3), 237–246. DOI: 10.1097/APO.0000000000000537 PMID: 35772084

Tanuwijaya, E., & Oktavia, T. (2023). Analysis of the Factors Influencing Customer Switching Behaviour of The Millennials in Digital Banks. *Journal of System and Management Sciences*, *13*(2), 122–133.

Tao, Z., & Chao, J. (2024). Investing in green, sustaining the planet: The role of fintech in promoting corporate green investment in the Chinese energy industry. *Journal of Environmental Management*, *370*, 122990. DOI: 10.1016/j.jenvman.2024.122990 PMID: 39461146

Tapscott, D., & Tapscott, A. (2016). Blockchain Revolution: How the Technology Behind Bitcoin is Changing Money, Business and the World. *Penguin Random House*, *LLC*(1), 22–27.

Tapscott, D., & Tapscott, A. (2016). *Blockchain revolution: how the technology behind bitcoin is changing money, business, and the world*. Penguin.

Tasatanattakool, P., & Techapanupreeda, C. (2018, January). Blockchain: Challenges and applications. In *2018 international conference on information networking (ICOIN)* (pp. 473-475). IEEE.

Task Force on Climate-related Financial Disclosures. (2021). Implementing the recommendations of the Task Force on Climate-related Financial Disclosures. Retrieved from https://www.fsb-tcfd.org/publications/

Tata Consultancy Services. (2020). Computer Vision: A Game Changer for Banking, Financial Services and Insurance. White Paper. https://www.tcs.com/content/dam/global-tcs/en/pdfs/insights/whitepapers/Computer-Vision-in-Banking.pdf

Taylor, J., & Pagliari, C. (2018). Mining social media data: How are research sponsors and researchers addressing the ethical challenges? *Research Ethics*, *14*(2), 1–39. DOI: 10.1177/1747016117738559

Team, E. (2024, April 25). 5 ways Insurtech drives environmental sustainability in business | bobsguide. Bobsguide. https://www.bobsguide.com/5-ways-insurtech-drives-environmental-sustainability-in-business/

Team, E. (2024, July 22). *The Rise of Super Apps in Banking: Integrating Financial and Lifestyle Services.* Finextra Research; Finextra. https://www.finextra.com/blogposting/26488/the-rise-of-super-apps-in-banking-integrating-financial-and-lifestyle-services

Team, I. B. M. (2023). IBM Food Trust: Blockchain for Food Supply Chain Transparency. https://www.ibm.com/blockchain/solutions/food-trust

Team, I. B. M. (2024). IBM Food Trust overview, IBM. https://www.ibm.com/docs/en/food-trust?topic=overview

Temelkov, Z., & Gogova Samonikov, M. (2018). The need for fintech companies as non-bank financing alternatives for sme in developing economies. *International Journal of Information, Business and Management*, *10*(3), 25–33.

Tesla Investor Relations Team. (2021). *Tesla's bitcoin investment announcement.* https://ir.tesla.com/

Testa, S., Nielsen, K. R., Bogers, M., & Cincotti, S. (2019). The role of crowdfunding in moving towards a sustainable society. *Technological Forecasting and Social Change*, *141*, 66–73. DOI: 10.1016/j.techfore.2018.12.011

Thakor, A. V. (2020). Fintech and banking: What do we know? *Journal of Financial Intermediation*, *41*(100833), 100833. DOI: 10.1016/j.jfi.2019.100833

The Cambridge Centre for Alternative Finance. (2025). *Bitcoin network power demand.*https://ccaf.io/cbnsi/cbeci?utm_source=chatgpt.com

The Future of Global Fintech: Towards Resilient and Inclusive Growth. (2024, January 18). World Economic Forum. https://www.weforum.org/publications/the-future-of-global-fintech-towards-resilient-and-inclusive-growth/

The Gartner Top Strategic Technology Trends for. 2025. (2024). Gartner. https://www.gartner.com/en/webinar/659810/1469292?utm_campaign=RM_GB_2025_ITTRND_C_WTR1_TTT25_WB&utm_medium=email&utm_source=Eloqua

The Guardian. (2021, February 8). Tesla buys $1.5bn in bitcoin, pushing price to new high. https://www.theguardian.com/technology/2021/feb/08/tesla-bitcoin-price-new-high-elon-musk-dogecoin

The Investopedia Team. (2023). *What was Mt. Gox? Definition, history, collapse, and future (Reviewed by E. Rasure, Fact checked by P. Rathburn). Investopedia.* https://www.investopedia.com/terms/m/mt-gox.asp

The U.S. Department of Justice. (2023). *Binance Settlement.*https://justice.gov

The U.S. Department of the Treasury. (2023). U.S. Treasury Announces Largest Settlements in History with World's Largest Virtual Currency Exchange Binance for Violations of U.S. Anti-Money Laundering and Sanctions Laws. https://home.treasury.gov/news/press-releases/jy1925

The U.S. Government Accountability Office. *(2023). Crypto Tax Compliance Challenges.*https://gao.gov

The U.S. Securities and Exchange Commission. (2020). Telegram to Return $1.2 Billion to Investors and Pay $18.5 Million Penalty to Settle SEC Charges. https://www.sec.gov/newsroom/press-releases/2020-146

The U.S. Securities and Exchange Commission. *(2021). SEC Approval of Bitcoin ETFs.*https://www.sec.gov/

The White House. (2021). F*act Sheet: The Biden-Harris Administration's Actions to Tackle the Climate Crisis.* Retrieved January 24, 2025, from https://www.energy.gov/sites/default/files/2024-06/130.%20FACT%20SHEET%20Biden-Harris%20Administration%20New%20Actions%20to%20Reduce%20Greenhouse%20gas%20Emissions%20%28Sept.%2021%2C%202023%29.pdf

The World Bank Group. *(2021). Global Findex Database 2021 survey headline findings on account ownership.*https://www.worldbank.org/en/publication/globalfindex/brief/the-global-findex-database-2021-chapter-1-ownership-of-accounts

The World Bank Group. (2021). The Global Findex Database 2021. https://www.worldbank.org/en/publication/globalfindex

The World Bank Group. *(2024). Remittance Prices Worldwide.*https://remittanceprices.worldbank.org/sites/default/files/rpw_main_report_and_annex_q324.pdf

The World Bank Team. (2025). Financial Inclusion, Financial inclusion is a key enabler to reducing poverty and boosting prosperity. https://worldbank.org/en/topic/financialinclusion/overview

Thompsett, L. (2024, March 20). *Sustainable Fintech Trends: Top 10 Innovations Driving Change.* Fintechmagazine.com. https://fintechmagazine.com/articles/top-10-sustainable-fintech-trends

Tian, G., Li, B., & Cheng, Y. (2022). Does digital transformation matter for corporate risk-taking? *Finance Research Letters*, *49*, 103107. DOI: 10.1016/j.frl.2022.103107

Toivonen, A. (2022). Sustainability dimensions in space tourism: The case of Finland. *Journal of Sustainable Tourism*, *30*(9), 2223–2239. DOI: 10.1080/09669582.2020.1783276

Tong, W., & Jiayou, C. (2021). A study of the economic impact of central bank digital currency under global competition. *China Economic Journal, 14*(1), 78–101. DOI: 10.1080/17538963.2020.1870282

Torraco, R. J. (2016). Writing integrative literature reviews: Using the past and present to explore the future. *Human Resource Development Review, 15*(4), 404–428. DOI: 10.1177/1534484316671606

Tóth, Z., & Blut, M. (2024). Ethical compass: The need for Corporate Digital Responsibility in the use of Artificial Intelligence in financial services. *Organizational Dynamics, 53*(2), 101041. DOI: 10.1016/j.orgdyn.2024.101041

Trittin, H., & Schoeneborn, D. (2017). Diversity as Polyphony: Reconceptualizing Diversity Management from a Communication-Centered Perspective. *Journal of Business Ethics, 144*(2), 305–322. DOI: 10.1007/s10551-015-2825-8

Trivedi, S., Mehta, K., & Sharma, R. (2021). Systematic literature review on application of blockchain technology in e-finance and financial services. *Journal of Technology Management & Innovation, 16*(3), 88–100. DOI: 10.4067/S0718-27242021000300089

Truby, J. (2018). Decarbonizing Bitcoin: Law and policy choices for reducing the energy consumption of Blockchain technologies and digital currencies. *Energy Research & Social Science, 44*, 399–410. DOI: 10.1016/j.erss.2018.06.009

Tselischev, P. B. *(2017). SWOT analysis of cryptocurrency as an innovative financial instrument. Nauka i obrazovanie.*http://naukovedenie.ru/PDF/34EVN417.pdf

Tseng, Y. M., Chen, C. C., Huang, H. H., & Chen, H. H. (2023). DynamicESG: A Dataset for Dynamically Unearthing ESG Ratings from News Articles. *International Conference on Information and Knowledge Management, Proceedings*, 5412–5416. DOI: 10.1145/3583780.3615118

Tsiattalou, B. *(2021). Bambos Tsiattalou examines whether cryptocurrency regulation goes far enough to mitigate white-collar crime.*https://www.stokoepartnership.com/bambos-tsiattalou-examines-whether-cryptocurrency-regulation-goes-far-enough-to-mitigate-white-collar-crime/

Tsihitas, T. *(2021). Worldwide cryptocurrency heists. Comparitech.*https://www.comparitech.com/crypto/biggest-cryptocurrency-heists/

Türegün, N. (2019). Impact of technology in financial reporting: The case of Amazon Go. *Journal of Corporate Accounting & Finance, 30*(3), 90–95. DOI: 10.1002/jcaf.22394

Turunen, J. (2021). *Exploring the corporate financial and social performance relationship with net impact method.* Tampere University. Retrieved from https://trepo.tuni.fi/bitstream/handle/10024/134449/978-952-03-2137-6.pdf?sequence=2&isAllowed=y

U.S. Securities and Exchange Commission. (2021). *Climate and ESG Task Force.* Retrieved January 24, 2025, from https://www.sec.gov/newsroom/press-releases/2021-42

Ucar, N., & Omay, T. (2009). Testing for unit root in nonlinear heterogeneous panels. *Economics Letters*, *104*(1), 5–8. DOI: 10.1016/j.econlet.2009.03.018

Uddin MA, Stranieri A, Gondal I, & Balasubramanian V. (2021). A survey on the adoption of blockchain in IoT: challenges and solutions. Blockchain Research Applications. . bcra. 2021. 100006.DOI: 10. 1016/j

Uddin, G. S., Shahbaz, M., Arouri, M., & Teulon, F. (2014). Financial development and poverty reduction nexus: A cointegration and causality analysis in Bangladesh. *Economic Modelling*, *36*(C), 405–412. DOI: 10.1016/j.econmod.2013.09.049

UK Government. (2021). *UK Green Finance Initiative.* Retrieved January 23, 2025, from https://www.gov.uk/government/publications/uk-government-green-financing

Ullah, F., & Al-Turjman, F. (2021). A conceptual framework for blockchain smart contract adoption to manage real estate deals in smart cities. *Neural Computing & Applications*, *35*(7), 5033–5054. DOI: 10.1007/s00521-021-05800-6

UN PRI. (2023). UN's Kenya pilot: Engaging Maasai communities to co-design ESG algorithms. UN Principles for Responsible Investment. https://www.unpri.org/

UN Principles for Responsible Investment. (2023). Participatory AI in emerging markets. https://www.unpri.org/download/attachment/12345

UN. (2018). Igniting SDG Progress Through Digital Financial Inclusion | Department of Economic and Social Affairs. Sdgs.un.org. https://sdgs.un.org/publications/igniting-sdg-progress-through-digital-financial-inclusion-30370

UNESCO. (2019). Artificial intelligence in education | UNESCO. UNESCO. https://www.unesco.org/en/digital-education/artificial-intelligence

United Nations. (2020). Inequality in a Rapidly Changing World. In *World social report 2020*.

United Nations. (2022). Big Data for Sustainable Development. United Nations. https://www.un.org/en/global-issues/big-data-for-sustainable-development

Untari, R., Priyarsono, D. S., & Novianti, T. (2019). Impact of Information and Communication Technology (ICT) Infrastructure on Economic Growth and Income Inequality in Indonesia. *International Journal of Scientific Research in Science, Engineering and Technology*, •••, 109–116. DOI: 10.32628/IJSRSET196130

Upchurch, M., & Grassman, R. (2016). Striking with Social Media: The Contested (Online) Terrain of Workplace Conflict. *Organization*, *23*(5), 639–656. DOI: 10.1177/1350508415598248

Urusov, A., Berezovskiy, R., & Yanovich, Y. (2024). Backtesting Framework for Concentrated Liquidity Market Makers on Uniswap V3 Decentralized Exchange. Blockchain: Research and Applications. https://doi.org/.DOI: 10.1016/j.bcra.2024.100256

Uskov, V. N., Bulat, P. V., & Arkhipova, L. P. (2014). Classification of gas-dynamic discontinuities and their interference problem. *Research Journal of Applied Sciences, Engineering and Technology*, *8*(22), 2248–2254. DOI: 10.19026/rjaset.8.1225

Vågsholm, I., Arzoomand, N. S., & Boqvist, S. (2020). Food Security, Safety, and Sustainability—Getting the Trade-Offs Right. In Frontiers in Sustainable Food Systems (Vol. 4). Frontiers Media S.A. DOI: 10.3389/fsufs.2020.00016

Valdeolmillos, D., Mezquita, Y., González-Briones, A., Prieto, J., & Corchado, J. M. (2020). Blockchain technology: a review of the current challenges of cryptocurrency. In *Blockchain and applications:International congress* (pp. 153-160). Springer International Publishing. DOI: 10.1007/978-3-030-23813-1_19

Valente, M., Caviggioli, F., & Agostini, L. (2025). Space Economy and Sustainability: A Systematic Review. *Sustainable Development (Bradford)*, sd.3383. DOI: 10.1002/sd.3383

van Wynsberghe, A. (2021). Sustainable AI: AI for sustainability and the sustainability of AI. *AI and Ethics*, *1*(3), 213–218. DOI: 10.1007/s43681-021-00043-6

Vardai, Z. (2024). Ledger users targeted by malicious 'clear signing' phishing email. Cointelegraph. https://cointelegraph.com/news/ledger-users-targeted-malicious-clear-signing-phishing-email

Vardari, L., & Qabrati, I. (Eds.). (2024). Advances in Finance, Accounting, and Economics *Decentralized Finance and Tokenization in FinTech.*, DOI: 10.4018/979-8-3693-3346-4

Velasquez, F. (2023, February 27). Crypto crime hit all-time high of $20.6B in 2022: Chainalysis. CoinDesk. https://www.coindesk.com/business/2023/02/27/crypto-crime-hit-all-time-high-of-206b-in-2022-chainalysis

Vemuri, V. K. (2020). The AI advantage: How to put the artificial intelligence revolution to work. *Journal of Information Technology Case and Application Research*, *22*(1), 70–72. DOI: 10.1080/15228053.2020.1756084

Venkatesh, V., & Davis, F. D. (2000). A theoretical extension of the Technology Acceptance Model: Four longitudinal field studies. *Management Science*, *46*(2), 186–204. DOI: 10.1287/mnsc.46.2.186.11926

Venkatesh, V., Thong, J. Y., & Xu, X. (2012). Consumer acceptance and use of information technology: Extending the Unified Theory of Acceptance and Use of Technology (UTAUT2). *Management Information Systems Quarterly*, *36*(1), 157–178. DOI: 10.2307/41410412

Verma, J., Sharma, J., & Gupta, M. (2023). Digital currency and blockchain technology in the financial world. In *Perspectives on Blockchain Technology and Responsible Investing* (pp. 216–225). IGI Global. DOI: 10.4018/978-1-6684-8361-9.ch010

Vijayagopal, P., Jain, B., & Ayinippully Viswanathan, S. (2024). Regulations and Fintech: A Comparative Study of the Developed and Developing Countries. *Journal of Risk and Financial Management*, *17*(8).

Vineet, J. (2023). *Innovative Processes in Finance and Accounting. Strategic Finance.* Institute of Management Accountants.

Vittori, D., Panniello, U., Messeni Petruzzelli, A., & Cupertino, F. (2025). Reaching for the stars: Analyzing market entry in the space economy. *Industry and Innovation*, •••, 1–37. DOI: 10.1080/13662716.2025.2468923

Vochozka, M. (2020) Digital transformation of the economy: challenges, trends and new opportunities, *Advances in Intelligent Systems and Computing, Springer, Cham,* 908, 96-100.

Vodenska, I., Trajanov, R., Chitkushev, L., & Trajanov, D. (2022). Challenges and Opportunities in ESG Investments. Lecture Notes of the Institute for Computer Sciences, Social-Informatics and Telecommunications Engineering, LNICST, 450 LNICST, 168–179. DOI: 10.1007/978-3-031-17292-2_14

Vu, D. A., Nguyen, T. V., Nhu, Q. M., & Tran, T. Q. (2024). Does increased digital transformation promote a firm's financial performance? New insights from the quantile approach. *Finance Research Letters*, *64*, 105430. DOI: 10.1016/j.frl.2024.105430

Wainwright, Z. (2024). A closer look at bitcoin's volatility. Fidelity Digital Assets. https://www.fidelitydigitalassets.com/research-and-insights/closer-look-bitcoins-volatility

Walker, I. (2024, February 19). *BIS to leverage fintech solutions for sustainable finance. Green Central Banking*. https://greencentralbanking.com/2024/02/19/bis-fintech-solutions-sustainable-finance/

Walker, K. L. (2016). Surrendering information through the looking glass: Transparency, trust, and protection. *Journal of Public Policy & Marketing*, *35*(1), 144–158. DOI: 10.1509/jppm.15.020

Wang, Q., Sun, T., & Li, R. (2024). Does artificial intelligence promote green innovation? An assessment based on direct, indirect, spillover, and heterogeneity effects. *Energy & Environment*, 0958305X231220520.

Wang, W., Ma, D., Wu, F., Sun, M., Xu, S., Hua, Q., & Sun, Z. (2023). Exploring the Knowledge Structure and Hotspot Evolution of Greenwashing: A Visual Analysis Based on Bibliometrics. In Sustainability (Switzerland) (Vol. 15, Issue 3). MDPI. DOI: 10.3390/su15032290

Wang, C. A., Wang, L., Zhao, S., Yang, C., & Albitar, K. (2024). The impact of Fintech on corporate carbon emissions: Towards green and sustainable development. *Business Strategy and the Environment*, *33*(6), 5776–5796. DOI: 10.1002/bse.3778

Wang, Q., Qu, J., Wang, B., Wang, P., & Yang, T. (2019). Green technology innovation development in China in 1990-2015. *The Science of the Total Environment*, *696*, 134008. DOI: 10.1016/j.scitotenv.2019.134008 PMID: 31465930

Wang, W. (2025). Digital finance and firm green innovation: Thc rolc of media and executives. *Finance Research Letters*, *74*, 106794. DOI: 10.1016/j.frl.2025.106794

Wang, Y., & You, J. (2022). [Retracted] The Operation Mode of Agricultural Supply Chain Finance Using Blockchain. *Computational Intelligence and Neuroscience*, *2022*(1), 3338030. PMID: 36188687

Wegrzyn, K. E., & Wang, E. (2021). Types of Blockchain: Public, Private, or Something in Between. https://www.foley.com/insights/publications/2021/08/types-of-blockchain-public-private-between/

Wei, L., & Li, M. (2024). Digital transformation, financing constraints and firm growth performance–From the perspective of financing channels. *Finance Research Letters*, *63*, 105272. DOI: 10.1016/j.frl.2024.105272

Weinzierl, M. (2018). Space, the final economic frontier. *The Journal of Economic Perspectives*, *32*(2), 173–192. DOI: 10.1257/jep.32.2.173

Wei, Y., Zhu, M., Lam, H. K. S., Liu, F., Yeung, A. C. L., Cheng, T. C. E., & Zhou, H. (2024). Protagonists in digital transformation: The impact of chief digital officers on firms' financial performance. *Journal of Industrial Information Integration*, *41*, 100678. DOI: 10.1016/j.jii.2024.100678

Wendl, M., Doan, M., & Sassen, R. (2022). The environmental impact of cryptocurrencies using proof of work and proof of stake consensus algorithms: A systematic review. Journal of environmental management, 326 Pt A, 116530. https://doi.org/.DOI: 10.1016/j.jenvman.2022.116530

Wen, F., Li, C., Sha, H., & Shao, L. (2021). How does economic policy uncertainty affect corporate risk-taking? Evidence from China. *Finance Research Letters*, *41*, 101840. Advance online publication. DOI: 10.1016/j.frl.2020.101840

Westerlund, J. (2007). Testing for Error Correction in Panel Data. *Oxford Bulletin of Economics and Statistics*, *69*(6), 709–748. DOI: 10.1111/j.1468-0084.2007.00477.x

Whiffin, A. (2024). Faster, Cheaper, Safer: How Tokenisation can change investing, Financial Tİmes. https://www.ft.com/content/dd020b5d-d031-4122-a488-ba7241b7f70d?utm_source=chatgpt.com

Wilson, L. (2020). Key benchmarks in the U.S. bond markets: An analysis of the Bloomberg U.S. Aggregate Bond Index. *Journal of Fixed Income*, *30*(1), 35–50.

Wilson, M., Paschen, J., & Pitt, L. (2022). The circular economy meets artificial intelligence (AI): Understanding the opportunities of AI for reverse logistics. *Management of Environmental Quality*, *33*(1), 9–25. DOI: 10.1108/MEQ-10-2020-0222

Wong, L. W., Tan, G. W. H., Ooi, K. B., Lin, B., & Dwivedi, Y. K. (2024). Artificial intelligence-driven risk management for enhancing supply chain agility: A deep-learning-based dual-stage PLS-SEM-ANN analysis. *International Journal of Production Research*, *62*(15), 5535–5555. DOI: 10.1080/00207543.2022.2063089

Wood, G. (2023). Polkadot Whitepaper, Web3 Foundation. https://polkadot.network

World Bank. (2020). World Bank Fintech. World Bank. https://www.worldbank.org/en/topic/fintech

World Bank. (2020a). How Regulators Respond to Fintech Evaluating the Different Approaches-Sandboxes and Beyond Public Disclosure Authorized Public Disclosure Authorized Public Disclosure Authorized Public Disclosure Authorized. 2020. https://documents1.worldbank.org/curated/en/579101587660589857/pdf/How-Regulators-Respond-To-FinTech-Evaluating-the-Different-Approaches-Sandboxes-and-Beyond.pdf

World Bank. (2022). Fintech and the Future of Finance. *World Bank*. https://www.worldbank.org/en/publication/fintech-and-the-future-of-finance

World Bank. (2023). Water access in Kenya: AI-driven solutions. https://www.worldbank.org

World Economic Forum. (2021). AI and market volatility. https://www.weforum.org/reports/ai-and-market-volatility-2021

World Economic Forum. (2023). Post-pandemic ESG risks. https://www.weforum.org/reports/post-pandemic-esg-risks-2023

World Wildlife Fund WWF. (2023). Real-time environmental monitoring and enforcement: The role of technology in detecting illegal deforestation. World Wildlife Fund. https://www.worldwildlife.org

Wu F., H. Hu, H. Lin, X. Ren (2021). Enterprise digital transformation and capital market performance: empirical evidence from stock liquidity, *Management World,* 37, pp. 130-144+110, DOI: 10.19744/j.cnki.11-1235/f.2021.0097

Wu, H., & Wang, Y. (2024). Digital transformation and corporate risk taking: Evidence from China. *Global Finance Journal*, *62*, 101012. DOI: 10.1016/j.gfj.2024.101012

Wu, Q. (2024). From bits to emissions: How FinTech benefits climate resilience? *Empirical Economics*, *67*(5), 1–29. DOI: 10.1007/s00181-024-02609-9

Wu, T. (2011). *The Master Switch: The Rise and Fall of Information Empires*. Vintage.

Wu, Y., & Huang, S. (2022). The effects of digital finance and financial constraint on financial performance: Firm-level evidence from China's new energy enterprises. *Energy Economics*, *112*, 106158. DOI: 10.1016/j.eneco.2022.106158

Wylde, V., Rawindaran, N., Lawrence, J., Balasubramanian, R., Prakash, E., Jayal, A., Khan, I., Hewage, C., & Platts, J. (2022). Cybersecurity, data privacy and blockchain: A review. *SN Computer Science*, *3*(2), 127. DOI: 10.1007/s42979-022-01020-4 PMID: 35036930

Xie, S. (2016). Advanced robotics for medical rehabilitation. *Springer tracts in advanced robotics, 108*(1), 357.

Xue, L., Dong, J., & Zha, Y. (2023). How does digital finance affect firm environmental, social and governance (ESG) performance? — Evidence from Chinese listed firms. *Heliyon*, *9*(10), e20800. DOI: 10.1016/j.heliyon.2023.e20800 PMID: 37867793

Xu, G., & Pan, X. (2024). Can digital transformation reduce the cost of debt capital? Empirical analysis based on the perspective of financial situation. *Applied Economics Letters*, •••, 1–6. DOI: 10.1080/13504851.2024.2351592

Xu, M., Chen, X., & Kou, G. (2019). A systematic review of blockchain. *Financial Innovation*, *5*(1), 1–14. DOI: 10.1186/s40854-019-0147-z

Yadav, H., Kar, A. K., & Kashiramka, S. (2022, September). Artificial intelligence adoption for FinTech industries-an exploratory study about the disruptions, antecedents and consequences. In *Conference on e-Business, e-Services and e-Society* (pp. 3-12). Cham: Springer International Publishing.

Yang, Y.-J., & Hwang, J.-C. (2020). Recent development trend of blockchain technologies: A patent analysis. *International Journal of Electronic Commerce Studies*, *11*(1), 1–12. DOI: 10.7903/ijecs.1931

Yan, J. (2025). How the use of alternative information in risk management fintech platforms influences SME lending: A qualitative case study. *Qualitative Research in Financial Markets*, *17*(1), 1–20. DOI: 10.1108/QRFM-08-2023-0198

Yan, R. S., & Qian, X. Y. (2018). Strategic analysis of digital transformation of Chinese operators in the era of digital economy. *China Soft Science*, *328*, 172–182.

Yaqoot, M., Diwan, P., & Kandpal, T. C. (2016). Review of barriers to the dissemination of decentralized renewable energy systems. *Renewable & Sustainable Energy Reviews*, *58*, 477–490. DOI: 10.1016/j.rser.2015.12.224

Ye, F., Ouyang, Y., & Li, Y. (2023). Digital investment and environmental performance: The mediating roles of production efficiency and green innovation. *International Journal of Production Economics*, *259*, 108822. DOI: 10.1016/j.ijpe.2023.108822

Yermack, D. (2017). Corporate governance and blockchains. *Review of Finance*, *21*(1), 7–31.

Yespergenova, L., Dosmanbetova, A., Baisheva, Y., Zhakipbekov, D., Zhumabayeva, M., & Faizulayev, A. (2023). The Factors that Drives the Cost Management Efficiency of Oil and Gas companies in Emerging Markets: The Case of Eurasian Economic Union. *International Journal of Energy Economics and Policy*, *13*(1), 328–334. DOI: 10.32479/ijeep.13711

Yonzan, N., Lakner, C., Mahler, D. G., Aguilar, R. A. C., & Wu, H. (2020). The impact of COVID-19 on global poverty under worsening growth and inequality. *World Bank Data Blog, 19.*

Young, A., Selander, L., & Vaast, E. (2019). Digital organizing for social impact: Current insights and future research avenues on collective action, social movements, and digital technologies. *Information and Organization, 29*(3), 100–157. DOI: 10.1016/j.infoandorg.2019.100257

Young, M. M., Bullock, J. B., & Lecy, J. D. (2019). Artificial discretion as a tool of governance: A framework for understanding the impact of artificial intelligence on public administration. *Perspectives on Public Management and Governance*, *2*(4), 301–313. DOI: 10.1093/ppmgov/gvz014

Yunis, M., Mirza, N., Safi, A., & Umar, M. (2024). Impact of audit quality and digital transformation on innovation efficiency: Role of financial risk-taking. *Global Finance Journal*, *62*, 101026. DOI: 10.1016/j.gfj.2024.101026

Z/Yen Group & University of Cambridge. (2021). 3rd global cryptoasset benchmarking study. Cambridge Centre for Alternative Finance. https://www.jbs.cam.ac.uk/wp-content/uploads/2021/01/2021-ccaf-3rd-global-cryptoasset-benchmarking-study.pdf

Zhang, M., Wang, D., & Zhang, B. (2022). Evolutionary machine learning for robust financial systems. Springer. https://link.springer.com/book/10.1007/978-981-12-4567-8

Zhang, A. Y., & Zhang, J. H. (2023). Renovation in environmental, social and governance (ESG) research: the application of machine learning. In *Asian Review of Accounting*. Emerald Publishing., DOI: 10.1108/ARA-07-2023-0201

Zhang, D. (2023). Can digital finance empowerment reduce extreme ESG hypocrisy resistance to improve green innovation? *Energy Economics*, *125*, 106756. DOI: 10.1016/j.eneco.2023.106756

Zhang, P., Cheng, Q., Zhang, M., & Luo, X. (2022). A group covert communication method of digital currency based on blockchain technology. *IEEE Transactions on Network Science and Engineering*, *9*(6), 4266–4276. DOI: 10.1109/TNSE.2022.3196816

Zhang, X., Zhang, J., & Zhao, K. (2020). A sustainable approach to financial inclusion: Understanding the impact of financial literacy and trust on FinTech adoption. *Sustainability*, *12*(1), 139. DOI: 10.3390/su12010139

Zhang, X., Zhao, Y., & Liu, X. (2024). Can fintech promote environmental investment in heavily polluting enterprises? evidence from China. *Frontiers in Environmental Science*, *12*, 1373685. DOI: 10.3389/fenvs.2024.1373685

Zhang, Y. (2023). Impact of green finance and environmental protection on green economic recovery in South Asian economies: Mediating role of FinTech. *Economic Change and Restructuring*, *56*(3), 2069–2086. DOI: 10.1007/s10644-023-09500-0

Zhang, Y., Cheng, X., & Wang, J. (2023). Systemic risks in AI-driven markets. *Journal of Financial Stability*, *45*, 101203. DOI: 10.1016/j.jfs.2022.101203

Zhang, Y., Chen, M., Zhong, S., & Liu, M. (2024). Fintech's role in carbon emission efficiency: Dynamic spatial analysis. *Scientific Reports*, *14*(1), 23941. DOI: 10.1038/s41598-024-74834-2 PMID: 39397038

Zhang, Y., Sun, J., Yang, Z., & Wang, Y. (2020). Critical success factors of green innovation: Technology, organization and environment readiness. *Journal of Cleaner Production*, *264*, 121701. DOI: 10.1016/j.jclepro.2020.121701

Zhang, Y., & Wang, X. (2023). Interdependence of DeFi and global stock markets: A data-driven analysis. *Journal of Economic Dynamics*, *15*(2), 98–112.

Zhao, G., Xin, Z., & Wang, Y. (2024). Effect of the sci-tech finance pilot policy on corporate environmental information disclosure—Moderating role of green credit. *Finance Research Letters*, *105177*, 105177. Advance online publication. DOI: 10.1016/j.frl.2024.105177

Zhao, S., Zhang, L., Peng, L., Zhou, H., & Hu, F. (2024). Enterprise pollution reduction through digital transformation? Evidence from Chinese manufacturing enterprises. *Technology in Society*, *77*, 102520. DOI: 10.1016/j.techsoc.2024.102520

Zhao, X., Sun, X., Zhao, L., & Xing, Y. (2022). Can the digital transformation of manufacturing enterprises promote enterprise innovation? *Business Process Management Journal*, *28*(4), 960–982. DOI: 10.1108/BPMJ-01-2022-0018

Zheng, C., Zhang, M., Zeng, C., Xiao, F., & Liu, M. (2023, June 27). Analysis of the Influential Factors and Prediction of Corporate ESG Performance under Multi-source Data Fusion - Based on Frontier Machine Learning. DOI: 10.4108/eai.2-12-2022.2328731

Zheng, F., Jiang, J., Ma, J., & Liou, J. (2024). Capital market opening, digital transformation, and corporate risk-taking: Empirical evidence from Shanghai Shenzhen-Hong Kong stock connect. *International Review of Economics & Finance, 96*(B), 103680. https://doi.org/DOI: 10.1016/j.iref.2024.103680

Zheng, Z., Xie, S., Dai, H., & Wang, H. (2017) An Overview of Blockchain Technology: Architecture, Consensus, and Future Trends, *Conference:6th IEEE International Congress on Big Data.* DOI: 10.1109/BigDataCongress.2017.85

Zhili, D., Boqiang, L., & Chunxu, G. (2019). Development path of electric vehicles in China under environmental and energy security constraints. *Resources, Conservation and Recycling*, *143*, 17–26. DOI: 10.1016/j.resconrec.2018.12.007

Zhong, J., Zhong, Y., Han, M., Yang, T., & Zhang, Q. (2024). The impact of AI on carbon emissions: Evidence from 66 countries. *Applied Economics*, *56*(25), 2975–2989. DOI: 10.1080/00036846.2023.2203461

Zhou, C., Zhou, J., Chen, G., & Xu, C. (2024). Digital finance and greener emissions: Evidence from China. *Applied Economics*, *56*(49), 5830–5844. DOI: 10.1080/00036846.2023.2266599

Zhou, Y., Chen, X., & Lin, W. (2025). Bank digital transformation and rural economic growth: Evidence from China. *Journal of Rural Studies*, *103*, 187–200. DOI: 10.1016/j.jrurstud.2025.000000

Zhu, Y., Lin, Y., Tan, Y., Liu, B., & Wang, H. (2024). The potential nexus between fintech and energy consumption: A new perspective on natural resource consumption. *Resources Policy*, *89*, 104589. DOI: 10.1016/j.resourpol.2023.104589

Zouridis, S., Van Eck, M., & Bovens, M. (2020). Automated discretion. *Discretion and the quest for controlled freedom*, 313-329.

Zuiderwijk, A., Chen, Y. C., & Salem, F. (2021). Implications of the use of artificial intelligence in public governance: A systematic literature review and a research agenda. *Government Information Quarterly*, *38*(3), 101577. Advance online publication. DOI: 10.1016/j.giq.2021.101577

About the Contributors

Muhammad Arslan (also known as Dr Arslan) is a dedicated academic and active researcher, currently serving as a senior lecturer (associate professor) of accounting and finance at the Open Polytechnic of New Zealand. Holding a PhD from Lincoln University, New Zealand, he specialises in corporate governance, financial reporting, and emerging financial technologies. He is a Certified Practising Accountant (CPA) and holds full membership in CPA Australia. He also has certifications in Xero Advisor and Payroll. Arslan is the recipient of the dean's award for his PhD thesis from Lincoln University, New Zealand, and has also received two Emerald Literati Awards for "Highly Commended Paper." He has also completed the New Zealand Certification in Adult and Tertiary Teaching (NZCATT) to develop his pedagogical skills further. With over 12 years of experience in higher education across New Zealand, China, Kazakhstan, the UAE, Mexico, and beyond, Arslan has taught and developed courses at both undergraduate and postgraduate levels. He has instructed diverse cohorts of students across face-to-face, blended, and online platforms. He also gave seminars and guest lectures at universities in Russia and Colombia. He employs active learning, experiential learning, and learner-centred approaches. His teaching portfolio includes a wide array of courses at undergraduate and postgraduate levels, including but not limited to Management Accounting, Financial Accounting, Introduction to Accounting and Taxation, Analysis of Financial Information, Business Research Methods, Advanced Corporate Finance, Advanced Financial Management, Corporate Finance, Principles of Finance, Corporate Reporting, Financial Reporting and Management Control, Quantitative Methods, Portfolio Management & Behavioural Finance, International Finance, Financial Statement Analysis and Reporting, and Financial Management. A prolific researcher, Arslan has established an international reputation with over 35 publications, including 26+ Scopus-indexed publications (14 Web of Science (WoS) with papers featured in SSCI, SCIE, ESCI, ABS, ABDC, and Q1/Q2). His contributions include six book chapters (five Scopus-indexed) and presentations at leading international conferences such as the Academy of Management (AOM), the International Conference of the Journal of International Accounting Research, the Annual New Zealand Finance Colloquium, and the ICGS Conference, among others. Arslan holds prominent editorial roles, serving as an Academic Editor at PLOS One (eISSN: 1932-6203), Guest Editor for the Journal of Risk and Financial Management (ISSN: 1911-8074), and Co-Editor-in-Chief of the International Journal of Business and Management (eISSN: 2815-9330). Additionally, he examines the funding applications and research articles for other leading journals in his field. He is a certified peer reviewer and mentor through Publons Academy and Web of Science (WOS) Clarivate. His standing as a globally recognised specialist and the cumulative influence of his work is reflected in an overall h-index of 16 and an i-index of 20 (Google). His research work has received over 970 citations (Google). He is an efficient, well-organised, and responsible scholar who understands how to get things done. A committed mentor, Arslan has already supervised several master's and PhD students. He regularly examines theses for several educational institutes. Beyond academia, Arslan has engaged in industry consultancy, working as a business advisor and contributing to EFMD Global accreditation efforts for KIMEP University's BFIN program. He is open to funding opportunities, visiting positions, research collaboration, and supervision.

Alimshan Faizulayev is an Associate Professor and Research Director in the Bang College of Business (BCB) at KIMEP University, Kazakhstan. His research interests include but not limited to financial technology in conventional and Islamic banking, blockchain technology, machine learning, and energy economics. He has published several papers in international journals.

Abdulkadri Toyin Alabi holds a BSc and MSc in accounting from Kwara State University. Abdulkadri is an associate member of professional bodies like the Institute of Chartered Accountants of Nigeria and the Chartered Institute of Taxation of Nigeria. As a dedicated and detail-oriented chartered accountant, he has been meeting the accounting needs of end users in the practical field of accounting in addition to in the academic setting. He currently works as a senior accountant for the Kwara State Government and as a researcher affiliated with Kwara State University. His research interests lie in the broad areas of financial and corporate reporting as well as forensic accounting and auditing. His recent publication features in Global Journal of Accounting.

Shahbub Alam is a Lecturer in the Department of Business Administration at Sonargaon University. He has extensive experience in university-level teaching and a strong academic foundation in business studies. He secured first place in both his

MBA and BBA programs and was awarded the prestigious Prime Minister Gold Medal. His research interests include financial analysis, and business management.

Aysegul Kirkpinar is an Assistant Professor and currently working in Izmir Katip Celebi University, Department of Business Administration. She got her PhD degree from Dokuz Eylul University, Institute of Social Sciences, Department of Business Administration and her master's degree in Accounting and Finance. Her research interests include financial management, financial modeling and forecasting, portfolio management and behavioral finance.

Halil Kiymaz is the Bank of America Professor of Finance at the Crummer Graduate School of Business at Rollins College in Winter Park, Florida, USA. He also holds the Chartered Financial Analyst (CFA) designation and is a Visiting Professor of Finance at KIMEP University. Professor Kiymaz maintains an extensive research agenda and has published more than 100 articles in scholarly and practitioner journals. He also has seven books titled "The Art of Capital Restructuring: Creating Shareholder Value through Mergers and Acquisition (John Wiley & Sons, Inc., 2011), "Market Microstructure in Emerging and Developed Markets" (John Wiley & Sons, Inc., 2013), "Private Equity: Opportunities and Risks" (Oxford University Press, 2015), "Mutual Funds – Building Blocks for Investment Portfolios," (Oxford University Press, 2016) and "The Savvy Investor's Guide to Pooled Investments (Emerald Publishing, 2019), Equity Market Valuation and Analysis (John Wiley & Sons, Inc. 2020), and Social Finance: Mobilizing Markets for Good (Palgrave Macmillan, 2025, Forthcoming). His research has appeared in the Journal of Banking and Finance, Financial Review, Global Finance Journal, Journal of Applied Finance, International Journal of Emerging Markets, Journal of Economics and Finance, Review of Financial Economics, Quarterly Journal of Business and Economics, Pacific-Basin Finance Journal, International Journal of Economics and Finance, Borsa Istanbul Review, and European Journal of Management. He has received several research awards, including the McGraw-Hill Irwin Best Paper Award and Outstanding Research Awards at the Global Conference on Business and Finance. He also serves on the editorial board of numerous journals. He is the former Academy of Financial Services president and served as the International Journal of Emerging Markets finance editor.

Carlos Muñoz Valdez is a visiting professor at IPADE Business School, where he specializes in Corporate Finance, Portfolio Management, and Fintech. He currently serves as Chief Financial Officer (CFO) of RBM Group, a leading conglomerate in Northwest Mexico with operations spanning retail and real estate. Additionally, Dr. Muñoz holds the role of Chief Investment Officer (CIO) for a prominent Family Office. Dr. Muñoz earned his degree in Finance from Universidad Panamericana and completed his MBA at IPADE Business School. He holds a doctoral degree from Rollins Business School, where his research focused on Volatility Models and Blockchain Technology. Residing in Northwest Mexico, Dr. Muñoz balances his professional endeavors with family life alongside his wife and their three daughters.

Muhammad Nauman is a academic and researcher with practical experience as an Inspector Inland Revenue at the Federal Board of Revenue, Pakistan. He brings a strong theoretical foundation, evidenced by his MBA (Banking and Finance) from the University of the Punjab and a Certificate in Accounting and Finance (CAF) from ICAP, Pakistan. His research and expertise focus on taxation, financial reporting, emerging financial technologies, and the dynamics of the Pakistan stock market.

Mira Nurmakhanova is a Chair of Finance and Accounting Department at Bang College of Business, KIMEP University. Mira is a Professor of Finance and she holds a PhD degree from Iowa State University.

Anand Patil, an Associate Professor at Christ University, Bangalore, holds a Ph.D. from Shivaji University, Kolhapur, and boasts over 23 years of teaching experience, coupled with three years in industry as an accounts and audit assistant. With expertise in Accounting, Finance, and Taxation, Dr. Patil actively contributes to research, presenting at international conferences and publishing in esteemed journals. His administrative roles include Head of the Curriculum Development Committee and Coordinator of IQAC, alongside mentoring students in finance and management, cementing his invaluable role in the university's academic community.

Lakshmi R B is an Assistant Professor in the Business and Management Department at Christ University in Bangalore. She is a highly distinguished academician with expertise in a wide range of areas. Dr. Lakshmi played a pivotal role in steering university-related activities, including securing approvals for new courses, coordinating with the university, and navigating the LIC and AICTE approval processes. Her commitment to academic excellence is evident in her contribution to crafting the PGPM curriculum. She is an active participant in organizing industry exchange programs, facilitating final placements, and orchestrating internships for students, creating a holistic learning environment. Dr. Lakshmi is a lifelong learner, actively engaging in various national and international Faculty Development Programs. Her research prowess is evidenced by her publications in UGC Care, Scopus indexed, and ABDC journals, contributing significantly to the academic discourse. Her engagement transcends borders, as she has coordinated international webinars, conducted virtual and offline outreach programs for students at premier institutes, and moderated national and international panel discussions. She has served as a resource person for several Faculty Development Programs (FDPs) on pedagogy and andragogy. As a recognized expert, she served as a resource person for a national-level webinar on "Design thinking," reflecting her commitment to innovative pedagogical approaches. Her multifaceted contributions to academia have earned her the prestigious AVLG Women's Achievers Award, further solidifying her position as a trailblazer in the realm of Business and Management education.

Batani Raghavendra Rao, Professor of Finance at CMS Business School, JAIN (Deemed-to-be University) and served as Area Chair of Finance from 2011 to 2020. He has 4 years of Corporate and 33 years of Academic experience including two years (Visiting) at IIM Indore. He is the Research Guide for PhD Programme of JAIN (Deemed-to-be University). He is a qualified Cost and Management Accountant, Fellow of the Institute of Cost Accountants of India, a Chartered Financial Analyst and PhD in Finance. He is a Chartered Management Accountant from CIMA, UK and a Chartered Global Management Accountant (AICPA, USA). As a visiting faculty at IIM Indore during 2013 and 2014, he earned a feedback rating of 9.01 on 10. He completed course on 'Decision Making Using Financial Accounting' from IIT Madras with 97% [Elite + Gold Class]. Dr. Rao is an IBM Data Science Professional. His areas of research interest are ESG, Financial Reporting and Analysis, Capital Markets, Financial Modelling, Derivatives, Forex, Risk Management and Income Tax. Case studies on 'Infosys Segment Reporting', 'Fundamental Analysis of Canara Bank' and 'EVA - Construction and Reporting at Pidilite Industries' authored by Dr. Rao are published with the Case Center, UK. He has publications indexed in Scopus and Web of Science. His article on MBO in Chartered Accountant Journal was adjudged as the best article. He conducted corporate training for GMR, HP, Sodexo and TVS Mobility. He earned a high rating feedback in the corporate training.

Safae Ismaili, a renowned scholar in economics and management, specializes in financial innovation and organizational resilience. She holds a Doctorate in Economics and Management from the Faculty of Legal, Economic, and Social Sciences in Fez and is currently a Professor at Fez Business School, Private University of Fez, where she leads flagship programs in Finance, Accounting, and Auditing. Her research focuses on blockchain integration in finance, economic crises' impact on resilience, and consumer behavior in financial technologies, with publications in top-tier journals. Dr. Ismaili also serves as Associate Editor of the Journal of Contemporary Issues in Business and Government. Known for her innovative teaching, she employs immersive technologies and interactive methods to prepare students for the complexities of modern finance. She has supervised numerous theses and dissertations, fostering academic excellence. An active member of professional associations, Dr. Ismaili is committed to advancing knowledge and interdisciplinary collaboration, contributing significantly to the global academic community.

Akilesh V.K.R is an Associate Consultant at KPMG Global Services, specializing in finance, data analytics and restructuring. He holds an MBA in Finance and International Business from Christ University, Bangalore, and a BTech in Mechanical Engineering from the same institution. With prior experience as a project intern at Symmetrix Computer Systems and a summer intern at Tech Innovations, he has honed his skills in investments and investor relations. Akilesh is passionate about business strategy and financial consulting, leveraging his expertise to drive value in a dynamic corporate environment.

Zazli Lily Wisker is a Programme Coordinator and a Senior Academic at Business School and School of Computing, EIT, New Zealand. Dr. Wisker's research interests are diverse, encompassing sales management, cultural marketing and diversity, sustainability and green marketing, internal branding, and destination marketing. She has published her work in several high-impact journals, including the Journal of Product & Brand Management, Marketing Intelligence & Planning, and the Journal of Islamic Marketing.

Aizada Zhakupova is currently serving as the Academic Dean of the School of Economics and Finance at Almaty Management University (AlmaU). With over five years of academic experience, she previously held the position of Senior Lecturer in Finance at Narxoz University, where she taught undergraduate and graduate courses in corporate finance, financial management, and investment analysis. Ms. Zhakupova is actively pursuing her Ph.D. in Finance, with research interests focused on financial markets, capital structure decisions, and the integration of sustainable finance principles in emerging economies. Her academic background is complemented by a strong commitment to curriculum development, academic leadership, and student mentorship. As a rising scholar and experienced educator, she continues to contribute to Kazakhstan's higher education landscape through research, teaching excellence, and institutional development.

Index

A

B

C

D

E

F

G

I

N

R

S

T

V

W

www.ingramcontent.com/pod-product-compliance
Lightning Source LLC
LaVergne TN
LVHW080614300926
849162LV00005B/1156